Technology for Chemical Cleaning of Industrial Equipment

2nd Edition

Wayne W. Frenier, FNACE

NACE INTERNATIONAL

NACE International
The Worldwide Corrosion Authority

Printed in the United States of America

ISBN: 978-1-57590-367-5

NACE International
The Worldwide Corrosion Authority
15835 Park Ten Place
Houston, TX 77084
nace.org

Preface

This book will update the first edition[1] and discuss new developments that have affected the commercial use of chemicals and devices to clean industrial equipment. This edition will emphasize the mechanisms of important cleaning processes and solvents and will give an overview of the science and technology of the formation and removal of fouling deposits in the industrial equipment environment. This publication is directed to a general technically-trained audience, with emphasis placed on the basic chemical and mechanical principles of industrial cleaning (IC).

The introduction describes the environment in which damaging deposits occur and reviews the technical aspects of IC in the wide variety of industrial plants where fouling may impede production or compromise safety. Subsequent chapters discuss aspects of the formation of inorganic scales and organic process deposits, including the driving chemical forces for deposit formation and the methods for deposit removal. Successful treatment design requires the operator to

- Understand the problem
- Evaluate solution options
- Design and execute cleaning procedures
- Perform a safe treatment
- Evaluate the results

Each major section and most subsections include reviews and the consensus of the current literature.

The scope of this book is limited to fouling phenomena in a wide range of industrial equipment, including power generation, pulp and paper, chemical production, and downstream petroleum refineries and gas plants. This discussion will not specifically include cleaning or inhibition of scale in transmission pipelines, but many of the techniques and technologies needed are like those described here and could be applied with appropriate modifications. See Frenier[2] for a review of cleaning pipelines in the upstream and midstream sectors.

Design of this Book

Chapter 1 "Introduction" describes the scope of the book and introduces the types of industrial equipment that require chemical or mechanical cleaning.

Chapter 2 "Review of Basic Chemical and Physical Processes Required in Industrial Cleaning Operation" describes the most important basic technologies needed to understand chemical and mechanical methods of scale removal, including chemical and hydrodynamic processes. This chapter also reviews the literature on scale testing protocols.

Chapter 3 "Formation and Deposition of Inorganic Scales" reviews the key chemical principles of the formation and deposition of inorganic scales that foul different types of industrial equipment. This includes scales that form in power boilers, cooling systems, and nuclear generating units.

Chapter 4 "Removal of Inorganic Deposits with Aqueous Solvents" comprehensively reviews the types of chemicals that have been used to dissolve the solids described in Chapter 3. This includes mineral acid-based solvents as well as those based on organic acids and chelating agents. Because there is extensive literature on the dissolution of magnetite, a separate section is devoted to these mechanisms.

Chapter 5 "Cleaning Nuclear Plant Equipment" is devoted to the cleaning of units in nuclear plants. The bulk of deposits that foul nuclear generating plant surfaces (or contain harmful radioactivity) are very similar (or identical) to those described in Chapter 3. Most of the cleaning chemicals are also described in Chapter 4. However, the operations in nuclear plants are highly controlled, and Chapter 5 focuses on the unique procedures in these plants.

Chapter 6 "Corrosion Inhibitors for Aqueous Cleaning Solvents" describes the wide range of corrosion inhibitors, or chemicals that block a metal's surface from attack, used in cleaning fluids and the uses for each type; addresses other chemicals that are used to control corrosion during cleaning activities; and describes lab and field methods for evaluating corrosion inhibitors for industrial cleaning solvents.

Chapter 7 "Formation and Removal of Organic and Mixed Deposits from Chemical Plants and Refineries" reviews the formation of organic solids in different types of units. While refineries and the various chemical plants have units that are fouled with inorganic deposits, many surfaces are also affected by a wide range of organic solids. This chapter also reviews the wide range of chemicals that may dissolve or flush out organic and mixed solids from these units.

Chapter 8 "Mechanical Equipment and Special Fluids for Descaling Units with/without Chemicals" describes the process of descaling units and unique fluids that form foams for applying cleaning chemicals. Many types of pumps, tanks, and high-pressure water jetting units are used with or without chemicals to remove fouling deposits. All the chemical solvents require large units (frequently mobile) for delivery and application. High-pressure water and other mechanical processes may "blast" the surfaces clean. Some chemicals may be used in these processes.

Chapter 9 "Disposal of Chemical Cleaning Wastes" reviews current chemical and mechanical disposal methods. All the services described in this book produce waste streams that amount to several to many times the volume of the units that are cleaned. Most of the fluids will contain hazardous products that must be removed or neutralized before final disposal or reuse.

Chapter 10 "Health, Safety, and Environmental Issues in IC" reviews the wide range of activities to safely and ethically apply cleaning chemicals and mechanical processes for deposit removal in various industries.

Chapter 11 "General Conclusions" briefly summarizes the book and reviews the key principles of safe IC activities.

Each chapter will include a summary. Chapters 3–8 also have a section with references and short abstracts of case histories that illustrate the principles described in these chapters.

Acknowledgments

I acknowledge the help and encouragement of HydroChem Industrial Services (Deer Park, TX). HydroChem Industrial Services provided images and important information on current cleaning operations.

I especially thank Jack Horvath, Chip Glover, Danny Foster, Ginger Curtis, Scott Holland, and Frank Moll of HydroChem Industrial Services for their consultations. I also acknowledge Metal Samples Company (Munford, AL) for permission to use images in this book.

I acknowledge the help and access to documents provided by Mike Little and Bob Varrin of Dominion Engineering (Reston, VA). Their knowledge about cleaning nuclear-related units worldwide has been very useful for making this book more complete. They also reviewed parts of the book.

Clean Harbors Energy & Industrial Services (Acheson, Alberta, Canada) permitted images of cleaning equipment to be used in this book. Tom McCartney and Rosanne Shank helped me to understand some technical issues.

Dedication

To Dolores, our children, and our grandchildren for the inspiration to write this book.

Contents

1 Introduction

Industrial cleaning (IC; also known as industrial services) uses reactive chemicals and specialized and, frequently, mobile equipment to remove unwanted deposits from the surfaces of various pieces of process equipment. High-pressure fluids also may be used to blast off deposits.

Included in the array of industrial equipment needing cleaning are components of power generating units, such as boilers and condensers, as well as heat exchangers and towers in refineries and petrochemical plants. Fouling deposits also affect the operation of other industrial equipment such as digesters and heat exchangers in paper mills. Nuclear power generating equipment including pressurized water reactors (PWR)[3] and boiling water reactors (BWR)[3] also have components that require cleaning. Most of these plants have cooling water and heat transfer units that can become fouled and require periodic cleaning.

This business segment is frequently lumped into the industrial and institutional cleaning sector, which generates multibillions in yearly revenue. However, the exact value of the services provided by the industrial sector is unknown to this author.

Clean Harbors[4] lists the following industries serviced by IC companies:

- Refineries
- Chemical and specialty chemical
- Transport of liquids, solids, and gases
- Power utilities (fossil fuel and nuclear)
- Pulp and paper
- Pharmaceutical and biotechnology
- Steel manufacturing

Further definitions of the scope of this industry (and thus this book) follow in later sections.

The removal of unwanted surface deposits is conducted for many reasons:

1) **To eliminate scales that contribute to increased corrosion.** Examples of these types of deposits include iron oxides, copper, and alkaline earth salts found on the watersides of many types of process equipment and steam generators. Organic deposits also can contribute to under deposit corrosion.

2) **To increase heat transfer**. Although steel may have a thermal conductivity of 52W/m-K (30 BTU/hr-ft-F), a highly conductive deposit such as magnetite (Fe_3O_4) has a thermal conductivity of only 5 W/m-K (3 BTU/hr-ft-F). Very refractory deposits (various silicates) have conductivity values of 3.97 W/m²K (0.7 Btu/hr×ft²×°F; for example, fire brick).
3) **To increase fluid flow rates and decrease pressure drops**. The pressure drop (Δp) of a fluid in a pipe at a constant flow rate is inversely proportional to the fourth power of the pipe diameter.

According to the Hagen-Poiseuille law,[5,6]

$$\Delta p = \frac{128\mu q l}{\pi g_c d^4}$$ (Equation [1.1])

g_c = Gravitational acceleration at the center of mass, m/sP²
q = Flow rate m³/s
l = Length of pipe, m
d = Pipe inside diameter (for round pipe), m
v = Flow velocity, m/s
ρ = Density, kg/m²
μ = Dynamic viscosity, kg/(m-s)

Thus, small changes in pipe diameter may have large effects on flow rate and pressure drops. Utility power plant surface condensers are regularly cleaned to increase the vacuum caused by condensing the steam. A clean condenser may have a vacuum that increases the boiler output by as much as 50 MW compared with a fouled unit.

4) **To remove spalling deposits**. Spalling deposits may be removed to reduce damage to downstream components. Superheat and reheat sections of power generating plants are cleaned to reduce damage to turbines from sloughed magnetite. Migrating deposits also foul flow channels in nuclear steam generators.
5) **To "safe" vessels by removing toxic (i.e., benzene) or flammable (hydrocarbon) vapors before vessel entry.** Refinery towers and tanks are frequently cleaned for these reasons. Nuclear equipment also may require cleaning to reduce radiation exposure for personnel. This is then called "decontamination-decon." Deposits also may contaminate the various products of a plant.

1.1. Background of Industrial Cleaning and Scope of this Book

Industrial chemical cleaning developed as an offshoot of technology devised to increase the flow from oil, gas, and brine from wells (see discussion of acidizing in Frenier and Ziauddin,[7] Ali et al.,[8] and Kalfayan[9]). Hydrochloric acid was used before 1930 to dissolve deposits from oil wells. The development of the first corrosion inhibitors by John Grebe[10] in 1932 allowed acid to be used (circa 1939[11]) to remove deposits from industrial boilers. The first utility power boilers were cleaned in about 1941 using inhibited hydrochloric acid. Since then, technology has been created to remove iron oxides and copper from boilers, to dissolve carbonate and silica deposits from various systems, and to remove many types of refinery/petrochemical foulants.

Methods have been invented to clean nuclear steam generators, to clean components in paper mills, and to chemically and mechanically treat the resulting wastes. Important advances have been made in the science of corrosion inhibition as applied to cleaning solvents and to the use of low-toxicity chelating agents and organic acids.

Several books cover aspects of chemical cleaning technologies. McCoy reviewed the technology up to 1981.[12] Chapter 9 of Cowan and Weintritt[13] has an excellent review of scale removal up to 1976 and will be cited where appropriate in this book.

Wackenhuth[14] produced a how-to manual covering chemical cleaning of fossil-fueled steam generating equipment. NACE International[15] published the *Industrial Cleaning Manual.* The Chemical and Mechanical Cleaning Unit committee (T-3M, now STG-6) of NACE's Technical Practice Committee produced three volumes of bibliographic information on cleaning technology.[16] Gutzeit[17] reviewed methods for cleaning refinery and process equipment that stressed the use of good planning and control procedures.

This book is an expansion of previous NACE International publications by Frenier[1, 18] and will review chemical as well as mechanical cleaning technology developed since about 1965. The emphasis will be on understanding the chemical/mechanical basis of the processes used to clean industrial equipment. Each section contains a review of the literature for a major technology area, with emphasis on technology that reached or affected the marketplace. There will be some duplication; for example, aspects of corrosion, passivation, and inhibitors also apply to technology for removing inorganic deposits such as iron oxide and copper. For clarity, technology developed before 1965 will be noted to facilitate comparison with more recent developments. The references, tables, and figures are contained within the 11 chapters.

1.2. Definitions and Terms Used in this Book

Some definitions of terms that apply to chemical/mechanical cleaning operations will be helpful.

Anode. This is the area of an electrochemical cell where oxidation takes place.

Anaerobic. Free of air or uncombined oxygen. This term frequently applies to a type of microorganism as well as the conditions in a specific environment.

Cathode. This is the area of an electrochemical cell where reduction takes place.

Cathodic Protection. This is an important corrosion protection process usually applied to external metal surfaces. The application of DC voltage polarizes the metal as a cathode and thus impedes corrosion.

Corrosion Inhibitor. A chemical agent that reduces the rate of various corrosion reactions by protecting the surfaces of a corrodible metal with a thin film.

Corrosive Agent Scavenger. This is a chemical or process that removes corrosive agents from a solution.

Crude Oil. This is a general term to describe unrefined petroleum products and specifically petroleum liquids. The chemical compositions of crude oil streams vary from formation to formation but usually contain a wide variety of hydrocarbons. Some crude oil streams also contain various amounts of chemicals containing nitrogen, sulfur, or oxygen.

Decontamination. This term usually refers to processes that remove harmful chemicals (solids, liquids, or vapors) from equipment. Decontamination is used in both the chemical processing and nuclear power industries to describe selected processes.

Deposit (Fouling). This refers to any solid that may form from a chemical reaction at a surface or may be transported to a surface by any chemical/mechanical process. This term may include inorganic scales, organic compounds, or mixtures of the two.

Dipole Moment. This is the measure of polarity of a polar covalent chemical bond. It can also be defined as the measure of strength of an electric dipole. It is defined as the product magnitude of charge on the atoms and the distance between the two bonded atoms. This is an important characteristic of a molecule and may affect adsorption on a surface.

Hydraulic Jetting. This term describes processes that use high-pressure water or water-containing particles to physically remove solids from surfaces.

Low Oxidation State Metal Ion (LOMI) Agents. These are metal complexes that stabilize a lower oxidation state of a metal ion in which a higher oxidation state is possible. Examples include Fe(II), Cr(III), and V(II) containing ions. They can be part of a solvent for catalyzing the dissolution of different metal oxide scales.

Microbiologically Induced Corrosion (MIC). This is corrosion caused or promoted by microorganisms, usually chemoautotrophs. It can apply to both metals and nonmetallic materials.

Molecular Modeling. This term encompasses a large range of physical, but mostly computational, activities to produce useful information about molecules or groups of molecules.

Natural Gas. A naturally occurring mixture of hydrocarbon gases that is highly compressible and expansible. Methane (CH_4) is the chief constituent of most natural gas (constituting as much as 85% of some natural gases), with lesser amounts of ethane (C_2H_6), propane (C_3H_8), butane (C_4H_{10}), and pentane (C_5H_{12}). Impurities also can be present in large proportions, including carbon dioxide, helium, nitrogen, and hydrogen sulfide.

Natural Gas Liquids (NGL). This refers to the heavier weight gaseous hydrocarbons: ethane (C_2H_6), propane (C_3H_8), normal butane (n-C_4H_{10}), isobutane (i-C_4H_{10}), pentanes, and even higher molecular weight hydrocarbons (sometimes called natural gasoline). When processed and purified into finished by-products, these are collectively referred to as NGL. Note that the terms "condensate" and "NGL" are sometimes used interchangeably, but NGL usually refers to the finished product.

Nondestructive Testing. Nondestructive testing is material testing in which the part being tested is not destroyed or rendered unusable.

Passivation. An electrochemical state in which an adherent film of a reaction product forms on the surface of a metal that reduces the corrosion rate to a low value.

Pig. The term comes from "Pipeline Inspection Gauges" and comprises a large variety of mechanical devices (see Frenier[2] for details) for inspecting, cleaning, or treating pipelines. In some parts of the world, they are called pipeline scrapers or inspection line devices.

Refinery. This is a very general term for a production facility composed of a group of chemical engineering unit processes and unit operations that refine certain materials or convert raw material into products of value. A petroleum refinery is one example, but other chemical plants refine/react raw materials into various useful products.

Reductive Dissolution. This is a process in which an iron oxide is dissolved from a steel surface and electrons are produced, which reduce another element. Ferric ions in scale are usually the reduced species. This process facilitates dissolution of some scales.

Scale. In the context of this book, a scale is a predominantly inorganic compound that forms in an aqueous solution owing to a chemical reaction. The result is usually a crystalline solid (frequently a well-defined and identifiable mineral). The scale may deposit on a surface or may remain suspended in the fluid.

Surface Facilities (Facilities). Devices or multiple devices and chemical operations that separate or treat the mixed hydrocarbon/water stream from a well or group of wells. Some parts of the treating process may be on the seafloor, on a platform, or on land. They also may be called separation facilities.

1.3. Types of Industrial Equipment Requiring Cleaning

This section will briefly review the types of industrial equipment that may require periodic or acute (off schedule) cleaning to remove fouling deposits. These include

- Non-nuclear power and steam generating process equipment, with an emphasis on the removal of inorganic solids.
- Chemical processing plants of all types, with an emphasis on organic/mixed foulants. Petroleum refineries, a wide range of chemical production facilities, and paper mills are included in this category. In some cases, these plants are part of petroleum production processes. Cleaning the numerous types of heat exchange units in these plants will be a focus of this book.
- Cooling and service water systems. These also have inorganic foulants, but these form at lower temperatures compared with steam generators and may require low-temperature solvents for cleaning.
- Nuclear power generation equipment. These units (the water sides) have challenges similar to those of conventional power generators. However, the radiation risks and nuclear regulations require different chemistries and processes. For this reason, Chapter 5 describes deposit and removal processes in nuclear plants.

Similar chemicals (and sometimes processes) are employed in a wide range of consumer-related situations, including cleaning shower stalls and sinks. Although the scope of IC is much larger, some of the chemicals and processes are familiar to many people, and this may aid in understanding IC operations.

1.3.1. Non-Nuclear Power and Steam Generating Processes

A vast number and many types of pressurized units produce and use (or recycle) steam for various purposes.

Utility power boilers and ancillary equipment generate steam for production of electricity. See the Department of Energy's annual list of all electricity generating boilers.[19] Cleaning of these (often very large) plants has been a major part of the IC industry. Because of the number, importance, risks, and costs associated with servicing these plants, a major fraction of technologies developed in laboratories and engineering groups has addressed problems in these facilities. Several different designs are in use and may require different techniques for cleaning. This is due to a range of deposits that are caused by different aqueous fluids, temperatures, and metallurgies and circulation processes. Refer to the Babcock & Wilcox website[20] for details of current designs. Several examples and characteristics are as follows.

Drum boiler. Here liquid water is circulated through a steam drum through the system and recycled directly. Characteristics/designs include

- Natural circulation units in which firing the unit causes circulation. Figure 1.1 shows the layout of a natural circulation coal-fired drum boiler plant. This diagram also shows several components that may need cleaning: the tubes and steam drum, the economizer, superheater and reheater, surface condenser, and possibly the turbine blades.
- Forced circulation units in which an internal pump helps circulate the boiler water. Figure 1.2 shows a drawing of the water circuit of a forced circulation boiler.
- Both boiler types use lower temperatures and pressures than supercritical units.

Once-through boiler. These do not have a steam drum, and most produce supercritical steam (Figure 1.3). These types of plants do not directly recirculate boiler water (since it becomes a supercritical fluid), but water is reused after condensation and treatment if necessary. A supercritical steam generator operates at such a high pressure (over 22 MPa or 3,200 psi) that the physical turbulence that characterizes boiling ceases to occur; the fluid is neither liquid nor gas but a supercritical fluid. There is no generation of steam bubbles within the water because the pressure is above the critical pressure point at which steam bubbles can form. As the fluid expands through the turbine stages, its thermodynamic state drops below the critical point as it does work turning the turbine, which turns the electrical generator, and power is ultimately extracted. The fluid at that point may be a mix of steam and liquid droplets as it passes into the condenser. Additional information on various boiler types can be found at Babcock & Wilcox[21] and Wikipedia.[22]

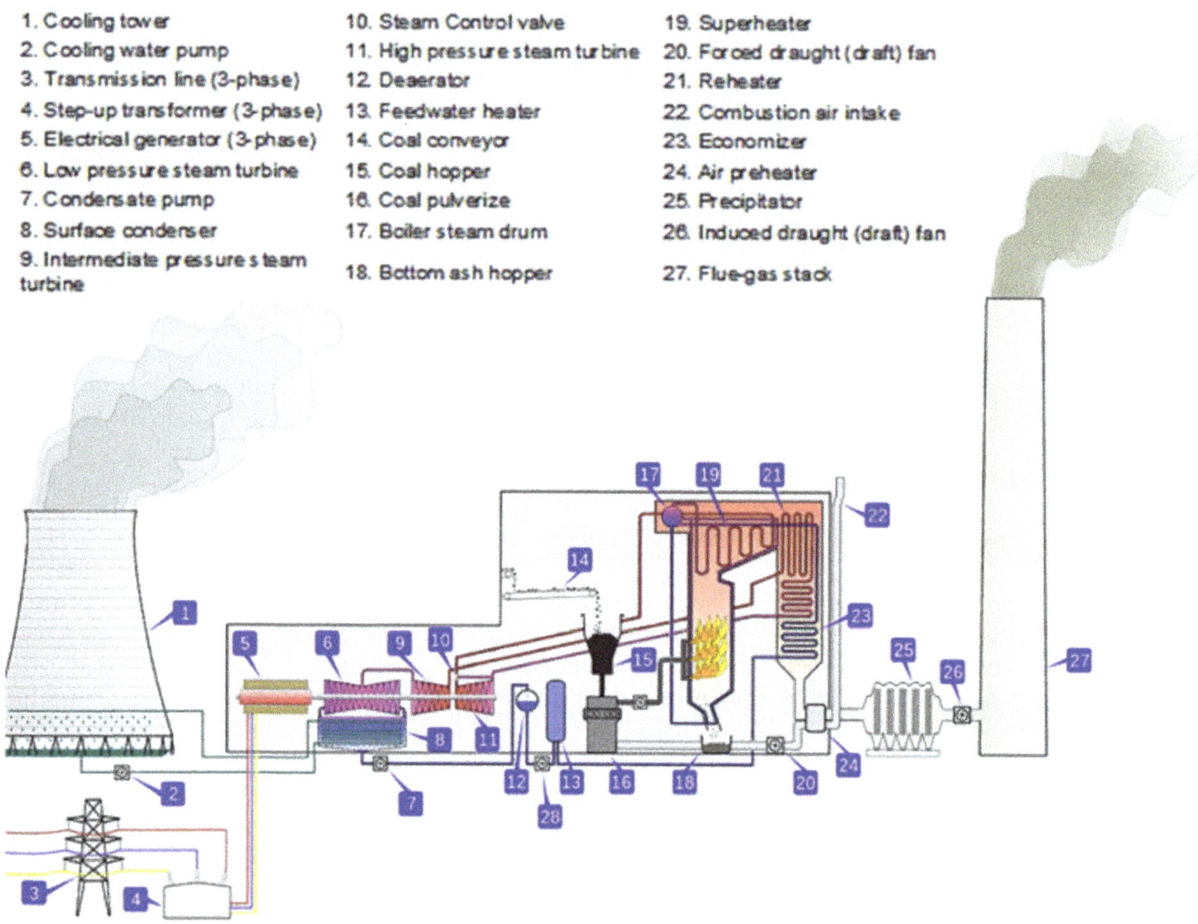

Figure 1.1 Layout of a Coal-Fired Drum Boiler Power Plant[23]

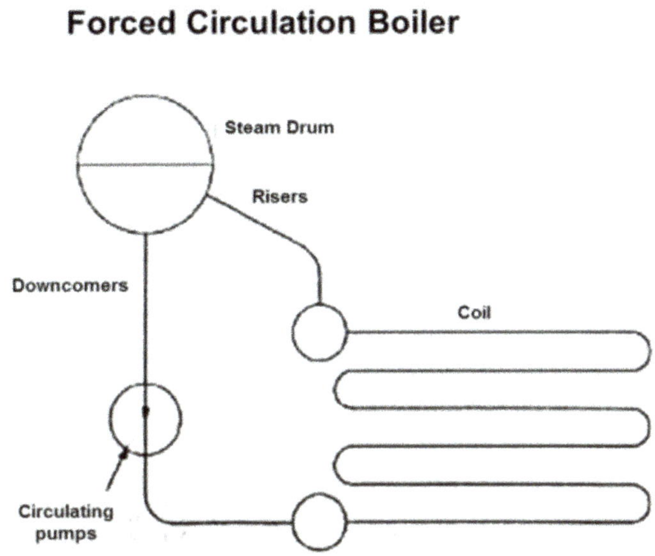

Figure 1.2 Drawing of Forced Circulation Drum Boiler Components and Flow Paths

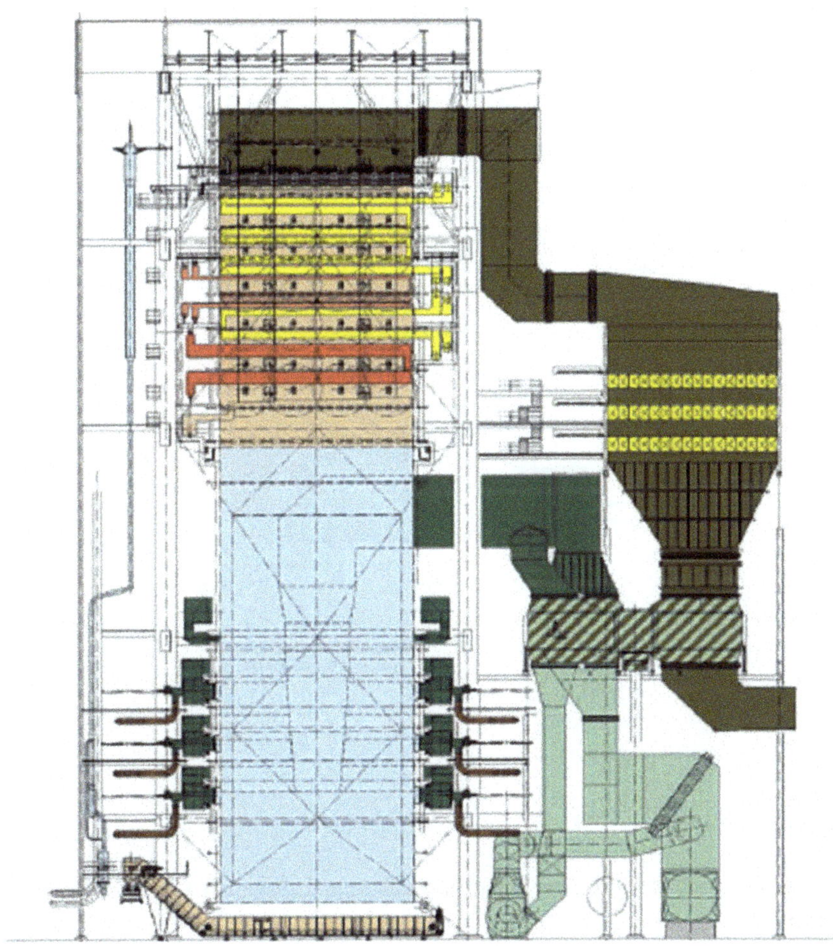
Figure 1.3 Diagram of a Supercritical Boiler[22]

The importance of the boiler types includes the different metals of construction, types of deposits to be removed, and the physical layouts that may require different equipment for circulation during cleaning.

Combined cycle power generating plants. These employ gas turbines to generate electricity and then use the hot exhaust gases in a heat recovery steam generator to produce steam that then powers an additional turbine. Since these units are more economical to build than other types of power generating units, many have been constructed in the last 20 years. A diagram of one design is shown in Figure 1.4.

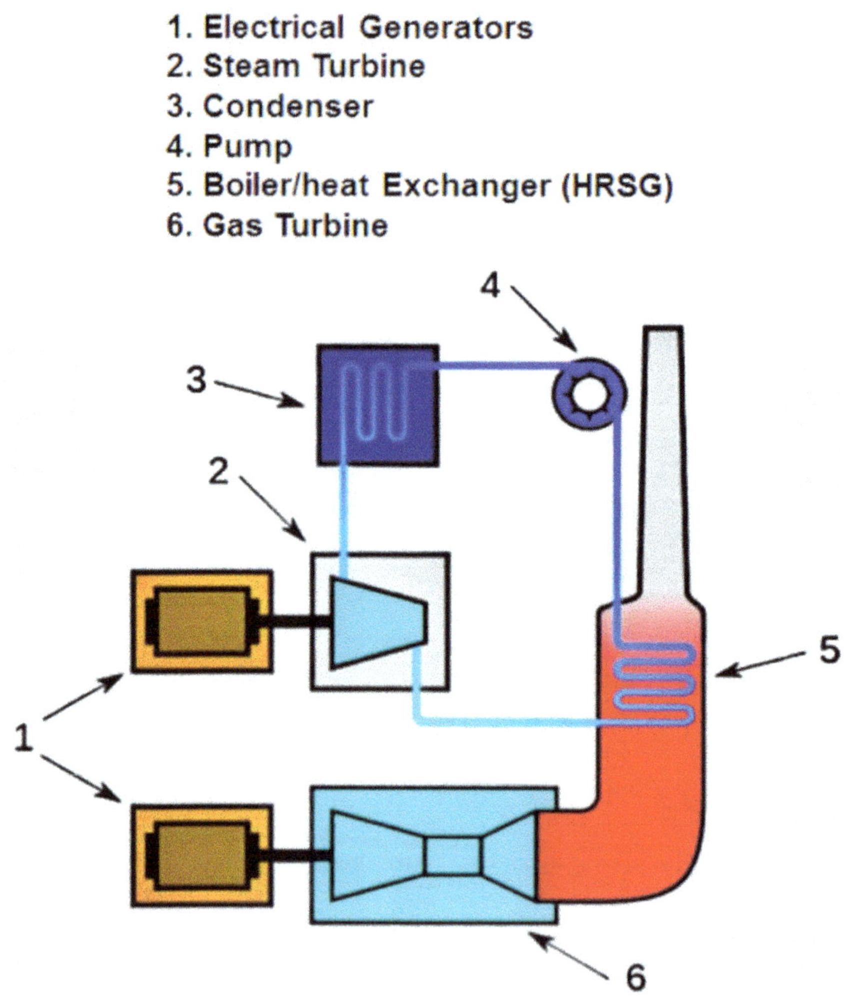

Figure 1.4 Diagram of Combined Cycle Power Plant

Many types and designs of industrial boilers are used in

- Refineries/chemical plants
- Paper mills
- Any industry that uses heat or steam or generates electrical power for internal use.

These units are usually smaller than utility boilers but have similar deposit problems (see Sections 1.4 and 3.2). They can be used to generate steam for processes and can also generate electricity for the plant.

Steam turbines of many sizes are used to convert the steam from a boiler or another source into electrical energy or for some other purpose. Figure 1.5 shows the turbine blades in an electrical generating steam turbine. These critical surfaces can become fouled, thus affecting power generation, and need periodic cleaning. See Section 8.4.3 for cleaning methods.

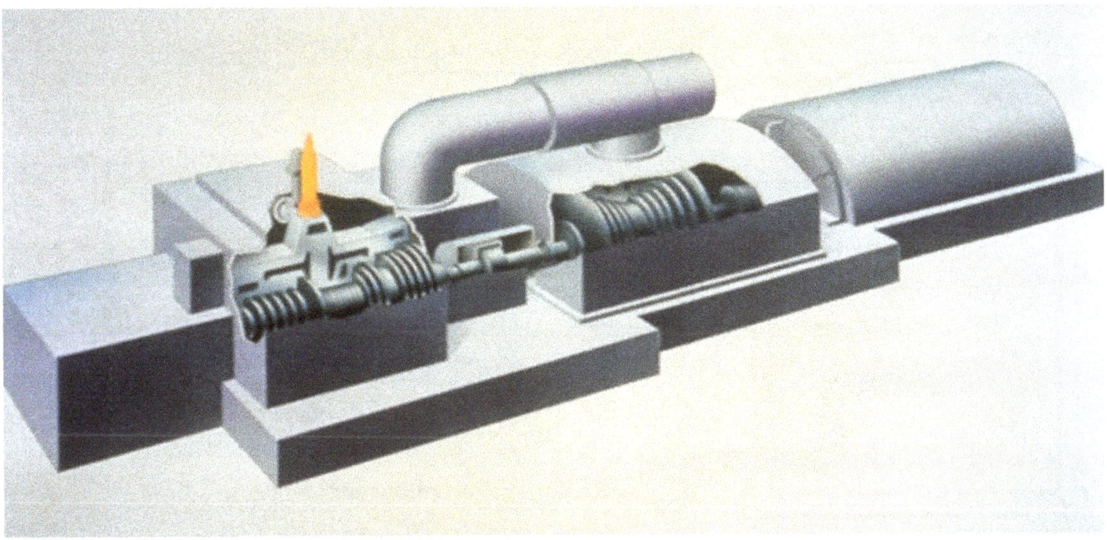

Figure 1.5 Steam Turbine and Turbine Blades

1.3.2. Nuclear Power Generating Plants

Unlike conventional power/steam generating plants, nuclear power plants generate heat by nuclear fission processes instead of by burning fossil fuels. The steam produced in these units is then used to run turbines and produce electricity. Wikipedia[3] notes that the plants can be categorized not only by the moderator (controller) of the nuclear fission reactions but also by the steam production design. In the United States, the reactors use light water (H_2O) as a moderator. Plants in Canada or in other countries may use deuterium oxide (D_2O) as a moderator and are called heavy water reactors.

Steam can be produced directly from the reactor heat (a boiling water reactor- BWR). See Figure 1.6, with details from the U.S. Nuclear Regulatory Commission,[24] who outline the process of a typical design concept of a commercial BWR:

1) The core inside the reactor vessel creates heat.
2) A steam-water mixture is produced when very pure water (reactor coolant) moves upward through the core, absorbing heat.
3) The steam-water mixture leaves the top of the core and enters the two stages of moisture separation, in which water droplets are removed before the steam can enter the streamline.
4) The streamline directs the steam to the main turbine, causing it to turn the turbine generator, which produces electricity.

The unused steam is exhausted to the condenser, where it is converted into water. The water is then pumped out of the condenser by a series of pumps, is reheated, and is pumped back to the reactor vessel. The reactor's core contains fuel assemblies that are cooled by water circulated using electrically powered pumps. These pumps and other operating systems in the plant receive power from the electrical grid. If offsite power is lost, emergency cooling water is supplied by other pumps, which can be powered by onsite diesel generators. This design is similar to a drum boiler in a conventional plant (except for the radioactivity). There are several operating BWRs in the United States.

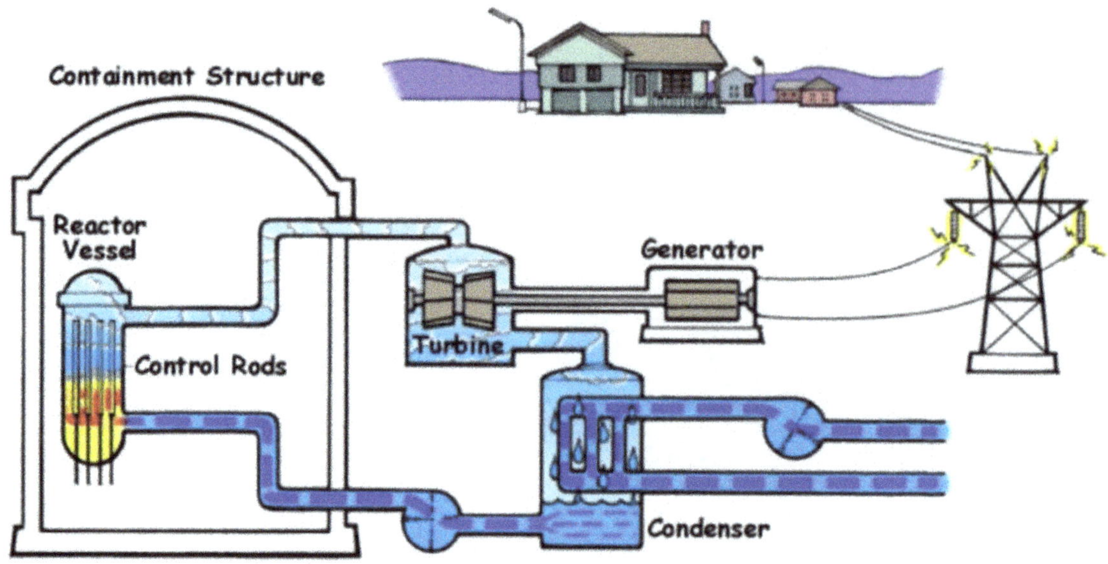

Figure 1.6 Layout of a Boiling Water Reactor (BWR)

Another type of nuclear reactor contains a secondary heat exchange loop, in which the water heated in the *primary* circuit takes heat from the radioactive pile and then heats water in the *secondary* loop. This produces the steam that drives the turbines. Because the primary loop is pressurized, these are called pressurized water reactors (PWR).

Reactors in Canada (CANDU units) use heavy water as the fission moderator but are pressurized and use steam generators. In other countries, many different designs are used. See publications from the World Nuclear Association[25] and the U.S. Nuclear Regulatory Commission[26] for more details of different designs and the fissionable fuels in use. Section 3.8 presents information about the deposits that form, and Chapter 5 discusses cleaning processes.

CANDU reactors and many light water reactor systems use a large heat exchanger, called the steam generator. See Figure 1.7 for a possible layout of the components. Figure 1.8 is a schematic of a steam generator with a drawing of a support plate that shows the annulus clogged with a deposit.

Figure 1.9 shows one of the several designs of a steam generator. Gay[27] shows additional designs. Note that designs will have changed over the many years of the development and use of PWRs.

In a typical design concept of a commercial PWR, the following process occurs:

1) The core inside the reactor vessel creates heat.
2) Pressurized water in the primary coolant loop carries the heat to the steam generator.
3) Inside the steam generator, heat from the primary coolant loop vaporizes the water in a secondary loop, producing steam.
4) The streamline directs the steam to the main turbine, causing it to turn the turbine generator, which produces electricity.

The unused steam is exhausted to the condenser, where it is condensed into water. The water is then pumped out of the condenser by a series of pumps, is reheated, and is pumped back to the steam generator. The reactor's core contains fuel assemblies that are cooled by water circulated using electrically powered pumps. These pumps and other operating systems in the plant receive power from the electrical grid. If offsite power is lost, emergency cooling water is supplied by other pumps, which can be powered by onsite diesel generators. Other safety systems, such as the containment cooling system, also need electric power. PWRs contain between 150-200 fuel assemblies. In the United States and the rest of the world, the PWR design is dominant.

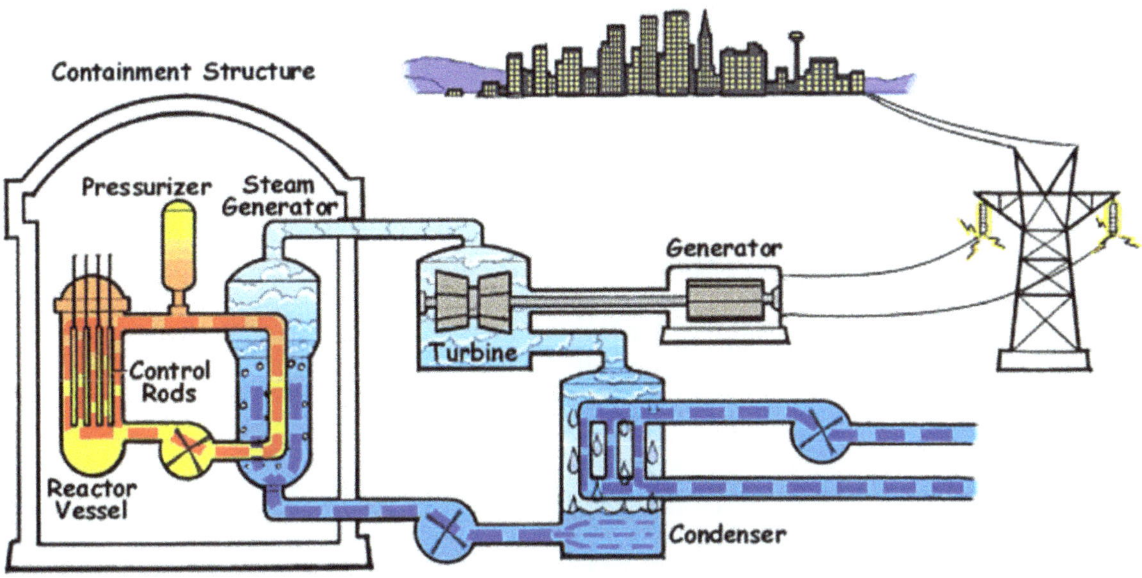

Figure 1.7 Pressurized Water Reactor (PWR) Plant Layout[26]

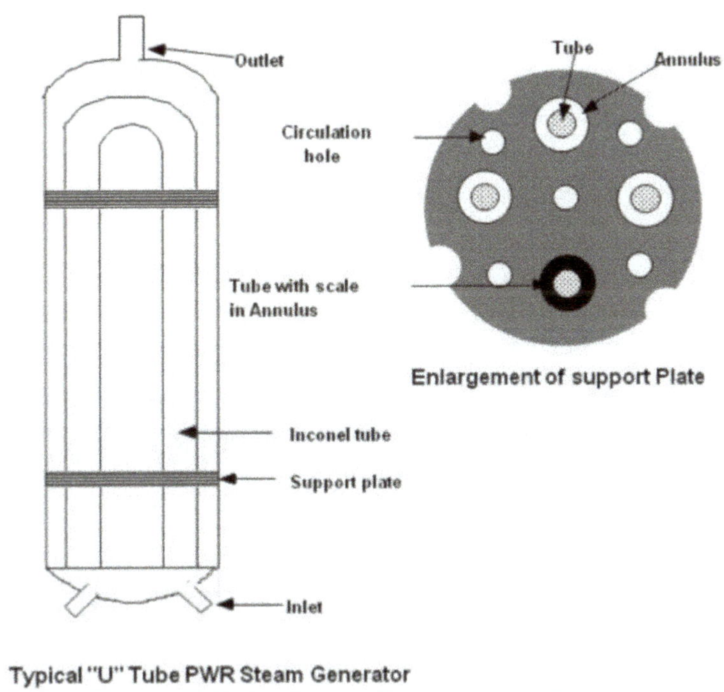

Figure 1.8 Pressurized Water Reactor (PWR) Steam Generator and Support Plate Detail

The metallurgy of the parts of the steam generator that contact water is important for the types of deposits that may form. Note in Figure 1.8 that the details of the support plate show that scale has filled the annulus and may damage the tube.

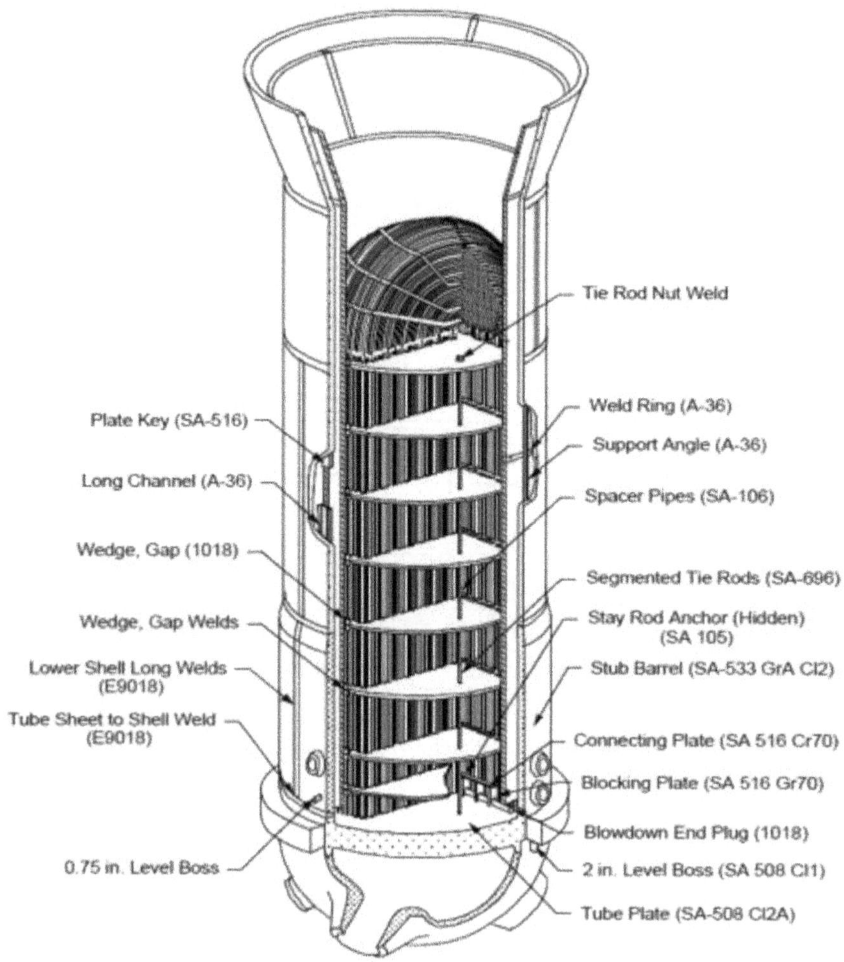

Figure 1.9 Details of a Steam Generator with Metal Designations[28]

Table 1.1 shows the nominal chemical specifications for some of the alloys in the steam generator shown in Figure 1.9. These metals have different corrosion characteristics and produce scaling deposits that differ from those of some non-nuclear boilers. These alloys must be considered when planning cleaning activities.

Table 1.1 Nominal Metal Chemical Specifications for Figure 1.9 Alloys

Code	C	Mn	Si	P	S	Cr	Mo	NI	Cu
E9018	0.07	.66	.48	0.19	.014	2.25	1.05		
AISI 1018	0.18								
SA 516	0.27-0.31	0.79 - 1.3	0.13-0.45	0.035 max	0.035 max				
A-36	.26	.75		<.04	<.04				.2
SA 106	.25-.35	.27-1.07	.10 min	.035 max	.035 max	.40 max	.15 max	.40	.40 max
SA 696									
SA 508	.19	1.35	.08	.006	.002	.17	.51	.82	
SA 533	.25	1.07-1.62		<.035	.15-.45		.45-.64	0-1	
SA 105	.35	.6-1.5	.1-.35	<.035	<.04	.31	<.12	<.12	

1.3.3. Cooling and Service Water Systems

The ability to cool process water and recycle it for reuse (or possible disposal) is integral to most industrial operations. The function of a cooling system is to remove heat from processes or equipment. Heat removed from one medium is transferred to another medium, or process fluid. Most often, the cooling medium is water. Figure 1.10 shows a simplified open cooling water system with a tower.

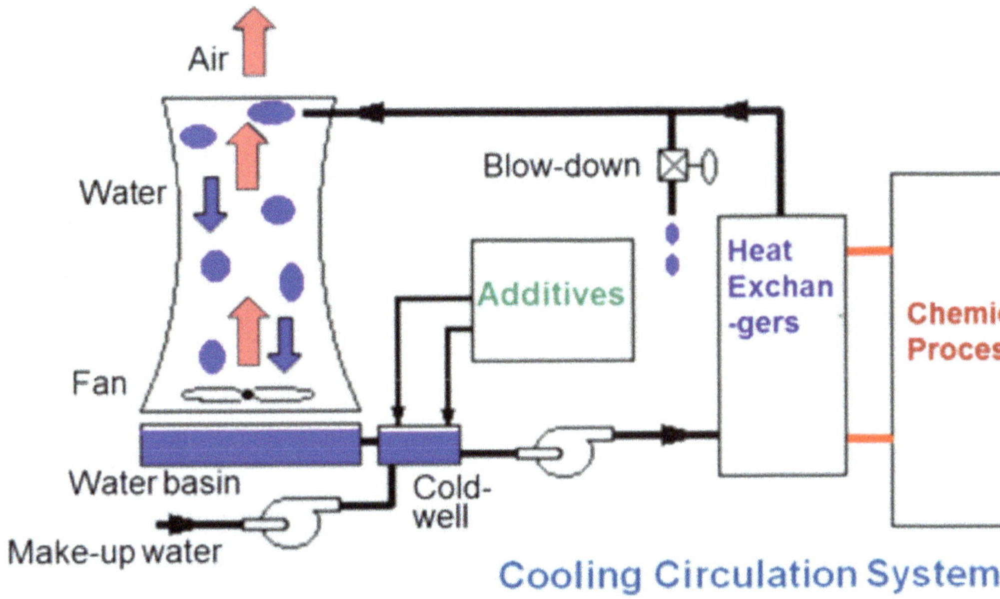

Figure 1.10 Diagram of a Cooling Water System

However, heat transfer concepts also can be applied to other fluids. These processes frequently cycle part of the water through open cooling towers where evaporative processes cool the water (see General Electric's *Handbook of Industrial Water Treatment*[29]). NACE[30] also provides comprehensive information about scaling and corrosion in cooling water systems.

Because the towers in many systems are excellent aerators, the water will become saturated with O_2 and possibly will be contaminated with CO_2 and other atmospheric gases. Contact with microbes also produces organic slimes and possibly MIC.

The gases make the water corrosive and capable of forming iron-containing scale. CO_2 also can be a source of ions that can result in calcium-based deposits. McCann[31] notes that the use of evaporative cooling, as in a tower, causes buildup of salt concentrations in the cooling water. Drawing off some of this briny water and making up the losses control how salty it becomes. Such blow down, of course, also removes some of the chemical additives used to inhibit corrosion and fouling. See Frenier[32] for examples of scale and corrosion control chemicals that may be useful.

Service water is a term that may include any industrial water use that does not need to meet drinking water standards. As well as being used in cooling systems, service water systems may be part of fire suppression or for any other requirements of the plant for large volumes of water.

Oil and gas facilities and most industrial plants frequently have systems where water is used to cool or to act as a heat transfer medium. These water-based heat transfer processes occur in many onshore and offshore facilities and especially in the various gas-treating systems described in Frenier.[2]

Since thousands of gallons of water can flow in these systems, corrosion is a constant issue. Various facilities that use water for cooling or for steam generation will develop corrosion scales if oxygen levels exceed about 0.5 ppm. For boiler feed water, levels above 20 ppb may not be acceptable.[33] Figure 1.11 from Xie et al.[34] shows the fouled surfaces and cleaned tubes of a heat exchanger where O_2 leaked into the shell side (water) and produced both iron scale and pitting.

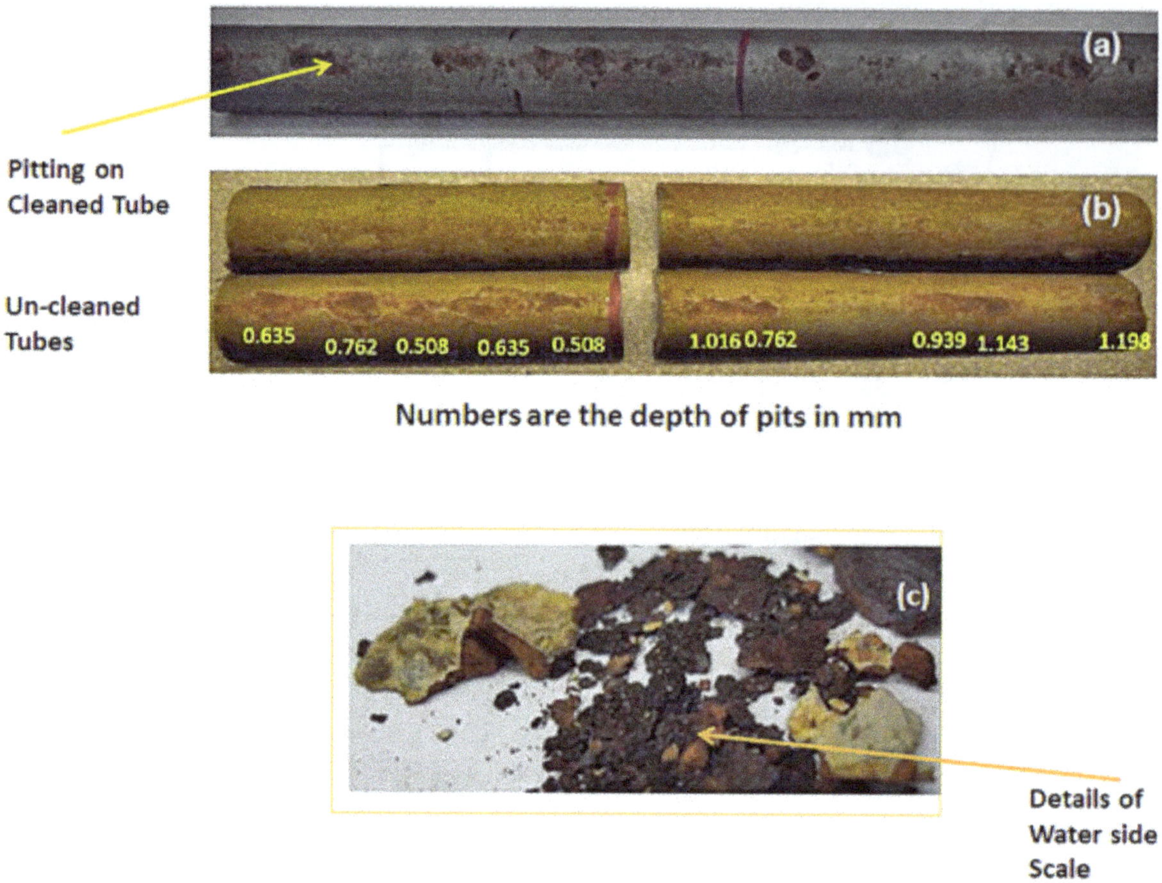

Figure 1.11 Oxygen-Related Corrosion of a Heat Exchanger. (a) HCl cleaned, (b) uncleaned, and (c) magnified scale[34]

The reactions that take place in these systems produce low-temperature solids but also necessitate control or removal processes.

Cooling water and service water systems are integral to the long-term operations of nuclear power systems. A report by Licina[35] claims that the service water system is a critical part of the plant infrastructure, is crucial to achieving a plant life of 40 years, and is also key to life extension. The primary mode of degradation for service water piping is general corrosion. A typical corrosion allowance of 0.102–0.305 cm (0.040–0.120 inches; 0.0254–0.0762 mm/y [1–3 mpy]) is applied for uncoated carbon steel piping to account for degradation in service. Service water systems may attain 25–40 years' life without much special attention; however, for longer lives, some repair or replacement should be expected. The keys are to take appropriate mitigation actions early to avoid surprises or to do the required repair or replacement once and no more.

Because the plant is built over piping or around it and because piping is not intended to be replaced or even worked on very much, access for repairs, refurbishments, or replacements that may be required to achieve design life or plant life extension can be difficult. Critical issues are prediction and evaluation of the condition of the piping; inspections to verify piping condition; and materials selection, fabrication, and inspection methods for temporary and permanent repairs, for refurbishments (e.g., cleaning, coatings, and linings), and for replacements (metallic and nonmetallic). Figure 1.12 shows an example of extreme corrosion in a plant's water system.

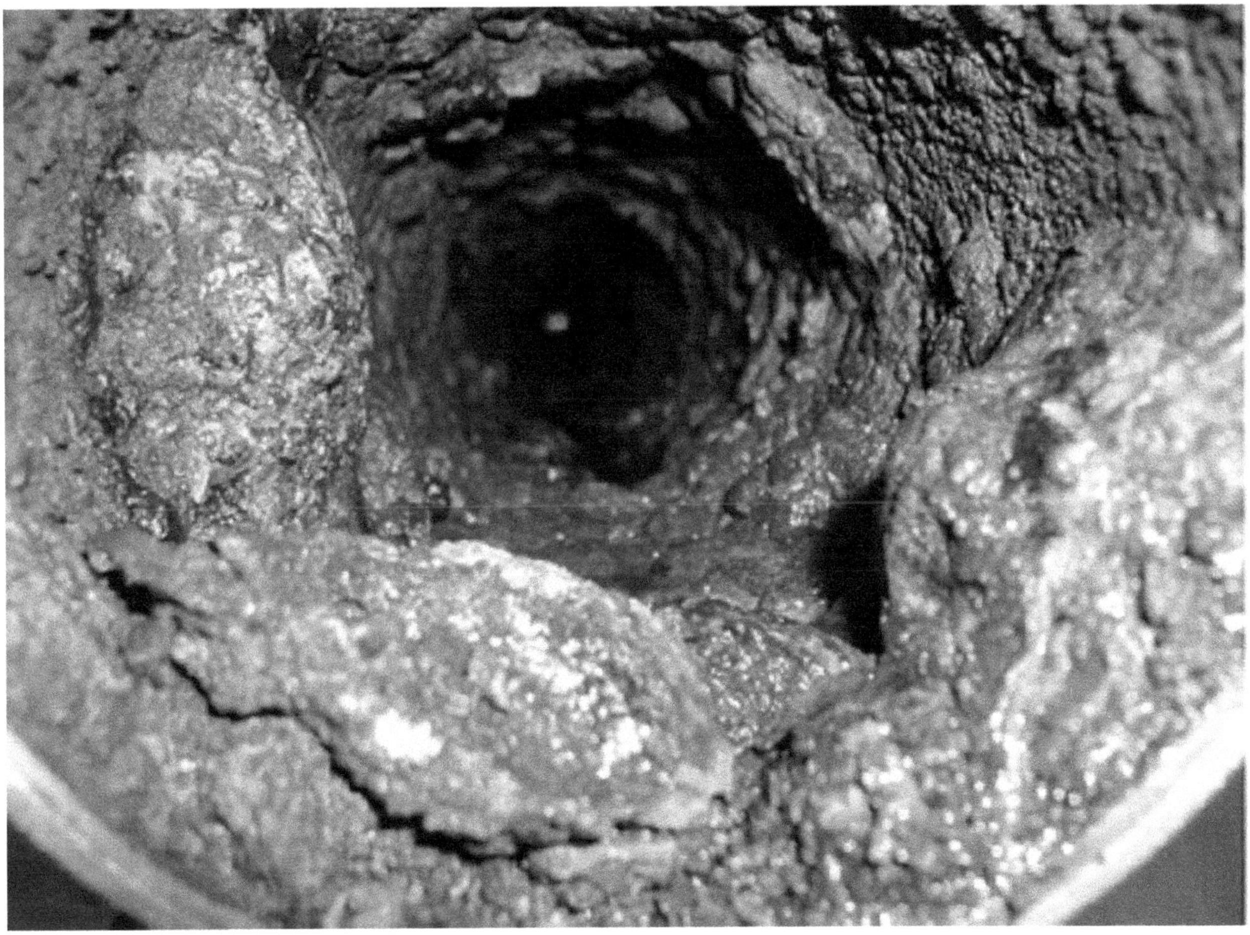

Figure 1.12 Service Water System Corrosion Scale

See Frenier and Ziauddin[36] and Davies and Scott[33] for more information on aqueous corrosion issues. Some closed systems rely entirely on heat transfer from one process to another. These systems also will develop scale because leaks and makeup water can introduce contaminants.

1.3.4. Chemical Processing Plants (Refineries, Chemical Plants, and Paper Mills)

This category includes essentially "everything else" in the non-nuclear industrial sector that may require cleaning. These plants have industrial boilers and cooling water systems, but they also process various chemical products. Raw materials (such as crude oil, refined products, and other chemicals as well as wood chips) come into the plant. Various chemical reactions are performed that usually require the application of heat, pressure, and various heat transfer operations. Various finished (or partially finished) products are sent to consumers or to other industries. Compared with the chemical reactions in power utility plants, the chemical reactions in chemical processing plants produce a large variety of organic or mixed organic/inorganic byproducts that may foul important units. Several examples of commonly treated industrial systems are discussed below.

Refineries process and produce products from crude oil, and chemical plants process thousands of different chemicals and produce more chemicals that are used by direct consumers and other industries. For the purposes of this discussion, paper mills are categorized as chemical plants because raw products (wood products) are processed into a wide variety of finished products, and many chemical engineering unit operations take place. The boilers and other parts of the mills require periodic cleaning.

Figure 1.13 shows a simplified layout of a petroleum refinery whose main function is to process crude oil into various useful chemical streams using furnaces, separation towers, pumps, and dozens of heat exchangers (not shown). Each refinery will have a specific set of unit operations and equipment based on the crude stream and the final products. Figure 1.14 and Figure 1.15 show two different types of heat exchange equipment in which a process stream is heated or cooled.

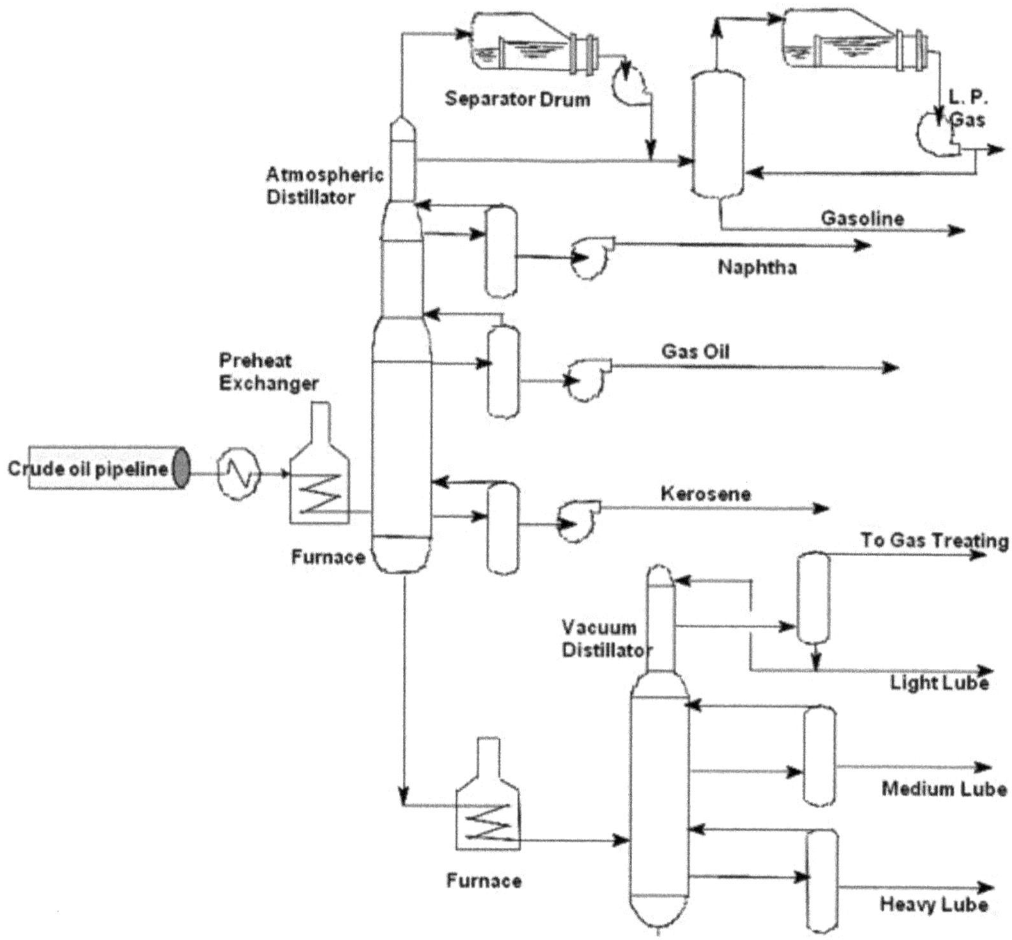

Figure 1.13 Basic Refinery Processes

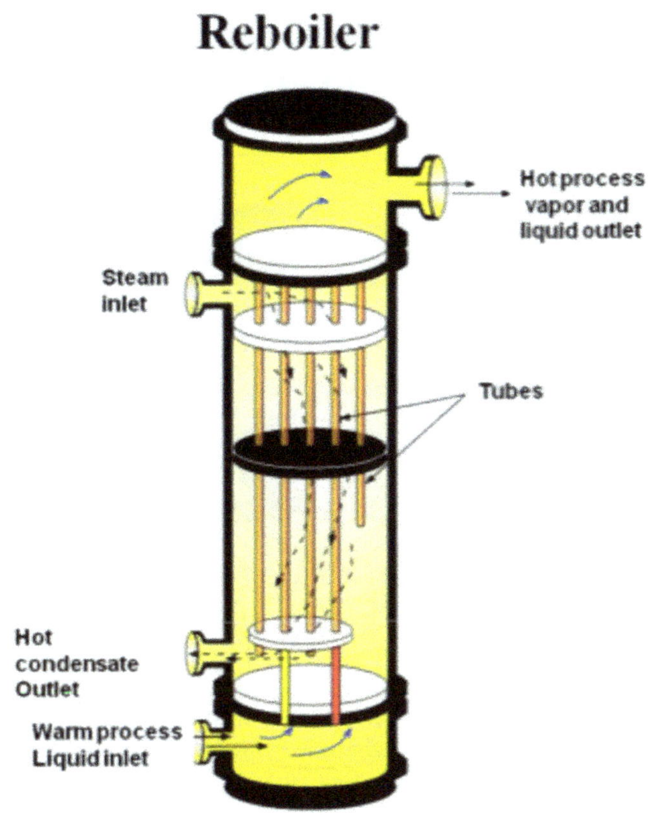

Figure 1.14 Diagram of a Reboiler

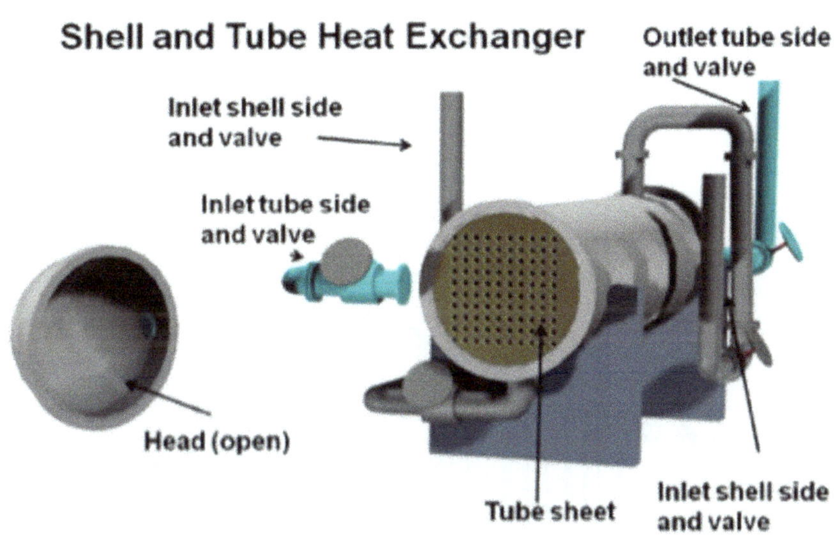

Figure 1.15 Shell and Tube Heat Exchanger

Figure 1.16 is a diagram of a bubble-cap counterflow tower that could be used for many purposes (in this case, for acid gas removal). Paper mills are another type of chemical plant where wood chips are reacted and turned into paper products, and other chemicals from the wood also are produced.

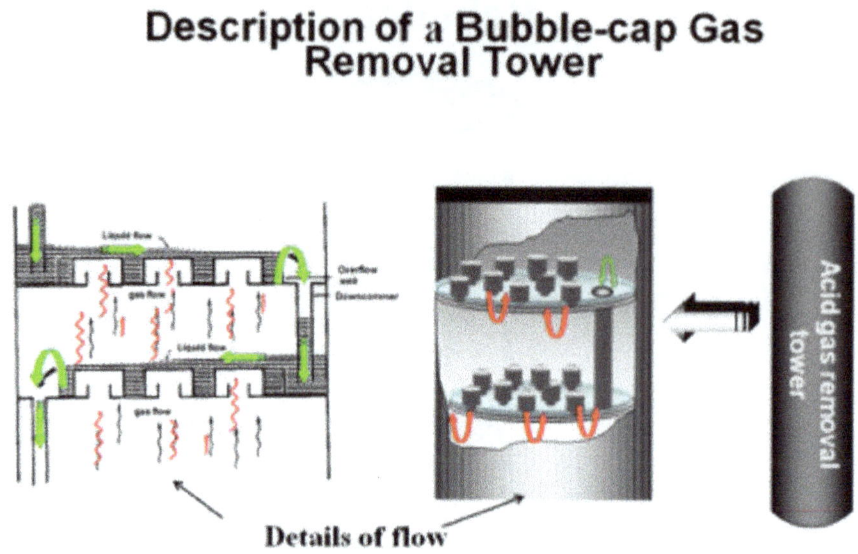

Figure 1.16 Details of an Acid Gas Removal Tower

An example of a paper mill is provided by Perdomo et al.,[37] who notes that Kraft digesters and evaporators are typically acid cleaned to remove scales that precipitate from chemical species present in liquors. Figure 1.17 shows a diagram of a Kraft-type paper mill.[38] The first phase uses an alkaline fluid (white liquor) in the digesters, producing a corrosive byproduct called black liquor. Both fluids can produce scale. Perdomo et al.[37] claims that the solids are mixtures of burkiete ($Na_2CO_3 2Na_2SO_4$) and carbonate scales on heat exchanger surfaces.

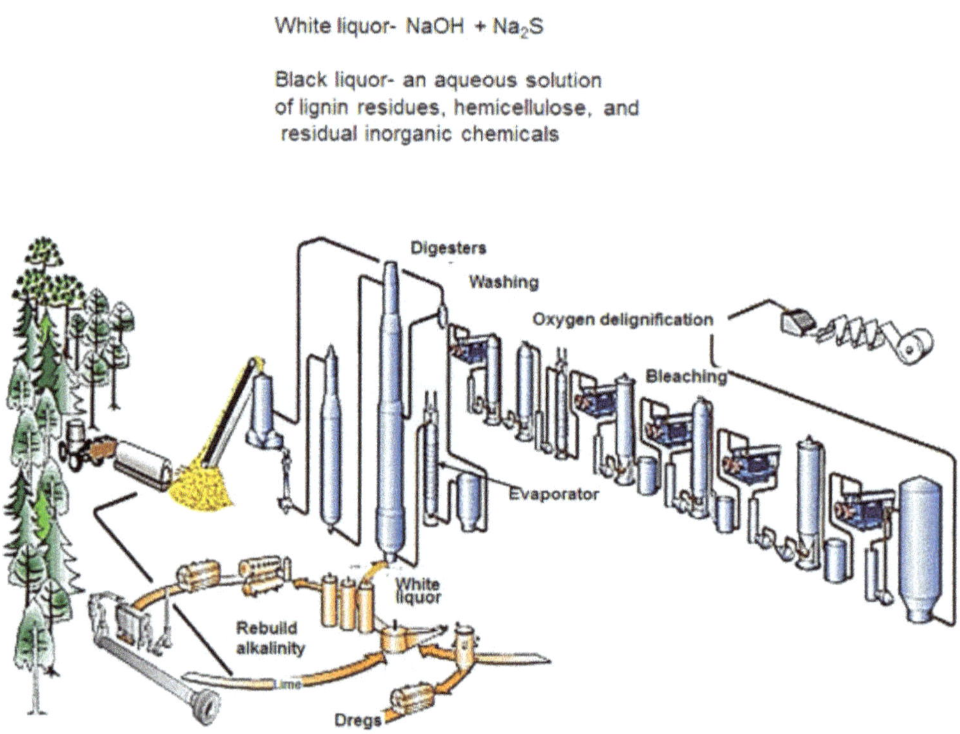

Figure 1.17 Diagram of a Kraft-Process Paper Mill[38]

Though such scales may not be harmful from the corrosion point of view, their presence and removal may affect the base metal. Scale removal may in turn affect the condition and integrity of the equipment by causing general or localized corrosion, which can lead to unexpected failures.

Note that some of these types of plants will become fouled by organic and inorganic solids that interfere with smooth and safe production.

1.4. Fouling Deposits Found in Industrial Equipment

This section introduces the wide variety of fouling solids that require removal. More details are in Chapters 3 and 7. Corrosion-related fouling deposits and equilibrium-related solids are the major reasons for performing chemical and mechanical cleaning operations. Most fouling deposits that form in process equipment have their origins in the process fluids themselves. Water-based fluids can transport and deposit a wide variety of minerals. In addition, corrosion products form due to the reaction of aqueous fluids with the metals of construction.

Hydrocarbon and petrochemical fluids transport and deposit various organic compounds, most of which include polymeric materials. Hydrocarbons also can transport sulfides and mercaptans, which can react with the construction materials to form iron or copper sulfides. Table 1.2 summarizes common inorganic scaling compounds. A more extensive list of inorganic scale is in Cowan and Weintritt.[13]

These inorganic scales can be further defined as corrosion product scales or salts that form due to changes in the equilibria of ions in an aqueous fluid. Figure 1.18 shows examples of corrosion product scales: industrial tubing fouled with black iron oxide and copper, red iron oxide, and iron sulfide (another salt).

Table 1.2 Summary of Common Inorganic Scale-Forming Compounds

Mineral Name	Common Name	Chemical Formula
Acmite	Sodium iron silicate	$NaFe(SiO_3)_2$
Analcite	Sodium aluminum silicate	$NaAlSi_2O_6 \cdot H_2O$
Anhydrite	Calcium sulfate	$CaSO_4$
Aragonite	Calcium carbonate-rhombic	$CaCO_3$
Barite	Barium sulfate	$BaSO_4$
Brucite	Magnesium hydroxide	$Mg(OH)_2$
Calcite	Calcium carbonate-hexagonal	$CaCO_3$
Chalcocite	Copper sulfide	Cu_2S
Chalcopyrite	Copper iron sulfide	$CuFeS_2$
Chromite	Iron chromium spinels	$CrFe_2O_4$
Copper	Copper	Cu
Covellite	Copper sulfide	CuS
Cuprite	Copper oxide	Cu_2O
Gibbsite	Aluminum hydroxide	$Al(OH)_3$
Goethite	Hydrous ferric oxide	FeOOH
Gypsum	Calcium sulfate dihydrate	$CaSO_4 \cdot 2H_2O$

Halite	Salt	$NaCl$
Hematite	Ferric oxide	Fe_2O_3
Hydromagnesite	Magnesium carbonate and hydroxide	$3MgCO_3 \cdot Mg(OH)_2 \cdot 3H_2O$
Hydroxyapatite	Calcium phosphate	$Ca_{10}(OH)_2(PO_4)_6$
Mackinawite	Iron sulfide	FE_xS_y
Magnesia	Magnesium oxide	MgO
Magnetite	Ferric-ferrous oxide	Fe_3O_4
Magnesite	Magnesium carbonate	$MgCO_3$
Montmorillonite	Aluminum silicate	$Al_2O_3 \cdot 4SiO_2 \cdot 4H_2O$
Noselite	Sodium aluminum silicate	$Na_8Si_2O_7O_{24} \cdot H_2O$
Pyrite	Iron sulfide	FeS_2
Pyrolusite	Manganese dioxide	MnO_2
Pyrrhotite	Iron sulfide	FeS_2
Serpentine	Magnesium silicate	$Mg_3Si_2O_7 \cdot 2H_2O$
Siderite	Iron carbonate	$FeCO_3$
Silica	Quartz	SiO_2
Silica, amorphous	Hydrated silica	$SiO_2 \cdot 2H_2O$
Sodalite	Sodium aluminum silicate	$Na_8Al_6Si_6O_{24} \cdot 2Cl$
Wüstite	Ferrous oxide	FeO

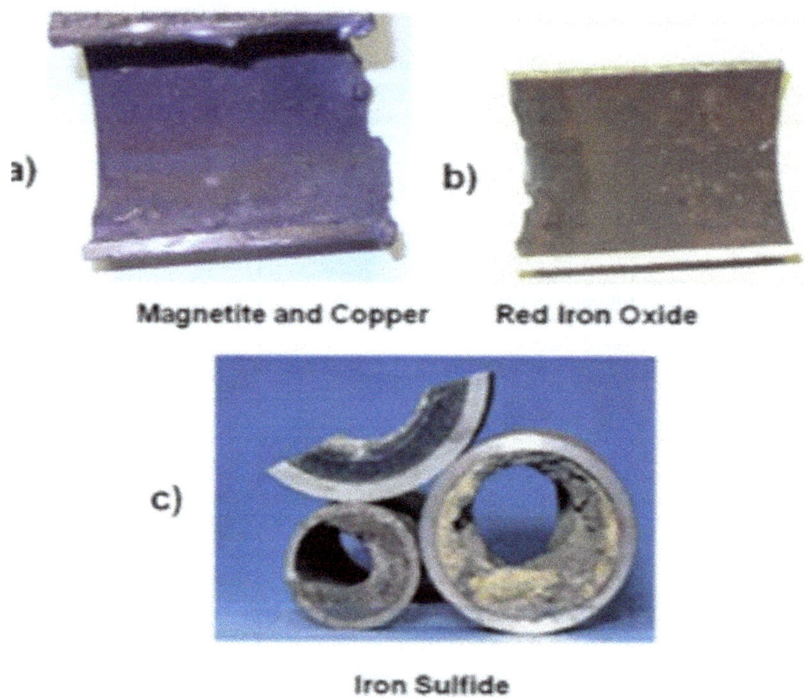

Figure 1.18 Corrosion Product Scale. Industrial tubing fouled with (a) black iron oxide and copper, (b) red iron oxide, (c) and iron sulfide (another salt).

Heifner[39] claims that deposits and/or scale within boiler tubes interrupt heat transfer, creating an insulating barrier resulting in higher tube metal temperatures and a slight loss in the thermal efficiency of the boiler. Higher tube metal temperatures first lead to accelerated material degradation by creep. Thus, it is desirable to remove such insulating deposit buildup before creep occurs and degrades boiler tubes. With deposits greater than 0.16 cm [1/16 in], under deposit hydrogen damage, acid phosphate corrosion, and caustic gouging can occur. Over time, such degradation will cause tube failures. See an example in Figure 1.19.

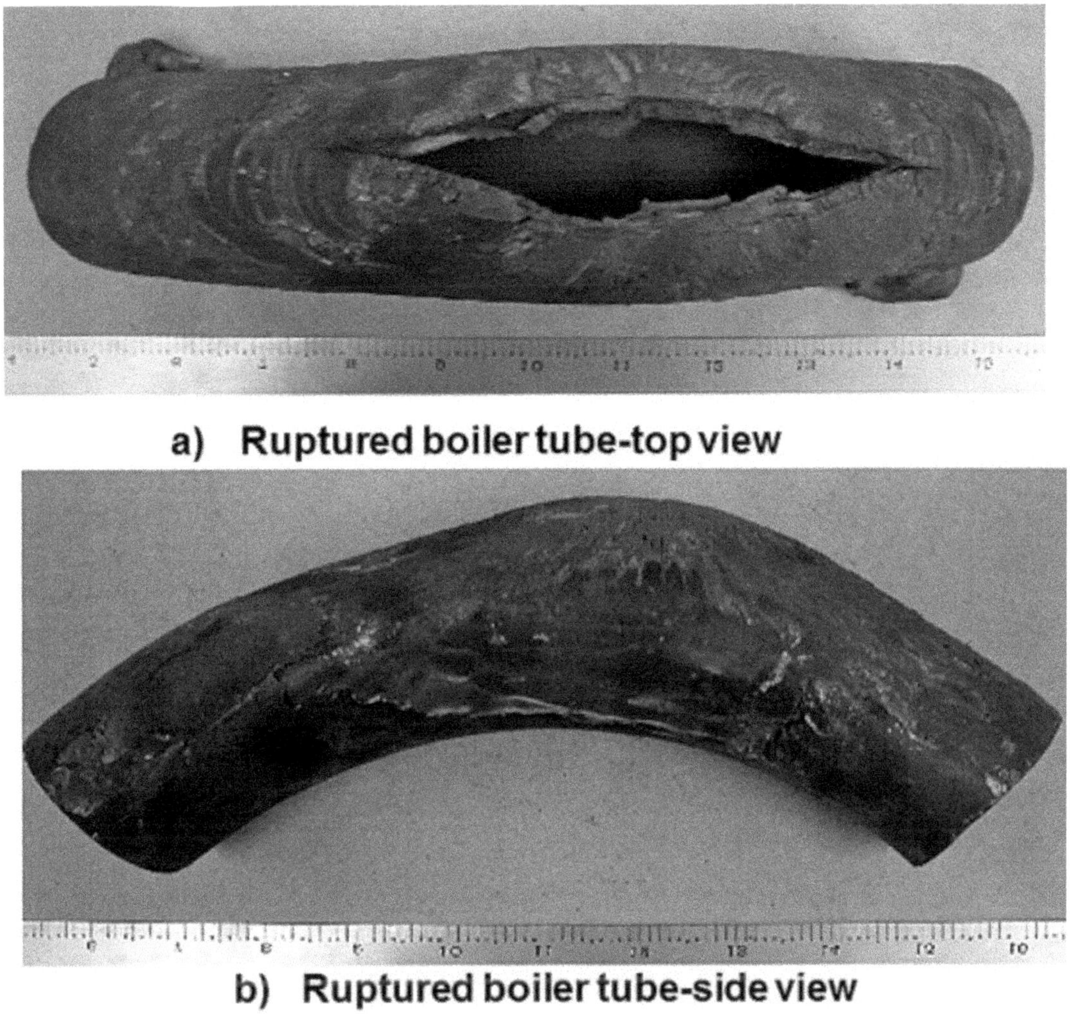

Figure 1.19 Ruptured Boiler Tubes. (a) Top view and (b) side view.[39]

Figure 1.20 shows two other types of precipitation (equilibrium change) deposits: calcite and wax. Other organic deposits also can form from the process fluids and may exacerbate pitting corrosion and cause fouling in chemical plants and refineries. Figure 1.21 shows a fouled heat exchanger and pitting of the tubes.[40]

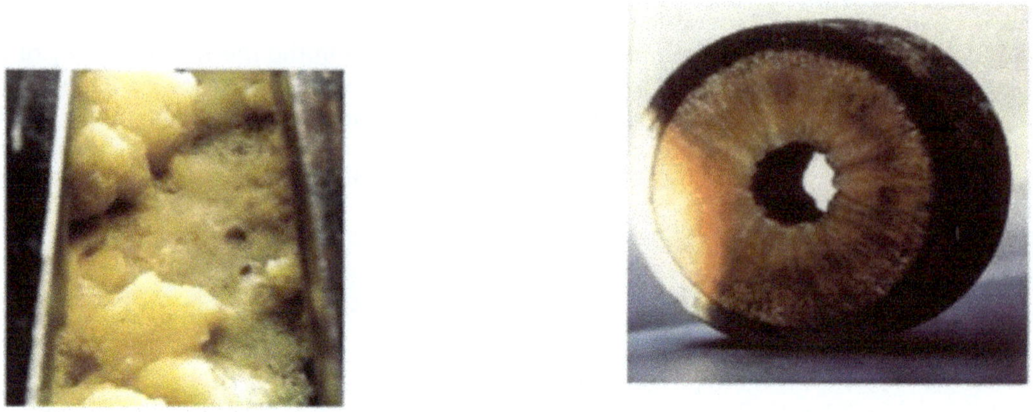

ı) Inorganic (Calcite) Deposit b) Organic (Wax) Deposit

Figure 1.20 Precipitation (Equilibrium Deposits): (a) Calcite and (b) wax

Details and formation mechanisms of various classes of deposits are covered in subsequent chapters: for inorganic solids, see Sections 3.3, 3.4, and 3.5; for nuclear plant solids, see Section 5.1; and for organic/refinery solids, see Section 7.4.

Figure 1.21 Fouled Heat Exchanger Bundle and Pitted Tubes[40]

1.5. Summary

This chapter introduced some of the technologies involved in chemical and mechanical cleaning of industrial equipment and

- Defined important IC terms
- Identified types of equipment that need to be cleaned
- Identified the types of fouling deposits

These aspects affect the planning and execution of the job and will be described in much more detail in subsequent chapters.

References

[1] W.W. Frenier, *Technology for Chemical Cleaning of Industrial Equipment* (Houston, TX: NACE International, 2001).

[2]W.W. Frenier, *Chemical and Mechanical Methods of Pipeline Integrity* (Richardson, TX: Society of Petroleum Engineers, 2017).

[3]"Nuclear Reactor," Oct. 27, 2014, Wikipedia, http://en.wikipedia.org/wiki/Nuclear_reactor (Nov. 2, 2014).

[4]Clean Harbors, "Industrial Services," 2013, Clean Harbors (Lowell, MA), http://www.cleanharbors.com/browse_by_service/industrial_services/ (Nov. 21, 2014).

[5]G. Hagen, "Uber Die Bewegung Des Wassers in Engen Zylinderschen," *Ann. Phys. Chem.* 46 (1839): 423-442.

[6]J.L. Poiseuille, *Compte Rendus* 11 (1840): 961, 1041.

[7]W.W. Frenier, M. Ziauddin, *Chemistry for Enhancing the Production of Oil and Gas* (Richardson, TX: Society of Petroleum Engineers, 2014).

[8]S.A. Ali, L. Kalfayan, C. Montgomery, *Acid Stimulation* (Richardson, TX: Society of Petroleum Engineers, 2016).

[9]L. Kalfayan, *Production Enhancement with Acid Stimulation*, 2nd ed. (Tulsa, OK: PennWell, 2008).

[10]J. Grebe, "Treatment of Deep Wells," U.S. Patent 1877504, 1932.

[11]A.D. Nesbitt, "A History of the Chemical Cleaning Business," CORROSION/93, paper no. NACE 93362 (Houston, TX: NACE International, 1993).

[12]J.W. McCoy, *Industrial Chemical Cleaning* (New York, NY: Chemical Publishing Co., 1984).

[13]J.C. Cowan, D.J. Weintritt, *Mineral & Organic Scale Deposits* (Lafayette, LA: Cowan & Weintritt Publishing Company, 2004).

[14]E.C. Wackenhuth, "Manual of Chemical Cleaning of Fossil-Fueled Steam Generation Equipment" (Palo Alto, CA: Electric Power Research Institute, 1984).

[15]NACE International, TPC Publication 8, "Industrial Cleaning Manual" (Houston, TX: NACE, 1982).

[16]NACE International, *Bibliography of Chemical Cleaning of Metal, Vols. I-III* (Houston, TX: NACE, 1985).

[17]J. Gutzeit, "Cleaning of Process Equipment and Piping," MTI Publication No. 51 (St. Louis, MO: Materials Technology Institute of the Chemical Process Industries, 1999).

[18]W.W. Frenier, "Review of Hse Innovations in the Chemical Cleaning Industries," CORROSION/2015, paper no. NACE C2015-5553 (Houston, TX: NACE, 2015).

[19]U.S. Department of Energy, "Annual Electric Generator Report" (Washington DC: U.S. Department of Energy, 2009).

[20]"Boiler Designs," Babcock & Wilcox Company (Charlotte, NC), http://www.babcock.com/Pages/results.aspx?k=boiler (Nov. 4, 2014).

[21]"Universal Pressure Boiler," Babcock & Wilcox Company (Charlotte, NC), http://www.babcock.com/products/Pages/Universal-Pressure-Boiler.aspx (Oct. 19, 2016).

[22]"Boiler," Wikipedia, Oct. 6, 2016, https://en.wikipedia.org/wiki/Boiler (Oct. 24, 2016).

[23]"Typical Coal Thermal Power Station," Wikipedia, Sept. 14, 2016, https://en.wikipedia.org/wiki/Thermal_power_station (Sept. 19, 2016).

[24]U.S. Nuclear Regulatory Commission, "The Boiling Water Reactor," U.S. Nuclear Regulatory Commission (Washington, DC), 2015, http://www.nrc.gov/reading-rm/basic-ref/students/animated-bwr.html
[25]World Nuclear Association, "Nuclear Power in the World Today," World Nuclear Association (London, U.K.), April 2014, http://www.world-nuclear.org/info/Current-and-Future-Generation/Nuclear-Power-in-the-World-Today/ (Nov. 5, 2014).
[26]Nuclear Regulatory Commission, "Pressurized Water Reactor," Nuclear Regulatory Commission, (Washington, DC), 2016, http://www.nrc.gov/reading-rm/basic-ref/students/animated-pwr.html (June 16, 2016).
[27]J.M. Gay, "The Role of Visual Inspection in Planning for Chemical and Mechanical Cleaning," CORROSION/98, paper no. NACE 98329 (Houston, TX: NACE, 1998).
[28]M.J. Little, R.D. Varrin Jr., A.T. Pellman, M.A. Kreider, "Advanced Scale Conditioning Agent (ASCA) Applications: 2010 Experience Update," EPRI SGMP 2010 SG Secondary Side Management Conference, San Antonio, TX, March 2–4, 2010.
[29]"Handbook of Industrial Water Treatment," General Electric Company (Fairfield, CT), 2012, http://www.gewater.com/handbook/index.jsp (Nov. 4, 2014).
[30]NACE Publication 46107, "Control of Corrosion, Deposition, and Microbiological Growth in Recirculating Water Systems" (Houston, TX: NACE International, 2007).
[31]M.J. McCann, "Cooling System Dynamics and Control," M.J. McCann Consulting, (Chadds Ford, PA), Aug. 19, 2014, http://www.mccannscience.com/cooling.htm (2016).
[32]W.W. Frenier, *Chemical Methods for Controlling Corrosion in Oil and Gas Activities* (Houston, TX: NACE International, 2017).
[33]M. Davies, P.J.B. Scott, *Oilfield Water Technology* (Houston, TX: NACE International, 2006).
[34]J. Xie, K. Yazdanfar, M. Butcher, T. Rockwell, "Corrosion of a Shell-and-Tube Heat Exchanger," CORROSION/2013, paper no. NACE-2013-2118 (Houston, TX: NACE, 2013).
[35]G. Licina, "Corrosion of Service Water Piping in Nuclear Systems," CORROSION/2007, paper no. NACE-07590 (Houston, TX: NACE, 2007).
[36]W.W. Frenier, M. Ziauddin, *Formation, Removal and Inhibition of Inorganic Scale in the Oilfield Environment* (Richardson, TX: Society of Petroleum Engineers, 2008).
[37]J. Perdomo, P.M. Singh, P. Conde, J. Mahmood, "Corrosion Prevention During Acid Cleaning in Digesters and Evaporators," CORROSION/2004, paper no. NACE-04246 (Houston, TX: NACE, 2004).
[38]"Kraft Process," Wikipedia, Nov. 4, 2016, https://en.wikipedia.org/wiki/Kraft_process (Nov. 5, 2016).
[39]S.C. Heifner, "Deposit Loading Characterization to Enhance Boiler Chemical Cleaning Decisions," CORROSION/2006, paper no. NACE 06460 (Houston, TX: NACE, 2006).
[40]M.A. Moore, M.M. Qarni, G.R. Lobley, "Corrosion Problems in Gas Treat Systems," CORROSION/2008, paper no. NACE 08419 (Houston, TX: NACE, 2008).

2 Review of Basic Chemical and Physical Processes Required in Industrial Cleaning Operations

This section introduces the important chemical and physical processes that form and remove fouling deposits, including surface chemical reactions, fluid mechanical concepts, corrosion reactions that generate fouling scales, and equilibrium scale deposition processes. Different methods for removing scales are introduced, and the final section (2.6) describes methods for evaluating field scales and testing solvents to remove them. More details and applications of these processes are discussed in Chapters 3–8.

2.1. Introduction to Surface Chemistry

Surface chemical reactions are at the core of all scale formation and industrial cleaning (IC) activities. The chemistry and physics of surfaces affect many processes and services performed in IC operations. These ideas are fundamental for understanding scale formation, corrosion, corrosion inhibition, and dissolution of scale on a surface.

The processes of "wetting" a surface are fundamental: if a surface is not in chemical contact with solvent, corrodent, or inhibitor, no reactions will take place.

The *surface* of any solid or liquid is an *interface* between that phase and others that could be a solid, a liquid, or a gas. In many situations, there will be three interfaces. The chemistry of the *interface* is affected by both surfaces. In one example (Figure 2.1), a drop of oil is placed on a solid surface in a jar of water. The forces at that surface (interfacial surface tension [IST-γ]) are caused by the cohesion of the molecules in that surface and depend on the molecules in all surfaces.

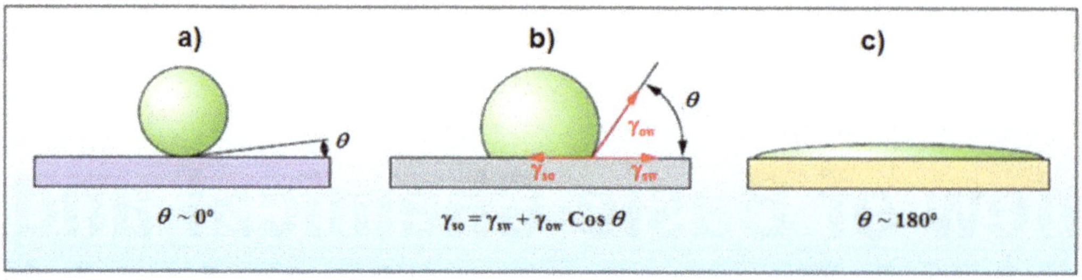

Figure 2.1 Wetting Angles of Water/Oil/Mineral Surface

Therefore, it is not a property of the liquid alone but of the liquid's interface with another material. In Figure 2.1(a), the surface is covered with a water film (hydrophilic), and the oil drop is repelled. At the point where the two surfaces meet (and three phases), they form a contact angle (θ), which is the angle that is tangent to the surface, with the solid surface. In this case, the contact angle is close to 0°. In Figure 2.1(b), the surface is partially *oil wet*, and in Figure 2.1(c), the surface is completely *oil wet*, and the oil droplet spreads out on the surface.

If a liquid is in a container (such as a partially filled pipe or tank), then there will be a liquid/gas interface at the liquid's surface and an interface between the liquid and the walls of the container. The IST between the liquid and gas is usually different than its IST with the container walls. Where two surfaces meet, all forces must be in balance.

Figure 2.2 shows an example of solid, liquid, and gas interfaces. Here, tension forces (*f*) are shown for the liquid-air interface, the liquid-solid interface, and the solid-air interface. Both the vertical and horizontal forces must cancel exactly at the contact point. The horizontal component of f_{la} is canceled by the adhesive force, f_A.

$$f_A = f_{la} \sin \theta \qquad \text{(Equation [2.1])}$$

The additional balance of forces is in the vertical direction. The vertical component of f_{la} must exactly cancel the force, f_{ls}.

$$f_{ls} = f_{sa} + f_{la} \cos \theta \qquad \text{(Equation [2.2])}$$

Since the forces are in direct proportion to their respective surface tensions, we also have:

$$\gamma_{ls} = \gamma_{sa} + \gamma_{la} \cos \theta \qquad \text{(Equation [2.3])}$$

Here, γ_{ls} is the liquid-solid IST, γ_{sa} is the solid-air IST, and γ_{la} is the liquid-air IST. This is known as Young's equation.

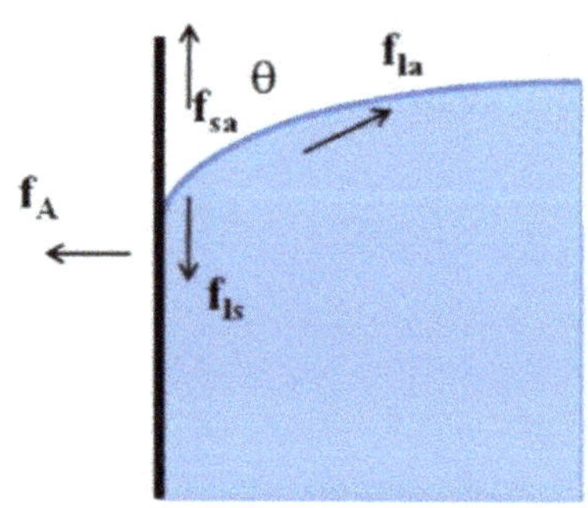

Figure 2.2 Forces at Surfaces

Table 2.1 lists some contact angle data for common fluids, and Table 2.2 shows some IST values for several liquids with air. The values of the contact angles and the IST control many of the properties of processes that will be described in subsequent chapters. Note in Table 2.2 that the IST of pure water is much higher than that of solutions of ethanol and water. Many surface-active agents (surfactants) also lower the IST of water and thus affect the contact of water with surfaces. The convention used in the oil and gas industry is that a near 180° contact angle indicates that the liquid will spread ("wet") on that surface. The reader should note which convention is being used when comparing data, as some conventions require wetting to have a 0° contact angle.

Table 2.1 Contact Angle Values Solid-Liquid (Sears and Zemanski[1]) Converted to Oilfield Convention

<table>
<tr><th>Liquid</th><th>Solid</th><th>Contact Angle</th></tr>
<tr><td>Water</td><td rowspan="6">Soda-lime glass
Lead glass
Fused quartz</td><td rowspan="6">180°</td></tr>
<tr><td>Ethanol</td></tr>
<tr><td>Diethyl ether</td></tr>
<tr><td>Carbon tetrachloride</td></tr>
<tr><td>Glycerol</td></tr>
<tr><td>Acetic acid</td></tr>
<tr><td rowspan="2">Water</td><td>Paraffin wax</td><td>73°</td></tr>
<tr><td>Silver</td><td>90°</td></tr>
</table>

The addition of other substances can greatly affect the wetting angles and make the surface hydrophobic or hydrophilic. Corrosion is a surface reaction, and the corrodent must wet the surface. Corrosion inhibitors are designed to migrate to a surface to adsorb and affect wetting and other surface reactions. Solvents for both inorganic and organic deposits must wet the surface (now covered by scales) to affect removal.

Measuring Contact Angles. The contact angle and IST can be measured in several ways. The sessile drop method is illustrated in Figure 2.3(a). Here, a drop of liquid has been placed on a surface in air, and the three IST vectors are noted. The contact angle (θ) can be measured visually by taking a photograph of the surface, usually using a microscopic device. See Howard et al.[2] for a specific example. Figure 2.3(b) and (c) illustrate the importance of the solid surface itself. In these examples, a water drop has been placed on two surfaces in an oil medium. When bronze (b) is the solid, the water does not displace the oil, whereas on glass (c), the water does displace the oil and wets the surface.

Table 2.2 IST of Various Liquids Contacting Air (Dean[3])

Liquid	Temperature (°C)	IST, γ (dyne/cm)
Acetic acid	20	27.6
Acetic acid (40.1%) + water	30	40.68
Acetic acid (10.0%) + water	30	54.56
Acetone	20	23.7
Diethyl ether	20	17.0
Ethanol	20	22.27
Ethanol (40%) + water	25	29.63
Ethanol (11.1%) + water	25	46.03
Glycerol	20	63
n-Hexane	20	18.4
Hydrochloric acid 17.7M aqueous solution	20	65.95
2-propanol	20	21.7
Methanol	20	22.6
n-Octane	20	21.8
Sodium chloride 6.0M aqueous solution	20	82.55
Sucrose (55%) + water	20	76.45
Water	0	75.64
Water	25	71.97
Water	50	97.91

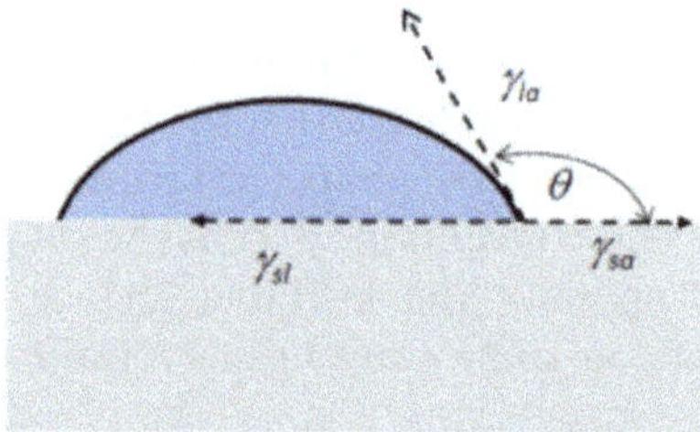

(a) Illustration of sessile drop method with liquid droplet partially wetting a solid substrate

(b) Water droplet immersed in oil resting on a brass surface

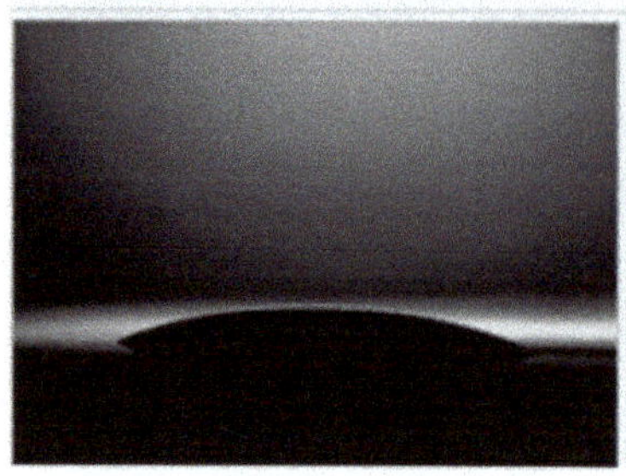

(c) Water droplet immersed in oil resting on a glass surface

Figure 2.3 Sessile Drop Method for Measuring Contact Angles

Figure 2.3 also illustrates a possible caution when interpreting information from the literature, since either the acute or the obtuse angle could be reported. If water is used to test the "wetting" of a surface, low numbers refer to a hydrophilic surface, and high values refer to a hydrophobic surface. For more information, see Tadmor.[4]

The IST of a liquid vs. air can be measured using the traditional Du Noüy ring method (see Frenier and Ziauddin[5]). Wetting properties of the surface or interface have little influence on this measuring technique. Maximum pull exerted on the ring by the surface is measured.[6] The most important experimental issues are using very clean equipment and fire cleaning the platinum ring used in the test.

Another method especially suited to check surface tension over long time intervals is the Wilhemly plate method that is described in Frenier and Ziauddin.[5] This method uses the interaction of a test plate with the liquid being tested: A vertical plate with a known perimeter is attached to a balance, and the force owing to wetting is measured.

Pendant drop test. In this method (used with oils), a drop of liquid is suspended from the end of a tube by surface tension (Figure 2.4).

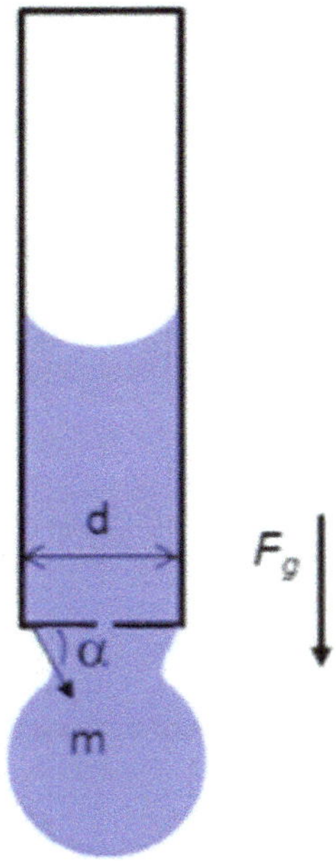

Figure 2.4 Surface Tension Measured by Pendant Drop Method

The force due to surface tension is proportional to the length of the boundary between the liquid and the tube, with the proportionality constant usually denoted as γ (the interfacial surface tension). Because the length of this boundary is the circumference of the tube, the force due to surface tension is given by Equation (2.4), where *d* is the tube diameter:

$$F_\gamma = \pi d \gamma \qquad \text{(Equation [2.4])}$$

Here, the mass (*m*) of the drop hanging from the end of the tube can be found by equating the force due to gravity:

$$F_g = mg \qquad \text{(Equation [2.5])}$$

The component of the surface tension in the vertical direction is $F_\gamma = \sin\alpha$. Thus,

$$mg = \pi d \gamma \sin\alpha \qquad \text{(Equation [2.6])}$$

Here, α is the angle of contact with the tube, and *g* is the acceleration due to gravity. The limit of this formula, as α goes to 90°, gives the *maximum* weight of a pendant drop for a liquid with a given surface tension,

$$\gamma = \frac{mg}{\pi d} \qquad \text{(Equation [2.7])}$$

Several important properties and uses of cleaning chemicals are related to wetting and capillary forces and inform[2] both the Laplace and Washburn equations, which can be of importance[7] in surface cleaning/scaling processes. The Laplace equation relates capillary pressure (the difference between the phase pressures of a nonwetting [P_2] and a wetting phase [P_1], such as water and oil [P_c]) to surface tension (γ) and contact angle (θ). Here, *r* is the radius of a capillary tube, *L* is the height of the fluid rise in the tube, ρ is the density of the fluid, and *g* is the gravitational acceleration (980 cm/sec^2).

$$(P_2 - P_1) = P_c = L\rho g = 2\gamma \cos\theta / r$$ (Equation [2.8])

To evaluate wetting of two fluids in the same geometry, the ratio of two Laplace equations can be used. For example, using water and a solution with unknown properties,

$$L_w \rho_w g = 2\gamma_w \cos\theta_w / r$$ (Equation [2.9])

and

$$L_u \rho_u g = 2\gamma_u \cos\theta_u / r$$ (Equation [2.10])

Dividing one by the other and removing the constants gives

$$\frac{L_u}{L_w} = \frac{\gamma_u \cos\theta_u}{\gamma_w \cos\theta_w}$$ (Equation [2.11])

The manipulation of ISTs of liquids, solids, and gases are used to protect the surfaces (inhibition) as well as to wet and penetrate a scale during dissolution.

2.2 Fluid Mechanics Effects on Deposition and Cleaning

The mechanics of fluid flow in industrial plant piping affects and is affected by the changing nature of the fluid phases. The piping and heat exchangers themselves are also affected by the production of scale and inorganic deposits (See Chapters 3, 5, and 7). The fluid mechanics at any point in a piping system and the presence of deposits will affect the causes of corrosion damage (Chapter 3), and pressure/flow changes affect deposition of scale.

Thus, there is a feedback loop in which both mechanical and chemical conditions influence each process. The entire chemical cleaning process system depends on flowing solvents through the units. Therefore, flowing conditions must be understood and anticipated for successful cleaning operations. This short section describes key ideas and references for more information.

The following subsections introduce single phase and multiphase flow equations in pipes. For further information, see Brill,[8] Asante,[9] Shoham,[10] and Falcimaigne and Decarre.[11]

2.2.1 Single Phase Flow

Single phase flow equations and concepts provide a starting point for the much more complex multiphase flow conditions found in many pipeline segments.

The pressure drop over a distance (L) of a single phase incompressible fluid can be obtained from the mechanical energy balance equation:[12]

$$\Delta p_L = \Delta p_{PE} + \Delta p_F + p_{KE}$$ (Equation [2.12])

In an expanded form,

$$\Delta p_L = \left(\frac{g}{g_c}\right)\rho l sin\theta + \frac{2 f_f \rho v^2 l}{g_c d} + \frac{\rho}{2 g_c}\Delta v^2$$ (Equation [2.13])

Δp_L is the difference between the upstream and the downstream pressure. Δp_{PE} is the pressure drop due to potential energy change and is the hydrostatic head of the fluid and accounts for the pressure change due to the weight of the column of fluid. p is the fluid density, and θ is the pipe deviation from the horizontal direction. Thus, the hydrostatic head for a horizontal pipe (θ = 0°) is zero. For vertical tubing, θ = 90° for upward flow and θ = -90° for downward flow. For fresh water, the potential energy pressure drop per foot of vertical distance is 0.433 psi/ft (1 N/cm^2/m). Δp_F is the pressure drop due to pipe friction and is obtained from the Fanning equation:

$$\Delta p_F = \frac{2 f_f \rho v^2}{g_c d} \qquad \text{(Equation [2.14])}$$

Here, v is the fluid velocity, and d is the pipe diameter. For a pipe with a constant cross section, the fluid velocity can be expressed in terms of flow rate (q):

$$v = \frac{4q}{\pi d^2} \qquad \text{(Equation [2.15])}$$

The Fanning friction factor (f_f) for laminar flow is

$$f_f = \frac{16}{N_{\text{Re}}} = \frac{4\pi d\mu}{q\rho} \qquad \text{(Equation [2.16])}$$

The Reynolds number of the flow is

$$N_{\text{Re}} = \frac{4q\rho}{\pi d\mu} = \frac{qd}{vA} \qquad \text{(Equation [2.17])}$$

Substituting v and f_f in the expression for Δp_F yields the Hagen-Poiseuille law for pressure drop in a pipe.[13, 14]

$$\Delta p_f = \frac{128\mu q l}{\pi g_c d^4} \qquad \text{(Equation [2.18])}$$

From these definitions and equations, we can see that, for a given flow rate, the pressure drop is proportional to $1/d^4$ for laminar flow; for turbulent flow in a smooth pipe, the pressure drop is proportional to $1/d^{4.75}$. Hence, any reduction in effective pipe diameter owing to deposit buildup can lead to a drastic increase in pressure drop and, hence, a similar decrease in the production from the reservoir. For turbulent flow in a smooth pipe, $f_f = 0.079/R^{0.25}$.

In Equation (2.12), Δp_{KE} is the pressure drop due to change in kinetic energy between various positions in the pipe. Generally, it is much smaller compared to Δp_{PE} and Δp_F. It is equal to zero if there is no change in fluid velocity between the two points of measurement. For example, for an incompressible fluid flowing through a pipe of uniform cross sectional area, the velocity does not change, and Δp_{KE} equals zero. In cases of high gas volumes or high gas-oil ratios, velocity may rapidly change, but even then Δp_{KE} generally accounts for less than 10% of the pressure loss.

In Equations (2.12–2.18):

g = Gravitational acceleration, m/s^2
g_c= Gravitational acceleration at the center of mass, m/s^2
f_f= Fanning friction factor (dimensionless)
q= Flow rate m^3/s
l =Length of pipe, m
A= Area of pipe, m^2
d = Pipe inside diameter (for round pipe), m
v= Flow velocity, m/s

ρ= Density, kg/m²
μ= Dynamic viscosity, kg/(m-s)
ν= Kinematic viscosity (μ/ρ), m²/s

Note that in various parts of the pipeline industry, different expressions are used in pipeline models; however, R_e is universally important since the transitions from laminar to turbulent flow is generally described by this number, with laminar flow conditions <2000, turbulent flow >4000, and a transition range between these values. The Darcy friction factor (f_D, also known as the Darcy-Weisbach friction factor) is used in some applications and equals $4(f_f)$ or $64/R_e$.

In addition to velocity, pipe diameter, viscosity, and other factors also can affect the flow regime. Shannon[15] reports that debris or scale in a pipe can change flow patterns. The equations above assume a smooth pipe; Figure 2.5 shows three possible conditions for the same velocity and viscosities.

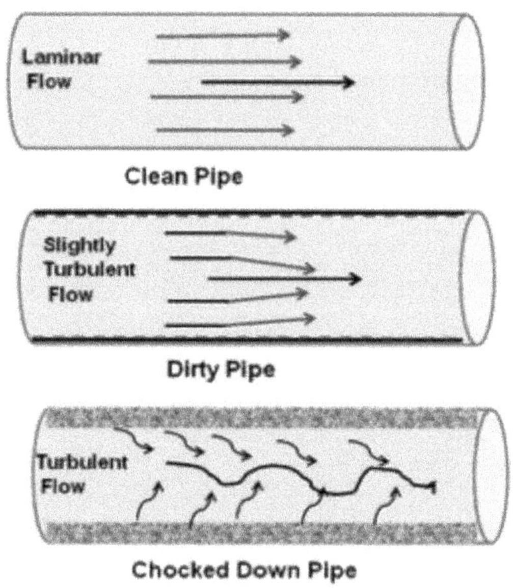

Figure 2.5 Effects of Deposits on Flow

In this set of conditions, there is laminar flow for a smooth pipe, slightly turbulent flow for a dirty pipe, and fully turbulent flow when the diameter has been constricted. The friction factor equation (Equation [2.16]) also will change for a dirty, rough pipe. To calculate the change,

$$\frac{1}{\sqrt{f}} = -2\log_{10}\left(\frac{\varepsilon}{3.7d} + \frac{2.51}{\mathrm{Re}\sqrt{f}}\right) \qquad \text{(Equation [2.19])}$$

In this equation, ε is the height of the roughness (mm). While this variable can be measured in the laboratory, it usually is calculated from the observed pressure drop in a line.

2.2.2. Multiphase Flow

Understanding multiphase flow conditions in the lines is important for predicting corrosion, solids accumulation, and distribution of inhibitors and cleaning fluids in piping. Shoham[10] states that the hydrodynamics of single-phase flow is well understood, and pressure drop vs flow rate and heat transfer can be calculated (see flow rate equations) in a straightforward manner. However, adding a second (or third) phase greatly complicates the analyses. Several modeling methods are used to predict hydrodynamic problems in the different flow regimes found in piping. See Shoham[10] and Kee et al.[16] for more details about methods for calculating important variables such as the friction factor and the pressure drop during multiphase flow conditions. Different multiphase flow combinations in piping include

- Hydrocarbon liquid/water
- Hydrocarbon gases/ hydrocarbon liquids
- Hydrocarbon liquid/water/gases
- Combinations of phases

Details are beyond the scope of this book; see Shoham[10] and Kee et al.[16]

2.3. Introduction to Corrosion in Industrial Units and Formation of Corrosion-Related Scales

Corrosion products constitute a significant portion of fouling scales that form on industrial surfaces, and corrosive agents are used to remove them. Thus, various corrosion processes degrade the metals of construction and may cause fouling scales to form.

A short review of corrosion chemistry is provided in this section. For more information, see Fontana,[17] Heidersbach,[18] and Roberge.[19] Additional details of the deposition of the scales onto iron surfaces are in Section 3.3.

All aqueous electrochemical corrosion processes require three agents:

- An oxidizing agent that accepts electrons (e^-)
- A reducing agent that gives up the electrons and is corroded (the metal)
- An aqueous electrolyte to complete an electrochemical circuit

Figure 2.6 summarizes the function of these chemicals to complete the electrochemical corrosion cell for iron (Fe). Stopping/reducing corrosion means breaking the circuit by using a coating or removing the oxidization agent.

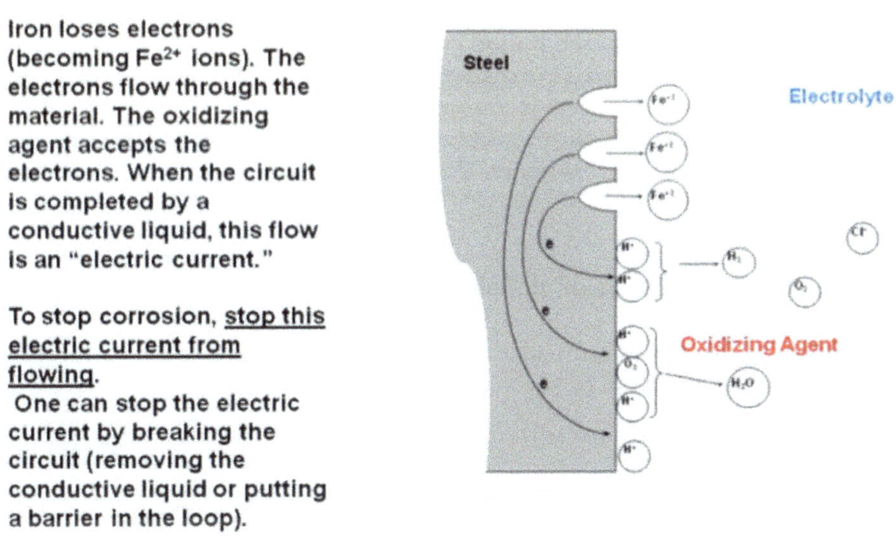

Figure 2.6 Electrochemical Corrosion on Steel

An example of a generalized electrochemical corrosion half-cell reaction system for a divalent metal in an acid is

$$\begin{aligned} &M \rightarrow M^{2+} + 2e^- \text{(oxidation at the anode)} \\ &2H^+ + 2e^- \rightarrow H_2 \text{(reduction at the cathode)} \\ &M + 2H^+ \Leftrightarrow M^{2+} + H_2 \uparrow \text{(equilibrium reaction)} \end{aligned} \qquad \text{(Equation [2.20])}$$

Construction materials in industrial systems usually are composed mostly of iron, but similar electrochemical reactions occur with nickel, manganese, molybdenum, and chromium. These elements are frequently added to the components of a production unit to change mechanical or chemical characteristics. Copper and its alloys with tin or zinc also may make up portions of water-contact systems. Aluminum, magnesium, and zinc and alloys with these metals may be part of cathodic protection devices (see Lazzari and Pedeferri[20]) or as surface protections films. These metals may be in the generalized industrial production system in some form. The exact arrangement and concentrations of these elements constitute the alloys of the units, and these alloys can have different corrosion properties. See Heidersbach[18] for brief reviews and references.

Corrosive agents (oxidizing agents) that may be in the fluids that contact industrial systems include CO_2, H_2S, O_2, and low molecular weight organic acids. When part of a cleaning solution, mineral acids (such as HCl) and organic acids also will corrode metals (See Section 4.1 and Chapter 6). Except for oxygen corrosion, all corrosion processes involve acid (H^+) attack in some form (See Equation [2.21]). In water, these agents react with iron to produce the reaction products that can form fouling scales:

- Hematite (Fe_2O_3)
- Hydrated iron(III) oxides (FeOOH)
- Magnetite (Fe_3O_4)
- Siderite ($FeCO_3$)
- Iron sulfides (Fe_yS_x)

These may constitute a significant portion of the inorganic fouling agents found in industrial units. The details of the corrosion product scales are provided in Section 3.3.

Note that the area of the metal from which the metal ions depart is called the anode, and the cathode is the area where the oxidizing agent accepts electrons. At equilibrium, half-cell reactions must proceed at the same rate (called the corrosion current, I_{cor}) and potential (the corrosion potential, E_{cor}). Also, the anodes and cathodes on the metal surface are separated from each other. This can be a micro-separation or a macro-separation, and the degree and type of separation can affect the corrosion of the metal. See Section 6.6.3 for electrochemical methods for monitoring corrosion.

This Section (2.3) describes general corrosion mechanisms and the chemical substances in the metal that can affect corrosion in the industrial environment. The consequences and manifestations of corrosion are discussed in Section 2.3.3.

2.3.1. Corrosion of Metals Owing to Cleaning Solvents

In most cleaning operations, the corrosion reaction of the solvents with the base metals is an acid attack of some type. The electrochemical equation is

$$2H^+ + Fe^0 = H_2 + Fe^{2+} \qquad \text{(Equation [2.21])}$$

The most common cleaning solvents are HCl, organic acids, and chelating agents such as EDTA. These chemicals will be discussed in Sections 2.5.1 and 4.1. This type of acid attack occurs in hydrochloric acid, organic acids, and chelating agents, including formulations that have pH values as high as 9, especially if they contain ammonium (NH_4^+) ions. The basic corrosion reactions are noted.

$$\text{HCl}: 2\text{HCl} + \text{Fe}^0 = \text{H}_2 + \text{FeCl}_2 \quad \text{(Equation [2.22])}$$

$$\text{Formic acid}: 2(\text{HCOOH}) + \text{Fe}^0 = \text{H}_2 + \text{Fe(HCOO)}_2 \quad \text{(Equation [2.23])}$$

$$\text{Ammonium EDTA}: (\text{NH}_4)_4\text{EDTA} + \text{Fe}^0 = \text{H}_2 + (\text{NH}_4)_2(\text{Fe(II)EDTA}) + 2\text{NH}_3 \quad \text{(Equation [2.24])}$$

The rate of corrosion can vary by several orders of magnitude depending on the acid strength and the temperature. In very general terms, the amount of damage potential increases with acidity and temperature, but most chemical scale removers are corrosive and require use of corrosion inhibitors (See Chapter 6).

2.3.2. Effects of Metallurgy on Corrosion Processes

In industrial equipment environments, corrosion reactions do not take place on pure iron but on a wide variety of alloyed steels. This Section (2.3.2) summarizes metallurgical concepts that may affect various types of corrosive attacks. For a more detailed discussion, see Chapters 4 and 8 of Heidersbach.[18] The alloy steels used in industrial units are solid solutions of iron with various other elements to give needed physical properties as well as corrosion resistance (in some cases). The production processes that affect the final alloy properties include a specific chemical recipe and the temperatures of formation as well as later heat treatments (such as annealing, quenching and tempering, normalizing, hot rolling, and pressing; see Chapter 4 in Heidersbach[18]).

The chemical recipes and manufacturing processes are designed to produce a specific microstructure in the steel to yield the physical and chemical properties required for a pipeline's or tubing's contents, physical location, and conditions (especially pressure).

The many parts of industrial units are constructed of various grades of carbon steel. Tubing and pipelines are specified in American National Standards Institute/American Petroleum Institute 5L.[21] These specifications include tensile stress requirements and list some ranges of the chemical constituents of the alloy. Note, however, that different ways of achieving a physical specification may result in different metal chemistries.

Boiler tubing metallurgies for carbon steels are specified by ASTM International A129[22] and may have chemical composition limits. Additional alloys that have small amounts of chromium, nickel, and other elements may be used in higher temperature sections of the boiler's water/steam systems. Some of the chemical specifications for steam generator construction are described by a range of codes noted in Table 1.1. Figure 2.7 shows typical microstructures for different carbon steels.

- In Figure 2.7(a), the white structures are α-ferrite, a crystal structure of very pure iron. The gray structures are called pearlite, which consists of bands of ferrite and cementite (Fe_3C).
- In Figure 2.7(b), another specimen that has low carbon content and a specific heat treatment (quenching and tempering) produces very fine-grained martensite structures (a highly strained body-centered cubic form of ferrite that is supersaturated with carbon).

The exact compositions and orientations of the microstructures give the metal its properties and can affect corrosion susceptibilities, and the various structures have different chemical potentials in different fluids. Figure 2.6 shows that the anodic and cathodic reactions take place on different locations on the metal's surface, and the actual differences in potential drive the redox reactions, thus affecting the corrosion rate.

Thus, the microstructures on the surface can change reaction rates, formation of protective scale, and the ways in which chemical corrosion inhibitors affect the corrosion reactions.

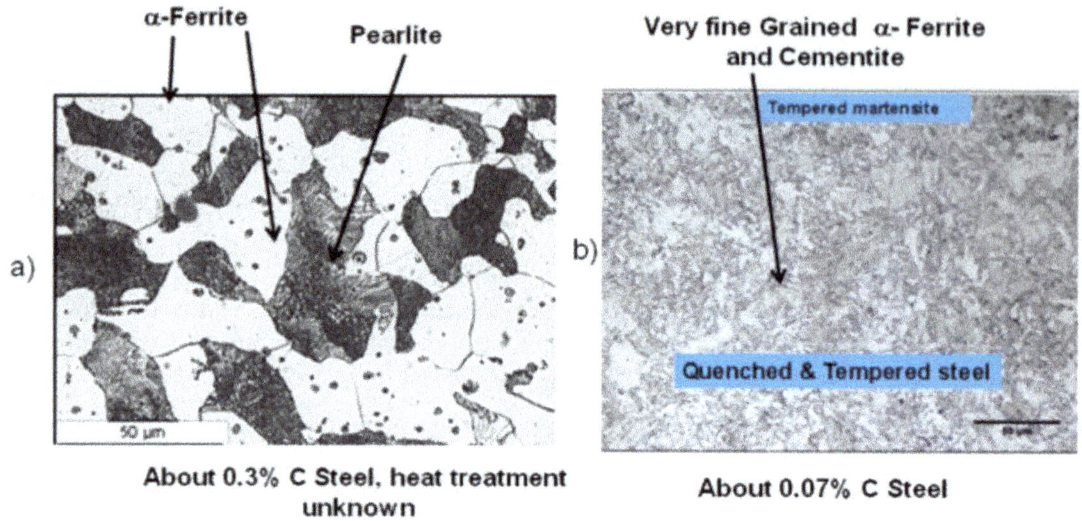

Figure 2.7 Microstructure of Carbon Steel[23]

In some operating conditions, including very hot CO_2 and/or H_2S environments, corrosion resistant alloys (CRAs), which contain much higher concentrations of chromium, nickel, and molybdenum may be required.[24] Examples of CRAs found in the pipeline sector include the martensitic (a different microstructure of iron[18]) stainless steels that have about 13% chromium (e.g., UNS J91150). These alloys are much more expensive than carbon steels but may be required under conditions in which corrosion inhibitors are ineffective.

Overall, specific metallurgies (from the chemical compositions as well as for forming and heat treatments) may substantially affect the corrosion susceptibility each type of unit made of steel. These metallurgies affect the scales formed and the type of cleaning solvent systems needed (including corrosion inhibitors).

Sections 2.3.3 and 2.3.4 describe specific manifestations of corrosion that can lead to scales of different morphologies. That is, these scales may contain mostly iron oxides/sulfides/carbonates, but how they are formed can greatly affect the surface conditions and their compositions. The chemical/mechanical cleaning processes must be designed specifically to remove these scales.

2.3.3. Localized Corrosion

Pitting and under deposit corrosion (UDC) are two of the most common manifestations of damaging corrosion to steel goods and may result from specific flowing conditions and solids formation processes. Because scales form in pits and crevices, the cleaning process may require modification to remove scale from the pits.

Pitting Corrosion. Pitting corrosion usually can be contrasted with uniform corrosion. In Figure 2.6, the active anodes (where metal is lost) and the cathodes (where electrons are lost to balance the charges) appear to be close together. In these conditions, corrosion is uniform (at least at the macro level), and large pits do not form. Very aggressive corrodents, such as HCl or concentrated organic acids, frequently cause uniform corrosion, especially when adequate acid corrosion inhibitors are present (See Section 6.1). In chemical or physical conditions in which the anodes are separated from the cathodes, various pitting processes may cause uneven corrosion, resulting in a corrosion rate that is much higher in a localized area than the average corrosion rate. Pitting can cause loss of integrity much faster than an even corrosion process.

The conditions that initiate pitting are difficult to determine in individual cases. However, the exact condition of a surface can have a large influence on the pitting behavior of a material. Generally, surfaces prepared with a rough surface finish are more susceptible to pitting and exhibit more pitting. Small amounts of scale or ions (such as Cl^-) can disrupt a protective coating and initiate pitting. Fontana[17] and Frankel[25] report that once a pit forms, pit development can be an autocatalytic process.

$$O_2 + 2H_2O + 4e^- \rightarrow 4OH^-$$
$$M \rightarrow M^{2+} + 2e^-$$
$$O_2 + 2H_2O + 2M \Leftrightarrow 2M^{2+} + 4OH^- \qquad \text{(Equation [2.25])}$$

In Figure 2.8, the pit bottom is the anodic area, and the surface is the cathodic area. In this figure, protons (H^+) are shown with the other ions. This can form when the metal ions are hydrolyzed, thus forming acid (Equation [2.26]). This acid in the pit bottom increases the corrosion process and the pit depth. If the pit then fills with the metal hydroxide salts (especially at the surface), the solids help maintain the low pH by preventing exchange with the bulk fluids and by accelerating the corrosion process.

$$M^{2+} + 2H_2O \Leftrightarrow M(OH)_2 + 2H^+ \qquad \text{(Equation [2.26])}$$

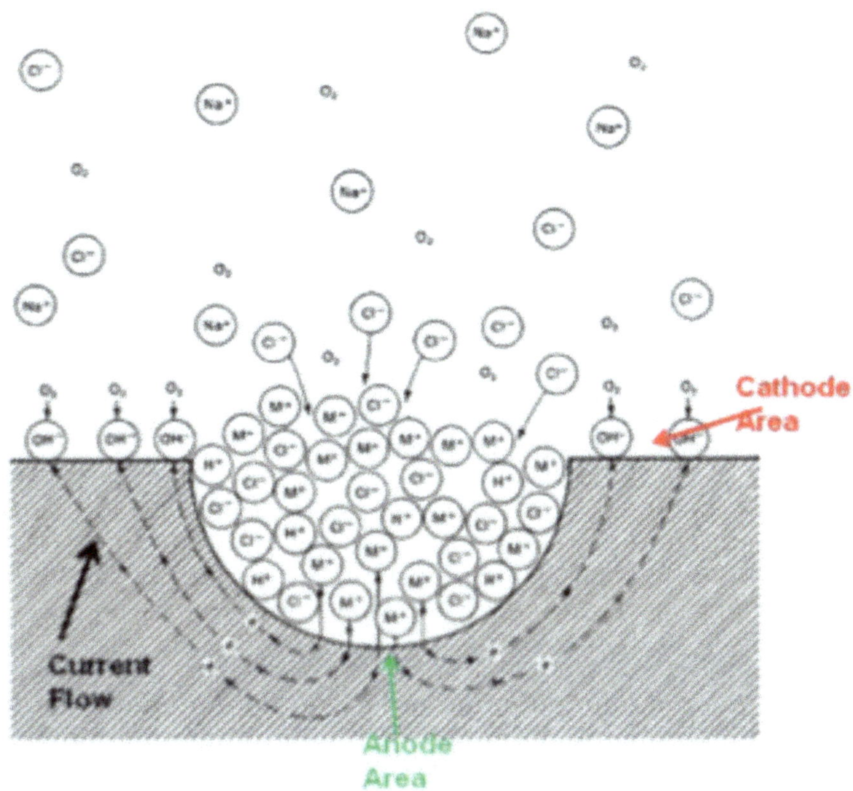

Figure 2.8 Autocatalytic Process in Oxygenated Brine

In Figure 2.9, corrosion occurs above or slightly below the water line (in a multiphase flow condition) in an area not contacted by soluble corrosion inhibitors. Heidersbach[18] reports that changes in the waterline level and the composition of the phases can exacerbate pitting corrosion.

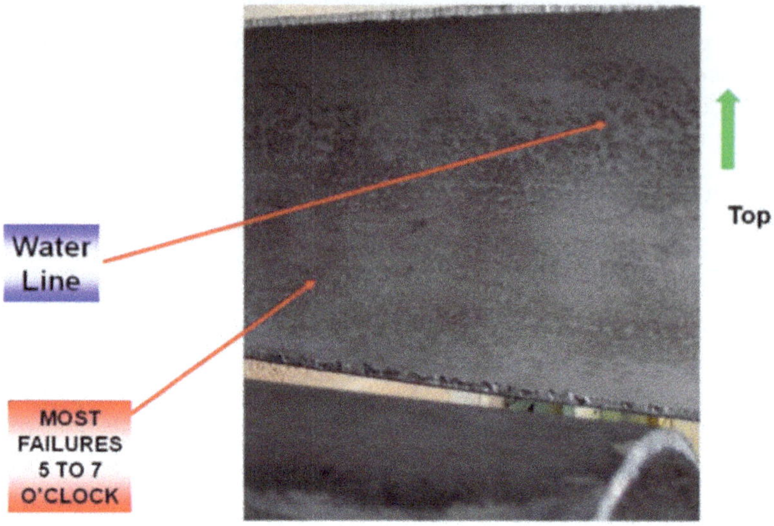

Figure 2.9 Pitting Corrosion below the Waterline

Crevice corrosion and pitting can initiate in welds. El-Sayed[26] reports a particularly extreme case of weld-related corrosion in which electric resistance seam-welded carbon and low-alloy steel pipes underwent preferential corrosion attack in the weld area, especially close to the fusion line, when exposed to neutral, salt-containing waters. In this case, a poor, incompletely penetrating weld allowed corrosive fluids to enter the weld, causing leaks.

During cleaning operations using chemical solvents, the goal usually is to remove all the scale from the pits to prevent this type of corrosion from continuing. Since this may account for a very small amount of metal in solution, a job design may include an additional contact time after the metal/time curve is stable. Some processes (especially nuclear cleaning [see Section 5.3]) may use a separate solvent for this purpose.

Pitting corrosion also may occur during chemical cleaning without the proper corrosion inhibitor. Section 6.6.4 describes tests for evaluating pitting during cleaning operations.

Microbiologically induced corrosion. Microbiologically induced corrosion (MIC) can be an issue in many water lines as well as in nuclear power utilities.[27] This is especially true when there is a lot of water (waste lines), when the flow is low, or when there are dead spots[28] in a line. MIC can manifest in the service water systems of nuclear plants and in piping in other types of units. Microbes of various types must be present, but they are common in many environments.

This type of corrosion, one of the most important topics in the corrosion protection field, is very complex because it involves reactions that take place in and under a mass of bioproducts. A compilation of NACE International conferences papers[29] covers several aspects of MIC, including threat assessment, prevention and mitigation, monitoring, and management.

Although the actual amount of MIC is difficult to determine in the pipeline environment, Jenneman[30] suggests that as much as 90% of oil pipeline corrosion may originate as MIC. Even so, most of the actual corrosion (oxidation) half-cell reactions can be described by Equations (2.1–2.9). This section will review current thinking on this very complex corrosion. For more information, see Augustinovic.[31]

Colonies of microorganisms can produce a scale-like mass under deposit corrosion (see next subsection). Jenneman[32] also has shown that microorganisms can generate acid that actually causes corrosion. The production of numerous byproducts of the microbiological reactions produce a sticky mass called a biofilm as well as inorganic solids. Figure 2.10(a)[33] shows the formation of biofilms on the surface, and several possible reactions under the film are noted in Figure 2.10(b). Because microorganisms also can cause sulfate to be converted to H_2S, treatment with biocides is usually recommended to reduce the effects. This type of scale/corrosion is shown in Figure 2.11:[34] (a) MIC in a water pipe, (b) a pipe that has a penetration, and (c) accumulation of biofilm inside a deaerator.

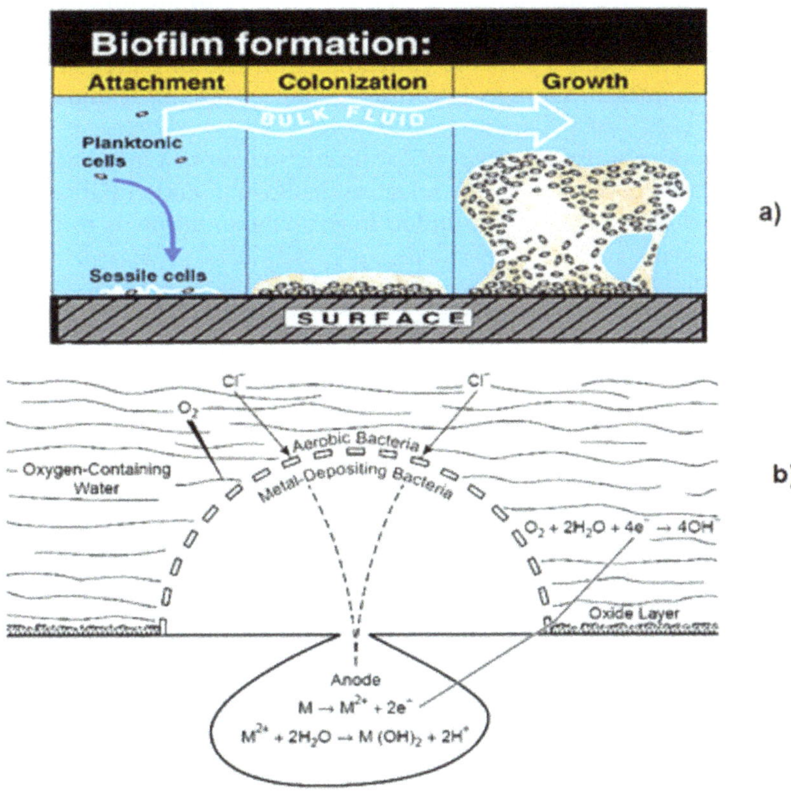

Figure 2.10 Biofilm Formation and Microbiologically Induced Corrosion[33]

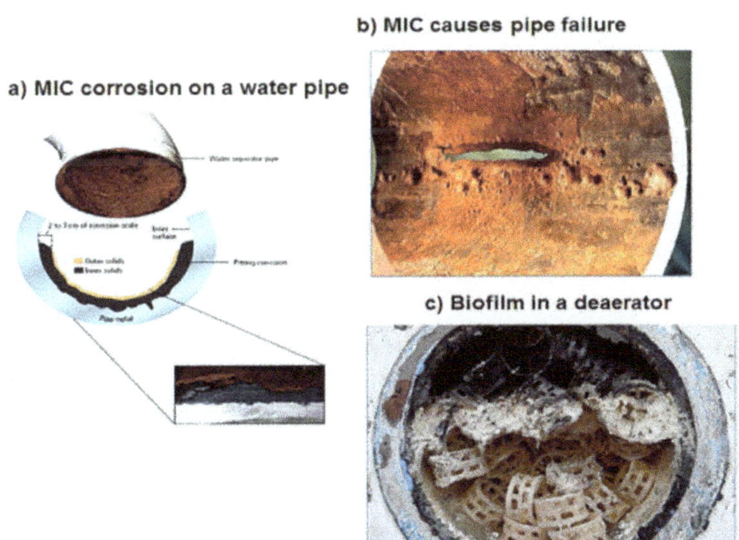

Figure 2.11 Microbiologically Induced Corrosion Scale Production[34]

Little and Wagner[34] discuss the mechanisms and consequences of MIC formed in pipelines. They suggest that the most aggressive MIC occurs when many physiological types of bacteria, including metal-oxidizing bacteria, sulfate-reducing bacteria (SRB), acid-producing bacteria, and metal-reducing bacteria, interact in complex ways within the structure of biofilms (Figure 2.10[33]).

Little et al.[35] note that identifying MIC after it has occurred requires a combination of microbiological, metallurgical, and chemical analyses. MIC investigations typically attempt to identify causative microorganisms in the bulk medium or in corrosion products, identify a pit morphology consistent with an MIC mechanism, and identify a corrosion product chemistry that is consistent with the causative organisms. The following sections discuss available techniques, advantages and disadvantages, and, most importantly, limitations.

MIC does not necessarily produce a unique form of localized corrosion. Little and Wagner[34] suggest that MIC can result in pitting, crevice corrosion, UDC, selective dealloying, and enhanced galvanic and erosion corrosion. Thus, the principal effect of bacteria under anaerobic conditions is to increase the probability that localized corrosion will occur.

Little and Wagner[34] suggest that bacteria can cause ideal conditions for pitting or crevice corrosion. Once localized corrosion has initiated, microbial reactions can maintain conditions (e.g., low oxygen) for continued pit/crevice growth. The rate at which pits propagate can depend on organic acid production by fungi in aerobic environments and by certain bacteria in anaerobic environments. Under anaerobic reducing conditions, aggressive MIC may occur. The following discussion about individual MIC mechanisms relates to carbon steel.

Augustinovic et al.,[31] Larsen et al.,[36] and Larsen et al.[37] highlight the use of molecular microbiology methods to determine the complex distribution of SRB and other microorganisms in oil field systems. These techniques are based on characterization of DNA and/or RNA in water samples rather than on counts of viable microorganisms. For details of analytical methods, see Whitby and Lund.[38]

These authors have identified SRB Prokaryotes and Archaea (called sulfate-reducing Archaea) in oil field waters. Note that Archaea are on a separate branch of the Earth's tree of life that also includes bacteria and more complex organisms (including mammals). Specific molecular microbiology methods are beyond the scope of this book; see Rassenfoss[39] for more information.

These authors[31,36,37] suggest that older methods of identifying microorganisms (i.e., culturing) identify only 10% of corrosion-causing microorganism species.

The various mechanisms of MIC depend on the flow conditions and microorganism species.[30, 34, 40]

UDC. UDC is like pitting but can occur in the presence of corrosion scale, other particles such as sand, or MIC. This type of corrosion happens when the deposit is permeable and nonprotective. The manifestation frequently is a large scale-filled pocket, whereas pitting (Figure 2.9) also can form on a surface that appears to mostly scale free. The conditions for the formation of H_2S corrosion (Section 3.3.2) and FeS scale frequently cause UDC.

The UDC processes cause a concentration cell (which causes an electrochemical polarization) that enhances corrosion and pitting.[41] Cervantes et al.[42] state that corrosion inhibitors should be tested for their ability to penetrate moderate solids deposits and to provide protection. A recent publication describes techniques for assessing inhibitor efficiency below sand deposits.[43] The best operational practice to protect a pipeline is to minimize low flow or stagnant conditions during operations and periodically remove solids with a maintenance pig.[44]

Two examples of pitting and UDC are barnacle-type corrosion and rust tubercle under deposit corrosion. Barnacle-type corrosion (Figure 2.12) forms when a tubing corrodes in an H_2S and water environment, and

FeS type scale builds up. If the scale is porous and is cathodic to the steel tubing, a barnacle forms. Figure 2.13 shows UDC/scale. An acidic layer of $FeCl_2$ prevents more FeS precipitation next to the steel, and a pit forms.[45] Rust tubercle UDC is a complex, localized process that forms a nodule-like structure. This may occur in a region of low fluid velocity in which solids can form and shield part of the metal and reduce the available oxygen. The portion of the steel exposed to low oxygen becomes anodic and corrodes faster than the rest of the surface. This is also called oxygen cell corrosion.

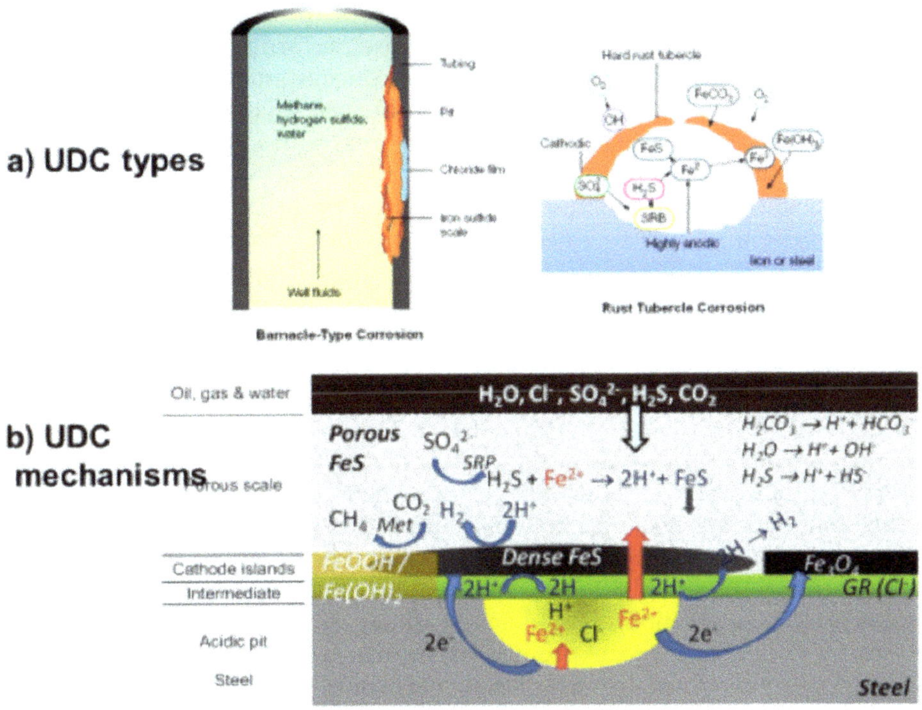

Figure 2.12 Under Deposit Corrosion (UDC)[46]

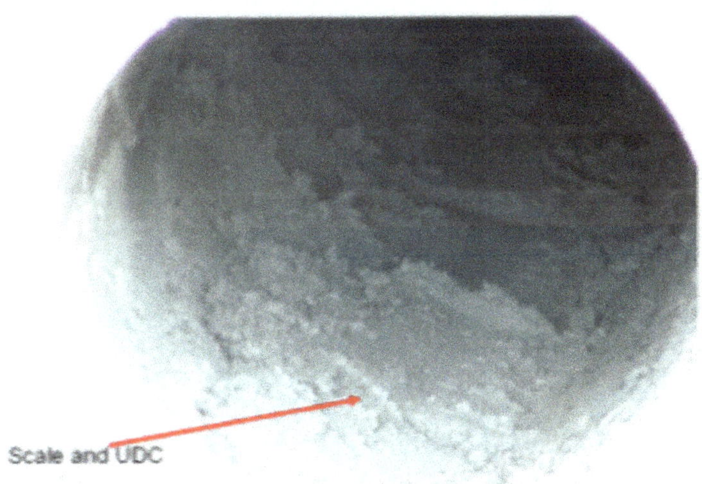

Figure 2.13 Scale and Under Deposit Corrosion

A major cause of UDC is MIC.[32] Heidersbach and Roodselaar[46] claim that water injection lines often do not last their design life owing to corrosion problems.

Vera et al.[47] note that in tubing UDC can be caused by inorganic scale/deposits:

- Deposits induce erosion or corrosion erosion at high velocities and preferentially in areas of high turbulence or changes in flow direction (e.g., elbows).
- Depending on the conditions, impacting particles can erode the base metal, increase corrosion rates by removing protective corrosion products, and/or decrease performance of filming inhibitors.
- Deposits decrease inhibition efficiency by reducing the concentration of corrosion inhibitor owing to its adsorption onto the solids, which typically have a high surface area.
- Corrosion can be induced under deposits, typically in areas with low flow velocities or where solids accumulate owing to changes in inclination.
- The use of traditional corrosion inhibitors for the mitigation of UDC is not always successful because of the vast variety of deposits and conditions that the corrosion engineer may encounter in the field.

Three possible mechanisms of formation of corrosion owing to surface deposits have been suggested.[47]

- A dominant internal anode (beneath the deposit) is coupled to a dominant external cathode (uncovered metal). This would cause corrosion under the deposits, starting preferentially from areas close to the edges of the deposit. This mechanism is claimed to be like crevice corrosion, in which the galvanic driving force originates from different chemistry inside and outside the deposit, and requires an ionic path. Also, this would typically occur in fluids with relatively high conductivity (e.g., saline water).
- If there is a separation of anodes and cathodes beneath the deposit only (no external cathode), the deposit acts mainly as a means of producing a chemistry suitable for pitting corrosion. The anode/cathode separation may develop from small differences in local chemistry arising from deposit irregularities.
- Where there is no local separation of anode and cathode, which would induce uniform corrosion beneath deposits, UDC can occur in the presence of hygroscopic salts or clays. This will cause corrosion where uncovered steel would not normally sustain corrosion.

Hydrogen Embrittlement and Corrosion-Related Cracking. Several types of corrosion-related cracking of high strength steels can lead to a failure. H_2S also can contribute to sulfide stress cracking (SSC).[24] MR0175 recognizes the risk of SSC and provides material requirements for H_2S partial pressures more than 0.05 psia. When H_2S is present, corrosion cells generate FeS and atomic hydrogen (H). The atomic hydrogen atoms may recombine on the surface to form H_2 gas.

The H_2S corrosion and the FeS layer can promote the movement of hydrogen into the metal and generate pressure. This leads to hydrogen embrittlement (HE; also called hydrogen-induced cracking) and, if combined with stress, may lead to SSC.

Stress corrosion cracking (SCC) can start with a pit or crevice. The zone at the tip of the crevice becomes plastic under the stress, allowing the crack to develop. Chloride ions can catalyze the corrosion (especially with stainless steels). Because these conditions can also cause HE, these related failure modes may be difficult to distinguish.

These conditions may occur in many units. The differences between the various modes of HE and SCC are thus complex. SCC can be caused by the combined action of a localized tensile stress and a cracking environment on susceptible steel at a near neutral pH[48] as well as in acidic conditions. Cracking and/or pitting during chemical cleaning are significant issues to be investigated before choosing a solvent.

2.3.4. Erosion/Corrosion (EC) and Impingement Damage

EC is a corrosion-inducing/enhancing process that usually requires two (or more) phases in the flow to enhance the attack of a corrosive fluid in the piping. It is the action of a mechanical force on the surface that may increase the corrosion rate or cause direct metal loss. The multiphase fluid can include liquid/gas, solid/liquid, or liquid/gas/solids. EC may include a tribocorrosion material loss mechanism caused by flowing fluid (with or without solid particles) that disrupts or thins the protective film of a corrosion product[49] scale or inhibitors. Tribology is the interaction of a solid surface's exposed face with interfacing materials. The interactions can be from the mass of the fluid passing the surface and/or from solids or a gas phase. This type of damage frequently occurs in pipe bends and other areas where there is an increase in velocity. A probe in a line also could be subject to EC or impingement damage. Such a disturbance in the flow also may cause excessive corrosion downstream of the probe.

For fluids that do not contain solids, Heidersbach[18] notes that rule 14E[50] provides an estimate of the critical velocity of a fluid before EC can occur:

$$V_{max} = \frac{C \bullet A}{\rho^{0.5}} \quad \text{(Equation [2.27])}$$

Here V_{max} is the maximum velocity (m/s), ρ is the density (kg/m^3), A is a conversion constant (1.23), and C is a service factor that is empirically determined. This equation *does not* consider the presence of solids such as sand or corrosion product debris.

Barker et al.[51] describe tests conducted in corrosive (CO_2) solutions using submerged impinging jets to simulate different flow regimes. The tests included addition of two different inhibitors as well as particulates (sand). They found that both inhibitors reduced corrosion under flow conditions, but there were differences in effectiveness. One inhibitor was affected by the flow (shear stress), but the other was not significantly affected when used at the same concentration.

Corrosion-caused scales of the variety describe above can be found in many industrial units, and deep cleaning to remove all of the deposit is necessary to prevent more damage and possible unit rupture.

2.4. Equilibrium-Produced Salt Scales

Apart from the iron-based scales, essentially all the other inorganic fouling deposits (Table 1.2) are formed when ions already in a solution combine to form a salt. Of course, the corrosion-based deposits also form due to the very rapid reactions of the iron from the substrate reactions with the anions in solution. These non-iron processes produce important industrial scales, such as calcite ($CaCO_3$), gypsum ($CaSO_4 \cdot 2H_2O$), and silicates (see Sections 3.1, 3.5, and 3.6). These fouling scales can form only when the solution becomes supersaturated. However, this condition can happen rapidly at the right temperature or pressure or when incompatible water sources are mixed.

2.5. Introduction to Cleaning Processes

Fouling deposits (Sections 1.4, 2.3, and 2.4) can be removed using chemical solvents, high pressure jetting, or a combination of the two. All cases have a mechanical component because the solvents often must be heated to react rapidly with the scale and must be transported, mixed, and circulated through the units being cleaned. This section briefly describes the solvents and mechanical units used in IC. More details of cleaning chemicals and commercial formulations are in Chapters 4, 5, and 7.

2.5.1. Chemical Solvents

The chemical solvent required depends on the type of deposit and the conditions and location of the equipment to be cleaned. Table 2.3 lists deposit types that require removal, current solvent classes, and usual application conditions. Details of the specific chemicals and mechanism of action are in Chapters 4 and 7.

Table 2.3 Cleaning Solvents in Current (or Past) Use

Deposit	Solvents	Use Conditions
Iron oxide	5% to 15% HCl (w/w/o HF) Hydroxyacetic/formic acids (and other organic acids) Other inorganic acids (H_2SO_4, sulfamic acid) Oxalic acid Mono-diammonium citrate Mono-sodium citrate Ammonium citrate Diammonium and tetra ammonium EDTA EDTA-organic acid mixtures Sodium HEDTA, pH 1.4 Tri-ammonium EDTA/N_2H_4 Monosodium HEDTA	150-180°F Fill/dump, circulating 180-200°F, circulating Ambient-150°F, circulating 100-200°F, circulating 180-275°F, circulating 180-200°F, circulating 180-250°F, circulating 200-300°F, circulating 150-200°F, circulating 100-150°F, circulating 180-200°F, circulating 180-200°F, circulating 140-200°F, circulating
Copper or copper oxides	Thiourea and derivatives in HCl Ammonium bromate Ammonium persulfate Ammonium carbonate/O_2/NH_3 Ammonium EDTA (with Fe and/or oxidizer) Ammonium citrate/oxidizer	150-170°F 120-170°F below 100°F below 150° 120-170°F 120-170°F
Calcium/magnesium carbonate	5% to 15% HCl Tetrasodium EDTA, trisodium HEDTA, tetrasodium GLDA 7% to 10% Sulfamic acid	below 150°F 180-300°F, circulating below 140°F
Calcium sulfate	Tetrasodium EDTA, trisodium HEDTA 1% NaOH, then 5% HCl EDTA-organic acid mixtures	180-300°F, circulating 100-150°F, circulating 120-150°F, circulating
Calcium phosphate compounds	5% to 10% HCl 7% to 10% Sulfamic acid Tetrasodium EDTA	100-150°F below 140°F 180-300°F, circulating
Silica and silicate compounds	Alkaline boil with soda ash or NaOH, high pH fluids (amines or alkali)+ chelating agents HCl with fluoride	Alkaline boil at 50-100 psi 150-170°F 150-180°F, fill/dump or circulate
Sulfides	5% to 10% HCl Alkaline $KMnO_4$ 5% to 10% HCl with aldehydes 5% to 15% H_2SO_4 with aldehydes	Heat slowly and scrub or flare H_2S 100-160°F 100-160°F
Disulfides	Chromic acid Stabilized H_2O_2	Hot chromic acid, 7%-10% 100-140°F
Organic despots	Detergents Terpene or emulsions Organic solvents (including chlorinated at one time) Alkaline $KMnO_4$, then HCl and oxalic acid	150-250°F, circulating 150-250°F, circulating 150-250°F, circulating 150-250°F, circulating

2.5.2. Mechanical Methods of Cleaning

Totally mechanical as well as chemical with mechanical methods are used to clean various types of process units. High pressure water cleaning of process equipment, especially heat exchangers, has a long history.[52, 53] Figure 2.14 shows a water lance that is inserted into a tube to blast out deposits. Although hand-lances are dangerous, very high pressure water (as well as water with particulates) has been used for this operation[54] for many years. Environmentally friendly particulate such as sodium bicarbonate[55] can be added to the water stream to improve removal effectiveness (see Chapter 8, which also reviews the use of trucks, pumps, and other equipment needed for IC).

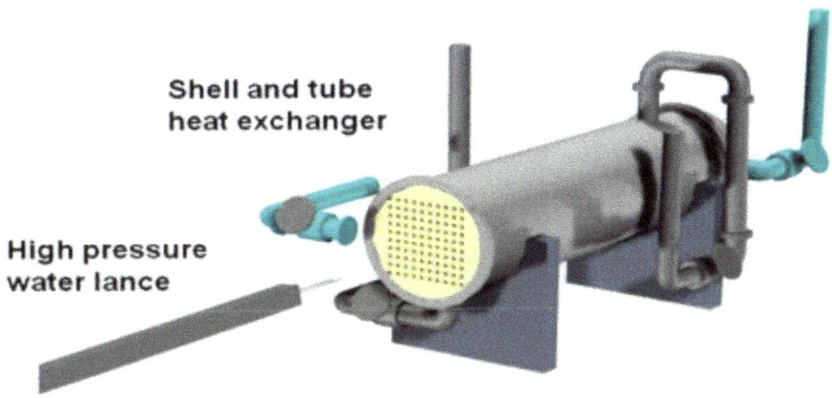

Figure 2.14 Water Lance and Shell and Tube Heat Exchanger

2.6. Testing of Deposits and Solvents

This section reviews the methods used to identify and quantify deposits that foul industrial units of all types. Once the deposit and the deposit's mass are determined, then several different solvents are evaluated and chosen on the basis of unit conditions, the site, and economics. On the job, various analyses must be performed to track and possibly modify the cleaning plan.

These evaluations must be performed before cleaning various units present in non-nuclear power generation, in refinery/petrochemical plants, and in various components of nuclear plants. Details of the solvents are in Chapters 4, 5, and 7.

Evans et al.[56] report that qualification programs are conducted for different chemical cleaning scenarios for maintaining steam generators. These programs vary from condition evaluations to extensive laboratory testing and qualification. The evaluations include material surveys and reviews for corrosion susceptibility, corrosion allowance estimates, deposit loading estimates, corrosion monitoring plans, and definition of technical requirements for the chemical cleaning application. Laboratory testing consists of dissolution and corrosion testing that incorporates actual and simulated steam generator deposits and critical materials of construction.

2.6.1. General Deposit Analysis Methods

Several levels of deposit analysis may be necessary.

The first step is to acquire a representative sample of a fouling deposit. Deposits for analyses can come from various parts of a plant or unit, and each sample represents only the point sampled. Cowan and Weintritt[57] note that scale deposits are seldom homogeneous. In many cases, they vary in composition not only along the length of the system but also with depth, perpendicular to the deposit (particularly with water-formed deposits). This is called a banding or layering effect. Layers and bands are probably the single most distinctive structural feature of scaling deposits. This reference has other useful information about collecting scale samples. As stated in Section 2.3, corrosion-based scale can differ greatly on the basis of the corrosion manifestation that causes surface scale, so sampling carefully is a paramount step.

A representative analysis of the unit's problems can be based on as many samples as is practical. Organic foulants and alkaline earth salts may be obtained by scraping a heavy deposit. However, assaying foulants on steel piping usually requires a section of the piping with a deposit in situ, which may not be possible but is the best method.

Sample requirements vary depending on the type of analysis and testing needed. Normally, a 50- to 100-gram sample is satisfactory for analysis. Larger samples are required if extensive solubilizing, disintegration, or mechanical removal studies are necessary. Once collected, the scale sample or tube length should be protected from further degradation. The labeling of the sample is critical. Every sample should be clearly labeled to show (if possible)

- The date of sampling
- The location of the sample in the unit
- Method of sample removal
- The time interval during deposition
- Any known water analysis data or process stream composition
- The pressure and temperature throughout the system (particularly at the point of deposition)
- History of scale removal
- Any changes in water or fluid flow in the system
- A complete system diagram and notation as to materials of construction in the system that might be affected by either mechanical or chemical removal
- Any information about previous cleanings or sampling

Various methods to assay the type and amount of scale on boiler tubing has a significant literature because the amount of deposit (as well as chemistries) greatly affect the solvent types used as well as the mass of fluid required to clean some potentially large (>20,000 gallons in most cases) volumes of these units.

Examples of sampling methods include

- Glass bead cleaning blasting (GBB) is reported[58] to be the most reliable of methods tested.
- Another report by NACE[59] describes a test method that uses GBB equipment to remove boiler waterside deposits on a piece of tubing from a representative area of a boiler. The test specimens are cut from a sample tube and weighed before and after the cleaning process, and the amount of deposit per surface area is estimated by measuring the weight loss of the tube sample test piece after deposit removal via GBB and dividing by the surface area of the test piece. The deposit weight density value is typically expressed in mg/cm^2 or g/ft^2 units and is defined as the weight of the deposit per boiler tube surface area. Procedures for test specimen processing, dimensional analysis techniques, sources of potential interferences, and a sample deposit weight density report calculation sheet are included in the method. Because of the velocity of the beads, a protective cabinet is required.
- An ASTM standard[60] describes several methods for determining boiler tube scale mass, including
    - Test Method A, a procedure using a scraper or vibrator, is used for deposits ranging from 1 g/ft^2 to 100 g/ft^2. The mass of the tube section is determined before and after the tests to give the mass of the deposit. The precision was determined in the collaborative study over the range of 16 g/ft^2 to 76 g/ft^2. This procedure allows for discretionary selection of the tube sample area. The removed deposit allows for further chemical analysis.
    - Test Method B uses inhibited HCl. The mass of the tube section is determined before and after the tests to give the mass of the deposit. It is generally used for deposits ranging from 1 g/ft^2 to 100

g/ft^2. The precision was determined in a collaborative study over the range of 28 g/ft^2 to 73 g/ft^2. The procedure averages out the heavier and lighter deposited areas. The solvent solution produced allows for further chemical analysis.

- Test Method C uses GBB. The mass of the tube section is determined before and after the tests to give the mass of the deposit. It is generally used for deposits ranging from 1 g/ft^2 to 100 g/ft^2. The precision was determined in a collaborative study over the range of 17 g/ft^2 to 88 g/ft^2. The procedure averages out the heavier and lighter deposited areas. The removed deposit does not allow for further chemical analysis because it is mixed with silica. Thus, any chemical analyses must be performed using surface methods that do not destroy the solids before cleaning.

The collaborative studies are referenced in the standard.[60] I have used methods A and B. Method A allows the scale to be analyzed on the basis of information in the next section. Method B does not assay organic components and produces a higher Fe result owing to corrosion of the tube but would more accurately describe the iron concentration if HCl is used as the solvent for cleaning the unit.

Hansen[61] provides more details of the sampling and analyses of boiler tube scales. Sampling is just one criterion for determining if cleaning is required. Other criteria include

- Tube sample deposit weights
- Tube sample deposit constituents
- Time of operation between cleanings
- Chordal thermocouple readings
- Tube failures
- Pressure drop (once-through units)

The determination of compositions is described in the next subsection.

Heifner[62] suggests that collecting a broad range of samples and collecting a broad range of information using other deposit loading characterization approaches would allow boiler operators to make more timely and economic deposit chemical cleaning decisions. Background and ideas on collecting this broad range of information to enhance boiler chemical cleaning are also described.

2.6.2. Chemical Analyses of a Deposit

Cowan and Weintritt[57] note that scale samples can be initially analyzed in the field using simple chemical tests (such as contact with an acid); however, laboratory processing is preferred.

Once a representative sample is obtained, the chemicals in the mass must be assayed. Figure 2.15[5] shows a flow diagram of one process (arrows) using chemical tests to determine the composition of inorganic scale and organic deposits. The following test protocol has been used for many years to analyze deposits from a range of industries, including pipeline, power production, refineries, and petrochemical plants.

1. A weighed sample of the deposit is ashed in a high-temperature furnace to burn off any hydrocarbon-containing chemicals. The weight loss is the loss on ignition (LOI).
2. The remaining solids are extracted with high-concentration HCl (at least 28%) at 200°F to determine the maximum solubility (MPS) in HCl. Analysis of the metals in HCl gives the total amount of metals in the sample. Any remaining solids are extracted with HCl/HF to give the silica content (SiO_2). Analysis of the metals in solution from the acid dissolutions can help determine the composition of the scale. Inorganic analyses can be performed using various chemical methods including atomic absorption spectroscopy, inductively coupled optical emission spectroscopy, or wet chemical tests[63] to determine the mass of the metals in the sample.

3. If the LOI is 10% or less, acidification of the solids will give a qualitative assessment of carbonates or sulfides on the basis of the formation of gas. X-ray analysis of the scale also can give a quantitative assessment of the elemental (x-ray fluorescence) composition or a semiquantitative assessment of crystalline compounds (x-ray diffraction [XRD]). This type of analysis also may identify specific crystalline minerals in the scale. Usually, iron oxides, sulfates, sulfides, and silicates can be identified using these techniques.
4. If the LOI is more than 10%, then extraction with organic solvents also is required to determine a process needed to dissolve the deposit. For large masses of organic deposit, separate hot extractions using alternate volumes of toluene and hexane can give a qualitative estimate of the chemical composition of the organic deposit. Subjecting the solution to Fourier transform infrared spectroscopy (FT-IR) or gas chromatography mass spectroscopy (GC-MS) also provides qualitative information that can facilitate a solvent choice (see the next subsection). If the MPS in HCl is more than 20%, multiple stages of an organic/inorganic solvent may be required or an acid emulsion or dispersion with an organic phase may be needed for complete removal.

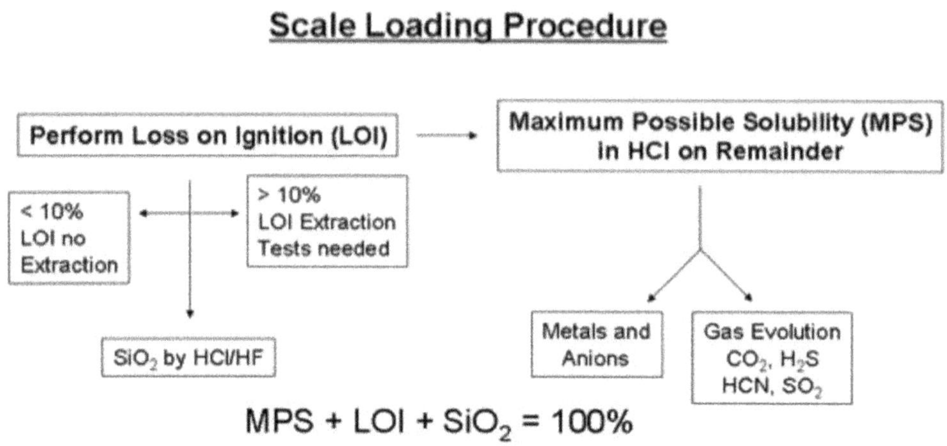

Figure 2.15 Deposit Analysis Process

Heifner[62] discusses additional methods for analyzing deposits on tubes, including metallographic examinations of the tube/scale interface using light microscopy and scanning electron microscopy (SEM). This is especially important if a tube failed (see Figure 1.19). A micrograph (Figure 2.16[62]) shows two boiler tube sections and the analysis of scale thickness and layers. This information can improve the development and choice of solvents and cleaning methods.

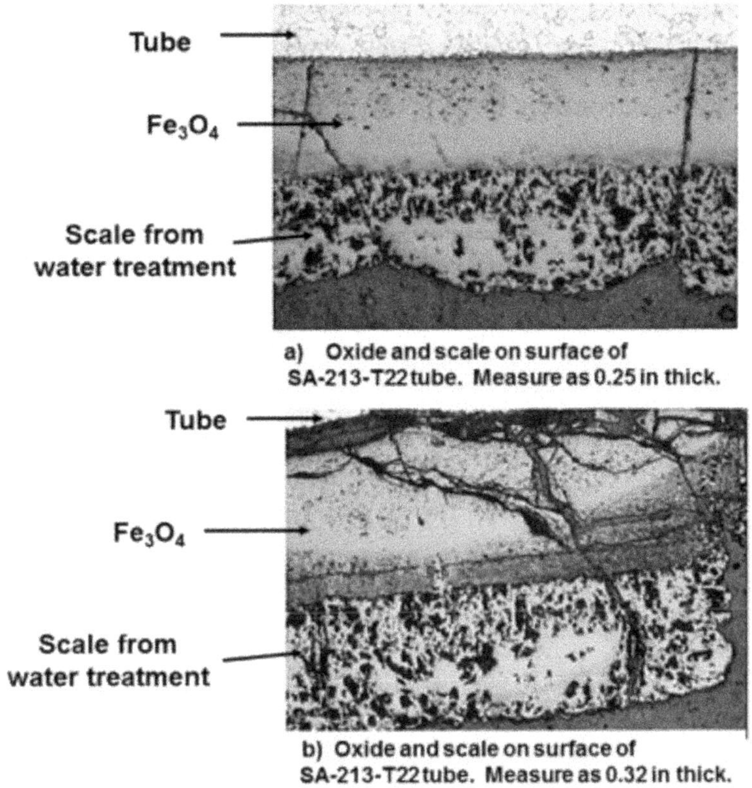

Figure 2.16 Micrograph of Scale Layers on Boiler Tubes (Nital Etch at 100× Magnification)[62]

2.6.3. Determining Organic Components in the Scale

It is usually much easier to determine the inorganic component of a deposit using the methods described in Section 2.6.2. This section describes chemical means for determining the composition of mostly organic deposit, which can be useful for choosing the appropriate solvent. This section will discuss the use of FT-IR to help identify organic deposits. This information may be used in the development of chemical dissolution formulations and methods for the subsequent removal of deposits.

Curtis and Weaver[64] described the uses of this technique. This method was originally developed to identify deposits in refinery equipment and other surface facilities but also is useful for determining the composition of pipeline deposits.

- **Identifying Classes of Organic Compounds using FT-IR.** The first step in this fingerprinting process is either physically or chemically separating component materials from the composite sample. Once separated, the functional groups or class of the component materials may be identified by FT-IR.
- **Separation by Physical Methods.** One of the more common examples of physically separating components in a refinery or petrochemical soil is based on the material's volatility relative to temperature. By increasing the temperature using an infrared lamp, heat gun, or other heat source, materials with a low boiling point will volatilize from the sample. At temperatures near or below the freezing point of water, materials with a high boiling point often will gel or solidify. Once solidified, these materials can be removed by filtration or other suitable means. Analysis of physically separated materials provides a better understanding of the organic materials within the sample. Electronically subtracting the FT-IR spectra of each component from those of the composite provides insights for the entire sample.

- **Separation by Chemical Methods.** A common method of chemically separating refinery or petrochemical soil components uses their solubilities. Some components are soluble in a polar solvent, whereas others are soluble in nonpolar solvents. Solvents are classified as polar and nonpolar by their molecular structure. Water is the most common polar solvent. Additional polar solvents are linear alcohols, such as methanol or ethanol. Traditional benchtop methods to evaluate organic soils can be time consuming and occasionally ineffective. FT-IR analysis of physically or chemically separated materials not only reduces analysis time but also provides unique insights.

An example from Curtis and Weaver[64] demonstrates this process. A sample of a foulant appeared to be black sooty metallic shavings. Following solvent extractions of this material with xylene and Freon, the solvents were evaporated, and the residues were analyzed using FT-IR. Because chlorinated solvents attack the epoxy that secures the zinc selenide crystal to the horizontal attenuated total reflectance boat, Freon was air dried from the residue. The residue was resolved in acetone for mounting purposes.

A spectral library search, restricted to the fingerprint region, strongly suggested the residue from each extraction was a defoamer (antifoaming agent). The class of defoamer was narrowed to a poly (alkyl methacrylate) copolymer in hydrocarbon oil. There were subtle differences between the residues from an acetone extraction and those from Freon and xylene. The appearance of peaks at 1852 cm^{-1}, 1774 cm^{-1}, and 1104 cm^{-1} strongly indicated a high molecular weight carboxylic anhydride as a component in the acetone-extracted material. A spectral library search confirmed this by identifying the compound class as di- and tetra-substituted carboxylic anhydrides. These materials commonly occur in emulsifiers and lubricants.

FT-IR analysis established that there were organic materials present in the form of surfactants and defoamer additives. These additives were believed to be the origin of the problem.

Several test methods for specific types deposits are listed in this section to illustrate additional methodologies.

Venditti et al.[65] describes the application of several chemical analytical methods to determine the composition of a mixed deposit fouling a refinery heat exchanger. The sample was extracted with various solvents, and the solutions were analyzed using size exclusion chromatography to determine the molecular weights. In addition, thermo gravimetric analysis, Fourier transform infrared spectroscopy, scanning electron microscopy/energy dispersive x-ray, and x-ray diffraction methods were used to identify some of the organic and inorganic parts of the scale.

Del Bianco and Stroppa[66] used the following method to test and custom-blend solvents for heavy deposits that may contain asphaltenes. The following operations were necessary to perform the measurements correctly:

1. Solutions with a known content of asphaltene material were dissolved to obtain calibrated straight lines to correlate absorbance and concentration at three selected wavelengths to perform the measurements (400 nm, 600 nm, and 800 nm). Such solutions were prepared from a mother solution obtained by filtering a mixture of approximately 100 mg of deposit in 100 mL of solvent. The test asphaltene material is a compositional continuum, and therefore the spectrum at UV-visible wavelengths is a function of the amount and of the quality of the dissolved material. The concentration of the dissolved product during the determination of the solubility curves was obtained by calculating the average values from the absorption values measured at the three wavelengths.
2. The measurement wavelength range is the widest possible for absorptions by the solvent under 400 nm, given the instrument limits. In most cases, the calibration straight lines showed good linearity within the examined concentration range. Via linear regression calculations, the extinction coefficients relevant to each individual wavelength were computed and were subsequently used to compute the concentrations when the solubility curves were determined.
3. Solubility curves for evaluating the solvent capacities were determined by measuring the amount of asphaltene material dissolved by mixtures at different values of the deposit:solvent ratio.

4. Over 15 experiments, the researchers prepared a set of mixtures with a known amount of deposit and increasing solvent volumes. These mixtures were sonicated for 20 minutes and then were mechanically stirred overnight. The resulting suspension was filtered under pressure with syringes equipped with 0.5-µm Teflon filters, the absorbance of the filtrate was measured, and the concentrations of dissolved organic material were obtained.

In this example (Del Bianco and Stroppa[66]), the solubility kinetics were reported. These values were determined using UV-visible absorption methods. The concentration of asphaltene material dissolved from the specimen (as pellets) of deposit that was soaked in the solvent was tested as a function of soaking time. The pellets were prepared by pressing an exactly weighed amount of 100 mg of sample with a press.

During the test, the solution was weakly stirred by a magnetic anchor to ensure the homogeneity of the solution and to avoid fragmentation of the pellet.

Ivanova and Shitz[67] developed a method to evaluate the efficiency of solvents with additives for removing mixed organic deposits in a Russian oil and gas field. The deposit sample is heated to the softening temperature, is thoroughly mixed, and is shaped in the form of a cylinder (12 mm × 20 mm). Then it is cooled and placed into a previously weighed basket (70 mm × 15 mm × 15 mm) made of brass (steel) gauze with 1.5 mm × 1.5 mm cells. The basket with the samples is weighed and placed in a glass hermetic cell with 100 mL of the test solvent. After 4 hours at 10°C, the basket with the residual nondestructed deposit is removed and dried to a constant weight. The destructed but undissolved deposit that had fallen from the basket in the cell is filtrated, dried to a constant weight, and weighed.

It is important to apply as many analytical methods as practical to determine the chemical composition of a deposit, especially if the deposit may contain many different organic solids and inorganic components. This information will help determine the appropriate solvent or mechanical treatment (see Chapters 4, 7, and 8).

2.6.4. Evaluation of Solvents

This section describes low-temperature and high-temperature (and pressure) methods to evaluate solvents for pipelines (usually at low temperatures), deposits, and boilers and chemical plants (frequently at higher temperatures). On the basis of information from previous cleanings, one or several solvents may require testing. The goal is to determine the most efficient solvent (dissolves most solids/unit weight) and possibly to indicate the dissolution kinetics to predict actual cleaning times.

Low-Temperature Evaluation of Solvents.

Frenier et al.[68] described the following procedure for measuring the effectiveness of different organic/inorganic deposit solvents. If the samples can be observed, an estimate of the cleaning time is possible.

A. Equipment Needed

- 4-ounce glass bottles
- Balance
- Hot water bath
- Wrist shaker

B. Procedure

1. Place approximately 10 g of sample into a clean, pre-weighed, 4-ounce bottle. Prepare one bottle for each solution to be tested.
2. Place the test bottles in hot to boiling water. Completely melt any organics if possible.

3. Carefully remove the bottles from the hot water bath, and roll the bottles horizontally as the organics begin to solidify.
4. Repeat the heating, cooling, and rolling procedure until a uniform, solid coating of organics is deposited on the inside of each test bottle.
5. Allow the paraffin bottles to cool to room temperature. Weigh each bottle.
6. Place 100 mL of the appropriate organic solvent or emulsion in each bottle. Cap and seal each bottle.
7. Place the bottles in a wrist shaker and shake for the duration of the test (up to 24 hours).
8. Visual results should be noted.
9. Carefully decant the solvent from the bottles, and dispose of the decanted solvent appropriately. Gently rinse the bottles with water and acetone, and allow them to dry.
10. Re-weigh the bottles.
11. Calculate the percent solid removed.

$$\text{Deposit Removed} = \frac{([P+B]2\ -B1)-([P+B]3\ -B1)}{([P+B]2-B1)}x100 \qquad \text{(Equation [2.27])}$$

[P + B]2 = Original weight of organics and bottle from step 5.
B1 = Original weight of bottle.
[P + B]3 = Final weight of paraffin and bottle from step 10.

12. Record the relative efficiency of each solvent system tested.

Casey[69] described the following low-temperature test procedure for organic solids. This procedure uses a gravimetric method to test solvents for removal of organic deposits, such as greases, on surfaces and uses the bob of a rheometer, which can be rotated to improve the rate of solids removal.

1. Pre-weigh a viscometer rotating bob (the rotor) to the nearest 0.01 g, and record as "Weight Before" (approximately 150 grams).
2. Lightly smear a uniform layer of axle grease on the exterior surface of the lower two-thirds of the rotor.
3. Weigh and record as "Before + Solids." Take care not to disturb the layer of grease (Figure 2.17) after this step.
4. Carefully attached the grease-smeared sleeve on the viscometer apparatus.
5. Prepare 500 mL of the test solvent in a glass beaker. The normal solvent concentration is 10% v/v.
6. Place the beaker of test solution on the viscometer fluid shelf. Adjust the fluid depth on the sleeve until the fluid meniscus and the etched ring on the rotor coincide.
7. Start the viscometer rotor at 300 rpm. Initialize/reset the stopwatch to 0.
8. Continue rotating the sleeve smeared with grease in the test solution for 30 minutes.
9. Stop the viscometer rotor. Lower the liquid-shelf, and properly discard the used solvent.
10. Carefully, remove the rotor from the viscometer to allow for air-drying.
11. Allow the rotor to air-dry overnight. Weigh the rotor and record as "Weight After."
12. Use the formula to determine the w/v% of grease removed by the solvent.

$$\textbf{\textit{Wt. \% Removed}} = \frac{(\text{Weight After})-(\text{Weight Before})}{([\text{Before}+\text{Solids}]-\text{Weight Before})}x100 \qquad \text{(Equation [2.28])}$$

Figure 2.17 Rotator with Organic Deposit[69]

This test can be modified for testing water-based solvent on inorganic minerals. The modification would involve skipping steps 2–5.

Additional methods for low-temperature dissolution of deposits are cited but not abstracted. Cowan and Weintritt[57] have a very detailed set of procedures that contains some analytical processes used before 1976.

See Spurrell and Bibbs[70] for evaluating detergent formulations for removing organic deposits from pipelines and facilities. Campbell[71] also lists simple methods for evaluating pipeline deposit cleaners.

The next subsections describe tests in which the temperatures of the units being cleaned will be >200°F, such as power boilers and many refinery units.

Evaluating Solvents for Refinery Equipment Cleaning. Although several of the processes described above can be used to evaluate solvents for various refinery deposits, Ramirez and Viloria[72] have developed specific processes to test solvents for some unusual conditions.

- **Efficiency of chemical treatments in cleaning aluminum fins**
  Aluminum fins covered with inorganic scales were cut in equal pieces to obtain the required number of samples for the study. These specimens were identified and weighed later and were exposed, at room temperature conditions, to various tests. The procedure followed in this case was based on the ASTM G1 standard and consists of obtaining weight loss of the scale-substrate system as a function of time. Exposure time periods in this evaluation were fixed in 3-minute intervals up to a total immersion time of 25 minutes.
- **Corrosion rate of aluminum in chemical treatments tested**
  Because 15 minutes is the average time established in cleaning procedures for aluminum fins, this period was used for corrosion tests. Aluminum coupons were degreased, marked, and weighed. Then they were subjected to the test solutions at room temperature.
- **Efficiency of chemical treatments in cleaning stainless steel heat exchangers**
  A solid deposit sample from a stainless steel heat exchanger was divided in portions of 20 g, which were immersed in 100%, 75%, 50%, 25% aqueous solutions of product for 22 hours. Then the solutions were filtered, dried, and weighed. Product efficiency was determined from differences in weight.

High-Temperature and High-Pressure Procedures. This procedure was modified from Frenier and Ziauddin[5] for testing predominately mixed scales. Tests should be performed in cells pressurized to approximately 600 psig. This equipment can be used for high-temperature and high-pressure simulations. This test is most appropriate for evaluating inorganic or possibly mixed deposits.

1. Prepare the scale dissolver (solvent).
2. Pour 100 mL of the scale dissolver into a test cell.
3. Place approximately 12 g of a scale sample in a scale basket (record exact mass).
4. Lower the scale basket into the test cell.
5. Pressurize the test cell to 500 psig.
6. Place the test cell in a temperature bath for the desired time (2–24 hours)
7. Remove the test cell, and release the pressure.
8. Filter the contents onto a pre-weighed filter paper.
9. Rinse the solid particles with water then with acetone to remove organics, and place the filter containing solids in a drying oven.
10. Weigh the dry solids and filter paper.
11. Test the dry solids for acid (HCl) soluble material, and develop a solvent for that portion.
12. The amount of scale removed by the organic dissolver is determined by steps 2–9.

C. Required parameters

- Piping Interval length to be treated, h (ft.)
- Internal radius of tubing, T_{ubing} (in.)
- Internal radius of scale, r_{scale} (in.)
- Scale density, ρ_s (lbm/ft^3)
- Dissolution capacity, G (lbm/gal)

The calculations for determining the needed volume of solvents are in Figure 2.18.

1) Volume of Scale $$V_s(ft^3) = 3.1416 * h * \left[\left(\frac{R_{tubing}}{12}\right)^2 - \left(\frac{R_{scale}}{12}\right)^2\right]$$

2) Volume of Chemical to remove all scale, $$V_c(gal) = \frac{V_s * \rho_s}{G}$$

3) Initial pipe volume, $$V_i(ft^3) = h * 3.1416 * \left[\frac{r_{scale}}{12}\right]^2$$

4) Estimated number of soaks $$= \frac{V_c * \rho_s}{h * G * 176.715 * \left[\dfrac{\dfrac{R_{tubing}}{12} + \dfrac{r_{scale}}{12}}{2}\right]^2}$$

Figure 2.18 Calculations for Solvent Volumes

2.6.5. Equipment for Testing the Removal of a Deposit from a Tubing Sample

If tubing samples with scale are available, different processes can test solvents and determine approximate cleaning times. Some of the equipment used in corrosion studies, especially the boiling flasks (Section 6.6.3) and the rotational equipment (Figure 2.19), can be used. This equipment also can be used to test the solubility of loose (removed) scale.

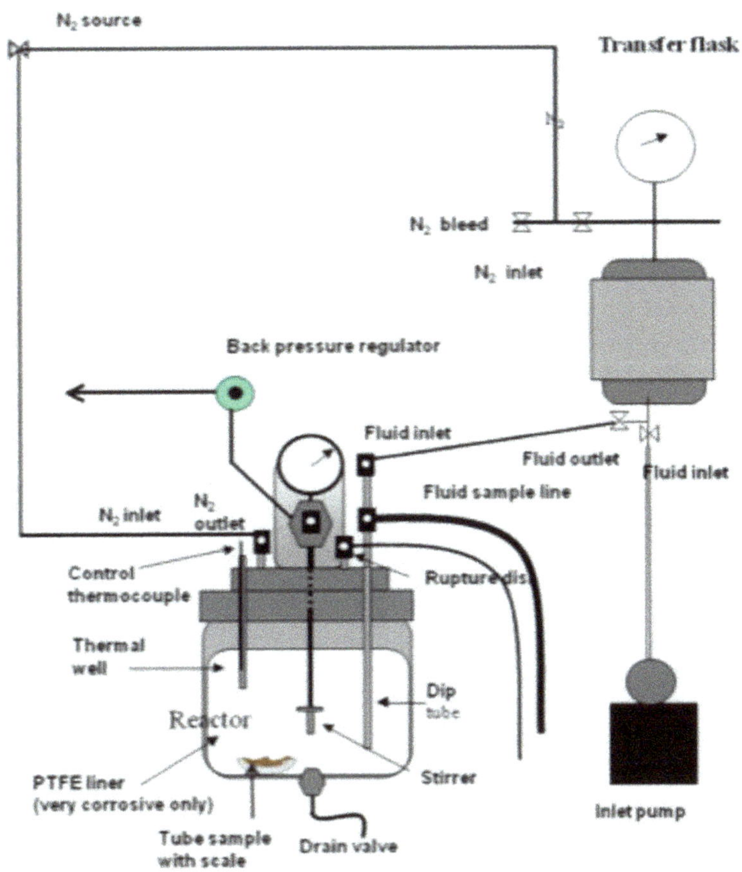

Figure 2.19 Autoclave Solvent Tester

The usual method is to stir a piece of tubing with scale in a boiling flask or autoclave and observe the buildup of the metal in solution from the scale, until a plateau is reached. The metric in most cases is the increase of [Fe] with time. Figure 2.20 shows the cleaning of boiler tube samples (scale in place) in inhibited tetraammonium EDTA at various temperatures. Note that this type of curve also is used onsite to follow the dissolution kinetics of an IC job. The results shown demonstrate that a temperature of at least 138°C will clean the tube efficiently.

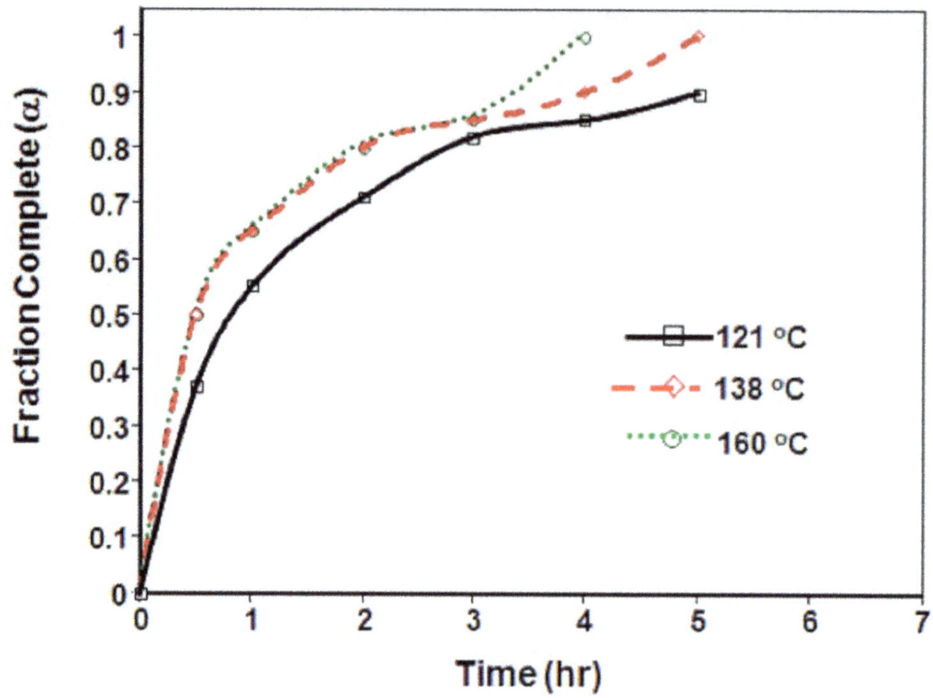

Figure 2.20 Metal Pick Up Curves for a Solvent at Different Temperatures

Details of an Autoclave Test. Many high-temperature, iron scale dissolution tests have been run in 1-L titanium autoclaves (such as Figure 2.19). Transfer flasks were fabricated from AISI 316 stainless steel so that a concentrated chelant solution could be injected into the autoclave when the test temperature had been reached. All the kinetic tests were conducted using the same batch of tubing obtained from an operational northeastern U.S. drum boiler. The tubes were machined to a constant outside diameter (3.34 cm) and length (4.74 cm), with a total surface area of 102.8 cm^2.

The interior surfaces were coated with about 1 g of oxide and 50 mg of copper. The composition of the scale as determined by x-ray diffraction and the chemical composition of the boiler tube are shown in Table 2. In each of the dissolution tests, two AISI 1010 carbon steel coupons in a PTFE mount were added to give a total wetted surface area of 190 cm^2.

For the scale dissolution test, the tube sample, 0.10% commercial corrosion inhibitor, corrosion coupons, and 250 mL of H_2O were put into a glass liner and then into the autoclave. The vessel was heated to the test temperature (usually 320°F [160°C]) and stirred at 200 rpm.

A total of 50 mL of the concentrate cleaning solution (a chelating agent) was drawn into an evacuated transfer bomb, which then was pressurized with N_2 to about 725 kPa. At the starting time, the chelant was blown into the reactor using a high-pressure nylon hose. The chelant concentration was 0.12M. At intervals, 1–2 mL of solution was withdrawn after the sample line was blown down with N_2. The samples were analyzed for [Fe] by atomic absorption spectroscopy. At the end of the test (4–6 hours), the samples were removed. The tube was rinsed with acetone and examined under low power magnification for scale and pitting. Figure 2.20 shows the data obtained by one set of tests.

Loop Tests. Large pieces of equipment (see Figure 2.21 and Figure 2.22) can be used to simulate cleaning a boiler tube. A particular unit[73] (Figure 2.21) was used to test solvents to clean once-through boilers, and it matched field results accurately, on the basis of total cleaning times. In this case, the temperature limit was about 200°F.

Much heavier loops have been constructed[74] that can test short sections (6–8 in) of scaled boiler tubes, where the treating temperatures will be >200°F and the pressures can range to about 80 psig. These units can predict cleaning time accurately, but the surface/volume ratio does not reflect the boiler, so the metal concentrations in the loop are much lower than found during the field cleaning. Figure 2.22 shows a high-temperature/high-pressure loop capable of >350°F cleaning temperatures. One of the key determinations of loop tests is an estimate of the cleaning time and the condition of the tube surfaces after cleaning. Solvents that cause pitting may be eliminated from consideration.

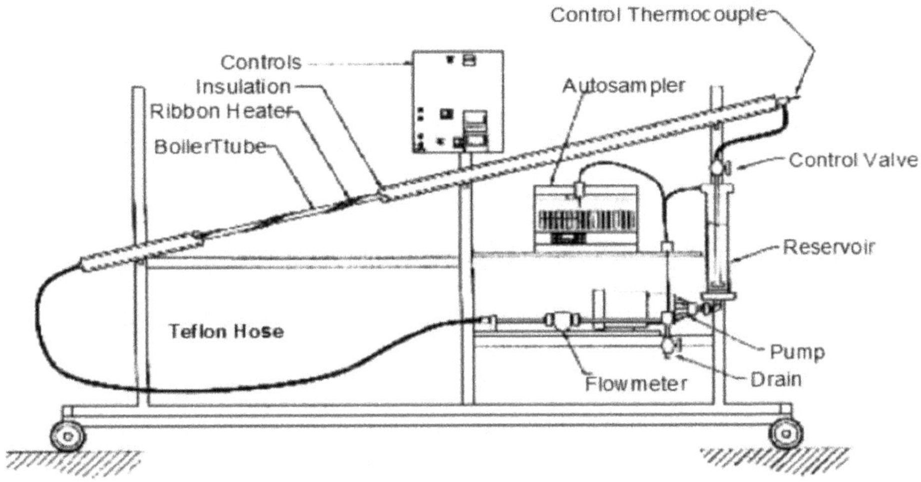

Figure 2.21 Low-Temperature Test Loop

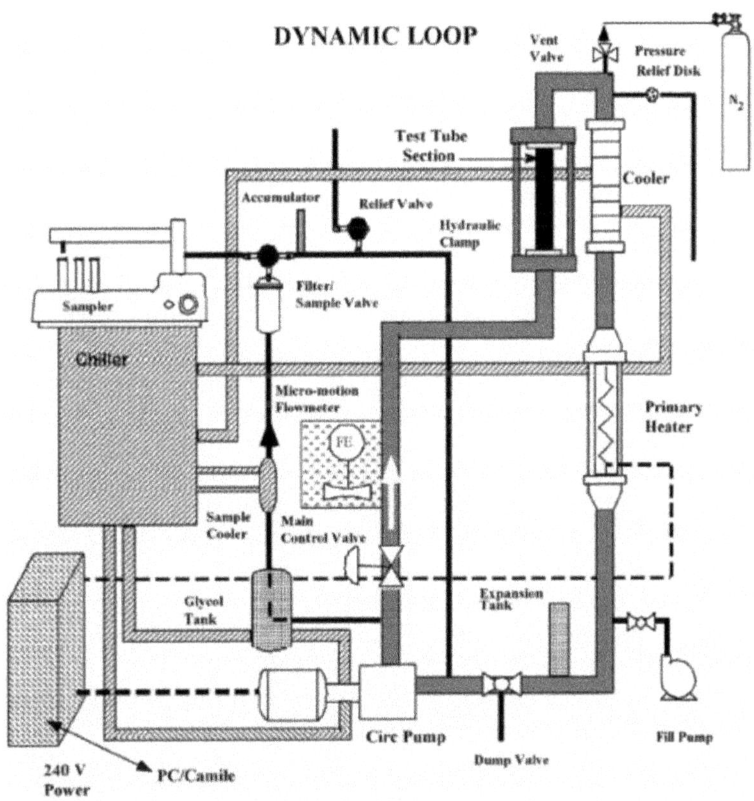

Figure 2.22 High-Temperature Flow Test Loop

2.6.6. Chemical Analyses of Cleaning Fluids

Onsite (and laboratory) analyses of the progress of a cleaning job was initially conducted from the tailgate of a pickup truck or on a lab bench using a kit of chemicals that included acid titration (to determine residual acid), chemical packets, and a battery-operated spectrophotometer to determine the iron and/or copper in the solutions. With the developments of ammonium EDTA[75] in the 1970s and the ability to remove tons of iron oxide from utility boilers, mobile laboratories were introduced. Often, a large trailer was modified with a range of titrations (chelating agent strength, acid strength, pH electrochemical properties [see Section 4.2]), and a small atomic absorption spectrophotometer for determining iron, copper, nickel, and chromium. The advantages were that the crews were protected from the weather and onsite hazards and had fresh coffee all night! Section 10.2.2 shows the interior of a modern lab. These types of automated metal analyses also are used in the research and development/support lab. Additional equipment may include inductively coupled plasma spectroscopy, x-ray, and other surface and organic chemical analytical devices.

2.7. Summary of Key Technologies

This chapter has introduced the technologies involved in chemical and mechanical cleaning of industrial equipment. Choosing the appropriate cleaning technology depends on several factors:

- Surface chemistry to identify the composition of deposits and removal methods
- Corrosion in the units that may affect the deposits and test methods for monitoring corrosion
- The types of fouling deposits in the units
- Introduction to chemical and mechanical cleaning processes
- Testing of deposits and solvent selection

These aspects affect the planning and execution of the job and will be discussed in much more detail in subsequent chapters.

References

[1]F.W. Sears, M.W. Zemanski, *University Physics*, 2nd ed. (Boston, MA: Addison Wesley, 1955).

[2]P.R. Howard, S. Mukhopadhyay, N. Moniaga, L. Schafer, G. Penny, K. Dismuke, "Comparison of Flowback Aids: Understanding Their Capillary Pressure and Wetting Properties," *SPE Production & Operations* 25, 3 (2010): 376-387.

[3]J. Dean, *Lange's Handbook of Chemistry*, 10th ed. (New York, NY: McGraw-Hill, 1961).

[4]R. Tadmor, "Line Energy and the Relation between Advancing, Receding, and Young Contact Angles," *Langmuir* 20, 18 (2004): 7659-7662.

[5]W.W. Frenier, M. Ziauddin, *Formation, Removal, and Inhibition of Inorganic Scale in the Oilfield Environment* (Richardson, TX: Society of Petroleum Engineers, 2008).

[6]PHYWE Systeme, "Surface Tension by the Ring Method (Du Nouy Method)," (Göttingen, Germany: PHYWE System GMBH, 2004).

[7]C.A. Grattoni, E.D. Chiotis, R.A. Dawe, "Determination of Relative Wettability of Porous Sandstones by Imbibition Studies," *Journal of Chemical Technology & Biotechnology* 64, 1 (1995): 12-24.

[8]J.P. Brill, "Multiphase Flow in Wells," *JPT* 39, 1 (1987): 15-21.

[9]B. Asante, "Two Phase Flow: Accounting for the Presence of Liquids in Pipeline Simulation," Paper Pipeline Simulation Interest Group 02W3, PSIG Annual Meeting, Portland, Oregon, October 23–25, 2002.

[10]O. Shoham, *Mechanistic Modeling of Gas-Liquid Two-Phase Flow in Pipes* (Richardson, TX: Society of Petroleum Engineers, 2006).

[11]J. Falcimaigne, S. Decarre, *Multiphase Production: Pipeline Transport, Pumping, and Metering* (Paris, France: IFP Publications, 2008).

[12]M.J. Economides, A.D. Hill, C. Ehlig-Economides, *Petroleum Production Systems* (Upper Saddle River, NJ: Prentice Hall Petroleum Engineering Series, 1994).
[13]G. Hagen, "Uber Die Bewegung Des Wassers in Engen Zylinderschen," *Ann. Phys. Chem.* 46 (1839): 423-442.
[14]J.L. Poiseuille, *Compte Rendus* 11 (1840): 961, 1041.
[15]D.R. Shannon, *The Art of Pigging* (Des Moines, IA: D.R. Shannon Company, 2010).
[16]K.E. Kee, M. Babic, S. Richter, L. Paolinelli, W. Li, S. Nesic, "Flow Patterns and Water Wetting in Gas-Oil-Water Three-Phase Flow - a Flow Loop Study," CORROSION/2015, paper no. NACE 2015-6113 (Houston, TX: NACE, 2015).
[17]M.G. Fontana, *Corrosion Engineering,* 3rd ed. (Boston, MA: McGraw-Hill, 1986).
[18]R. Heidersbach, *Metallurgy and Corrosion Control in Oil and Gas Production* (Hoboken, NJ: John Wiley & Sons, Inc., 2011).
[19]P. Roberge, *Corrosion Basics,* 2nd ed. (Houston, TX: NACE International, 2006).
[20]L. Lazzari, P. Pedeferri, *Cathodic Protection* (Milan, Italy: Polipress, 2006).
[21]American National Standards Institute/American Petroleum Institute Spec 5L, Specification for Line Pipe, 44th ed., Includes Errata and Addendums 1 (2009), 2 (2010), and 3 (2011) (Washington, DC: American Petroleum Institute, 2011).
[22]ASTM International A192/A192M-02(2012), Standard Specification for Seamless Carbon Steel Boiler Tubes for High-Pressure Service (West Conshohocken, PA: ASTM International, 2012).
[23]E. Akeer, B. Brown, S. Nesic, "The Influence of Mild Steel Metallurgy on the Initiation of Localized Co2 Corrosion in Flowing Conditions," CORROSION/2013, paper no. NACE 2013-2383 (Houston, TX: NACE, 2013).
[24]ANSI/NACE MR0175-2009, Materials for Use in H_2S-Containing Environments in Oil and Gas Production: Parts 1, 2, and 3 (Identical to ISO 15156-1:2009, 15156-2:2009, and 15156-3:2009) (Houston, TX: NACE International, 2009).
[25]G.S. Frankel, "Pitting Corrosion," in *Corrosion: Fundamentals, Testing, and Protection ASM Handbook*, vol. 13a (Materials Park, OH: ASM International, 2003).
[26]M.H. El-Sayed, "Grooving Corrosion of a 24-in Seam Welded Pipeline," *MP* 53, 6 (2014): 62-65.
[27]P. Angell, "Use of Microbial Kinetics to Control MIC in the Nuclear Industry," CORROSION/2002, paper no. NACE 02475 (Houston, TX: NACE, 2002).
[28]J.G. Stoecker, *Practical Manual on Microbiologically Influenced Corrosion,* vol. 2A (Houston, TX: NACE International, 2001).
[29]R.B. Eckert, *Introduction to Managing Microbiologically Influenced Corrosion* (Houston, TX: NACE International, 2015).
[30]G.E. Jenneman, "Biocorrosion: What's Old, What's New," Tulsa Section Meeting of International, Tulsa, OK, Nov. 19, 2013.
[31]A. Augustinovic, O. Birketveit, K. Clements, M. Freeman, S. Gopi, T. Ishoey, G. Jackson, G. Kubala, J. Larsen, B.W.G. Marcotte, J. Scheie, T.L. Skovhus, E. Sunde, "Microbes-Oilfield Enemies or Allies," *Oilfield Review* 24, 2 (2012): 4-17.
[32]G. Jenneman, "Microbiologically Influenced Corrosion in the Oil Field," Tulsa Section Meeting of NACE International, Tulsa, OK, Sept. 25, 2006).
[33]S. Seagraves, Y. Wu, "Comparison of Scale Index Calculations and Two Predictive Models," CORROSION/96, paper no. 96186 (Houston, TX: NACE, 1996).
[34]B.J. Little, P. Wagner, "Microbiologically Influenced Corrosion," in *Peabody's Control of Pipeline Corrosion,* 2nd ed., ed. R.L. Bianchetti (Houston, TX: NACE International, 2001).
[35]B.J. Little, J.S. Lee, R.I. Ray, "Diagnosing Microbiologically Influenced Corrosion: A State of the Art Review," *Corrosion* 62, 11 (2006): 1006-1017.
[36]J. Larsen, K. Sorenson, K. Hojris, T.L. Skovhus, "Significance of Troublesome Sulfate-Reducing Proaryotes (Srp) in Oilfield Systems," CORROSION/2009, paper no. NACE 09389 (Houston, TX: NACE, 2009).
[37]J. Larsen, K. Rasmussen, H. Pedersen, K. Sørensen, T. Lundgaard, T.L. Skovhus, "Consortia of Mic Bacteria and Archaea Causing Pitting Corrosion in Top Side Oil Production Facilities," CORROSION/2010, paper no. NACE10252 (Houston, TX: NACE, 2010).
[38]C. Whitby, S. Lund, *Applied Microbiology and Molecular Biology in Oilfield Systems* (New York, NY: Springer, 2011).

[39]S. Rassenfoss, "From Bacteria to Barrels: Microbiology Having an Impact on Oil Fields," *JPT* 63, 11 (2011): 32-38.

[40]D. Enning, H. Venzlaff, J. Garrelfs, H.T. Dinh, V. Meyer, K. Mayrhofer, A.W. Hassel, M. Stratmann, F. Widdel, "Marine Sulfate-Reducing Bacteria Cause Serious Corrosion of Iron under Electroconductive Biogenic Mineral Crust," *Experimental Microbiology* 14, 7 (2012): 1772-1787.

[41]H.G. Byars, *Corrosion Control in Petroleum Production* (Houston, TX: NACE International, 1999).

[42]V.J. Mena Cervantes, L.S. Zamudio Rivera, M. Lozada y Cassou, H.I. Beltrán Conde, E. Buenrostro González, S. López Ramirez, Y. Douda, A. Morales Pacheco, R. Hernández Altamirano, M. Barcenas Castañeda, "Composición De Aditivo Inhibidor Dispersante De Asfaltenos a Base De Oxazolidinas Derivadas De Polialquil O Polialquenil N-Hidroxialquilsuccinimidas," Mexico Application MX/E/2007/084388, 2007.

[43]W.H. Durnie, R. DeMarco, A. Jefferson, B.J. Kinsella, "Developments in Laboratory Techniques to Assess the Persistency of Carbon Dioxide Corrosion Inhibitor Films," 9th European Symposium on Corrosion Inhibitors (9SEIC), Ferrara, Italy, Sept. 4-8, 2000.

[44]W.W. Frenier, *Chemical and Mechanical Methods of Pipeline Integrity* (Richardson, TX: Society of Petroleum Engineers, 2017).

[45]D. Brondel, R. Edwards, A. Hayman, D. Hill, S. Mehta, T. Semerad, "Corrosion in the Oilfield Industry," *Oilfield Review* 6, 2 (1994): 4-18.

[46]K.L. Heidersbach, A.C.V. Roodselaar, "Understanding, Preventing, and Identification of Microbial Induced Erosion-Corrosion (Channeling) in Water Injection Pipelines," CORROSION/2012, paper no. NACE, 2012-1221 (Houston, TX: NACE, 2012).

[47]J.R. Vera, D. Daniels, M.H. Achour, "Under Deposit Corrosion (UDC) in the Oil and Gas Industry: A Review of Mechanisms, Testing and Mitigation," CORROSION/2012, paper no. NACE 2012-1379 (Houston, TX: NACE, 2012).

[48]M. Ajmal, B. Saleem, A.R. Saleemi, "Stress Corrosion Cracking of 18-Inch (45.72cm) Diameter Pipeline (X52)," CORROSION/2012, Paper NACE 2012-1130 (Houston, TX: NACE, 2012).

[49]X. Hu, A. Nevill, J. Wells, V.D. Souza, "Prediction of Erosion-Corrosion in Oil and Gas - a Systematic Approach," CORROSION/2008, paper no. NACE 08540 (Houston, TX: NACE, 2008).

[50]API 12.ARP 14E, *Recommended Practice for Design and Installation of Offshore Production Platform Piping Systems*, 5th ed. (Washington, DC: American Petroleum Institute, 1991).

[51]R. Barker, X. Hu, A. Neville, S. Cushnaghan, "Flow-Induced Corrosion and Erosion-Corrosion Assessment of Carbon Steel Pipework in Oil and Gas Production," CORROSION/2011, paper no. NACE 11245 (Houston, TX: NACE, 2011).

[52]W.W. Frenier, *Technology for Chemical Cleaning of Industrial Equipment* (Houston, TX: NACE International, 2001).

[53]W.W. Frenier, S.J. Barber, "Choose the Best Heat Exchanger Cleaning Method," *Chemical Engineering Progress* 94, 7 (1998): 37-45.

[54]"HydroChem Story," HydroChem (Deer Park, TX), 2014, http://www.hydrochem.com/ (January 18, 2014).

[55]A.C. Rubey III, A.M. Taylor, "Cleaning Method and Apparatus Using Sodium Bicarbonate Particles," U.S. Patent 5588901, 1996.

[56]S.E. Evans, J.F. Remark, S.L. Watson, F. Ammann, "Steam Generator Chemical Cleaning Experiences 1998-2002," CORROSION/2003, paper no. NACE-03579 (Houston, TX: NACE, 2003).

[57]J.C. Cowan, D.J. Weintritt, *Mineral & Organic Scale Deposits* (Lafayette, LA: Cowan & Weintritt Publishing Company, 2004).

[58]M.J. Esmacher, G.H. Bodman, "Use of Comprehensive Deposit Analysis Techniques to Evaluate Boiler Deposits Prior to Chemical Cleaning," Paper IWC-00-33, 2000 International Water Conference, Pittsburgh, PA, Oct. 22-26, 2000.

[59]NACE TM0199-2013, Standard Test Method for Measuring Deposit Mass Loading ("Deposit-Weight-Density") Values for Boiler Tubes by the Glass-Bead-Blasting Technique (Houston, TX: NACE International, 2013).

[60]ASTM D3483-14, Standard Test Methods for Accumulated Deposition in a Steam Generator Tube (West Conshohocken, PA: ASTM International, 2014).

[61]K. Hansen, "Chemical Cleaning of Boilers," Paper IWC 12-67, 73rd Annual IWC, San Antonio, TX, Nov. 4-8, 2012.

[62]S.C. Heifner, "Deposit Loading Characterization to Enhance Boiler Chemical Cleaning Decisions," CORROSION/2006, paper no. NACE 06460 (Houston, TX: NACE, 2006).
[63]Hach, "Test Kits" (Loveland, CO: Hach Company, 2005).
[64]G. Curtis, C.W. Weaver, "Use of FT-IR to Identify Organic Deposits," CORROSION/98, paper no. NACE 337 (Houston, TX: NACE, 1998).
[65]S. Venditti, C. Berrueco, P. Alverez, T.J. Morgan, M. Millan, A.A. Herod, R. Kandiyoti, "Developing Characterization Methods for Foulants Deposited in Refinery Heat Exchangers," International Conference on Heat Exchanger Fouling and Cleaning-VIII-2009, Schlandming, Austria, June 14-19, 2009.
[66]A. Del Bianco, F. Stroppa, "Effective Hydrocarbon Blend for Removing Asphaltenes," U.S. Patent 5382728, 1995.
[67]I.K. Ivanova, E. Shitz, "Hydrocarbon Solvents on the Hexane Base for Oil Organic Deposits Elimination of the Irelyakh Gas and Oil Field," *Oil and Gas Business* 1 (2008): 1-7.
[68]W.W. Frenier, M. Ziauddin, R. Venkatesan, *Organic Deposits in Oil and Gas Production* (Richardson, TX: Society of Petroleum Engineers, 2010).
[69]G. Casey, "Evaluating Cleaning Solutions for the Removal of Hydrocarbon Deposits in Pipelines and Process Equipment," CORROSION/2006, paper no. NACE 06531 (Houston, TX: NACE, 2006).
[70]C.H. Spurrell, M. Bibbs, "Refinery Chemical Cleaning: Criteria for Detergent Selection," CORROSION/2000, paper no. NACE-00446 (Houston, TX: NACE, 2000).
[71]S. Campbell, "The Ins and Outs of Pipeline Cleaners: Testing and Evaluation for Chemical Cleaners," CORROSION/2000, paper no. NACE-00434 (Houston, TX: NACE, 2000).
[72]M. Ramirez, A. Viloria, "A Methodology for the Evaluation of Chemical Treatments Used in Cleaning of Refinery Components and Equipment," CORROSION/2003, paper no. NACE 03650 (Houston, TX: NACE, 2003).
[73]W.W. Frenier, D. Larson. "Development and Testing of a New Solvent for Cleaning Once-through Boilers," Paper IWC 94-74, 55th International Water Conference, Pittsburgh, PA, Oct. 31–Nov. 3, 1994.
[74]HydroChem, GA-CHEM-0613, Chemical Services, 2013, (Deer Park, TX), ChemicalServices_web.pdf.
[75]W.W. Frenier, W.C. Kennedy, "Mechanism of Copper Dissolution in Chelant Solvents," CORROSION/83, paper no. NACE 222 (Houston, TX: NACE, 1983).

3 Formation and Deposition of Inorganic Scales

This chapter reviews the formation and deposition of inorganic fouling scales in industrial equipment. However, general principles apply to inorganic deposits, organic deposits, and mixed deposits. Several major reviews of the literature of allied technologies include the books by Frenier[1,2,3,4] and the references in the Introduction. Cowan and Weintritt[5] dedicate several chapters to water-formed scales, such as sulfates and carbonates. Since chemistry works the same if the conditions are the same, this information also applies to industrial surfaces and oilfield surfaces. Many specific references in this chapter, especially on iron oxide formation and deposition, also can be found in Frenier.[6, 7]

It is important to note that solids may form from a saturated solution and not deposit. Also, in some cases, scale can form on a surface without detectable solids suspended in the aqueous phase. Thus, this chapter discusses homogeneous and heterogeneous scaling processes. In addition, the rate at which a deposit forms can greatly affect the need for chemical cleaning.

There are four distinct types of formation/deposition processes:

- Electrochemical processes (usually corrosion) with solution equilibrium changes, which form
  - Iron oxides
  - Iron sulfides
  - Iron carbonates
  - Copper and copper oxides
- Scale and depositions caused by additional changes in solution equilibria, including
  - Alkaline and alkaline earth solids, such as calcite, gypsum, and other sulfates and silicates
  - Iron sulfides
  - Organic deposit/mixed deposits (waxes, asphaltenes, and some refinery deposits)
- Depositions caused by organic chemical reactions (frequently at high temperatures), including
  - Complex compounds formed form the reactions in crude oil or in chemical process reactors
  - Reactions in acid gas removal solutions
- Deposition of "debris" produced in one section of a plant and then transported to another region where fouling deposits form. Changes in flow conditions can cause the deposit to form.

These processes follow the thermodynamic and kinetic laws reviewed in the following subsections. For precise nomenclature, keep in mind that an inorganic mineral (or scale) can form in a fluid but may or may not form a fouling deposit on a surface.

3.1. Simplified Aqueous Solution Thermodynamics

3.1.1. Introduction

The concepts of aqueous solution thermodynamics are crucial for understanding scale formation and removal. Because the topic is covered in great depth in textbooks on aqueous chemistry,[8,9] only a simplified description of important principles is presented here. Also see the discussions in Frenier and Ziauddin.[1]

A system at chemical equilibrium represents a dynamic state in which two or more opposing reactions are taking place at the same time and at the same rate. Figure 3.1 shows a chemical system approaching equilibrium. Early in the process, the rate of the forward reaction (r1), i.e., the formation of products from reactants, is large. As the system progresses, the rate of the forward reaction decreases, and the rate of the reverse reaction (r2) increases. When the rates are equal, the system is in equilibrium, and the net rate (i.e., rate of formation of products – rate of formation of reactants) is equal to zero. Once a closed system reaches chemical equilibrium, the chemical composition of the system is independent of time and previous history.

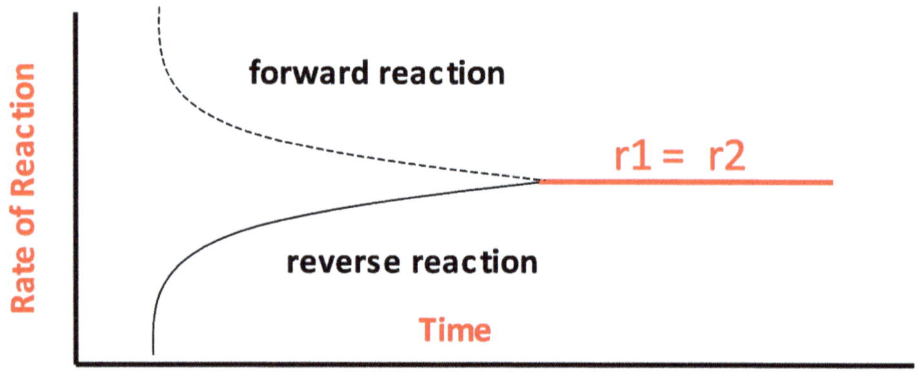

Figure 3.1 Approach of a Chemical System to Equilibrium

Equilibrium equations are used to relate the amounts of reactants and products at equilibrium. Consider, for example, a simple ideal chemical system consisting of reactants A and B and products C and D. The reaction stoichiometry is given by

$$aA + bB \leftrightarrow cC + dD \quad \text{(Equation [3.1])}$$

This implies that the forward and reverse reactions are

$$aA + bB \rightarrow cC + dD \text{ (forward reaction)} \quad \text{(Equation [3.2])}$$

$$cC + dD \rightarrow aA + bB \text{ (reverse reaction)} \quad \text{(Equation [3.3])}$$

If the system is ideal, then an *equilibrium constant* for the system can be expressed in terms of concentrations as

$$K_{eq} = \frac{[C]^c[D]^d}{[A]^a[B]^b} \quad \text{(Equation [3.4])}$$

Here, K_{eq} is the equilibrium constant, and [ ] denotes concentration of the species. The equilibrium constant also can be expressed in terms of the free energy change for the reaction (ΔG°) as

$$K_{eq} = \exp\left(\frac{-\Delta G^\circ}{RT}\right) \quad \text{(Equation [3.5])}$$

Here, R is the gas constant, and T is the temperature. K_{eq} is a function of temperature and pressure only and does not depend on the composition of the system. Therefore, if the composition of the system changes by subsequent reactions or addition of new reactants or products to the system, the same value of the equilibrium constant(s) can be used to calculate the equilibrium distribution in the new system as long as the temperature and pressure remain constant and the equilibrium assumption is valid.

Equilibrium constants are not constants in the true sense because they do depend on temperature and pressure. An increase in temperature may affect the forward and reverse reactions differently. The reaction that absorbs the most heat will increase that rate more than the other reaction. A new equilibrium constant will now represent the new situation. The simple equation by Van't Hoff can be used to compute the change in equilibrium constant owing to the change in temperature. The equation can be expressed as

$$\frac{d \ln K_{eq}}{dT} = \frac{\Delta H^\circ}{RT^2} \qquad \text{(Equation [3.6])}$$

Here, ΔH° is the standard enthalpy change of reaction, T is the temperature, and R is the gas constant. If the reaction is exothermic (i.e., if ΔH° for the reaction is negative), then the equilibrium constant decreases as temperature increases. Conversely, K_{eq} increases with temperature for an endothermic reaction.

If ΔH° is assumed independent of temperature, then integrating Equation (3.6) yields an even simpler result:

$$\ln \frac{K_{eq}}{K_{eq,ref}} = -\frac{\Delta H^\circ}{R}\left[\frac{1}{T} - \frac{1}{T_{ref}}\right] \qquad \text{(Equation [3.7])}$$

Here $K_{eq,ref}$ is the value of the equilibrium constant at the reference temperature $T_{eq,ref}$. This approximate equation implies that a plot of $\ln K_{eq}$ versus the reciprocal temperature gives a straight line. This equation is helpful in interpolating and extrapolating equilibrium constant data with reasonable accuracy.

The effect of pressure on equilibrium constants is typically smaller than the effect of temperature. However, for deep wells, this effect can be significant and needs to be considered along with the change in temperature. The pressure depending on the equilibrium constant can be calculated from

$$\left(\frac{\partial \ln K_{eq}}{\partial P}\right)_T = \frac{\Delta V^\circ}{RT} \qquad \text{(Equation [3.8])}$$

Here, ΔV° is the molar volume of change of the reaction with all reactants and products in their standard states.[9] If the molar volume change of the reaction is independent of pressure, then the integration of the above equation yields

$$\ln \frac{K_{eq}}{K_{eq,ref}} = -\frac{\Delta V^\circ (P - P_{ref})}{RT} \qquad \text{(Equation [3.9])}$$

Here, K_{eq} is the equilibrium constant at the desired pressure P. $K_{eq,ref}$ is the equilibrium constant at P_{ref}, which is typically 1 bar.

In very dilute aqueous solutions, the anions and cations behave in an ideal manner: each ion will act as if it is independent of all other ions in the solution. In real solutions, especially those with high concentrations of other ions (such as a brine), the ions are affected by the other ions in solution, and the fluid may not behave as if it has the same number of ions as described by the concentration. The equilibrium constants in such non-ideal systems are then expressed in terms of species activity. For example, if the chemical system considered previously with components A, B, C, and D is non-ideal, then the equilibrium constant is given by

$$K_{eq} = \frac{\{C\}^c \{D\}^d}{\{A\}^a \{B\}^b} = \exp\left(\frac{-\Delta G^\circ}{RT}\right) \qquad \text{(Equation [3.10])}$$

Here { }[4] denotes activity of the species. The activity of the species can be thought of as an effective concentration of the species in solution.

3.1.2. Formation of Inorganic Scale

Equilibrium constants are very helpful in describing a crucial concept in the formation of scale, which is the idea of solubility of a salt in water. The solubility (*S*) of a salt is the mass of the salt dissolved in a known weight (or volume) of fluid and can be expressed in molar as well as mass units; therefore, care must be exercised in taking data from the literature. Consider, for example, the dissolution of sodium chloride (NaCl). The chemical equation is

$$NaCl(s) \leftrightarrow Na^{+}(aq) + Cl^{-}(aq) \qquad \text{(Equation [3.11])}$$

This equation states that the dissolved sodium and chloride ions are in equilibrium with the solid sodium chloride salt. The forward reaction is the dissolution of NaCl(s) to produce one cation (Na^+) and one anion (Cl^-). The reverse reaction is the combination of the cation and anion to yield solid NaCl. When enough sodium and chloride ions are in the solution to exceed the solubility of the salt, a solid ($NaCl_{(s)}$) forms.

For some salts with different numbers of charges on the cations and anions, the generalized equation is

$$M_mX_x(s) \leftrightarrow mM^{z+}(aq) + xX^{z-}(aq) \qquad \text{(Equation [3.12])}$$

Here, *m* is the number of cations with a charge *z*+, and *x* is the number of anions with a charge of *z*-. An example of a more complex salt is $FeCl_3$. Thus,

$$FeCl_3(s) \leftrightarrow Fe^{3+}(aq) + 3Cl^{-}(aq) \qquad \text{(Equation [3.13])}$$

The equilibrium constant for dissolution of salts is of special importance in scaling studies and is given special notation. For example, consider the equilibrium constant for the general dissolution reaction expressed in Equation (3.12) above:

$$K_{eq} = \frac{\{M^{z+}\}^m\{X^{z-}\}^x}{\{M_mX_x\}} \qquad \text{(Equation [3.14])}$$

Since the activity of a pure solid is equal to one, the above equation can be simplified to

$$K_{eq} = K_{sp} = \{M^{z+}\}^m\{X^{z-}\}^x \qquad \text{(Equation [3.15])}$$

Here, K_{sp} is the equilibrium constant for the dissolution of the salt, or the *solubility product* (Table 3.1). If the salt is dissolving into a dilute solution (i.e., into an ideal solution), then activities of the species are equal to their concentration, and K_{sp} can then be expressed as

$$K_{sp} = [M^{z+}]^m[X^{z-}]^x \qquad \text{(Equation [3.16])}$$

For example, consider the dissolution of $CaSO_4$ in water:

$$CaSO_4(s) \leftrightarrow Ca^{2+}(aq) + SO_4^{\ 2-}(aq) \qquad \text{(Equation [3.17])}$$

K_{sp} for $CaSO_4$ is given by

$$K_{sp} = \{Ca^{2+}\}\{SO_4^{\ 2-}\} \qquad \text{(Equation [3.18])}$$

If $CaSO_4$ is dissolved in a dilute solution, the above equation can be further simplified to

$$K_{sp} = [Ca^{2+}][SO_4^{2-}] \quad \text{(Equation [3.19])}$$

An extensive compilation of solubility data is provided by Linke.[10] K_{sp} values for various ionic compounds can be obtained from the *CRC Handbook of Chemistry and Physics*[11] or from *Lange's Handbook of Chemistry*.[12] These data are helpful in assessing the scaling potential of oilfield brines.

In most brines, activities instead of concentration values must be used in equilibrium calculations. The activities of the dissolved ions at equilibrium are a function of temperature, pressure, and brine composition. Many models are available for calculating activities of species and are discussed in textbooks on aqueous thermodynamics; for example, see Zemaitis et al.[8] Various software programs described in Frenier and Ziauddin[1] (see Section 2.2.11[1]) can be used to automate these calculations. However, note that this is an evolving field, so additional programs will continue to be produced. For convenience, a few programs are described in Section 3.2.1 of this chapter.

A further complicating factor is that many salt solutions can exist in a state of supersaturation (i.e., the concentration of cations and anions in solution is much larger than that predicted by chemical equilibrium calculations). Supersaturated solutions can exist for a long time and may not produce a scale unless initiated by, for example, a seed crystal or an appropriate surface. A useful quantity for describing such solutions is the saturation ratio (SR):

$$SR = \{M^{z+}\}^m \{X^{z-}\}^x / K_{sp} \quad \text{(Equation [3.20])}$$

The saturation ratio for a salt measures the degree of its supersaturation (this is the thermodynamic driving force for scale formation). Precipitation can only occur when $SR > 1.0$. Much of the art and science of predicting scaling involves calculating the real *SR* in complex aqueous fluids. The log(*SR*) is called the saturation index (SI). This method is the most used in the literature since small numbers (0, 1, 1.5, or 2), compared with fractions of large numbers, indicate the degree of supersaturation. Various correlations are available in the literature for estimating SI in terms of easily measurable quantities, such as pH, alkalinity, and concentration of dissolved ions and ionic strength of the fluid. For example, the saturation index for calcite from Stiff and Davis[13] is

$$SI = pH - k - pCa - pAlk \quad \text{(Equation [3.21])}$$

Here, *pCa* is the negative log of [Ca], and *Alk* is a calculation of bicarbonate concentration determined by titration.[13]

This formula was developed for predicting precipitation of calcium carbonate from brines. The term "*k*" accounts for the ionic strength and species activities. Note that this is *not* the equilibrium coefficient. For complex fluids, such as oilfield brines, the *SR* values are usually calculated using geochemical models (described later in Section 3.2.1). Kan et al.[14] notes that solubility in concentrated brines often is calculated at standard temperatures and pressures, and this may lead to errors in estimating the SI values. Davies and Scott[15] describe several saturation indices or scaling indices for oilfield waters (see chapter 14.2 of that book). These are useful within a range of water conditions but are not as general as the models described in subsequent chapters. Table 3.1 compiles useful scale nomenclatures with an example of the value of the different terms.

Table 3.1 Scale Nomenclature

Term	Symbol	Definition	Example for $CaCO_3$
Solubility	S	Mass of a solid dissolved in a known weight (or volume) of fluid.	220 mg/L, 80°F, TDS=500, ALK=100, pH 7.5
Solubility Product	K_{sp}	$K_{sp} = \{M^{z+}\}^m \{X^{z-}\}^x$ at equilibrium	3.3×10^{-9} (ref16)
Saturation Ratio	SR	$\mathcal{R} = \{M^{z+}\}^m \{X^{z-}\}^x / K_{sp}$ in the real solution	SR < 1.0, no precipitate
Saturation Index	SI	$\log_{10}$(SR)	SI=0, no precipitate

3.2. General Mechanisms for Formation of Inorganic Scales

3.2.1. Thermodynamic Treatments of Scale Formation

This section applies the concepts described in Section 3.1. Several investigators have described scale formation and deposition mechanisms. From the standpoint of the equipment owners, scale is not a problem if the deposits do not block critical flow paths. Scale will not form if the local environment is not supersaturated, if the rate of formation of scale on a surface is less than the residence time of the water in a supersaturated state, or if it *may* not foul a critical surface. Thus, conditions for supersaturation, the rate of crystal growth, and surface factors determine the extent of possible damage. The generalized mechanistic steps are described to illustrate the complexity of controlling scaling in an aqueous environment. Specific examples of the wide range of inorganic scales are described in later sections. These ideas apply directly to the primary saturation-limited scales (calcite, gypsum, and other sulfates), but once corrosion-caused ions become mobile, these mechanisms also apply.

Crabtree et al.[17] proposed a mechanism of formation of the most common types of inorganic scale including calcium carbonate and alkaline earth sulfate deposits. These are categorized as salt formers. Although the driving force for scale formation may be a temperature or pressure change, outgassing, a pH shift, or contact with incompatible water, many produced waters that have become over saturated and scale-prone do not always produce scale.

Scale forms by growing from a solution. In a saturated fluid, the first step is formation of unstable clusters of atoms, a process called homogeneous nucleation. Figure 3.2 depicts this process from supersaturation, to formation of ion pairs, to the final deposition on a crystal. The final part of the process is *deposition* of scale crystals on a surface. This step can occur simultaneously with nucleation and crystal growth or after these steps. In some cases, visible scale may not appear in solution, but heterogeneous scaling on the surface may still occur.

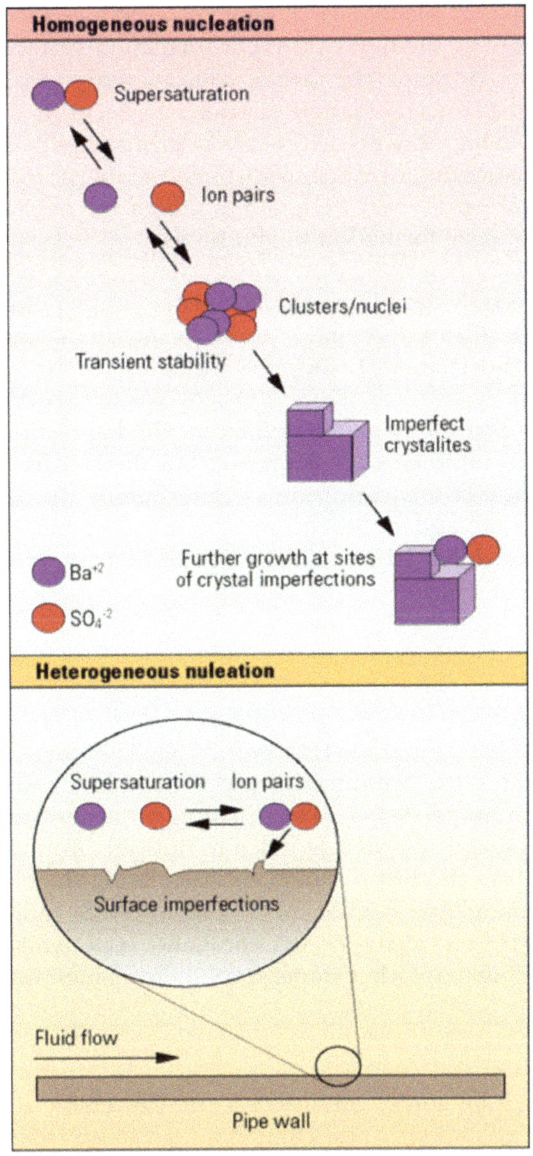

Figure 3.2 Process of Scale Formation

The atom clusters form small seed crystals triggered by local fluctuations in the equilibrium of ion concentration in supersaturated solutions. The seed crystals subsequently grow as ions are adsorbed onto imperfections on the crystal surfaces. The energy for seed crystal growth is driven by a reduction in the surface free energy of the crystal, which decreases rapidly with increasing radius after a critical radius is exceeded. Thus, large crystals favor continual growth, whereas small seed crystals may dissolve. On the basis of constant concentration experiments, Nancollas et al.[18] observed that redissolution of seed crystals may decrease and end when the crystals reach a nanoscale.

Thus, given a large enough degree of supersaturation, the formation of any seed crystal will encourage an increase in the growth of scale. The seed crystal, in effect, is a catalyst for scale formation. Crystal growth also tends to initiate on a pre-existing fluid-boundary surface, a process called heterogeneous nucleation. Heterogeneous nucleation sites include surface defects such as pipe surface roughness, existing scale or perforations in production liners, or even joints and seams in tubing and pipelines. Thus, corrosion is complexly linked with precipitation of alkaline earth scale. In these cases, nucleation is likely heterogeneous rather than homogeneous because of the many different formation surfaces. A high degree of turbulence also can catalyze scale deposition. However, the rate of scale formation may be higher under laminar flow. The practical observation is that the accumulation of scale can occur at the position of the bubble point pressure in the flowing system. This explains why scale deposits rapidly build on downhole completion equipment. Through this understanding of nucleation phenomena, scale inhibitors—discussed later—have been developed that use chemicals specifically designed to inhibit the nucleation and growth stages of scale formation and reduce the rate of scale formation to almost zero.

Because of the difficulties of precisely describing the chemistry at any point, empirical scaling indices have been developed. One of the first indices developed[19] does not predict scaling accurately. A modification by Stiff and Davis[13] yielded a more satisfactory SI (shown in Equation [3.21]).

A more general SI described by Tomson et al.[20] is defined as the log of the saturation ratio (see Table 3.1) Note that this SI does not include an ionic strength factor, but the activity coefficients are calculated using various software packages. Because of the complexity of determining the SI of a calcium carbonate system, the use of model package is recommended. Several models are available, but this book does not compare them or endorse a particular model. Model predictions should be compared with actual field results to calibrate the model.

ScaleSoftPitzer™

ScaleSoftPitzer™ is an Excel-based software program developed by M. Tomson and A. Kan as part of the Brine Chemistry Consortium at Rice University (Houston, TX). This consortium includes several major oil companies, large service companies, the National Science Foundation, and the U.S. Environmental Protection Agency. This software program predicts scale formation for 16 minerals, including calcite, dolomite, barite, three calcium sulfates, iron and zinc sulfide, calcium fluoride, iron carbonate, strontium sulfate, and sodium chloride. The program uses the Pitzer model for calculating the effects of high temperature, pressure, and total dissolved solids on activity coefficients.[20] The program can be used to predict the effect of methanol and mono ethyleneglycol on calcite, barite, and halite scale tendency and the effect of methanol on gypsum and celestine scale tendency. Other mineral solubility values are being added.

OKSCALE

OKSCALE was developed by K. Raju and G. Atkinson at the University of Oklahoma.[21,22,23,24] The development was supported by various oil and chemical companies. However, the program is currently not maintained, but copies of the program may still be available. This program considers the effect of temperature and pressure on solubility using classic thermodynamics. A modified Pitzer formulation is used for calculating activity coefficients. The effect of $MgSO_4$ ion association is considered explicitly.

DownHole

The DownHole series by French Creek Software includes various software products for modeling oilfield scale (http://www.frenchcreeksoftware.com). The basic DownHole SAT™ program uses an ion association model to provide scaling indices for about 19 scale forming species. It calculates scaling indices, saturation levels, momentary excess, and distribution of ions. Water can be evaluated as a single brine or mixed in any ratio as in an injection well. Various methods for pH prediction are also included. DownHole Rx Engineer's Edition adds the ability to select a water treatment product and obtain dosage recommendations. See discussions by Ferguson.[25]

MultiScale[26]

This multiphase equilibrium program considers the phase distribution of the scale active gases. Water evaporation is also included in the model. The program can be used to predict the formation of the most common and troublesome mineral scales: calcite ($CaCO_3$), siderite ($FeCO_3$), anhydrite ($CaSO_4$), gypsum ($CaSO_4 \cdot 2H_2O$), celestine ($SrSO_4$), barite ($BaSO_4$), halite (NaCl), and mackinawite (FeS). The current model claims to be very flexible with respect to input data and is well suited for pH and scale prediction in process plants, oil recovery, geothermal wells, water pipelines, and wastewater treatment plants.

3.2.2. Kinetics of Scale Formation

The word "kinetics" has several different meanings in scale formation/deposition. The current general discussion describes kinetic models for initial scale formation in fluids. Deposition models are more complex and will be described in Section 3.7. In one formation model, based on the work of Tomson et al.,[20] the scale formation rate is described in terms of the induction time (t_{ind}) after a supersaturated condition exists and before the appearance of scale. It is assumed that no scale will form in the system if the induction time is less than the residence time of the fluid in the supersaturated condition. However, Neville[27] has shown that this may not always describe the system accurately. Although induction time models do not predict when the scale buildup starts to affect operation of the system, they are critical starting points for an estimation. If the nucleation time is much greater than the time required for growth of crystal nuclei to a detectable size, the induction period can be expressed in terms of the primary nucleation rate, J, as

$$t_{ind} = \frac{1}{J} \qquad \text{(Equation [3.22])}$$

The primary nucleation rate is given by

$$J = A\exp\left[-\phi\beta\frac{\gamma^3 \upsilon^2}{(k_B T)^3(\ln SR)^2}\right] \qquad \text{(Equation [3.23])}$$

Here, A is the frequency factor, ϕ is the energy barrier factor ($\phi = 1$ for homogeneous nucleation, and $\phi < 1$ for heterogeneous nucleation), β is a shape factor, γ is crystal surface energy, υ is the molecular volume of the crystalline phase, k_B is the Boltzman constant, T is the absolute temperature, and S is the supersaturation ratio.[28] Combining Equation (3.22) with Equation (3.23) gives

$$\ln t_{ind} = \phi\beta\frac{\gamma^3 \upsilon^2}{(k_B T)^3(\ln SR)^2} + \frac{1}{A} \qquad \text{(Equation [3.24])}$$

Equation (3.24) can be applied to elucidate the nucleation mechanism. As can be seen in Figure 3.3, a change in slope of the $\ln(t_{ind})$ as a function of $(\ln SR)^{-2}$ reveals a transition from homogeneous to heterogeneous nucleation. Moreover, the surface energy can be determined by extracting slope values.

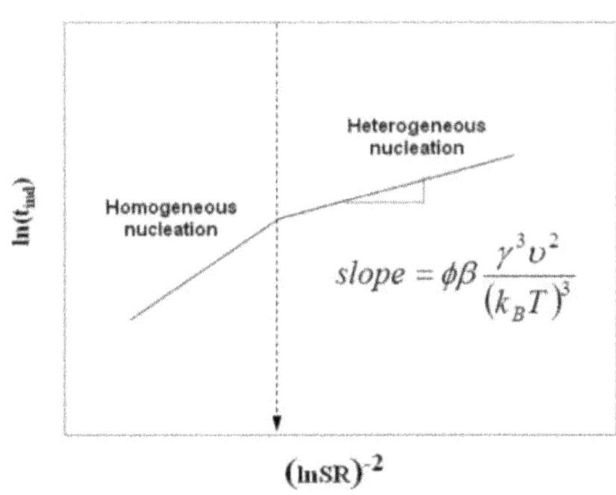

Figure 3.3 Nucleation Model[28]

For example, Tomson et al.[20] have found that the measured induction time for barite, in the absence of inhibitor (t^0_{ind}.), over about five orders of magnitude can be described by the following function:

$$Log(t^0_{ind.}, \sec) = \frac{2.2 + [1087 - .3T(^\circ K)]^2}{T^2 SI} - \frac{0.12 + [1087 - .3T(^\circ K)]^3}{T^3 SI^2} \qquad \text{(Equation [3.25])}$$

The term in brackets is related to the surface tension function from classic nucleation theory, and the calculated interfacial surface tension is close to that reported for barite and correspondingly for calcite. Several other researchers have attempted to predict the effect of inhibitors on mineral scale formation. One of the most successful theories of precipitation kinetics is the Burton-Caberrara-Frank spiral growth mechanism. This theory suggests that kink sites grow continuously in a spiral, thereby overcoming the expected activation energy for nucleation of a new site on a flat particle surface. A corollary of this is that inhibition is treated as an adsorbed inhibitor blocking the advance of a spiral. Only the active growth sites need be poisoned to prevent the spiral from continuing to grow. These types of equations have been used as the basis of calculating the minimum amount of scale inhibitor (microbiologically induced corrosion) required to prevent bulk scale formation.

The active growth sites can be poisoned by less than monomolecular coverage of the crystals by inhibitor. Although the information described above can be used to predict scale formation in bulk, scale formation on surfaces can be contradictory and appear to be random.

Additional details on heterogeneous (Figure 3.2 and Figure 3.3) scale formation are described in the papers by Zhang et al.[29] and Neville[27] but are beyond the scope of this book.

This section has described only the chemistries that control the appearance and formation of inorganic scales in a supersaturated solution. Corrosion product scales have not been studied in this manner because the corrosion and scale formation takes place on or near the surfaces. For most practical applications, the deposition of a fouling scale onto a surface is much more important and is a more complex physical/chemical process. Some additional details will be discussed in Section 3.7 after descriptions of different categories of scaling compounds that foul industrial equipment.

3.3. Iron Corrosion Product Scales

These primary corrosion product scales consist of the family of iron oxides, which are of concern to most owners of industrial equipment. Iron sulfides are associated with "sour" facilities, including most refineries and paper mills. Iron carbonates require the action of carbon dioxide as an oxidant and may be found in refineries and in connecting equipment. The resulting scale is siderite and is more likely to be found in pipelines and oil and gas producing facilities than in industrial equipment.

Iron scales form primarily as corrosion product deposits. This means that an oxidizing agent first corrodes iron and then the iron-containing salt precipitates. See the review of corrosion chemistries in Section 2.3. Because of the very low solubility of iron salts at pH values >3, the SR values will be high, and deposition will occur rapidly. Note also that the manifestation of the corrosion (pitting, under deposit corrosion, and MIC) can greatly affect the morphology and mixtures of oxides and other salts in the deposit on a steel surface. See Section 2.3.3.

3.3.1. Iron oxides

These scales form primarily from iron-based metals corroded by the oxygen in aqueous fluids or in very hot gases. Talbot and Talbot[30] describe various iron oxides that may form in different parts of an industrial environment. Common rust is not a determinate chemical substance but changes continuously as it develops. A generalized composition for most rust is FeOOH. This will be the predominant oxide in service water equipment. Iron can exist in oxidation states Fe(II), Fe(III), and Fe(I). The highest state Fe(VI), in the ferrate ion FeO^{2-}, is stable only in highly oxidizing alkaline media and is seldom significant in corrosion in recesses or in chemistry generally.

If O_2 is present, the chemical reactions depend on the pH. In slightly acidic fluids (such as those that contain carbon dioxide or the other components of water), the reaction is

$$2Fe + O_2 + 4H^+ \rightarrow 2Fe^{2+} + H_2 + 2H_2O \qquad \text{(Equation [3.26])}$$

The iron salts produced are a complex group of chemicals owing to different oxidation states of Fe and the ability to complex with many anions. The final compound depends on the amount of oxygen, the pH, and the temperature. The iron(II) ions are soluble only at pH values less than about 3. A common reaction product is the familiar red rust that is described approximately as FeOOH. The presence of oxygen in a sour system can be particularly problematic because it changes the protective coatings. New pipelines and other tubing will always have a coating of an iron oxide compound (called mill scale) owing to the manufacturing processes and environmental exposures, including a hydrostatic test. These coatings then can affect the formation of additional scale and organic solids.

The common oxidation states are iron(II) and iron(III), formerly called ferrous and ferric iron, respectively. Both form oxides and hydrous oxide phases, soluble cations, and corresponding series of salts. The redox potential of the Fe(III)/Fe(II) couple in dilute solution is E = 0.771 V (SHE), but this potential is sensitive to the chemical environment of the metal ions. Iron(III) is the stable oxidation state of simple compounds and aqueous solutions in equilibrium with atmospheric oxygen. There are two iron(III) oxides.

Mineral scales include red-brown oxide (α-Fe_2O_3; hematite) and γ-Fe_2O_3. Note that α-Fe_2O_3 has a corundum (Al_2O_3) structure with Fe^{3+} ion in the octahedral spaces of a hexagonal close-packed lattice of oxygen ions. The γ-Fe_2O_3 form, sometimes incorrectly called spinel like, has Fe^{3+} ions distributed randomly in the octahedral and tetrahedral interstices of a face-centered cubic lattice of oxygen ions. It is produced by dehydrating lepidocrocite (γ-FeO(OH)), or by oxidizing the lower oxide (Fe_3O_4). Iron(III) is precipitated from neutral or alkaline solution as a red-brown gelatinous material loosely described as ferric hydroxide, but there is no clear evidence that it contains discrete hydroxyl ions, and its constitution is uncertain.

Because the iron(II) ion has a lower charge density than the iron(III) ion, iron(II) salts are much less susceptible to hydrolysis. Iron(II) forms simple salts with most common anions except when they are oxidizing enough to raise the oxidation state from iron to iron(III). Iron forms an oxide (FeO; wustite) with a simple cubic structure, but it is unstable at temperatures below 570°C. Iron(II) does, however, participate in Fe_3O_4 (magnetite), an anhydrous mixed iron(II)/iron(III) oxide, stable at ambient temperature with an inverse spinel structure. It can form by oxidation of iron at temperatures below 570°C or by decomposition of FeO formed at higher temperatures:

$$4FeO = Fe + Fe_3O_4 \qquad \text{(Equation [3.27])}$$

Iron oxides containing various amounts of ferric(III) and ferrous(II) are present in many industrial units. From the previous discussion, we can see that the exact composition depends on the reaction conditions, including the presence of oxygen in the system, which will favor[30] production of Fe(III)-containing compounds.

Bornak[31] discusses the formation of magnetite on the surfaces of utility boiler tubes. He notes that the magnetite layer forms on the metal surface under normal boiler conditions in pure water per the following reaction:

$$3Fe + 4H_2O = Fe_3O_4 + 4H_2 \quad \text{(Equation [3.28])}$$

Ferrous hydroxide ($Fe(OH)_2$) is believed to be an intermediate in this process, converting to magnetite above 212°F (100°C):

$$3Fe(OH)_2 = Fe_3O_4 + 4H_2 + 2H_2O \quad \text{(Equation [3.29])}$$

The resulting metal oxide layer is protective in that it acts as a barrier to further corrosion. Bornak[31] and Shreir[32] claim that the growth of any protective layer involves two principal steps:

1) Surface reactions at the metal/oxide and oxide/environment interfaces
2) Transport of materials through the oxide

The growth rate of the oxide layer will be controlled by the slower of these steps. This is demonstrated by the parabolic rate law observed for the iron-magnetite system under boiler conditions. Mathematically, dx/dt = k2/x, where x is the thickness of the oxide layer and k2 is the parabolic rate constant. It is the transport step (2), or some portion of it, that appears to be rate-controlling. This conclusion is based largely on the observed morphology of the system and various proposals regarding magnetite growth.

The addition of oxygen or other ions into water results in the formation of other iron oxides/hydroxides that can alter and change the protective nature of magnetite.

Shreir[32] claims that further corrosion in the boiler to form unstable scales is caused by caustic attack: at a high pH, the magnetite becomes soluble in NaOH.

$$Fe_3O_4 + 4OH^- = 2FeO_2^- + FeO_2^{2-} + H_2O \quad \text{(Equation [3.30])}$$

Under these conditions, iron reacts rapidly with hydroxide ions to form the soluble FeO_2^{2-}:

$$Fe + 2OH^- = FeO_2^{2-} + H_2 \quad \text{(Equation [3.31])}$$

Chloride Ions

According to Potter and Dibble,[33] the acid-chloride concentration rises locally because of the flux of Cl- ion to the steel surface (anode) along with the oxygen carriers. At the anode, OH- disappears as it combines with Fe and further decomposes to the insoluble oxide. Together with local evaporation, this process results in a reduction of pH, thereby creating an acidic region with a high Cl- concentration. Further anodic action results in the formation of hydrolysable chlorides, whose ingress causes a degradation in these zones of the protective film.

A review by Hansen[34] of utility boiler scales and cleaning notes that the type and amount of deposits are also determined by

- Boiler design
- Boiler operating pressure
- Drum-type vs. once-through
- Base load vs. cycling operation
- Hours of operation
- Water treatment program
- Water chemistry upsets
- Condenser leakage
- Fuel
- Burner configuration

- Boiler and pre-boiler materials of construction
- Pre-boiler and boiler offline storage

Sarin et al.[35] describe the formation of iron oxides in low-temperature water systems. The investigators propose that corrosion-caused scale formation in corrosive water is like a layer cake. See Figure 2.8, Figure 2.12, and Figure 2.16 for various corrosion manifestations that cause tubercles. Inside a corrosion tubercle, the metal floor is corroding and producing the iron for the scale formation. The next layer is called the porous core. The next layer is a shell-like layer followed by a surface layer. Various iron oxides are proposed to be in the different layers. See Figure 3.4 for a depiction of this process.[35]

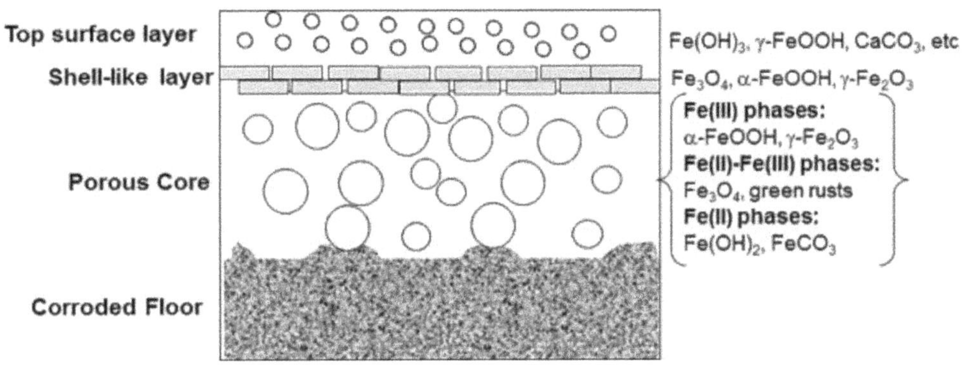

Figure 3.4 Schematic of Scale Structure Showing Characteristic Features and Composition of Corrosion Scales found in Iron/Steel Water[35]

3.3.2. Iron Sulfides

There is a family of iron sulfides that can be formed in several different ways. If a system has sour fluid (H_2S present), a corrosion reaction is the most likely method for forming iron sulfide scale.

The net corrosion reaction is

$$Fe + H_2S = FeS + H_2 \quad \text{(Equation [3.32])}$$

This is the result of the anodic partial reaction:

$$Fe = 2e^- + Fe^{2+} \quad \text{(Equation [3.33])}$$

and the cathodic partial reaction:

$$H_2S + 2e^- = 2H^+ + S^{2-} \quad \text{(Equation [3.34])}$$

Sulfate-reducing bacteria can generate the hydrogen sulfide:

$$SO_4^{2-} + 4H_2 = S^{2-} + 4H_2O \quad \text{(Equation [3.35])}$$

$$2H^+ + S^{2-} = H_2S \quad \text{(Equation [3.36])}$$

Some possible reactions at low temperatures and low concentrations of hydrogen sulfide can form the mackinawite (the most reduced) form of FeS from rust:

$$2FeO(OH) + 3H_2S = 2FeS + S + 4H_2O$$ (Equation [3.37])

This may be converted to greigite (Fe_3S_4) by further reaction:

$$3FeS + S = Fe_3S_4$$ (Equation [3.38])

Under hotter conditions with higher concentrations of gas, pyrite is formed:

$$FeS + S = FeS_2$$ (Equation [3.39])

Berner[36] reported that the product depends upon the environment. In weakly acidic solutions (pH = 4), mackinawite and pyrite form:

$$2FeO(OH) + 3H_2S = FeS + FeS_2 + 4H_2O$$ (Equation [3.40])

However, in neutral and alkaline mixture, mackinawite and sulfur are produced:

$$2FeO(OH) + 3H_2S = 2FeS + S + 4H_2O$$ (Equation [3.41])

Smith and Pacheco[37] describe the formation of the various phases of iron sulfide in aqueous environments. Different iron sulfide compounds form on the metal surface at different temperatures and H_2S activity values. This can affect the amount of sulfide formed and the ease of removal. Mackinawite forms on steel surfaces under conditions in which FeS is still soluble in the bulk solution (i.e., the concentrations of Fe^{2+} and H_2S are still below saturation for FeS). Mackinawite forms rather than the more stable pyrrhotite because pyrrhotite forms much more slowly. Smith[38] provides details of this process by way of a stability diagram (Figure 3.5) that shows the various forms of FeS as a function of temperature and activity of H_2S.

Dunlop et al.[39] proposed a simple order-of-magnitude correlation of the carbon dioxide (CO_2)/H_2S partial pressure ratio to define the coexistence of $FeCO_3$ and all forms of FeS. On the basis of tabulated data at 25°C, and specifically ignoring temperature and solution effects, Dunlop et al.[39] determined that FeO would be the corrosion product if the CO_2/H_2S partial pressure ratio were >500 and that FeS would form if the ratio were ˜500. Because Dunlop was looking at conditions in which both $FeCO_3$ and sulfide were stable, he described a very simple way to define the $FeCO_3$. Smith[38] derived the equilibrium constant, K, for this boundary that is similar in form to Dunlop but is not limited to a temperature of 25°C for mackinawite formation as the corrosion product. When the FeS levels in the bulk fluid meet or exceed the saturation levels, the reaction kinetics allows pyrrhotite to become the corrosion product. This boundary is defined by the following: mackinawite may form, but it will transform to pyrrhotite so rapidly that it is not detected.

Morse et al.[40] describe additional details about the formation of iron sulfide scales from hydrogen sulfide and other sources contained in natural waters.

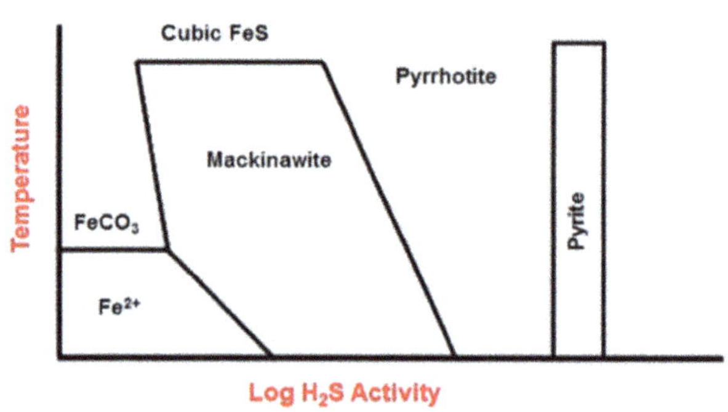

Figure 3.5 Iron Sulfide Stability Diagram[38]

Iron sulfides also can form when iron-containing fluids are mixed with sulfide-saturated liquids. Nasr-El-Din et al.[41] described the formation of various types of iron sulfide and use of scavengers to prevent the formation of this scale. These authors suggest that iron generated by acidizing fluids is a major cause of FeS-containing scales. This represents the formation of FeS owing to more conventional precipitation of a salt when the solubility of FeS is exceeded owing to changes in water quality and is analogous to the formation of calcite and sulfate scales.

In industrial equipment, acid soluble FeS is common. However, the morphology of the sulfide is important. If it has a very high surface area, it may be pyrophoric (easily combustible) when dry, and caution when opening a vessel thought to contain FeS is advised. Decontamination procedures (Sections 7.4.5 and 7.6) are performed to remove this danger from refinery units.

A brochure from Jeffries[42] notes the reactions that may occur in refinery vessels when FeS is exposed to dry air.

$$4FeS + 3O_2 = 2Fe_2O_3 + 4S + heat \qquad \text{(Equation [3.42])}$$

$$4FeS + 7O_2 = 2Fe_2O_3 + 4SO_2 + heat \qquad \text{(Equation [3.43])}$$

3.3.3. Iron Carbonate-Siderite

This salt is a scale produced by a corrosion product reaction that forms in a manner like iron sulfide (described above). Iron carbonate-siderite is common in topside oil and gas facilities that may be cleaned by IC companies or by local operators.

The formation reaction is

$$Fe + CO_2 + H_2O = FeCO_3 + H_2 \qquad \text{(Equation [3.44])}$$

This scale can be protective when it initially forms on the surface. Natural corrosion inhibitors in the crude as well as added materials influence the thickness of the scale and the protection provided by it. Siderite placed in carbonate-free water is extremely insoluble, but pH and other gases greatly change the overall solubility.

Palacios and Shadley[43] propose that a consequence of corrosion of steel in the presence of oxygen-free carbon dioxide and a water phase is the formation of iron carbonate ($FeCO_3$ corrosion scale). This scale provides some protection of the steel from further corrosion. The protectiveness of the scale depends on environmental factors and characteristics of the steel. Environmental factors include temperature, carbon dioxide partial pressure, solution chemistry, fluid velocity, single-phase or multiphase flows, geometry factors, and solution pH. Steel characteristics that affect corrosion include chemical composition and heat treatment, which determine the microstructure of the steel. These studies showed that iron carbonate scales are very brittle and, under some conditions, can be removed easily from metal surfaces. Adherence of the scale to the metal and the thickness of the scale depend on the microstructure of the metal specimen. For cases in which the metal specimen had a normalized microstructure (1018 CR and API N-80 normalized), the iron carbonate scale was adhered strongly and, for some cases, was thicker than for specimens having a quenched and tempered microstructure. Moreover, the crystals forming the scale on normalized steel specimens were larger and more densely packed than those for quenched and tempered specimens. The success rate in forming iron carbonate scales on specimens was greater for normalized steel than for quenched and tempered steel.

Petroleum well fluids often contain H_2S as well as CO_2. The ratio of these gases and the temperatures and pressures in the well control the deposition of iron-containing scales and the corrosion rates. Some details are in the section about formation of iron sulfides (Section 3.3.2). Additional information on Sweet-CO_2 corrosion and Sour-H_2S corrosion can be found in the collection of NACE papers compiled by Burke et al.[44] FeS-containing deposits also occur in hydrocarbon-processing equipment in refineries.

Kriel and Lacey[45] report that a rare iron carbonate scale can form from the precipitation in formations that have a high concentration of natural iron carbonate mineralogy. This is contrasted with the formation of the iron carbonate corrosion product noted above.

Most types of iron carbonate scale can be prevented using corrosion inhibitors that are described in Frenier.[4] The use of metallurgy compatible with the well conditions is another method for reducing corrosion and thus a control method of iron carbonate scale; however, this subject is beyond the scope of this book. See NACE recommended practices for sweet and sour service (NACE MR0175/ISO 15156[46] and MR0176-2000[47]) and data from the European Federation of Corrosion.[48] Also see discussions of corrosion of oil and gas production equipment in Chapters 1–3 of the book by Heidersbach[49] and the book by Frenier.[50]

3.4. Copper Deposits

Copper frequently enters a unit (especially a boiler) from the feed water train if there are any cupreous alloys in the system (such as surface condensers). Over many months and years of operation, even very small amounts (parts per billion) may cause many pounds of copper to plate onto the iron base surfaces.

$$Fe + Cu^{2+} = Fe^{2+} + Cu \qquad \text{(Equation [3.45])}$$

This plated copper (on the steel) and any copper oxides that also could form provide sites for galvanic corrosion of the steel substrates and should be removed during chemical cleaning operations. These deposits are not found often in newer plants, but removing them is part of the history and technology of chemical cleaning of industrial plants.

Copper metal and more often copper oxide (CuO) foul the blades and inlets of steam turbines. Even small amounts (25 lb.) can change the fluid flow enough to reduce the power output. Methods for dissolving copper are described in Section 8.4.4. Also see Foster and Trimble.[51] Copper deposits (Cu and CuO) also are found in some nuclear steam generators. The chemistry for removal is similar to methods described in Section 4.2, but additional details specific to steam generators are in Section 5.3.

3.5. Alkaline Earth Scales

This family of solids consists of the salts of calcium, barium, strontium, and radium that form in the presence of halide, hydroxide, carbonate, or sulfate ions. Note that the salts of Cl^- and Br^- are highly soluble and exist as scales only in the most concentrated brines. Also see Chapters 2, 3, and 4 of Cowan and Weintritt.[5]

3.5.1. Carbonates

Several types of carbonate scale are found in the industrial environment. These include calcium carbonate ($CaCO_3$; calcite, aragonite, and vaterite), magnesite ($MgCO_3$), and iron carbonate-siderite ($FeCO_3$; see Section 3.3.3). The various forms of calcium carbonate differ only in crystal structure; calcite is the most common. See the mineralogy database for crystal structures (http://webmineral.com). Magnesium and calcium carbonate can be formed when aqueous solutions of the metal salts are contacted by dissolved CO_2. See Table 3.2 for solubility data. Dolomite ($CaMg(CO_3)_2$) is a mixed mineral that usually only forms in the earth.

Alkaline earth carbonate scale formation is controlled by some of the most complex chemistry of all the minerals commonly encountered in industrial equipment. To predict carbonate precipitation, the chemistry of three phases must be understood:

1) Carbon dioxide dissolves in water to form carbonic acid.
2) Carbonic acid dissociates to carbonate and bicarbonate, which lowers the pH.
3) Carbonate ions interact with calcium to precipitate calcite (in most cases).

The overall equation is

$$Ca^{+} + 2HCO_3^{-} = CaCO_3 + H_2O + CO_2 \qquad \text{(Equation [3.46])}$$

All the chemical reactions are linked, and modifying any one (pH or partial pressure of CO_2) influences the others. An important contribution to scale formation can be the reduction of partial pressure of CO_2, which causes precipitation of calcium carbonate. This is called flashing or auto scaling.

Reduction of pH causes more bicarbonate to form and calcium carbonate to precipitate. Note that the general solubilities of the carbonates is $Fe^{2+} < Ca^{2+} < Mg^{2+}$, but the solubility of a given salt is dependent on $p[CO_2]$ and pH, and the presence of dissolved iron usually requires an active corrosion reaction. Note also that these units are moles per 1000 g of water. Unfortunately, there is no standard way of recording solubility values, and mixed units and differing experimental conditions are found in many compilations.

Table 3.2 Solubility of Alkaline Earth Salts[52]

M^{2+}	Radius (10^{-8} cm)	Fluorides	Hydroxides	Carbonates	Sulfates
Mg	0.31	V.S.	1.2×10^{-11}	1.2×10^{-3}	2.4
Ca	0.65	2×10^{-4}	2.1×10^{-2}	1.5×10^{-4}	1.5×10^{-2}
Sr	1.13	1×10^{-3}	6.5×10^{-2}	7.0×10^{-4}	5.0×10^{-3}
Ba	1.35	1×10^{-2}	2.8×10^{-1}	1.0×10^{-4}	1.0×10^{-5}

Moles per 1000 g water at 1 atmosphere and saturated with CO_2 for carbonates; V.S., very soluble.

Carbonate scale can form owing to changes in pH or pressure and by incompatible mixing of waters. Almost absolute accuracy in water analysis is required for predicting scaling because of the critical dependence of this reaction on pH. Calculation errors in pH, which is a function of all solution species, and in the pressure of CO_2 can give orders of magnitude error in saturation factors for carbonates. Physical scale samples are often required to confirm scale predictions from water analyses.

Typically, an increase in temperature increases the water solubility of a mineral. More ions are dissolved at higher temperatures. Similarly, decreasing pressure tends to decrease solubilities and, as a rule of thumb, the solubility of most minerals decreases by a factor of two for every 7000-psi [48-MPa] decrease in pressure.

Not all minerals conform to the typical temperature trend; for example, calcium carbonate shows the inverse trend of increasing water solubility with decreasing temperature. See Figure 3.6.

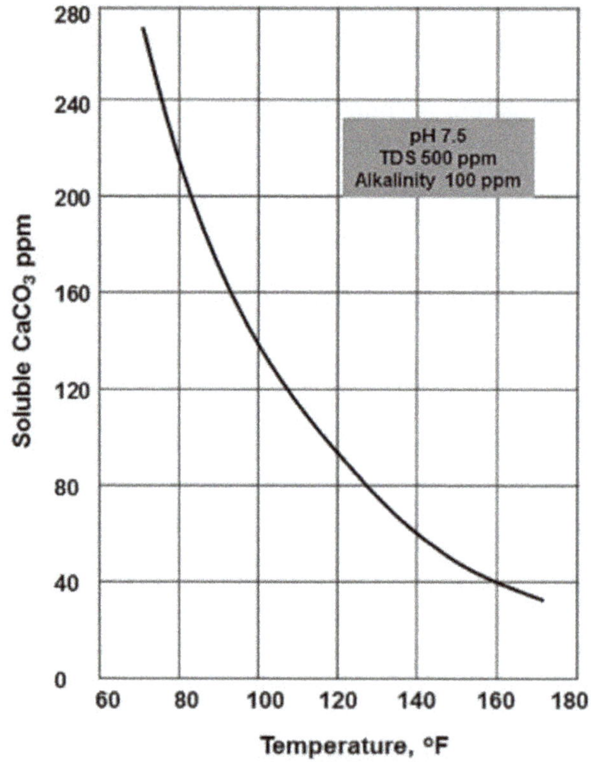

Figure 3.6 Solubility of Calcite (pH 7.5, TDS 500 ppm, Alkalinity 100 ppm)

By comparison, the solubility of barium sulfate increases by a factor of two in temperatures ranging from 25°C to 100°C (77°–212°F) and decreases by the same magnitude as temperatures approach 200°C (392°F). This trend is itself influenced by the background brine salinity. Additional trends are described in Section 3.5.2.

An example of the effects of pH on the precipitation of calcium carbonate is described by MacAdam and Parsons.[53] As the pH of water increases, so does the conversion of bicarbonate to carbonate and hence the potential (called the precipitation potential here, which is the same as the saturation ratio [SR] of calcium carbonate; see Figure 3.7). These authors also reported the effect of pH on deposition of a calcium carbonate scale and showed how increasing the pH from 8.8 to 10 increased the deposition of calcium carbonate from 2 mg/cm^2 to 12 mg/cm^2, after a 2-hour run.

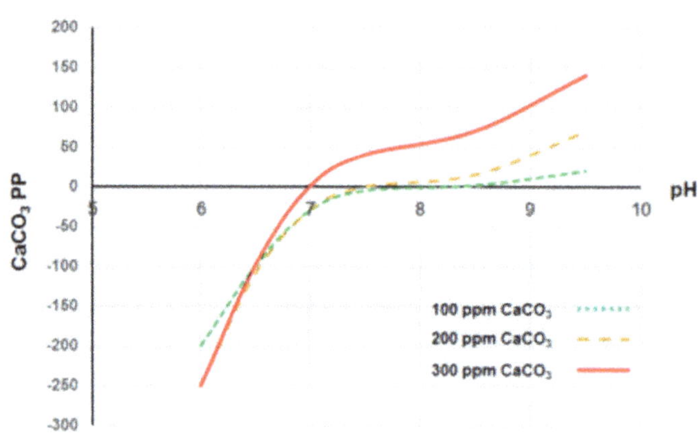

Figure 3.7 Effect of pH and Hardness Level on the Calcium Carbonate Precipitation Potential (PP)

An additional complexity is the solubility of carbonate minerals in the presence of additional acid gases such as hydrogen sulfide (H_2S). Carbonate solubility increases as fluid acidity increases, and CO_2 or H_2S at high pressure supply significant acidity. Consequently, waters in contact with both carbon dioxide and acid gases can be rich in dissolved carbonate. This trend has a complex nonlinear dependence on brine composition, temperature, and the pressure of the gas above the liquid phase; this gas pressure effect is orders of magnitude greater than the normal effect of pressure on the solubility of a mineral. Because of the complexity of the precipitation/scale formation process, a detailed knowledge of the downhole water chemistry is required to predict scaling/non-scaling conditions.

3.5.2. Sulfates

Since the basic chemical reactions for the precipitation of alkaline earth sulfates is much less complex than formation of the carbonates, the literature on formation of these salts has concentrated on the morphology, adherence to surfaces, and the rate at which crystallization is initiated. The formation of the scale on a surface also is important.

Three different calcium sulfates occur, each with different amounts of water of hydration:

- Anhydrite ($CaSO_4$)
- Hemihydrate ($CaSO_4 \bullet 1/2H_2O$)
- Gypsum ($CaSO_4 \bullet 2\ H_2O$)

Barite ($BaSO_4$) and celestine ($SrSO_4$) are not hydrated. Magnesium forms complex crystals with sulfate, but these have high solubility and usually are not problematic. Radium is an alkaline earth metal, exists naturally, and can substitute for barium in barite and cause the deposit to become radioactive. Additional details on various crystallographic structures are found in the Mineralogy Database (http://www.webmineral.com). The composition, hydration, and morphology can become critical when remediation methods must be employed because these factors can greatly affect the rate of dissolution or physical removal.

Calcium, barium, and strontium sulfate scale formation is usually the result of mixing incompatible waters (e.g., seawater with a high sulfate content and formation water with high barium content). As this water continues to be produced up the well bore and pressure and temperature fall, additional scale will form. See Table 3.2, which shows the solubilities of various salts. The solubilities of the various salts as a function of temperature, pressure, and salt content are a very complex subject. See discussions of models in Kan et al.[54] Note the changes with respect to salinity and temperature of $SrSO_4$ in that it is more soluble in some higher concentrations of brines. See Figure 3.8.

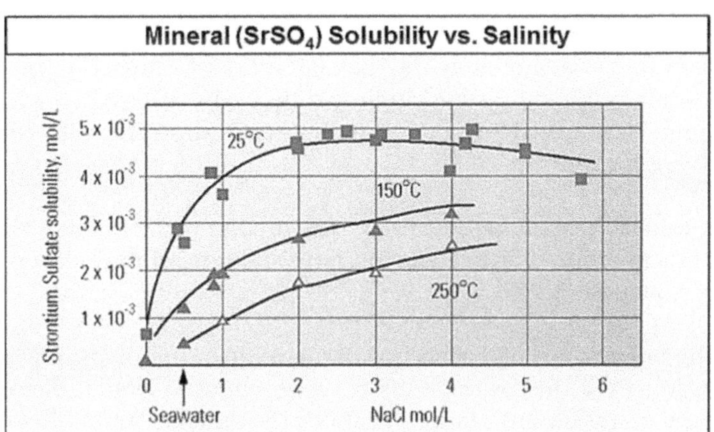

Figure 3.8 Mineral Solubility (Molality) and Salinity

Figure 3.9 shows plot of sulfate mineral solubilities (k_{sp} values from Langmuir[9]) for barite, gypsum, and anhydrite as a function of temperature. There are very large differences already noted for the sulfates. Gypsum is unstable at temperatures >50°C, and anhydrite is unstable at temperatures below that value.

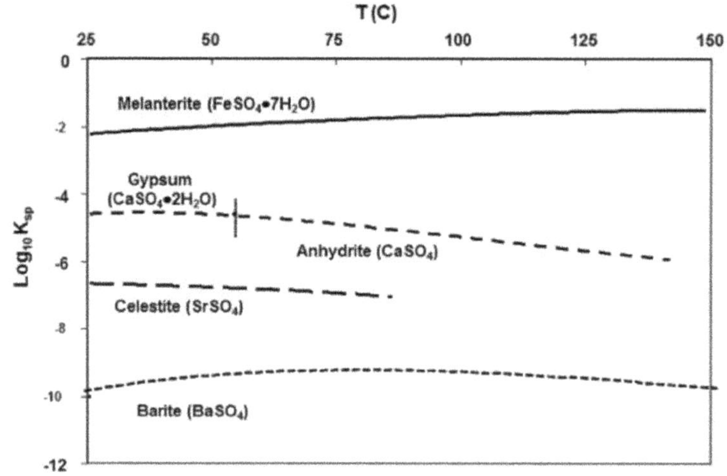

Figure 3.9 Solubility of Sulfate Minerals as a Function of Temperature (K_{sp} values from Langmuir[9])

The solubilities of these salts in water are a major determination of scaling. The reasons for different solubility are difficult to understand. Becker[52] relates the solubilities of a range of alkaline earth metals to crystal structure. He proposes that the coordination numbers of the central metal are not altered by the change of the anion from carbonate to sulfate. However, the metallic salt resulting from the neutralization of charge by the larger and stronger sulfate anion causes a greater distortion of the coordinated complex. Thus, the crystal forms of the multiple combinations of complexed salts present different morphologies when the individual unit cells are deformed. His theory is that the deformation of the complex also reduces the ability to accept water at the open ligand sites. This inability to accept coordination water changes the solubility characteristics of the salt complex. A general trend of increasing water solubility is observed with increasing coordination by water at the available ligand sites of the metallic salts. Therefore, decreases in the number of available ligand sites result in lower water solubility. Coordination numbers (ligand sites) assumed by the central metal depend on the metal's available energy levels. They are determined by the physical conditions to which the metal cation is exposed.

He notes that the highest coordination number is determined by the highest quantum state reached by the available metallic orbitals. Electronic effects exerted by ionic ligand groups also affect changes in the complex and the coordination number assumed by the metallic cation. Comparison of the alkaline earth metal salt complexes found in some crude oil systems indicates that salts of large ions follow trends expected if changes in hydration energy dominate. Thus, the solubility trends show the following order: $BaSO_4 < SrSO_4 < CaSO_4 < MgSO_4$. See Table 3.2.

Because solubility is the critical factor, the equations that predict the induction time of the start of scale formation are described as a function of the saturation ratio and temperature. It is useful to review the ideas from Tomson et al.[20] See Equation (3.25).

Scale morphology and the importance of surface factors have been the major focus of research on barium sulfate formation. $BaSO_4$ growth studies by the FAST-5[55] consortium show that the nature of a metal surface (e.g., roughness or chemical composition) can significantly affect the amount of heterogeneous scale surface nucleation and growth.

Amjad[56] elucidates the kinetics of gypsum scale formation on heat exchanger surfaces (brass or stainless steel) from aqueous solutions. They found that gypsum scale formation may take place *directly* on the surface of a heat exchanger without any bulk or spontaneous precipitation in the reaction cell. The kinetic data also suggest that the rate of scale formation is a function of solution supersaturation, heat exchanger surface area, and the metallurgy of the heat exchanger. The test setup (use of a loop of metal tubing that was heated internally by flowing water) allowed the testers to show a direct relationship between loss of calcium from the solution and mass of scale formation on the heat exchanger tubing. The test used a temperature differential of 6°C for the water and 67°C for the tubing. Note from Figure 3.9 that this is in the range where gypsum becomes unstable and may form anhydrite (which has a negative solubility/temperature profile.)

Nevile[27] also has found that calcite scaling may occur on a surface and not in solution under specific conditions.

3.6. Silicates

Silica (SiO_2), hydrated silica ($SiO_2.H_2O$), and silicates can form in surface waters that contain dissolved silica or from the side reactions of sandstone acidizing.[57] Also see Chapter 5 of Cowan and Weintritt.[5] Metal silicates (Na, Mg, and Al) form from the mixing of incompatible waters. These types of deposits usually form on surface equipment such as boilers and heat exchangers but occasionally foul other surfaces. Several metallic silicates are noted in Table 1.2.

Where the water supply contains high amounts of soluble silica (>150 ppm), polymerized silica (SiO_2) can be expected. If the pH is raised to 8.5 or greater, alkaline earth silicates will precipitate from solution. Silica or silicate deposits are a common problem for geothermal wells and generating equipment where the hot brine may contain large amounts of dissolved silica that precipitates when the temperature is lowered. The chemistry and kinetics of precipitation is very complex and depends on the pH as well as the initial concentration of soluble silica. Aluminosilicates are the building blocks of most clay and clay-like minerals. See the reference book by Iler.[59]

Silica scale growth is described by Demadis[59] as a complex process involving numerous equilibria. In evaporative cooling water systems (where the pH of interest is 7 to 9.5), colloidal silica usually is favored at a pH of 8.5; in contrast, magnesium silicate ($3MgSiO_3.5H_2O$) scale forms at a pH of more than 8. There is no clear cutoff, and silica and magnesium silicate usually form in random mixtures. Silica exhibits normal solubility characteristics, which increase proportionally to temperature. In contrast, magnesium silicate exhibits inverse solubility. Formation of colloidal silica results from silicate ion polymerization, contrasting with crystalline scales such as calcium carbonate ($CaCO_3$), calcium sulfate ($CaSO_4$), and barium sulfate ($BaSO_4$), which are products of cation-anion association.

When silicate ions polymerize, they form primarily crosslinked polymeric chains and oligomeric structures. Iler[58] also describes the very complex formation of silica polymers. Dissolved silica principally precipitates out of solution in three ways: by surface deposition, in living organisms, and through bulk precipitation. In surface deposition, forms of the silicate ion condense with any solid surface possessing OH groups.

If the surface contains MOH moieties (M = metal), this reaction is enhanced further. Such pronounced silica deposition phenomena in the water treatment industry are observed on severely corroded metallic surfaces covered with metal oxides/hydroxides. Once the receptive surface is covered with silica scale, additional silica is deposited on an already formed silica film. Silica in living organisms is called biogenic and appears in certain microorganisms that can deposit silica from highly undersaturated solutions. Bulk precipitation allows silicate ions to polymerize and form small colloidal silica particles. The particles grow and agglomerate, forming larger silica particles. The precise mechanism of silica formation is not well understood.

Gharaibeh et al.[60] proposed a mechanism for the formation of silica on a metal surface. The authors claim that silica deposits on surfaces via heterogeneous nucleation steps, where silica monomers deposit on the surfaces followed by a reaction with functional groups such as -OH, $-COO^-$, and $-NH_3^+$ (Figure 3.10). After the surface is fully covered by silica deposition, further silica deposition is then controlled by silica-silica interactions. The second step in silica scaling is controlled by a condensation process, where water molecules are expelled from the formed silica film, as shown in Figure 3.10(b). The gel in this step is called soft gel since it is not fully polymerized and has low water content. However, with aging, more water molecules are expelled from the film, resulting in a dry condensed polymerized silica film (Figure 3.10[c]). It is worth noting that the process of silica polymerization/condensation involves complex chemistry that is not yet fully understood. Previous studies have revealed that inorganic metal ions, such as Mg^{2+}, Ca^{2+}, Na^+, and Fe^{3+}, significantly affect silica scaling kinetics.[61,62]

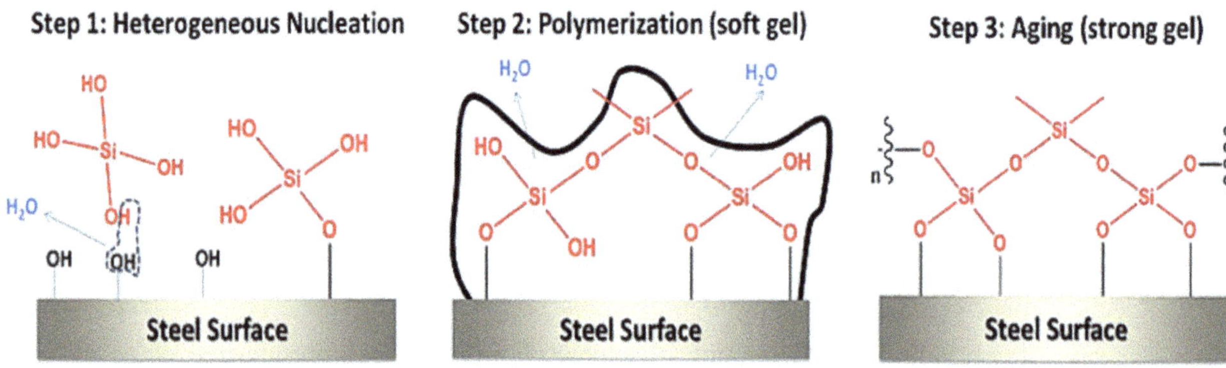

Figure 3.10 Schematic Illustration of Silica Scaling on a Steel Surface with Hydroxyl Functional Groups[60]

Silica and silicate scale formation is a significant problem for the boiler feed water and cooling water industries. Two industrial sources are noted (Kemmer[63] and Betz Handbook[64]). Silicate scales also can form in well tubulars, especially in geothermal wells, and in many different industrial units. Silicon-containing deposits also may be found in nuclear steam-generating equipment and can be very difficult to remove (see Section 3.8). Once formed, solvents containing HF or strong alkalis are the only solutions that will dissolve silica or silicates. See several alternate solvents in Sections 4.4.2 and 5.3.

3.7. Kinetics of Deposition of Scales

Precipitation of an inorganic solid and deposition are not the same, and deposition is more complex than solids precipitation. Because of the importance of carbonate and sulfate scaling, several research studies are abstracted; however, the principles can apply to other mineral scales that form from solution equilibrium changes such as silicates.

Important factors affecting surface scaling rates (deposition rates) include the specific mineral deposited, temperature, degree of saturation (SR), fluid composition (which also affects SR), and surface chemistries. In flowing systems, such as cooling water processes, pipelines, and heat exchangers, hydrodynamics and surface effects may especially control the amount (and rate) of surface scales. This section describes both hydrodynamic and surface chemistry effects.

The actual amount of scaling that occurs in an aqueous system has been related to the nucleation rate, as described by Tomson et al.[20] Models of nucleation kinetics are generally based on the classic theory of nucleation and are characterized by the logarithm of the induction time being proportional to the nuclei surface tension divided by the saturation index squared, to various exponents.

The models of Tomson et al.[20] are similar to the equation proposed by Kladi et al.[65] for the growth of calcium carbonate.

$$Logt_{(ind.)} = A + \frac{B\gamma_s^3}{((2.3kT)^3 \log^2 SR)} \qquad \text{(Equation [3.47])}$$

In these cases, the induction time is related to the surface energy (γ_s), which can be calculated from the slope of the log-log plot (Figure 3.3). The term "B" is related to the molecular volume of growing spheres, and "k" is the Boltzman constant. Different calcium carbonates form with time: vaterite, followed by aragonite, then calcite. These equations are based upon the work of Mullin.[66] They conclude that the crystallization of calcium carbonate in seawater is controlled by the diffusion of the scale crystals and is a largely homogeneous process. The process is second order with respect to the SR (or SI). But the presence of magnesium in seawater can substantially increase the induction times. Note that these models do not describe the actual deposition process onto external surfaces.

Neville et al.[67] discusses the potential use of atomic force microscopy to study the formation of electrodeposited calcareous scales in uninhibited and inhibited conditions. They note that an important feature of cathodic protection in seawater is the formation of calcareous deposits, a term that is often used to describe magnesium (e.g., $Mg(OH)_2$) and calcium-containing (e.g., $CaCO_3$) compounds. Deposition involves the nucleation and growth of minerals on a metallic surface from the interfacial solution, which is supersaturated with respect to certain compounds. They also claim that the electrodeposition method offers a controlled way to generate surface scaling deposits, and, as such, electrodepositing could be used to assess the efficiency of scale inhibitors and the rate of nucleation and growth of crystals on solid surfaces. In this study,[67] the development of scale on a stainless-steel surface through application of a cathodic potential in different brine systems was used to follow deposition of calcium- and magnesium-containing compounds on the surface in uninhibited and inhibited conditions. Chen et al.[68] concluded that bulk precipitation and surface precipitation are different processes and should be studied independently. This work described the scaling on the surface as well as the bulk precipitate.

Neville[27] provides information on the rate of the actual deposition of calcium carbonate scale on a surface compared with the rate of bulk scale formation. Neville claims that the quantification of scale formed on a surface and the visualization of scale as it forms in the bulk phase improve understanding of the field processes. This paper is focused on the calcium carbonate system, but much of the discussion is also relevant to $BaSO_4$ and other common inorganic scales (e.g., $CaSO_4$).

The work by Amjad[56] described in Section 3.5.2 supports the observations that heterogeneous scale deposition is a major mechanism in some situations. This study's setup encouraged surface deposition by using a heated surface for deposition of $CaSO_4$. The author proposed a rate law for deposition:

$$Rate = -dC/dt = ks(C - C_e)^2 \qquad \text{(Equation [3.48])}$$

Here, *C* is the total molar calcium ion (or total sulfate) concentrations at time *t*, *Ce* is the equilibrium solubility value, t is the time in minutes, *k* is the rate constant, and *s* is a function of active growth sites on the substrates. Their data showed that the $CaSO_4$ crystals formed on brass at a higher rate than on stainless steel.

Tests of a metal surface were performed using a rotating cylinder (of stainless steel) electrode (RCE). Rotating disks or cylinders have been used to study the kinetics of different surface controlled reactions. See Section 6.6.3, which describes the use of a rotating disk for passivation studies. The flow dynamics of the

RCE have been studied by Silverman[70] for corrosion evaluations. Using a RCE, the Reynolds number (Re; see Section 2.2) can be calculated to determine the shear stress at the vicinity of the surface. The Re of the RCE with outer diameter, d_{cyl} (cm), can be computed as

$$Re = U_{cyl} \bullet d_{cyl} \bullet \frac{\rho}{\mu} \qquad \text{(Equation [3.49])}$$

U_{cyl} (cm s^{-1}) is the linear velocity, ρ is the solution density (g cm^{-3}), and μ is the viscosity of the solution (g cm^{-1} s^{-1}). The linear velocity at the outer diameter (i.e., surface velocity) can be calculated as

$$U_{cyl} = d_{cyl} \bullet \frac{F}{60} \qquad \text{(Equation [3.50])}$$

Here, F is the RPM of the cylinder.

Determining the relationship of the R_e values of the RCE vs the scaling rates can provide mechanics information of the scale deposition processes.

In addition, Neville[27] described an in situ flow cell that was used to determine the rates of scale formation (using a turbidity meter), and a see-through in situ cell was used to observe deposition of solids on the surface. Figure 3.11 shows the layout of the system for determining bulk scale kinetics and surface scale deposition conditions.

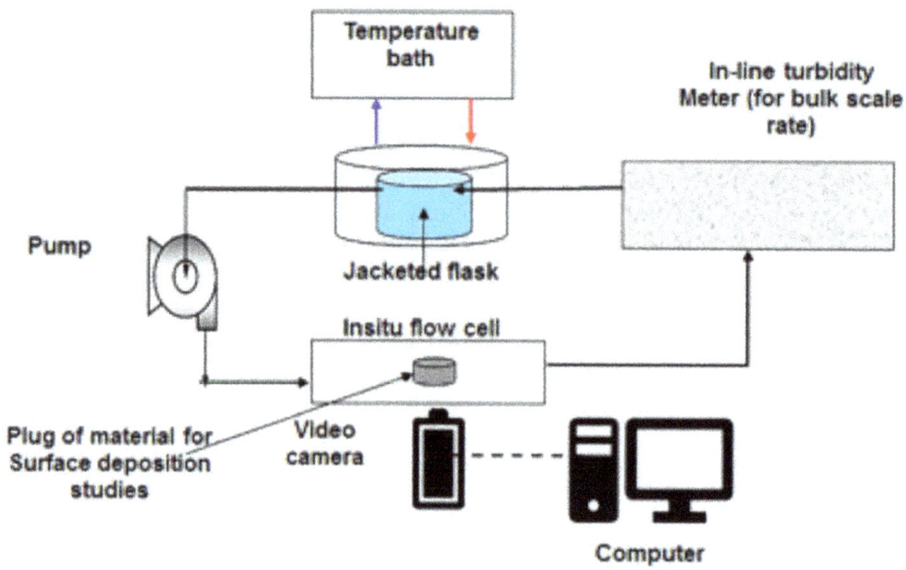

Figure 3.11 Loop to Study Bulk and Surface Scale Rate[27]

This study was designed to test the proposition that although the processes in the bulk solution will affect the scale deposition processes, surface reactions also are important. However, the study did not show that reducing the amount of bulk precipitation proportionally reduced the extent of surface deposition. There *was* a correlation of surface scaling and bulk precipitation, but it was not linear when the dependent variable was the SR (which can depend on several factors). Figure 3.12 depicts the correlation shown in Neville.[27] The conclusions of the paper are:

1) Surface deposition is a surface crystallization process, and strategies for its management, whether in critical or noncritical components, must be based on an understanding of the process and parameters at the surface, at the surface/fluid interface, and in the bulk fluid. Only then can efficient scale management be achieved at surfaces.

2) No single parameter accurately describes surface scaling tendency, and, as such, experimental evidence is normally needed to confirm the scaling tendency suggested by measurements of commonly used parameters, such as roughness and surface energy.

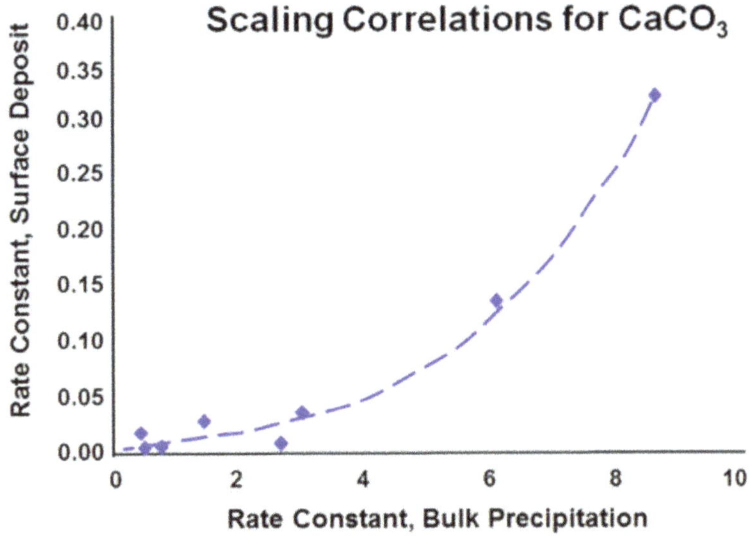

Figure 3.12 Rate Constant Correlation for Surface and Bulk Processes across the Temperature Range and Different SR Values[27]

Collins[70] describes details of the scaling process in a pipe:

- Brine enters the pipe in a supersaturated state; as it flows along the pipe, nucleation and growth occur, creating crystals in the bulk of the solution. The crystals increase in size in the direction of flow. The mechanism by which these crystals are deposited depends upon their size; therefore, the mechanism of deposition also depends on the position in the pipe.
- Crystal size may increase towards the walls of the pipe, especially in laminar flow, since the solution near the pipe wall has a lower velocity and therefore carries crystals in a later stage of growth. Owing to the removal of crystals by deposition, the concentration of crystals in the brine near the wall tends to be lower than in the center of the pipe.
- Transverse mixing and crystal migration, especially in turbulent flow, suppresses changes in crystal size and concentration near the wall.

The deposition of particles (in this case, crystals produced by precipitation) occurs in two steps (see Collins[70]): transport to a surface and attachment to that surface. Collins claims that attachment does not always occur when a particle reaches a surface because of the presence, in some cases, of an energy barrier to deposition. Data in the Neville[27] report also indicate that some scale can form by heterogeneous processes that do not involve transport of precipitates solids.

Collins[70] describes six transport mechanisms by which crystals may reach a surface:

- Particle motion owing to electrostatic forces
- Deposition owing to gravity forces (sedimentation)
- Deposition caused by shear dispersion and shear forces
- Brownian diffusion
- Transport owing to inertial forces
- Eddy diffusion only operates in the turbulent flow regime

According to Collins,[70] transport mechanisms are affected by the transition from laminar to turbulent flow because the eddies that occur in the latter can keep particles, especially smaller particles, in suspension. The actual calcium carbonate (and other) scale adhesion and accumulation (deposition) in a line or tube can be affected by the flow dynamics.

These steps also apply to the case described by Hamid et al.[71] who claim that there is a five-step process that controls the amount of deposition in a flowing pipe. In Figure 3.13(a), the flow in a clear (but unsaturated) solution over a rough surface is shown. In Figure 3.13(b), the solution is saturated with a salt, and micro-scale crystals start to form (see also Figure 3.2). In Figure 3.13(c–e), scale forms at the boundary area if the flow is above a minimum level and continues to accumulate until the flow is above this minimum. This is due to erosion of the scale particles.

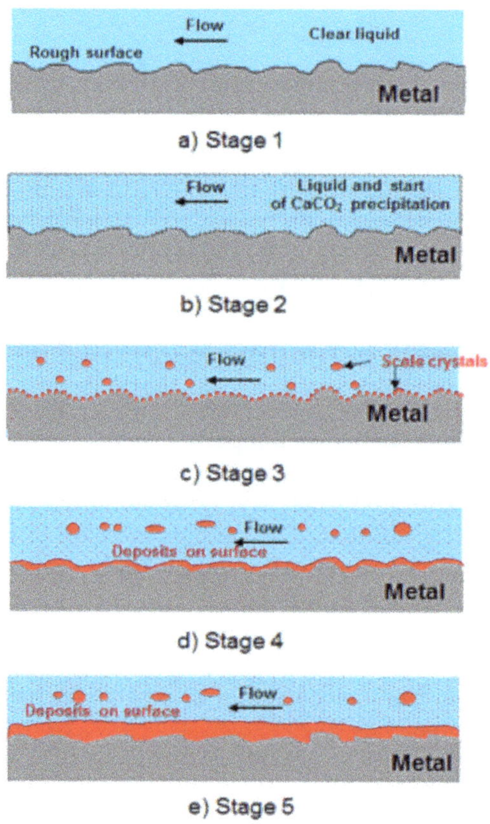

Figure 3.13 Scale Deposition in a Flowing Pipe[72]

Collins[70] also describes the theoretical prediction of scale deposition on surfaces. This theory proposes that there is a minimal surface free energy for each scale below which scale will not form, even at infinite induction times. Collins also provides a correlation between surface energy and deposition behavior (induction period and asymptotic deposition resistance), which is in satisfactory agreement with the experimental data. He claims this model also explains previous contradictory results on the effects of surface energy on deposition behavior. Although low-adhesion surfaces cannot replace chemical inhibition in the oilfield owing to the difficulty and cost of preparing the surfaces, effective anti-scaling surfaces could be crucial to protect key components both downhole and topside.

Parsons and MacAdam[72] report results of tests of carbonate scaling on heat transfer surfaces. In these tests, the metallurgy as well as the roughness affected the amount of deposit formed. Thus, stainless steel and a mirror-like surface resisted scale buildup. Note, however, that these tests are difficult to reproduce, and effects on downhole equipment are still being explored. More details of some anti-fouling surfaces as well as the Collins[70] theory will be described in the inhibition sections.

Neville[27] and Cowan and Weintritt[5] describe the effect of the surface on scale adhesion. These studies indicated that the amount of corrosion and use of the equipment (related to wear and roughness) resulted in increased scaling. The studies reviewed indicate that the surface/solution environment is very complex and difficult to predict. However, reducing corrosion, wetting angle, and roughness may reduce some scale adherence.

The composition and surface conditions (the real surface area and other deposits) of a metal also can affect the deposition and accumulation of fouling scales, such as calcium carbonate. MacAdam and Parsons[73] report results of tests of carbonate scaling on heat transfer surfaces. In these tests, the roughness and the metallurgy affected the amount of deposit formed. Doyle et al.[74] showed that the scaling rate for magnesium ammonium phosphate was dependent on the material on which the scale formed, but once the material was roughened, the impact of the initial material was significantly reduced. In this case, stainless steel and a mirror-like surface resisted scale buildup.

MacAdam and Parsons[73] reported that for calcium carbonate, increasing surface roughness from 0.2 μm to 0.8 μm led to a 30% increase in scaling rate. Surface material also influences the rate of scale formation, and this is often linked to surface free energy of the material. The lower the surface energy, the lower the adhesion strength between calcium carbonate crystals and surfaces.

Förster et al.[75] tested the effects of surface finish on calcium sulfate fouling. See Table 3.3. The results indicate that the induction period for copper scaling was significantly shorter than for stainless steel and followed the value of the surface free energies of each material.

Table 3.3 Effect of Surface Free Energy and Roughness on Induction Time of $CaSO_4$ Fouling (Surface Temperature, 75°C; Bulk Temperature 40°C); Adapted from Förster et al.[75]

Material	Surface free energy (nM m^{-1})	Mean Surface Roughness (μm)	Induction Time (h)
Copper	50	1	28.7
Aluminum	48	1	50.4
Steel (stainless)	40	5	139
Brass	35	1	177
Diamond Like Carbon	37	5	>475
Polytetrafluoro-ethylene (PTFE)	18	11	89

Quddus and Al-Hadhrami[76] have studied the effects of flow rates on deposition of two scales (calcium carbonate on stainless steel and calcium sulfate on aluminum) using RCE methods. The deposition rates for each scale/metal were plotted vs the square root of the Reynolds number ($Re^{0.5}$) for the rotating cylinder. See Figure 3.14. The data show that the $CaCO_3$ and $CaSO_4$ scaling rate increased with $Re^{0.5}$, suggesting that the process is diffusion controlled. According to Levich[77] and Fredd and Fogler,[78] the mass transfer coefficient should increase in proportion with $(Re)^{0.5}$. In addition, the figure shows that the weight gain of $CaCO_3$ scale on stainless steel specimens was less compared with $CaSO_4$ scale on aluminum specimens. The authors of this paper claim that the disparity is due to the difference in their solubilities and therefore the concentrations used. Note that the solubility of $CaSO_4$ is higher than the solubility of $CaCO_3$ by a factor of 10^2. See Table 3.2.

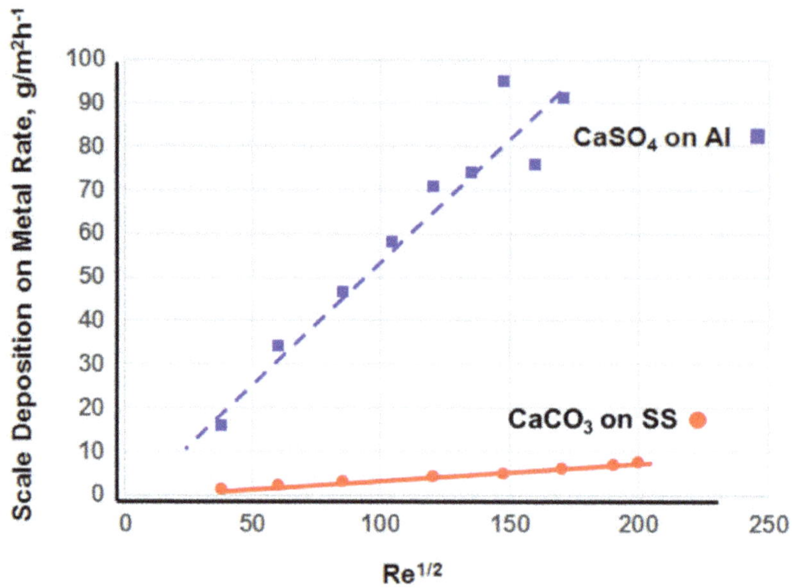

Figure 3.14 A Study of Scale Deposition in Relationship to a Flow Parameter ($Re^{0.5}$) for Calcium Carbonate and Calcium Sulfate

This type of information may be useful for determining details of the rate of scale deposition, since the diffusion coefficients for the processes may be calculated from the slope of a plot of this type. See De Rozieres.[79] Quddus and Al-Hadhrami[76] and Vazirian et al.[80] have shown that the actual amount of scale found on the surface can be affected by adhesion of the scale crystals to the surfaces.

The data from Amjad[56] for the deposition of $CaSO_4$ on different heat exchange alloys is in line with the information in Table 3.3. This study showed that copper-based surfaces are more active than stainless steel surfaces and that the surface free energy may be important in other crystalline deposition situations, in addition to calcite.

Østvold and Ranhold[81] studied $CaCO_3$ scaling kinetics and effects of other solids in the laboratory. The induction time for precipitation was determined in a series of experiments with various surface temperatures and the ionic compositions of the water from which the precipitation occurred. The effect of sand from a North Sea oilfield reservoir on the rate of $CaCO_3$ precipitation has been investigated by Østvold and Ranhold.[81] The data show that sand enhances the rate and reduces the induction time for calcite precipitation. At 100–120°C, precipitate was observed after a few seconds.

Sand suspended in solution acted as crystallization sites for the dissolved carbonate. Mg^{2+} and SO_4^{2-} ions in the solution increased the induction time and slowed the $CaCO_3$ crystallization. This implies that compounds that block active sites for calcite crystallization may inhibit $CaCO_3$ precipitation. With Mg^{2+} and SO_4^{2-} ions in the solution, aragonite was formed instead of calcite.

The presence of other scales also may increase or decrease the rate and amount of calcite deposited. This will cause a buildup of an inorganic scale such as calcite on top of an iron oxide layer and an organic solid on top of them. This leads to mixed deposits as described in Section 3.9.

A summary of all the reports in this section show that various processes can influence the actual amount of solids that form on a surface and thus must be removed. These include

- Scaling mineral type and the total brine composition that will control the SR
- The degree of supersaturation (SR), which can change with time within a flow line

- Hydrodynamic conditions
- Composition and condition of the surface
- Scale inhibitors (see Frenier and Ziauddin[1])

3.8. Formation of Deposits in Nuclear Power Plants

Although the deposits found in nuclear plants are like those found in non-nuclear units, the unique radioactive products require additional discussion.

The corrosion products, containing iron oxides as well as oxides of nickel and chromium and mixed oxides, are formed on the basis of the chemistries described in Sections 3.3.1, 3.4, 3.6, and 3.7. In addition, corrosion products and solids entering the secondary side of pressurized water reactors via feedwater can deposit on steam generator tubes and other internal surfaces. These tenacious deposits inhibit heat transfer, block tube supports, and create crevices where corrosive impurities can accumulate. If not corrected, these deposits may lead to stress corrosion cracking and tube failure.

According to Vepsäläinen,[82] the deposits are formed mainly from corrosion products that are transported from the secondary cycle. Iron oxides are the primary component in the deposits, but they may contain significant proportions of copper and copper oxides, depending on the construction materials of the secondary cycle. Soluble species, such as silicates and sulfates, can concentrate and precipitate within deposits.

Vepsäläinen[82] also claims that flow-accelerated corrosion (FAC) on the secondary cycle is the main source of iron deposit precursors. The FAC magnetite layer becomes thin due to enhanced dissolving at the high flow and turbulence. FAC occurs in alkalinized and deaerated water. Then, deposits and sludge form in the steam generator and involve a range of possible physical and chemical phenomena, such as turbulent deposition, boiling deposition, gravitational settling, and re-entrainment. The deposition process is a complex phenomenon in which multiple factors affect the deposit formation rate.

The chemistry of the solids includes magnetite, lepidocrocite, maghemite, hematite, silicon dioxide, apatites, copper, and copper oxides. The process from Jevec et al.[83] shows the deposition and spalling of tubing deposits (Figure 3.15) on tubes. The figures show at least two layers that come from corrosion of the base metal as well as deposition owing to scale formation and solids migration processes.

Details of the process (Vepsäläinen[83]) include

1) Incubation
2) Initiation
3) Growth
4) Growth-limiting stage
5) Spalling and re-deposition

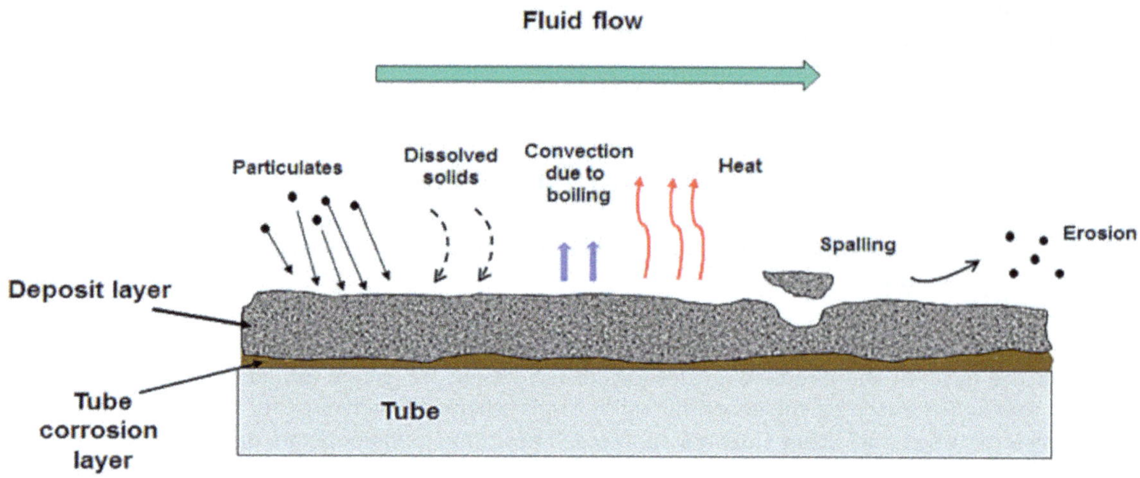

Figure 3.15 Pressurized Water Reactors Tube Scale Formation Mechanism[83]

In addition to scale on the tubing and bridging the tube sheets, sludge piles form on horizontal surfaces in steam generators owing to the gravitational settling, entrapment, and consolidation of particulate and chemical species from the feedwater. Sludge pile integrity is thought to be maintained and improved by reactions among the sludge pile species or by chemical reactions with dissolved species that tend to consolidate the sludge pile matrix. Two examples from Vepsäläinen[82] of such consolidations are

1) The cementation of iron oxide particles such as magnetite by ferrous hydroxide:

$$Fe(OH)_2(s) = Fe_3O_4(s) + 2H_2O + H_2 \qquad \text{(Equation [3.51])}$$

2) The reaction between the dissolved iron species or magnetite and phosphate to form sodium iron phosphate.

Semmler and Leinonen[84] reported that the scale and sludge that formed on the surfaces of secondary systems of CANDU reactor systems consisted of a powdery deposit and flake-like solids. Various analyses (SEM, XRD) showed that the powdery solids consisted of magnetite, metallic copper, hematite, and copper oxide (Cu_2O). In addition to the iron and copper-containing elements, the flakes also contained zinc silicate (Zn_2SiO_4). These solids were recovered after water lancing operations. Silicates may increase the difficulty of dissolving all the deposits. See Sections 4.1, 4.2, 4.4.2, and 5.3 for discussions of solvents for these types of deposits.

According to Little et al.,[86] industry experience has shown that most impurities and corrosion product material generated in the secondary systems of pressurized light water reactors and CANDU units accumulate in the steam generators. Most of these impurities and corrosion products deposit as scale on steam generator tubing, with smaller percentages depositing or accumulating on other internal surfaces such the top of the tube sheet and within tube support plate (TSP) crevices and flow passages. Depending on the exact location, chemistry, and structure of these secondary steam generator deposits, such accumulations can result in operational challenges, including

- Heat-transfer fouling, which degrades the secondary steam pressure, lowers thermal margins, and in some cases requires reductions in thermal power or other remedial measures. Heat-transfer fouling is attributed to insulating scale that deposits on tubing surfaces within the steam generator.
- Deposit-induced changes in flow patterns, some of which can cause thermal-hydraulic instabilities and steam generator water level oscillations, flow-accelerated corrosion of internal components such as the

TSPs, or flow-induced vibration that leads to tubing wear, fretting, and/or fatigue cracking. These problems are generally attributed to deposit accumulations formed within TSP flow passages.
- The formation of occluded regions in which corrosive species can be concentrated, which can promote or aggravate degradation of steam generator tubing (for example, by outer diameter stress corrosion cracking). Deposit-induced degradation is most often observed within crevices formed at tube-to-tube sheet or tube-to-TSP intersections, where deposit accumulations tend to consolidate. However, similar degradation has also been observed under scale formed on free span steam generator tubing surfaces.

Marks et al.[86] claim that steam generator tube deposit fouling occurs predominantly on the tube outer-diameter surfaces. These deposits may cause reductions in thermal efficiency (heat-transfer fouling), corrosion of tubing or other internal steam generator components, and other operational problems because of secondary flow blockage or redistribution. Two examples of steam generator tube deposits are shown in Figure 3.16. Each image represents an optical photograph of the deposit cross section obtained with a light microscope; the tube interface surface in each case is located on the left side of the image, whereas the secondary bulk flow interface is located on the right side.

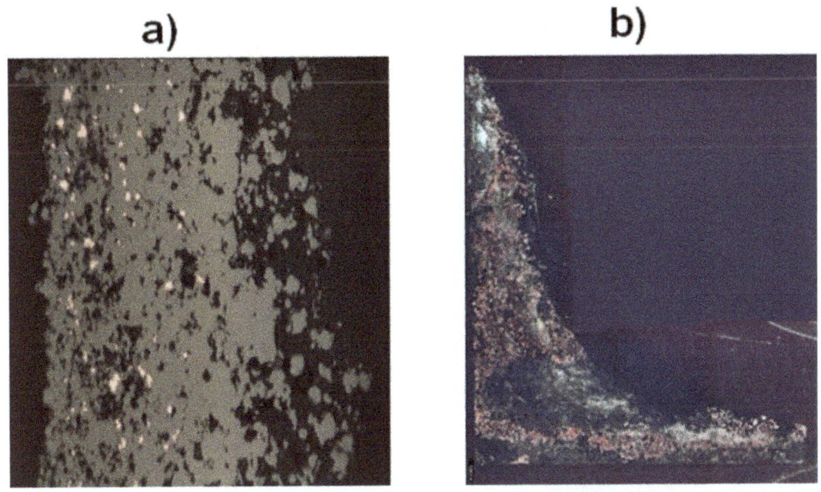

Figure 3.16 Optical Microphotographs of Steam Generator Deposit Cross Sections: Tube Scale (Left) and Hard Support-Structure Deposit (Right)[86]

In Figure 3.16, the image on the left (a) depicts the scale that forms on the tube free spans. This example is approximately 0.006 inches (6 mils = 152 μm) in thickness. Such deposits, which are typically composed mainly of magnetite (Fe_3O_4; the gray material in this figure) with some copper inclusions (visible as the small bright areas in the image), influence the thermal performance of the steam generators because they cover the tube heat-transfer surfaces.

These authors point out that the image on the right (b) reflects a much harder, consolidated deposit that formed at the intersection between a tube and a support structure within the steam generator (e.g., the tube sheet at the base of the tube bundle). Such hardened, concrete-like deposits are typically highly heterogeneous and contain exfoliated magnetite-bearing tube scale pieces as well as various other chemical species that cement the deposits into a large amalgamation with hard collars surrounding many tubes. These hardened deposit accumulations (often termed a sludge pile) lead to concentrations of deleterious chemical species, which in turn can exacerbate tube corrosion processes. The example shown in Figure 3.16(b) is more than 0.15 inches (3.8 mm) in total height and width. Additional background concerning steam generator tube deposits and their properties is included in other publications.[87, 88, 89] Additional details are in Section 5.1.

Inorganic deposits also form in waste storage tanks in major nuclear sites (for example, the site in Savannah River, SC). King and Hay[90] claim the sludge in the tanks contains the aluminum commands of gibbsite ($Al(OH)_3$) and the iron phases hematite, maghemite (Fe_2O_3, γ-Fe_2O_3), goethite (α-$FeO(OH)$), lepidocrocite (γ-$FeO(OH)$), magnetite, and wustite (FeO). However, no mechanisms of the formation of the minerals were provided. The removal of these types of scales is described in Chapter 5.

3.9. Field Scale Deposits

Earlier sections of this book discussed the formation of relatively pure, one-component deposits. However, field scales often are mixtures of iron oxides, silica, and calcium salts and iron sulfides. Figure 1.18 and Figure 3.17 show photos of tubes fouled with mixed iron oxides, as well as calcite and iron sulfides. In some boiler tubes, copper deposits also may be present. Depending on the industry/units that are fouled, the scales also may contain various organic solids or liquids. Figure 3.18 depicts the buildup of several layers of scale on a steel pipe. The first layer contains iron oxides or sulfides. Other water-precipitated scales, such as calcite, may deposit onto the iron salt, followed by an organic compound if these chemicals are flowing in the piping.

Chapter 7 describes the formation and removal of organic and mixed deposits. The dissolution of FeS, which frequently occurs in refineries, also is described in Chapter 7.

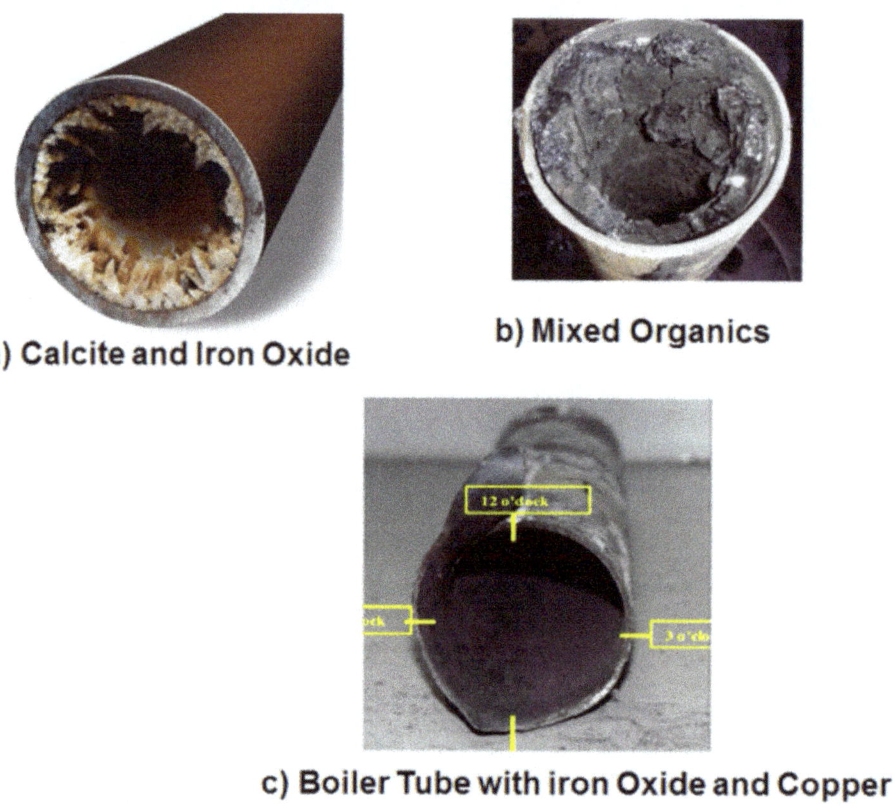

Figure 3.17 Mixed Field Deposits

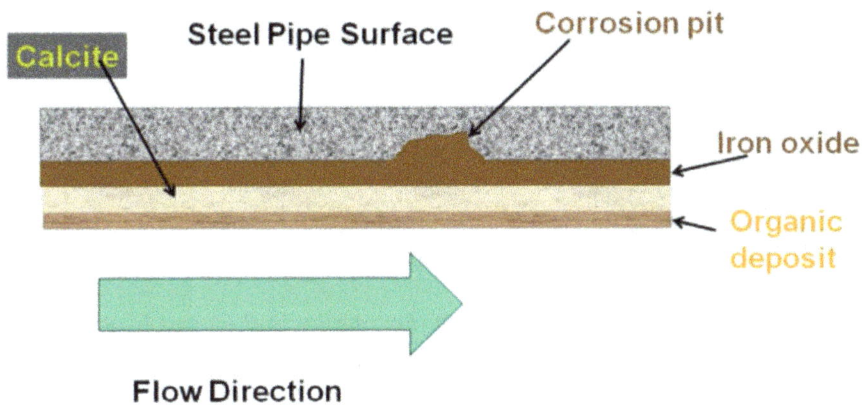

Figure 3.18 Mixed Scale in a Pipe

References

[1]W.W. Frenier, M. Ziauddin, *Formation, Removal, and Inhibition of Inorganic Scale in the Oilfield Environment* (Richardson, TX: Society of Petroleum Engineers, 2008).

[2]W.W. Frenier, M. Ziauddin, *Chemistry for Enhancing the Production of Oil and Gas* (Richardson, TX: Society of Petroleum Engineers, 2014).

[3]W.W. Frenier, "Review of HSE Innovations in the Chemical Cleaning Industries," CORROSION/2015, paper no. NACE C2015-5553 (Houston, TX: NACE, 2015).

[4]W.W. Frenier, *Chemical and Mechanical Methods of Pipeline Integrity* (Richardson, TX: Society of Petroleum Engineers, 2017).

[5]J.C. Cowan, D.J. Weintritt, *Mineral & Organic Scale Deposits* (Lafayette, LA: Cowan & Weintritt Publishing Company, 2004).

[6]W.W. Frenier, F.B. Growcock, "Mechanism of Iron Oxide Dissolution," *Corrosion* 40, 12 (1984): 663-668.

[7]W.W. Frenier, *Technology for Chemical Cleaning of Industrial Equipment* (Houston, TX: NACE International, 2001).

[8]J. Zemaitis Jr., D.M. Clark, M. Rafal, N.C. Scrivner, *Handbook of Aqueous Electrolyte Thermodynamics* (New York, NY: American Institute of Chemical Engineers, Inc., 1986).

[9]D. Langmuir, *Aqueous Environmental Geochemistry* (Englewood Cliffs, NJ: Prentice Hall, 1997).

[10]W.F. Linke, *Solubilities: Inorganic and Metal-Organic Compounds*, vol. I and II (Washington, DC: American Chemical Society, 1958).

[11]D.R. Lide, *CRC Handbook of Chemistry and Physics*, 87th ed. (Boca Raton, FL: CRC Press, 2006).

[12]J.A. Dean, *Lange's Handbook of Chemistry*, 14th ed. (New York, NY: McGraw-Hill, Inc., 1992).

[13]H.A. Stiff, L.E. Davis, "A Method for Predicting the Tendency of Oil Field Water to Deposit Calcium Carbonate," *JPT* 4, 9 (1952): 213-216.

[14]A.T. Kan, G. Fu, M.B. Tomson, "Prediction of Scale Inhibitor Squeeze and Return in Calcite-Bearing Formation," Paper SPE 93265, 2005 International Symposium on Oilfield Chemistry, Houston, TX, Feb. 2-4, 2005.

[15]M. Davies, P.J.B. Scott, *Oilfield Water Technology* (Houston, TX: NACE International, 2006).

[16]M.M. Benjamin, *Water Chemistry* (New York, NY: McGraw-Hill, 2002).

[17]M. Crabtree, D. Eslinger, P. Fletcher, M. Miller, A. Johnson, G. King, "Fighting Scale- Removal and Prevention," *Schlumberger Oilfield Review* 11, 3 (1999): 30-45.

[18]G.H. Nancollas, L. Wang, Z.J. Henneman, P.P. Emmerling. CORROSION/2007, paper no. 07050 (Houston, TX: NACE, 2007).

[19]W.F. Langelier, "The Analytical Control of Anti-Corrosion Water Treatment," *J. Am. Water Works Assc.* 28 (1936): 1500-1521.

[20]M.B. Tomson, A.T. Kan, G. Fu, M. Al-Thubaiti, "A Molecular Theory of Mineral Scale Inhibition," CORROSION/2004, paper no. NACE 04075 (Houston, TX: NACE, 2004).
[21]G. Atkinson, K. Raju, R.D. Howell. "SPE 21021: The Thermodynamics of Scale Prediction," SPE International Symposium on Oilfield Chemistry, Anaheim, CA, 1991.
[22]K. Raju, G. Atkinson, "Thermodynamics of Scale Mineral Solubilities 3. Calcium Sulfate in Aqueous NaCl," *J. Chem. Eng. Data* 35 (1990): 361-367.
[23]K. Raju, G. Atkinson, "Thermodynamics of Scale Mineral Solubilities 2. Srso4(S) in Aqueous NaCl," *J. Chem. Eng. Data* 34 (1989): 361-364.
[24]K. Raju, G. Atkinson, "Thermodynamics of Scale Mineral Solubilities 1. Baso4(S) in H_2O and Aqueous Nacl," *J. Chem. Eng. Data* 33 (1985): 490-495.
[25]R.J. Ferguson, "The Kinetics of Cooling Water Scale Formation and Control," Association of Water Technologies Annual Meeting, Sept. 14–17, Atlanta, GA, 2011.
[26]Multiscale, Multiscale 7.1 (Haugesund, Norway: Expro Petrotech, 2015).
[27]A. Neville, "Surface Scaling in the Oil and Gas Sector: Understanding the Process and Means of Management," *Energy and Fuels* 26 (2012): 4158–4166.
[28]V. Tantayakom, P. Charoensirithavorn, H.S. Fogler, "Kinetic Study of Scale Inhibitor Precipitation Squeeze Treatment," *Crystal Growth & Design* 5, 1 (2005): 329–335.
[29]Y. Zhang, H. Shaw, R. Farquhar, R. Dawe, "The Kinetics of Carbonate Scaling–Application for the Prediction of Downhole Carbonate Scaling," *Journal of Petroleum Science and Engineering* 29 (2001): 85-95.
[30]J. Talbot, D. Talbot, *Corrosion Science and Technology* (Boca Raton: CRC Press, 1997).
[31]W.E. Bornak, "Chemistry of Iron and Lts Corrosion Products in Boiler Systems," *Corrosion* 44, 3 (1987): 154-157.
[32]L.L. Shreir, *Corrosion: Metal/Environment Reactions*, vol. I (London, U.K.: Newnes-Butterworths, 1976).
[33]J.M. Potter, W.E. Dibble Jr., "Chemical Aspects of Iron Colloid Plugging in Quartz Sands and Implications for Formation Damage," *Journal of Petroleum Technology* 37, 9 (1985): 1682-1688.
[34]K. Hansen, "Chemical Cleaning of Boilers," Paper IWC 12-67, 73rd Annual IWC, San Antonio, TX, Nov. 4-8, 2012.
[35]P. Sarin, V.L. Snoeyink, D.A. Lytle, W.M. Kriven, "Iron Corrosion Scales: Model for Scale Growth, Iron Release, and Colored Water Formation," *Journal Of Environmental Engineering* 130, 4 (2004): 364-378.
[36]R.A. Berner, "Iron Sulfides Formed from an Aqueous Solution at Low Temperatures and Atmospheric Pressure," *Journal of Geology* 72, 3 (1964): 293-306.
[37]S.N. Smith, J.L. Pacheco, "Predicting Corrosion in Slightly Sour Environments," CORROSION/2002, paper no. NACE 02241 (Houston, TX: NACE, 2002).
[38]S.N. Smith, "Prediction of Minimum H2s Levels Required for Slightly Sour Corrosion," CORROSION/94, paper no. NACE 94011 (Houston, TX: NACE, 1994).
[39]A.K. Dunlop, H.L. Hassell, P.R. Rhodes, "Fundamental Considerations in Sweet Gas Well Corrosion," CORROSION/83, paper no. NACE 46 (Houston, TX: NACE, 1983).
[40]J.W. Morse, F. Millero, J.C. Cornwell, D. Rickard, "The Chemistry of Hydrogen Sulfide and Iron Sulfide in Natural Waters," *Earth-Science Reviews* 24 (1987): 1-42.
[41]H.A. Nasr-El-Din, B.A. Fadhel, A.Y. Al-Humaidan, W.W. Frenier, D. Hill, "An Experimental Study of Removing Iron Sulfide Scale from Well Tubulars," paper SPE 60205, International Symposium on Oilfield Scale, Aberdeen, U.K., Jan. 26-27, 2000.
[42]D. Jeffries, Pyrophoric Ignition Hazards in Typical Refinery Operations (San Ramon, CA: Chevron, 2010).
[43]C.A. Palacios, J.R. Shadley, "Characteristics of Corrosion Scales on Steels in a Co_2-Saturated Nacl Brine," *Corrosion* 47, 2 (1991): 122.
[44]P.A. Burke, A.I. Asphahani, B.S. Wright, *Advances in Co2 Corrosion*, vol. 2 (Houston, TX: NACE International, 1985).
[45]B.G. Kriel, C.A. Lacey, "The Performance of Scale Inhibitors in the Inhibition of Iron Carbonate Scale," Paper SPE 27390, SPE international Symposium on Formation Damage Control, Lafayette, LA, Feb. 7-10, 1994 (Society of Petroleum Engineers, 1994).
[46]NACE MR0175/ISO 15156, "Petroleum and Natural Gas Industries–Materials for Use in H2s-Containing Environments in Oil and Gas Production," (Houston, TX: NACE, 1975).
[47]NACE MR0176-2000, "Metallic Materials for Sucker-Rod Pumps for Corrosive Oilfield Environments" MR0176-2000 (Houston, TX: NACE, 2000).

[48]P.S. Jackman, L.M. Smith, *Advances in Corrosion Control and Materials in Oil and Gas Production (European Federation of Corrosion Report No. 26)* (London, U.K.: The Institute of Materials, 1999).
[49]R. Heidersbach, *Metallurgy and Corrosion Control in Oil and Gas Production* (Hoboken, NJ: John Wiley & Sons, Inc., 2011).
[50]W.W. Frenier, *Chemical Methods for Controlling Corrosion in Oil and Gas Activities* (Houston, TX: NACE International, 2017).
[51]C.D. Foster, J. Trimble, "Foam Cleaning Boosts Turbine Output," *Power Engineering* March (1999).
[52]J.R. Becker, *Corrosion and Scale Handbook* (Tulsa, OK: PennWell Books, 1998).
[53]J. MacAdam, S.A. Parsons, "Calcium Carbonate Scale Formation and Control," *Reviews in Environmental Science & Bio/Technology* 3, 2 (2004): 159-169.
[54]A.T. Kan, X. Wu, G. Fu, M.B. Tomson. "Validation of Scale Prediction Algorithms at Oilfield Conditions," Paper SPE 93264, SPE International Symposium on Oilfield Chemistry, The Woodlands, Texas, Feb. 2-4, 2005).
[55]"Flow Assurance and Scale Control 2001-2004," Report 6 (Glasgow, U.K.: Heriot-Watt University, 2004).
[56]Z. Amjad, "Kinetic and Morphological Investigation of Calcium Sulfate Dihydrate (Gypsum) Scale Formation on Heat Exchanger Surfaces in the Presence of Inhibitors," CORROSION/2017, paper no. NACE C2017-9020 (Houston, TX: NACE, 2017).
[57]A.D. Hill, R.S. Schechter, "Fundamentals of Acid Stimulation," in *Reservoir Stimulation*, 3rd ed., eds. M. J. Economides, K.G. Nolte (New York, NY: John Wiley and Sons, 2000).
[58]R.K. Iler, *The Chemistry of Silica* (New York, NY: John Wiley & Sons, 1979).
[59]K.D. Demadis, E. Nfofotistou, "Inhibition and Growth of Colloidal Silica: Designed Chemical Approaches," *MP*, April (2004): 38-40.
[60]S.A. Gharaibeh, T.R. McCartney, R.A. Shank, "A Novel Non-Toxic Method for the Decontamination of Silicate Scales: A Case Study," CORROSION/2017, paper no. NACE C2017-9087 (Houston, TX: NACE, 2017).
[61]I.A.-M. Sheikholeslami, "Pretreatment and the Effect of Cations and Anions on Prevention of Silica Fouling," *Desalination* 139, 1-3 (2001): 83-95.
[62]T.K. Sahachaiyunta, "Effect of Several Inorganic Species on Silica Fouling in Ro Membranes," *Desalination* 144, 1-3 (2002): 373-378.
[63]F.N. Kemmer, *The Nalco Water Handbook* (Naperville, IL: Nalco, 1988).
[64]Betz Laboratories, *Betz Handbook of Industrial Water Conditioning*, 9th ed. (Trevose, PA: Betz Laboratories, 1991).
[65]A. Kladi, P.G. Klepetsenis, T. Ostvold, C.G. Kontogiannis, P.G. Koutsoukos, "Crystal Growth of Calcium Carbonate in Seawater. The Effect of Temperature and of the Presence of Inhibitors," in *Advances in Crystal Growth Inhibition Technologies*, ed. Z. Amjad (New York, NY: Kluwer Academic/Plenum Publishers, 2000).
[66]J.W. Mullin, *Crystallization* (Oxford: Butterworth-Heinemman, 1993).
[67]A. Neville, M. Bepoix, G.M. Graham, A.P. Morizot, "Assessing the Potential of Atomic Force Microscopy to Study the Formation of Electrode Calcareous Scales in Uninhibited and Inhibited Conditions," SPE 74653, SPE International Conference on Health, Safety, and Environment in Oil and Gas Exploration and Production, Kuala Lumpur, Malaysia (2002).
[68]T. Chen, A. Neville, M.D. Yuan, "Effect of PPCA and DETPMP Inhibitor Blends on CO_2 Scale Formation," SPE 87442, SPE 6th International Symposium on Oilfield Scale, Aberdeen, U.K., May 26-27, 2004.
[69]D.C. Silverman, "Rotating Cylinder Electrode for Velocity Sensitivity Testing," *Corrosion* 40, 5 (1984).
[70]I.R. Collins. "A New Model for Mineral Scale Adhesion," Paper SPE 74655, International Symposium on Oilfield Scale, Aberdeen, U.K., Jan. 30-31, 2002).
[71]S. Hamid, O. DeJesus, C. Jacinto, R. Izetti, H. Pinto, E. Droguett, C. Edwards, J. Cassidy, H. Zhang, P. Dagenais, "A Practical Method of Predicting Chemical Scale Formation in Well Completions," Paper SPE 168087 SPE SAS Annual Technical Symposium and Exhibition, Khobar, Saudi Arabia, May 30-31, 2013.
[72]S.A. Parsons, J. MacAdam, "Scaling on Heat Transfer Surfaces: Chemical Versus Non-Chemical Control," CORROSION/2004, paper no. 04073 (Houston, TX: NACE International, 2004).
[73]J. MacAdam, S.A. Parsons. "Scaling on Heat Transfer Surfaces: Chemical Versus Nonchemical Control," CORROSION/2004, paper no. 073 (Houston, TX: NACE International, 2004).

[74]J.D. Doyle, K. Oldring, J. Churchley, S.A. Parsons, "Struvite Formation and the Fouling Propensity of Different Materials" *Water Research* 36 (2002): 3971–3978.
[75]M. Förster, W. Augustin, M. Bohnet, "Influence of the Adhesion Force Crystal/Heat Exchanger Surface on Fouling Mitigation," *Chemical Engineering and Processing* 38 (1999): 449-461.
[76]A. Quddus, L.M. Al-Hadhrami, "Hydrodynamically Deposited Caco3 and Caso4 Scales," *Desalination* 246 (2009): 526-533.
[77]V.G. Levich, *Physicochemical Hydrodynamics* (Englewood Cliffs, NJ: Prentice-Hall, 1962).
[78]C.N. Fredd, H.S. Fogler, "Influence of Transport and Reaction on Wormhole Formation in Porous Media," *AIChE Journal* 44, 9 (1998): 1933-1949.
[79]J. De Rozieres, F.F. Chang, R.B. Sullivan. "Measuring Diffusion Coefficients in Acid Fracturing Fluids and Their Application to Gelled and Emulsified Acids," Paper SPE 28552, SPE Annual Technical Conference and Exhibition, New Orleans, LA, Sept. 25-28, 1994.
[80]M. Mohsen Vazirian, T.V.J. Charpentier, M.D.O. Penna, A. Neville, "Surface Inorganic Scale formation in Oil and Gas Industry: Adhesion and Deposition Processes," *Journal of Petroleum Science and Engineering* 137 (2016): 22-32.
[81]T. Östvold, P. Randhol, "Prediction and Kinetics of Carbonate Scaling from Oil Field Waters," CORROSION/2002, paper no. 317 (Houston, TX: NACE International, 2002).
[82]M. Vepsäläinen, "Deposit Formation in PWR Steam Generators" (Espoo, Finland: VTT Technical Research Centre Of Finland, 2010).
[83]J.M. Jevec, S.J. Klimas, P. King, K. Fruzzetti, "Multivariable Assessment of Flow Accelerated Corrosion and Steam Generator Fouling: Literature Review" (Palo Alto, CA: Electric Power Research Institute, 2003).
[84]J. Semmler, P.J. Leinonen, "Characterization and Dissolution Studies of Bruce Unit 3 Steam Generator Secondary Side Deposits," CORROSION/98, paper no. NACE 98324 (Houston, TX: NACE, 1998).
[85]M.J. Little, R.D. Varrin Jr., A.T. Pellman, M.A. Kreider, "Advanced Scale Conditioning Agent (ASCA) Applications: 2010 Experience Update," EPRI SGMP 2010 SG Secondary Side Management Conference, San Antonio, TX, Mar. 2–4, 2010.
[86]C. Marks, M.A. Kreider, M.J. Little, R.D. Varrin Jr., "Benefits of Partial Removal of Corrosion Deposits from Nuclear Steam Generators: ASCA and Code Applications," CORROSION/2017, paper no. C2017-9481 (Houston, TX: NACE, 2017).
[87]C.R. Marks, C.E. Anderson, H. Feldman, "Sg Deposit Characteristics: Relationship to Tube Degradation and Management," SGMP 2010 Steam Generator Secondary Side Management Conference, Electric Power Research Institute, San Antonio, TX, Mar. 2–4, 2010.
[88]R.D. Varrin Jr., "Characterization of PWR Steam Generator Deposits" (Palo Alto, CA: Electric Power Research Institute, 1996).
[89]R.D. Varrin Jr., R. Stanley, G. Gary, "Recent Experiences with Secondary Side Deposit Characterization," Paper EPRI No. 1008468, 2003 Steam Generator Secondary Side Management Conference, Electric Power Research Institute, Savannah, GA, Feb. 2003.
[90]W.D. King, M.S. Hay, "Alternative and Enhanced Chemical Cleaning: Basic Study Results Fy 2010" (Aiken, SC: Savannah River National Laboratory, 2011).

4 Removal of Inorganic Deposits with Aqueous Solvents

Chapters 1, 2, and 3 of this book provide background information on the equipment that may require cleaning, important chemical and mechanical principles, and deposits that foul industrial units. Chapter 3 reviews the deposition to inorganic fouling scales. Chapters 4–8 describe the chemical and mechanical methods for removing these fouling solids.

Chapter 4 reviews the large amount of information on the dissolution of inorganic scaling solids from industrial equipment. Industrial chemical cleaning has been based on the use of powerful solvents to remove many of the fouling deposits that occur in a wide variety of plants. This chapter describes the chemicals and the known mechanisms of deposit dissolution of the inorganic scales that were described in Chapter 3. Table 1.2 lists the major inorganic scaling minerals. Note that the dissolution process always requires other equipment to place, circulate, and remove the solvents; thus, the processes are always chemical and mechanical. Some of the most used equipment types are described in Section 8.1.

An earlier book[1] focused on the removal of inorganic deposits, especially from utility power boilers. This is still important, but other parts of the industry (such as refineries) may now compose a large percentage of jobs performed by contractors. As the percentage of coal-fired units diminish and as improved water treatment processes are introduced, the cleaning cycles have been lengthened. Thus, the current book has more information on non-fossil fuel utility plant processes.

Section 2.5.1 introduced chemical cleaning solvents, and Table 2.3 shows an overview of solvent types and temperatures for removing deposits. Chapter 4 emphasizes what is known about the chemicals and chemical reactions for removing inorganic scales from metallic surfaces. Table 4.1 lists the chemical names and key chemical characteristics of the materials used in formulations to clean metal surfaces. The data include acidity values and formation (equilibrium) constants for many chemicals. Most of the chemical structures are in Figure 4.1. Also, see Section 9.3 for a discussion of additional biodegradable chelating agents that are not currently used in IC operations (as of the time of this book's publication) but may be in future formulations.

Solvents for iron oxides are described first (Section 4.1) followed by copper solvents (Section 4.2). Processes to remove both iron and copper are described in Section 4.2.3 along with passivation procedures (Section 4.2.4). Because of the significant amount of literature available, a separate Section (4.3) reviews detailed iron oxide dissolution mechanisms. Although this chapter is concerned primarily with non-nuclear equipment, the chemistries are also used to descale or safe nuclear plant units. See Chapter 5 for more details.

Section 4.4 reviews the chemicals and mechanisms for dissolving alkaline earth salts and silica/silicates. However, some these solvents will also dissolve iron oxides and alkaline earth oxides and carbonates.

The cleaning chemicals discussed in this chapter are usually dissolved in water to produce solutions that contain between 5% and 25% active chemicals by weight. Note that the cited literature may be confusing because commercial solvent chemicals frequently are concentrated aqueous solutions that are then diluted to produce the application formulation. Some reports will state the use concentration as a percentage of the commercial product. Except for cleaning solutions that are highly alkaline (pH 12), all the field formulations should contain a corrosion inhibitor (CI) to protect the base metal. CIs for corrosive solvents are described in Chapter 6.

Other additives, including surfactants, solvents, and dissolution "boosters," also may be in the commercial cleaning formulation and are described in Section 4.5.

Table 4.1 Chemical Characteristics of Reactive Chemicals[2,3]

Compound	Sol. Water, (wt.%)	MW		pKa Values					LogK Equilibrium Constant				
		Acid	Na salt	pK1	pK2	pK3	pK4	pK5	Fe (II)	Fe (III)	Ca (II)	Ba (II)	Cu(II)
Hydrochloric acid (HCl)	v.	36.5		-7									
Hydrofluoric acid (HF)	v.	20		3.2									
Nitric acid (HNO_3)	v.	63		-1.6									
Sulfuric acid (H_2SO_4)	v.	98		-3	2								
Phosphoric acid (H_3PO_4)	v.	98		2.1	7.2	11.9							
Sulfamic acid	97.1			1.0									
Formic acid	v.	46		3.75						3.1	1.43	1.38	2.0
Acetic acid	v.	60		4.8					1.4	3.4	1.18	1.07	0
Citric acid	50	192	258	5.7	4.4	2.9			4.4	11.5	3.5	2.6	6.1
Glycolic acid	v.	76		3.83						2.9	1.6	1.04	2.8
Oxalic acid	v.	90		1.27	4.28				4.7	9.4	3		6.3
Hydroxyethyl ethylenediamine-triacetic acid (HEDTA)	10	278	344	9.8	5.4	2.6			12.2	19.8	8.4	6.2	
Ethylenediamine-tetra-acetic acid (EDTA)	0.1	292	380	10.2	6.1	2.7	2.0		14.2	25.0	10.7	7.8	18.8
Diethylenetriamine-penta-acetic acid (DTPA)	0.5	393	503	10.5	8.5	4.3	2.6	1.8	16.5	28.0	10.9	8.9	
GLDA-L-glutamic acid, N,N-diacetic acid	40	263	351	9.36	5.0	3.5	2.6			11.7 (est.)	5.9		
Hydroxyethyliminodi-acetic acid (HEIDA)	15 ??	177	221	8.7	2.2					11.6	4.8	2.8	
Ethylenediamine disuccinic acid (EDDS)		292		9.8	6.8	3.9	2.4			22	4.6		
Iminodisuccinic acid (IDS)		249		10.1	4.8	3.8	3			15.2	4.3		
Methylglycine diacetic acid (MGDA)	Sol?	157		10	2.5	1.6				16.5	7		

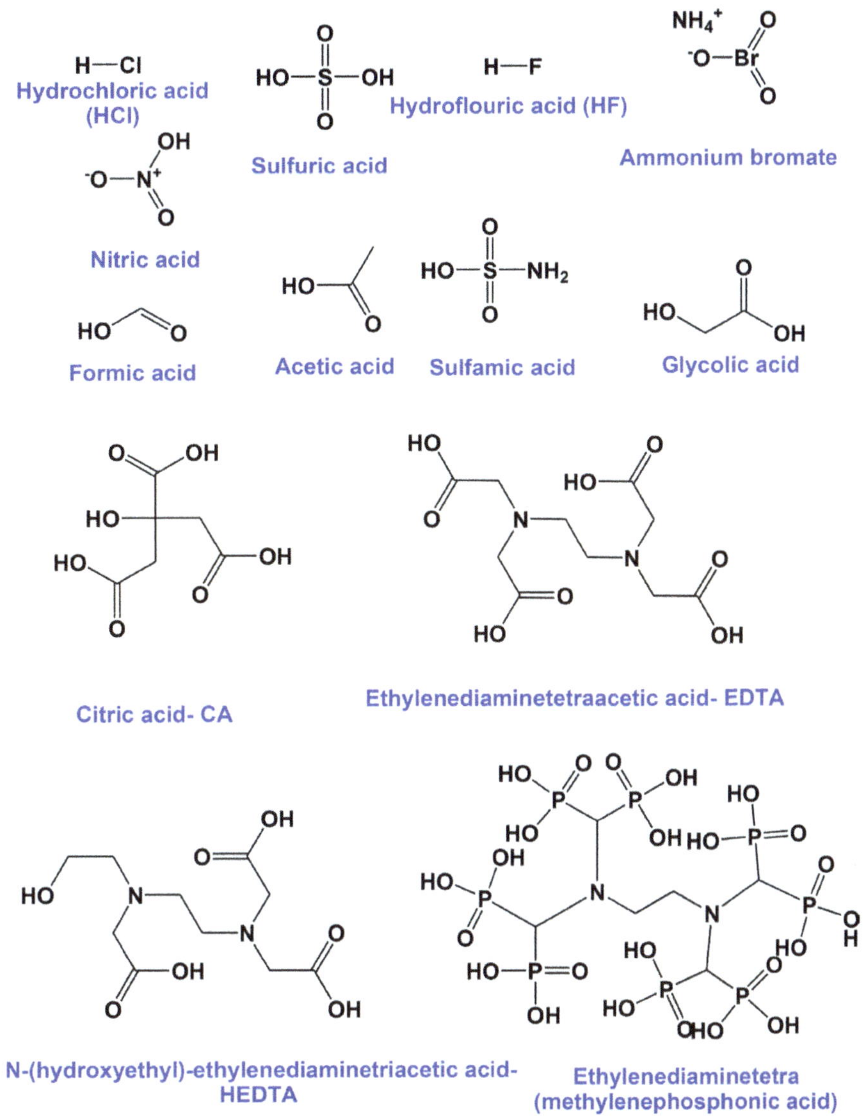

Figure 4.1 Cleaning Chemical Structures

4.1. Solvents for Iron Oxides

The inorganic deposits described in Table 1.2 in Chapter 1 and in Chapter 3 form and foul the surfaces of heat transfer and other classes of equipment. These include power and waste-heat boilers, heat exchangers, and various parts of cooling water systems. Iron oxides, copper, and calcium-containing and silicon-containing deposits make up the largest portion of the material that requires removal from metallic surfaces. Many of these units that require cleaning have very large volumes (sometimes >50,000 gal) and thus require equally massive amounts of chemicals for thorough descaling. Some of these chemicals and processes are also applied to decontaminate and remove scale from nuclear power generating plants. See Chapter 5.

This section describes the many solvents used for removing iron oxide deposits, including aqueous solvent-containing mineral acids, organic acids, chelating agents, and mixtures of these chemical classes. The choice of a specific solvent system depends on the metal of construction, the exact type and density of the oxide, the configuration of the unit, cost, and disposal considerations. This section describes the chemicals in current (or previous) use and the generally accepted mechanism. The science of iron oxide dissolution has been investigated in detail, and more information is in Section 4.3, which may be optional reading for some who use this book. That section (4.3) appears after section 4.2 (copper solvents) because some of the detailed chemistry also applies to combined iron/copper removal processes.

Because of the corrosivity of all iron oxide solvents, it is assumed that appropriate CIs are in the commercial formulations; however, lab test solvents may not contain a CI. Details of appropriate CIs are in Chapter 6. Additional details are in Frenier and Ziauddin[4] and Frenier.[5] Methods for analyzing the scale and selecting a solvent are in Section 2.5.

4.1.1. Mineral Acids

Mineral acids were the first chemicals used in IC operations and include

- Hydrochloric acid (HCl)
- Hydrofluoric acid (HF)
- Sulfuric acid (H_2SO_4)
- Phosphoric acid (H_3PO_4)
- Nitric acid (HNO_3)
- Sulfamic acid (H_2NSO_3H)

As noted previously, field solvents contain from about 5–10 wt.% of the active chemical. From an operational standpoint, most of the acids will be diluted from concentrated (but usually less than 100%) liquid commercial products. Therefore, the used concentration should reflect the absolute wt.% of the 100% active chemical, not the percent of the original delivered product. This is not consistent in the literature, so readers must be careful when making comparisons of commercial solvents.

Most of these acids have very low pK_a values (see Table 4.1) and are completely ionized at the use strength. An exception is HF, which has a pK_a value higher (less ionized) than formic acid, and phosphoric acid, which is about 10 times as ionized as HF.

These strong acids were some of the first materials used to remove fouling deposits from industrial equipment. The development of CIs by Grebe[6] to protect metal in contact with the various mineral acids (used initially in well stimulation) allowed the chemical cleaning industry to provide safe, useful services.

Hydrochloric Acid. Hydrochloric acid, the simplest, most common, and most versatile mineral acid, is used on many types of industrial process equipment. It is used at strengths from 5%–28% (5%–10% is the usual IC range) and can arrive on site in strengths of about 28–36 wt.%. These final solutions can be inhibited at temperatures up to about 180°F (82°C). Oilfield acids are used at temperatures up to about 350°F (177°C), but different standards of corrosion protection apply.

HCl solutions usually are not used to clean Series 300 stainless steel, free machining alloys, magnesium, zinc, aluminum, cadmium, or galvanized steel. Dangers to the metals of construction owing to general or localized attack[1] limit the use of HCl for cleaning the previously mentioned metals. Most other metals of construction can be adequately protected during cleaning using CIs (see Chapter 6). The HCl solutions will dissolve carbonates, some phosphates, some sulfate scales, ferrous sulfide, iron oxides, and copper oxides. Gypsum ($CaSO_4 \cdot 2H_2O$) is a common foulant in water-contacted units but is only sparingly soluble in HCl. See Section 4.4 for other solvents for low HCl soluble solids.

Using appropriate additives, iron oxides, carbonates, fluoride deposits, and copper oxide can be removed from surfaces with inhibited HCl.

The equations listed below represent the basic solvent reactions of several pure minerals with HCl.

$$CaCO_3 + 2HCl = CaCl_2 + CO_2 + H_2O$$ (Equation [4.1])

$$Fe_3O_4 + 8HCl = FeCl_3 + 2FeCl_2 + 4H_2O$$ (Equation [4.2])

$$CuO + 2HCl = CuCl_2 + H_2O$$ (Equation [4.3])

$$FeS + 2HCl = FeCl_2 + H_2S$$ (Equation [4.4])

Sulfate and phosphate scales may be converted to the more soluble chloride salts by the action of HCl if they are very thin, but a two-stage process of soda ash (Na_2CO_3) followed by HCl or chelating agents is needed for dense scales. See Section 4.4.

When an iron oxide (magnetite in this example) is dissolved from a steel surface with HCl, Equation (4.2), which is not balanced, becomes:

$$Fe_3O_4 + Fe + 8HCl = 4FeCl_2 + 4H_2O$$ (Equation [4.5])

Note that the Fe(III) iron was reduced to Fe(II), and additional iron is removed during a reductive dissolution process. This occurs as iron oxide is dissolved from a carbon steel surface and must be anticipated during the cleaning process to obtain an accurate estimation of the scale to be dissolved. More information on the importance of reductive dissolution processes is in Section 4.3.

Samuel[7] described the use of HCl and other solvents for dissolving various deposits. The author concludes that HCl is effective for removing carbonate, phosphate, copper oxide, silicate, and iron oxide deposits. Trabanelli et al.[8] used a dynamic loop to compare cleaning effectiveness and corrosion produced by inhibited 7.5% HCl at 167°F (75°C). Inhibited 1.5% formic acid:1.5% hydroxyacetic acid-formic acid mixtures (HyFor) were also used at 194°F (90°C). More details of organic acids are in Section 4.1.2.

On the basis of those tests, the HCl solvent was much more aggressive as a solvent than HyFor, especially for magnetite that was not electrically coupled to steel. Electrochemical tests performed during the acid cleaning indicated that initially the HCl solvent was highly corrosive. However, by the end of the cleaning test, the solvent was not substantially more corrosive than clean, inhibited acid. A likely cause of the initial corrosion could be the release of ferric ions from the scale on the tubes being cleaned.

Ruiz and Frost[9] reviewed several solvents for cleaning boilers. According to these authors, the advantages of HCl include low cost, ability to dissolve various deposits, and readily available CIs. However, HCl has limited ability to dissolve copper (without thiourea) and has limitations for cleaning Series 300 stainless steel. HCl cleanings also result in high water usage owing to requirements for many rinses. Rusch[10] made a similar review of solvents for use in cleaning paper mill equipment. The same considerations apply for the use of HCl in paper mills as in boilers and other equipment.

Hydrofluoric acid. Another important mineral acid is HF, which is used to remove silica-containing deposits. In the United States, HF usually is generated by adding ammonium difluoride (NH_4F_2H or NH_4F • HF-ABF) to HCl:

$$NH_4F_2 + HCl = NH_4Cl + 2HF$$ (Equation [4.6])

The generated HF reacts with silica to form a more soluble fluorosilicic acid:

$$SiO_2 + 6HCl = H_2SiF_6 + 2H_2O$$ (Equation [4.7])

This can involve very complex chemistries if the pH rises or if clays (aluminosilicates) are present. See the discussions of mud acid (HCl/HF) solutions in Frenier and Ziauddin.[11] As long as the pH is less than about 3, the reaction products will stay dissolved. Problems also can occur if the deposits contain calcium or magnesium salts that could precipitate CaF_2 or MgF_2.

Another claimed advantage of the use of HF or HF-containing solutions is the partial control of ferric iron corrosion. Fluoride ions form very stable complexes with ferric ions:

$$\begin{aligned} &\text{A. } 3Fe^{-} + Fe^{3+} = FeF_3, \log K = 11.5 \\ &\text{B. } 6Fe^{-} + Fe^{3+} = FeF_6^{3-}, \log K = 16.7 \end{aligned} \quad \text{(Equation [4.8])}$$

There are additional discussions of methods to control ferric corrosion in Section 6.4.2.

Bradley and Smith[12] reported that corrosion rates of exposed steel were much lower when mixtures of HF and HCl were used to dissolve magnetite, compared with HCl alone. In both cases, an organic CI was used to control the attack of the acid on the steel. Lower corrosion rates also were reported for HF solutions compared with similar concentrations of HCl. Similar results were noted by Alfandry et al.[13] Poff[14] also described the benefits of adding ammonium difluoride (ABF) to acidic solvents. This compound decreased the corrosion rates of the acidic solvents.

HF also has been added to organic acid solvents to catalyze the dissolution of magnetite from some power boiler surfaces. Spillner[15] claims that fluorides increased the rate of iron oxide dissolution in several solvents. See Section 4.1.2. Bieller and Brochardt[16] described the advantages of HF as a boiler solvent, including lower corrosion rates and increased cleaning rates compared with HCl solvents. Since F^- ions form strong complexes with Fe^{3+} species in solution, the complexation may explain part of the inhibition mechanism.

However, when testing HCl/HF solvents in the lab (see Section 6.6) more corrosion inhibitor may be required to obtain satisfactory corrosion weight loss values than the same solvent containing only HCl.

Phosphoric acid. Phosphoric acid solutions (inhibited) have been used for many years to clean surfaces. Rusch[17] patented a formulation containing phosphoric acid and a CI. Cowan and Weintritt[18] claim that use of this particular acid is rather limited, since problems caused by the reprecipitation of calcium phosphates can occur in waters containing high levels of calcium. Phosphoric acid is sold as an 85% by weight solution. It is sometimes compounded with water-miscible solvents and wetting agents and sold as a ready-to-use dilute cleaning solution.

After dissolving the iron oxide with phosphoric acid (Cowan and Weintritt[18]), sodium hydroxide and an oxidizer (sodium nitrite) may be added to passivate the surfaces. Mixtures containing HCl, phosphoric acid, citric and oxalic acid, and an amine (such as hexamethylenetetramine) were reported by Garcia[19, 20] to be a noncorrosive solvent for cleaning surfaces. Garcia also claimed that the buffered acid solvents were not harmful to skin or other low oxidation state materials and proposed that the conjugate bases from the organic acids control the amount of hydronium ions (H_3O^+) available to attack the substrates. Lau and Stadig[19] described similar technology. This study reports the use of highly buffered acids for cleaning and passivation of heat exchangers and claims that additional inhibitors are not needed during the acid treatment, which lasted for 2–3 hours at 140°F (60°C). The formation of insoluble iron oxalates and phosphates may cause some of the inhibition described. The exchangers were flushed with water and not given an additional passivation treatment.

Sulfuric acid. Sulfuric acid diluted to 10%–20% can be used to clean series 300 stainless steel as well as other metals (excluding aluminum galvanized steel or magnesium). This acid cannot be used to remove deposits that contain calcium or barium owing to the low solubility of $CaSO_4$ and $BaSO_4$.

Dilute sulfuric acid (10%) can be used with several aldehydes (Frenier et al.[21] and Frenier[22]) to dissolve FeS without evolving hydrogen sulfide (see Section 7.6) and has been employed in IC processes where HCl cannot be used. Note that CIs for sulfuric acid (and organic acids) require chemistries that are different from those needed for inhibiting HCl solutions. See Sections 6.2 and 6.3.

Sulfamic acid. This a much weaker acid compared with the other mineral acids. Its principal advantage is that it is easier and safer to handle and can be used by plant personnel for small cleaning jobs. Cowan and Weintritt[18] report that sulfamic acid is used to remove rust, algae, and hard water scales from air conditioning systems, cooling towers, evaporative condensers, seawater evaporators (distillation units), and heat exchangers. Sulfamic acid is also used to clean items such as copper-bottom kettles, steam irons, dishwashers, silverware, toilet bowls, bricks and tiles, and processing equipment in the food and dairy, chemical and petroleum, paper and pulp, sugar refining, and brewing industries.

Nitric acid. Nitric acid can be used to clean stainless steel (frequently with HF; see ASTM[23]). Carbon steel corrosion is very difficult to inhibit in this acid because it is an oxidizing acid even at low (5%) concentrations.

Examples of Applications of Mineral Acids. Tanis[24] applied the HCl acid technology described above to boiler cleaning. See Figure 4.2, which shows the layout of the unit and some ancillary equipment needed for cleaning. The suggested steps include

1) Preparation: Solvent selection (acids or chelating agents) as well as the physical layout of the unit
2) Temperature adjustment (to bring the boiler temperature to <170°F (<77°C) for HCl
3) Acid soak (usually lasts 6–8 hours): The boiler is drained under nitrogen and then filled with a hot (150°F [66°C]) rinse of condensate grade water
4) After boil: A solution of soda ash and sodium phosphate is heated to >212°F (>100°C) to neutralize any residual acid. Many boiler-cleaning sequences also include a passivation stage using one of the solutions described in Section 4.2.4.

Tanis[24] also included sections on inspections and field analytical methods for following the course of the job. Several analytical tests were used to determine the concentration of iron as well as free (unspent) acid. This type of setup could be used with many solvents that are circulated through a boiler.

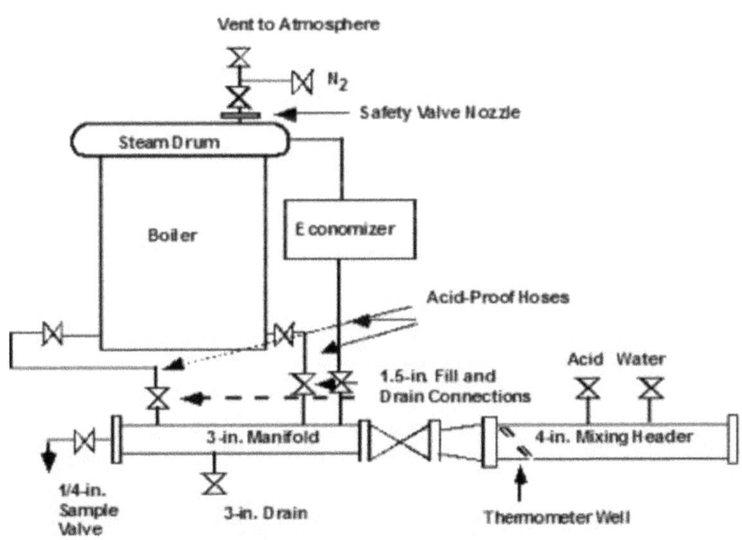

Figure 4.2 Boiler Cleaning Setup

4.1.2. Organic Acids

Deposits that foul industrial equipment, such as boilers, steam generators, and heat exchangers, consist of oxides of iron, copper, nickel, and various metal salts such as calcium carbonate and calcium sulfate. Mineral acids such as hydrochloric acid (HCl) can dissolve these deposits. However, it is not always desirable to treat the fouled metal with a strong mineral acid because of the danger of damage to the equipment during or after cleaning or if this solvent is difficult to circulate in a unit.

All the chelating agents described in the next subsection (4.1.3) also are organic acids. This section includes a review of the use of organic acid solvents that are not primarily chelating agents and solvents where the chelating agent component is a secondary constituent.

Four organic acids, the primary chemicals of this class, have been used in various IC applications. The structures of these acids are shown in Figure 4.1, and the properties are presented in Table 4.1:

- Formic acid (FA)
- Hydroxyacetic acid (glycolic acid- HA)
- Oxalic acid
- Citric acid (a chelate acid described in Section 4.1.3)

The basic reactions of formic (HCOOH) and hydroxyacetic acid (HO - CH_2COOH) with magnetite and, assuming reductive dissolution through contact, with an iron surface are

$$Fe_3O_4 + Fe + 8[HCOOH] = 4[Fe^{2+}(HCOO^-)_2] + 4H_2O \qquad \text{(Equation [4.9])}$$

$$Fe_3O_4 + Fe + 8[HO - CH_2COOH] = 4[Fe^{2+}(HO - CH_2COO^-)_2] + 4H_2O \qquad \text{(Equation [4.10])}$$

The reactions of magnetite or hematite with oxalic acid are more complex and depend on pH. This compound is a dibasic acid (Figure 4.1) that has applications in the nuclear cleaning industries (Panias et al.[25]). See Section 5.2 for more details.

Eberhard and Rosene[26] reported the use of solvents consisting of formic acid or citric acid for cleaning non-drainable tubes in superheaters and reheaters. The solvent (5% or 10%) also included a CI. Reich[27] developed a formulation for removing iron oxide from steel surfaces that consisted of a mixture of citric acid and formic acid. Reich claimed that the mixture of the two acids would hold more iron in solution than either of the acids alone. Citric acid used alone precipitated iron after several hours of contact.

McLaughlin[28] investigated a mixture of formic acid with hydroxyacetic acid (HA) and formic acid and a mixture of citric acid with hydroxyacetic acid. He found that HA with formic acid at a 2:1 ratio held more iron in solution than other mixtures. This work yielded the "HyFor" mixtures that have been the standard solvents to clean once-through (OT) and supercritical boilers (see the design in Figure 1.3). Sullivan and Horvath[29] described the testing and cleaning of different OT boilers using 6% HyFor (1 formic:2 hydroxyacetic acid ratio). They note that the deposit that occurs in these units is predominately magnetite (Fe_3O_4). However, operators of OT generating units that have converted over to oxygen treatments are seeing a change in the characteristics of deposits. This deposit is a mix of magnetite and hematite (Fe_2O_3) and is more tenacious and dense.[30]

By using scale dissolution tests like those described in Section 2.5.4, Sullivan and Horvath[29] found that HyFor alone would not dissolve all the deposit from the tube sections. They found that adding up to 1% ABF was required to dissolve all the oxygen-treated deposits at about 92°C. The suggested mechanism is that the F^- ions complex with the Fe^{3+} ions in the scale to produce a more stable FeF_3 ion. They discussed actual cleaning jobs using 3% HyFor/ABF that did not completely clean the unit. A later cleaning job used 6%

HyFor/0.5% ABF/CI for 24 hours of contact time at 200°F (93°C); the post cleaning tube analysis showed that all tube samples had less than 1 g/ft² (<1.07 g/cm²) remaining deposit.

Frenier and Larson[31] proposed a solvent that is predominantly formic acid, with iron stabilizers and CIs. This mixture was developed for cleaning OT boilers. Frenier[32] patented an aqueous cleaning solution containing formic acid and at least one carboxylic acid having at least two carbon atoms wherein the weight ratio of formic acid to higher carboxylic acid is greater than about 4:1. This solvent is contacted with the scale in the absence of an oxidizing agent. Preferred carboxylic acids are the mono-, di-, hydroxy-, and polyhydroxy-carboxylic acids with two to six carbon atoms. Citric acid is the most preferred chemical in this process.

Bomb cleaning tests are useful for preliminary screening. Dissolution and corrosion tests were conducted using a stirred pressure bomb (Figure 2.19) with a formic/citric (FC) solvent. In separate tests, the scale dissolution rate (using an OT boiler tube section) was estimated to be about 6 hours. See Figure 4.3. The corrosion rate was then calculated for the remaining time of the test (30 hours; see Section a) rate). The cleaned tube section was then (test 2) exposed to fresh solvent for an additional 30 hours to determine the corrosion rate for the uncontaminated solvent (Section b). Acceptable corrosion rates were found for both tests.

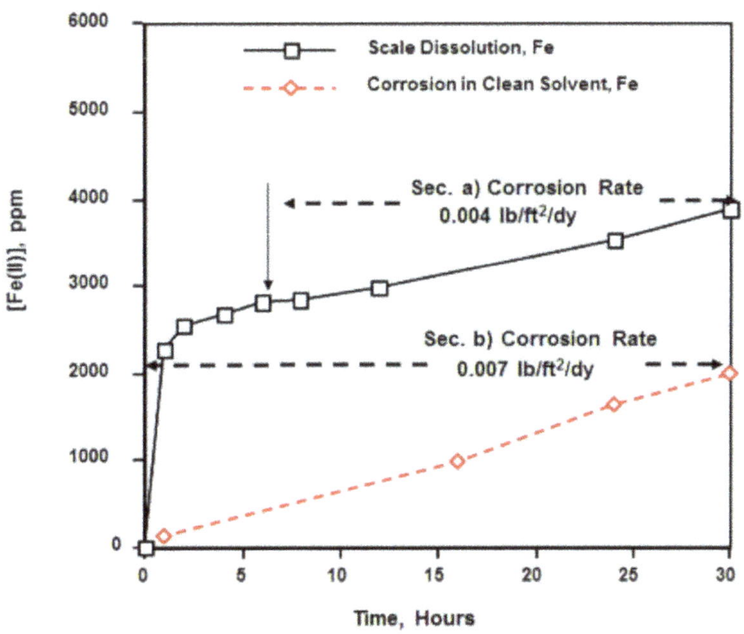

Figure 4.3 Pressure Bomb Iron Curves for F/C Solution, 0.25% Inh. at 200°F. OT Tube Cleaning a) and Fresh Solvent Corrosion Curves (b).

A dynamic loop (Figure 2.21) was constructed to compare the solvent (FC) with HyFor. The loop uses a 10-ft section of boiler tubing as the heat exchanger and matches the surface/volume ratio and flow rates observed during cleaning of the OT type of boiler. Oxide dissolution curves developed using the loop demonstrate that the FC solvent (Figure 4.4) cleaned the tubes almost as fast at 150°F (66°C) as HyFor when used at 200°F (93°C). Note that the pH increased as the iron dissolved and the solvent was spent.

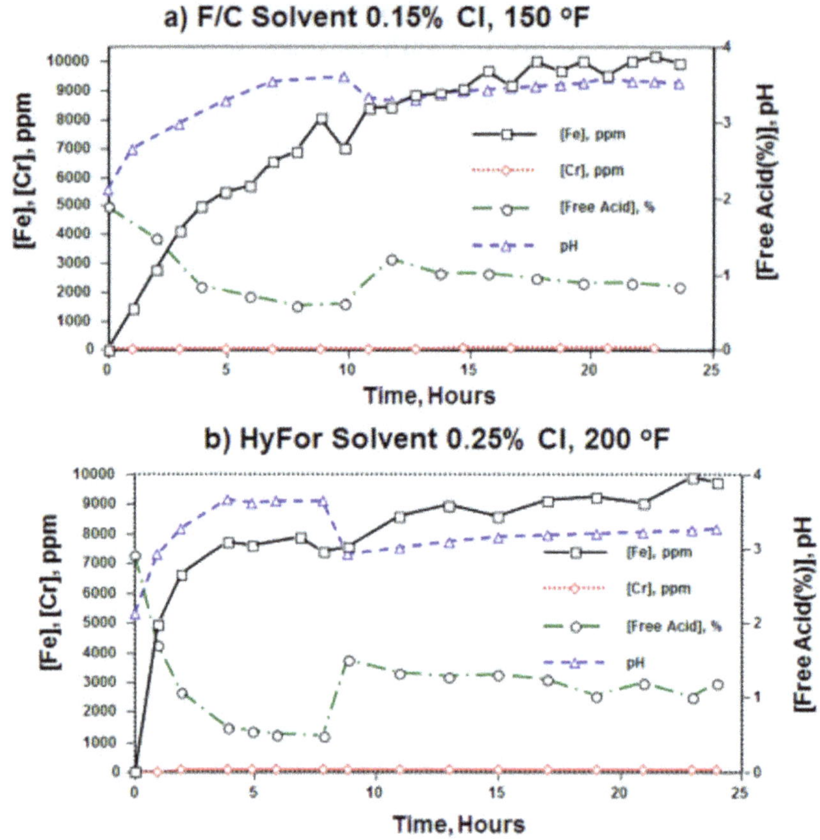

Figure 4.4 Loop Cleaning of OT Tube. a) F/C Solvent, 0.15% Corrosion Inhibitor at 150°F. b) HyFor, 0.25% Corrosion Inhibitor at 200°F

The kinetic profile for the operational cleaning of a 25,000-gal OT boiler using this solvent (formic/citric solution) is shown in Figure 4.5. The unit was cleaned in less than 24 hours despite scale loading approaching 80 g/ft^2.

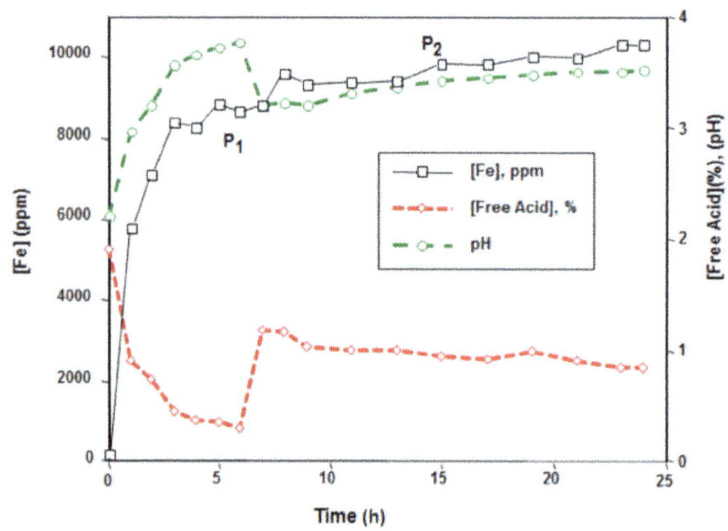

Figure 4.5 Cleaning Profile of Formic/Citric Acid Fluid at 200°F (93°C)

Poulos[33] described a study that compared ammonium citrate with several organic acid solvents. In this test, loose magnetite was dissolved in the solvents. Monoammonium citrate or EDTA solvents were much more effective than the organic acids (including HyFor). Because reductive dissolution mechanisms were not active, his conclusions are valid. However, when magnetite is deposited on carbon or alloy steel, reductive dissolution yields ferrous ions, which are very soluble in organic acids. Frenier and Larson[31] noted that loose magnetite dissolved much slower in the formic/citrate solvent than the magnetite on the tube surfaces.

A mixture of citric acid and gluconic acid (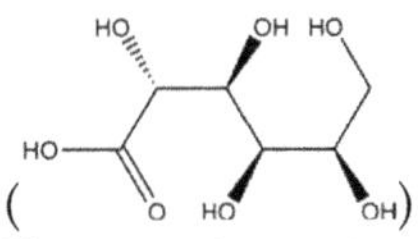) that contains ammonia and a corrosion inhibitor was investigated by Veysset and Saurin[34] as a solvent for removing sludge (iron oxide and copper) from PWR steam generators. The oxide dissolution is conducted at a pH of 3-3.5 and at 176°F (80°C). The copper removal stage is run at a pH of 9.5 and at 122°F (50°C). Hydrogen peroxide is added to improve copper dissolution. This process is claimed to be as effective as the EPRI-SGOG process that used EDTA. More details of these types of solvents are in Section 5.3. Mechanistic details of iron oxide and copper removal are discussed in subsequent chapters.

Additional organic acid formulations also have been proposed. Horner[35] described the use of citric acid, tartaric acid, and gluconic acids in abrasive and electro-polishing of metals. Gluconic acid will complex and prevent the precipitation of iron during cleaning operations. Cooling water systems were cleaned using this process. Oxalic acid has been proposed and used for removing sludge from nuclear waste tanks[36] and as part of nuclear steam generator decontamination solutions.[37] More details about these types of solvents are in Section 5.3.

4.1.3. Chelating Agents

An alternative solvent family consists of aqueous solutions of chelating agents with pH values of 2 to 12. A chelating agent is defined as a chemical that forms an equilibrium complex between the metal ion and the complexing agent. Chelating agents are characterized by the formation of more than one bond between the metal and the molecule of the complexing agent, resulting in the formation of a ring structure incorporating the metal ion (see Figure 4.6 for the sodium salt of calcium hydroxyethyl ethylenediamine triacetic acid [HEDTA]).

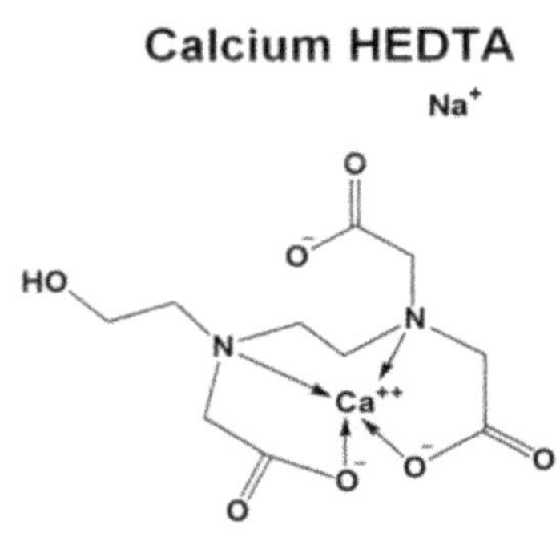

Figure 4.6 Chelate Structure CaHEDTA

This is contrasted with the reaction of a complexing agent such as ammonia or thiourea with a metal ion such as Cu^{2+} or Cu^{+}, in which the complexing agent coordinates with only one reactive site on the metal ion. The formation of rings by the multidentate (many reactive sites) ligand gives added stability to the complex.

$$M + nL = ML_n \quad \text{(Equation [4.11])}$$

The equilibrium for the reaction does not go as far to the right for monodentate ligands as it does when a chelating agent is present. This is known as the chelate effect. The equilibrium constants for these reactions, $K = [ML_n]/[M][L]^n$, define the strength of the chelate complex. For example, the equilibrium constant for Cu^{2+} reacting with four molecules of ammonia (NH_3) is 10^{13} (log K = 13), whereas for the complex of Cu^{2+} with the chelating agent ethylenediaminetetraacetic acid (EDTA), log K is 19 (see Table 4.1 and Figure 4.1).

Formulations containing chelating agents and organic acids have been used to clean fouling deposits from metal for more than 30 years. This section reviews the published literature on the use of chelating agents in chemical cleaning solvents, including formulations containing citric acid, EDTA, hydroxyethylethylenediaminetriacetic acid (HEDTA; see Figure 4.1 for structures), and organic phosphonates. Some details of mechanisms are in Sections 4.2.3 and 4.3. Note also that although chelating agent solvents (with 10%–25% active ingredients) are generally *less* corrosive than mineral acids, they are usually still corrosive enough to damage the base metal, and the additional iron dissolved would consume the solvent. Thus, CIs (with different chemical structures; see Section 6.3) are required.

Citric acid. In 1958, Loucks et al.[38] described the use of citric acid as a cleaning agent. Sodium citrate at a pH of 4.0 was used to clean a service water system that was heavily scaled with iron oxide.[39] The authors described the dissolution of iron oxide with either ammonium or sodium citrate. A precipitate of ferrous citrate was noted at a low pH (3.5), but this precipitate dissolved at higher pH values. A 65,000-gal (245,700-L) cooling water system was cleaned with 10% sodium citrate.

Citric acid was one of the first organic acids to be used in solvents to remove iron oxides from process equipment. Although inhibited citric acid was used to clean various power generating units in the 1950s, a gray precipitate also was formed. This problem is described in more detail because it is a central issue when citric acid is used in cleaning formulations. Morris[40] described inconsistent laboratory experiments in which precipitates formed with iron concentrations >1% and pH values from 4.0 to 4.85. However, he noted that many operational units had been cleaned without any evidence of a precipitate.

The author speculated that the precipitate was a ferrous (Fe[II]) salt. Hamm et al.[41] identified three Fe(II) complexes in aqueous solutions of citric acid and ferrous iron: FeHCit, $FeCit^-$, and $FeOHCit^2$. Because FeHCit is the only identified uncharged species, the precipitate probably contained this compound. The aqueous chemistry of ferric/citric acid is more complex than the ferrous chemistry. The known Fe(III) complexes include $FeHCit^+$, FeCit, $FeOHCit^-$, and $Fe(OH)_2Cit^{2-}$. Bobtelsky and Jordan[42] identified additional iron complexes (such as Fe_2Cit_3) in solutions containing sodium citrate and Fe(III). These authors also offer evidence that hydrated species may be present. One composition that solved the iron precipitate problem was patented by Reich.[27] He made mixtures of citric acid with formic acid that did not produce precipitates during removal of mill scale from steel pipes.

A process for cleaning iron oxide and copper from utility boilers was described by Alfano.[43] This method uses monoammonium citrate at pH values between 3 and 5 to remove the iron oxide. The ammonium citrate solved the precipitation problems encountered with the use of citric acid alone. The pH is then raised to 9 with ammonia or an amine, and an oxidizer is introduced. The oxidizer dissolves the copper. The iron oxide removal stage is conducted at 150°F (65°C) to 212°F (100°C), and the copper removal stage is conducted at 150°F (65°C). The oxidizers include ammonium persulfate, potassium perchlorate, potassium bromate, sodium nitrite, potassium permanganate, and nitro phenyl sulfonic acid. More details of copper removal processes are in Sections 4.2.1, 4.2.3, and 4.2.4.

Bell[44] also described a two-stage process for removing iron oxide and copper. The iron oxide stage is conducted at a neutral pH (7), and the copper removal stage is conducted at a pH of 9–10 with an oxidizer.

A large boiler was cleaned[45] with the two-stage Alfano process using 7500 lbms (3375 kg) of citric acid, 3000 gal (11250 L) of ammonium hydroxide, and an inhibitor. A total of 1560 lbms (702 kg) of iron and 30 lbms (13.5 kg) of copper were removed. The use of the corrosion inhibitor is significant because neither the Alfano nor Bell patents mention the use of an inhibitor. In fact, one advantage of the neutral-pH Bell process is claimed to be lower corrosion rates than found with low-pH ammonium citrate. Several boiler tubes were cleaned in a lab loop.[46] Ammonium citrate was included. The major result was that ammonium citrate required a velocity of at least 1 ft./sec (0.3 m/sec) to adequately clean that boiler tube.

Alfano[43] and Bell[44] described the mechanism for suppressing the precipitation of citrate salts, including the formation of iron ammonium citrate complexes such as $FeNH_4Cit$ and $Fe(NH_4)_2OHCit$. In addition, Bell[47] proposed that the formation of ferric ammonium citrate ($FeNH_4Cit^+$) takes place during the Citro-Solv (an accepted name for the process) oxidation stage. Blume[48] also proposed that the addition of ammonia and the formation of ammonium ferrous citrate suppressed the formation of the insoluble ferrous acid citrate (FeHCit). An extensive review by Blair et al.[49] (with 112 references) reiterated the use of citric acid and endorsed the use of the ammonium salts. Arrington and Bradley analyzed the precipitate formed in citric acid solutions and stated that it consisted of $FeHCit \bullet H_2O$ and that its formation was pH dependent. This observation led them to propose the use of sodium citrate (monosodium and disodium salts) as solvents for heavily scaled service water piping. A patent was also issued to Arrington and Bradley[50] for the use of sodium citrate solutions to remove iron oxides from systems that have both iron-base and copper-base metals in the water circuit being cleaned.

Further investigations by Frenier and Kennedy[51] corroborate Arrington's assertion that the complex that precipitates is the ferrous acid citrate species. The species that exist in solution as a function pH are shown in Figure 4.7. An insoluble FeHCit forms in the pH range of about 3.0 to 4.0. Raising the pH to >4 with either ammonia or sodium hydroxide eliminates this precipitate. Citric acid and especially ammonium citrate are still used in some IC applications. Daniels[52] notes that as of 2014, ammoniated citric acid is considered to be an excellent solvent and is often the solvent of choice for pre-commissioning cleanings, where deposits are anticipated to be light and composed exclusively of iron. It can also be used in a high-temperature (higher pH) and low-temperature (lower pH) scenarios, like EDTA. Chemical cleaning wastes containing citric acid are often easier to dispose of than EDTA-containing wastes. See Chapter 9.

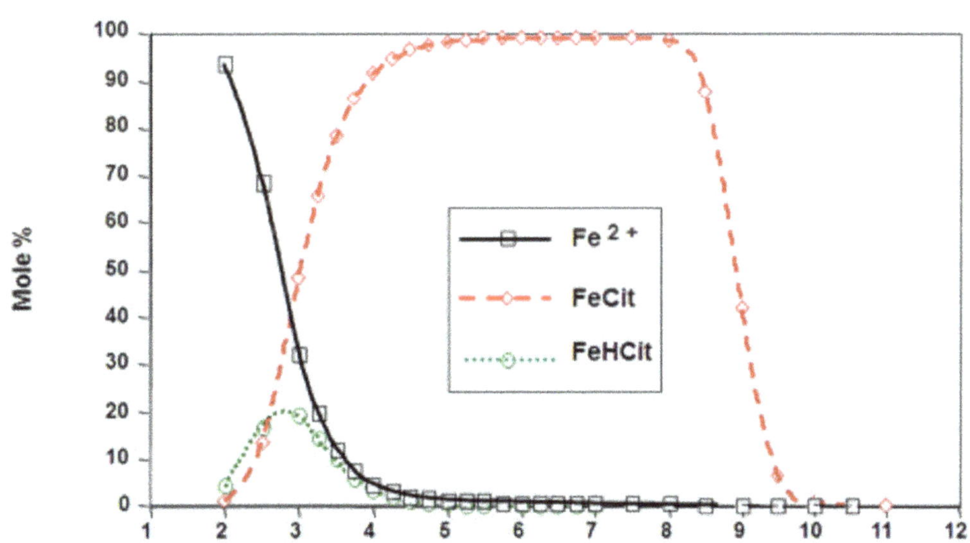

Figure 4.7 Ferrous/Citric Acid Speciation (10% Citric Acid Solution, Equimolar Iron)

EDTA and polyaminopolycarboxylic acids (PAPCA). This section introduces the use of PAPCA chemicals. EDTA, the most used chemical in this family, is a very versatile chemical that forms metal ion complexes that have higher equilibrium constants than citric acid (Table 4.1). Chemical cleaning solvents with pH values from 4.5 to 9.2 have been formulated, and they can remove iron and copper-containing deposits (as well as salts of calcium, nickel, and chromium).

Ammonium salts of PAPCAs, and especially EDTA, have been used to dissolve iron oxides from steel at a pH of 9.0 and at temperatures up to 300°F (150°C) and leave the steel in a passive condition (resists rusting). A high pH (~9.2) iron oxide and passivating solvent was described by Lesinski.[53,54] The copper in the scale is not removed by this process but can be dissolved by contacting the boiler surfaces with ferric EDTA produced by oxidizing the ferrous EDTA generated during the iron oxide dissolution. This process was patented by Harriman et al.[55] This copper removal stage is usually conducted at temperatures of 150–180°F (65–82°C). The ferric ion is generated by introducing an oxidizing agent such as air, sodium nitrite, hydrogen peroxide,[56] or gaseous oxygen.[57] The commercial application of ammonium EDTA for dissolving iron oxide and copper from utility and industrial boilers was described by Slicker.[58] The mechanism of dissolution of magnetite in alkaline chelating agents was described by Frenier[59] and Frenier and Growcock.[60] The iron oxide dissolves by a reductive dissolution mechanism:

$$Fe_2O_3 + Fe + 8NH_4^+ + 4chel = 4[Fe - chel]^{2+} + 8NH_3 + 4H_2O \qquad \text{(Equation [4.12])}$$

The same mechanism is active when ammonium citrate is the major solvent. More details on the iron oxide dissolution mechanisms are in Section 4.3, and other copper-removing processes are in Section 4.2.

Note that the rate of the reaction depends roughly on the strength of the iron chelate. These determinations were based on tests (Frenier[59]) run on boiler tubes at a pH of 9.2 and at 325°F (163°C). EDTA, diethylenetriaminepentaacetic acid (DTPA), and HEDTA all form strong chelate complexes with Fe(II) and are much more aggressive solvents than nitrilotriacetic acid (NTA) and citric acid, which form much weaker complexes with iron. See Figure 4.8.

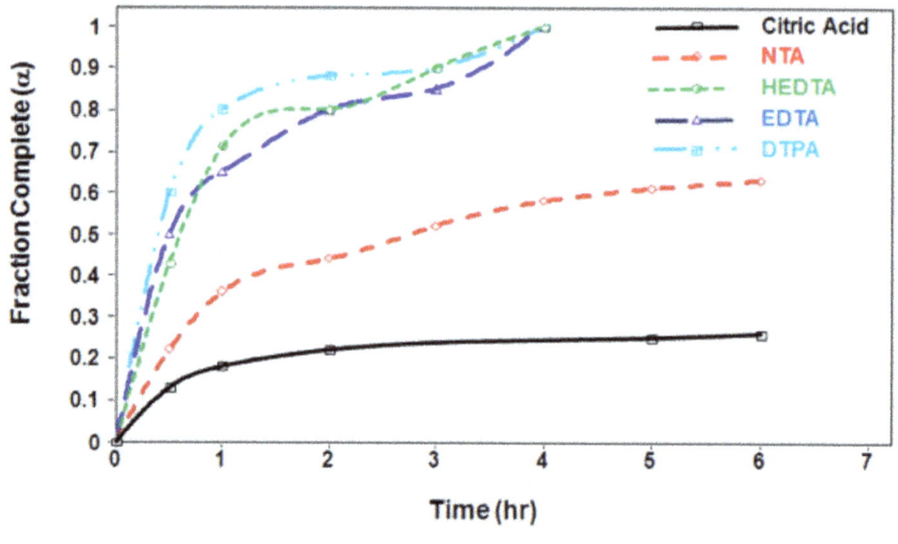

Figure 4.8 Iron Oxide Dissolution Tests in Ammonium Chelates at a pH of 9.2 (Tests were Run at 320°F [160°C])

This study also showed that the magnetite must be electrically coupled to the steel for the kinetics to be favorable. The reductive dissolution of the base iron is driving the reaction. In addition, sodium chelates are not effective at a high pH because these salts do not have protons to satisfy Equation (3.10). However, sodium EDTA also can be used to dissolve iron oxides at a pH of 4–5.[61]

Vinso[62] used sodium EDTA at a pH of 8–11 to remove hardness, then lowered the pH to 4–5 with acid, and used the same solution to dissolve iron oxides. More details about these dissolution mechanisms are contained in Sections 4.2 and 4.3.

If the boiler scale contains copper, a second copper removal stage must be conducted using ferric EDTA as the oxidizing agent (also see Section 4.2 for more information on alternative copper removal processes). Before the copper can be dissolved, the ferric EDTA must passivate the steel so that the copper will not replate.

Frenier and Kennedy[63] described the conditions under which passivation is most readily achieved. These include high flow rates, high concentrations of ferric iron, and low temperatures. High concentrations of free chelant slows passivation. The study recommends using strong oxidizers such as air plus $NaNO_2$ or H_2O_2 to generate the ferric EDTA. The dissolution of copper itself was studied.[51,63] For EDTA, the ferric chelate dissolves the copper, whereas in citric acid, the external oxidizer (used to generate the ferric chelate) dissolves the copper. Ferric EDTA is much more effective as a copper solvent than ammonium citrate with external oxidizers. This work also recommends using strong oxidizers, such as air plus $NaNO_2$ or H_2O (later work[57] also recommends O_2), to generate the ferric chelate. More details of the Fe/Cu solutions are in Section 4.2.3.

One advantage of the pH 9 tetraammonium EDTA (TAEDTA) solvent is that it can be employed to clean natural circulation boilers using thermal cycling to achieve flow in the tubes. Typically, at the start of a cycle, the steam drum temperature is about 260°F (127°C). By means of an igniter or some other auxiliary heat input, the temperature is raised to 300°F (150°C). By successive cooling and heating, the solvent flows, and fresh chelant is brought in contact with the scale. Shorthouse[64] calculated that the flow in the tubes may vary from 0.19 to 1.3 ft./sec (0.057 to 0.39 m/sec) and recommends using more but smaller sources of thermal input, positioned as low as possible within the furnace to promote circulation.

Several EDTA-based formulations were developed to dissolve magnetite (Fe_3O_4) found in nuclear steam generators. Although parts of a steam generator may be carbon steel, some of the scale also has been deposited on alloy steels (containing various amounts of nickel and chromium) that are resistant to reductive dissolution and requires dissolution of the steel substrate.

Pitzer[65] described a solvent that contains ammonium EDTA at a pH of 6.5 to 9 containing from 0.1 to 0.5 molar concentration of hydrazine (N_2H_4). Magnetite is dissolved at 203°F (95°C) in the absence of a metal surface. The reductive dissolution apparently is provided by the hydrazine. Pocock and Leedy[66] also developed a formulation of ammonia, EDTA, and hydrazine containing a corrosion inhibitor to remove magnetite from nuclear steam generators. The solvent was used at a pH of 7 and at 200°F (93°C). This mixture was much more effective than other solvents such as thioglycolic acid, citric/formic acid, ammoniated EDTA, or EDTA/citric acid. Additional testing of solvents for cleaning nuclear steam generators was conducted by Welty et al.,[67] who tested 50 solvents for removing sludge (mostly magnetite, copper, and copper oxide) from the secondary side of steam generators. The solvent chosen for removing magnetite was pH 7 EDTA with hydrazine and an inhibitor. The copper solvent was 5% EDTA adjusted to a pH of 7 with ammonia, then to 10 with ethylenediamine and 2% hydrogen peroxide, as the oxidizer to dissolve the copper. Additional details of nuclear plant solvents are in Chapter 5.

The EDTA formulations also can be used at temperatures from 100°F to 200°F (38–93°C) if the pH value has been adjusted to 4–5. The cleaning rate of magnetite from the boiler tube for tetraammonium EDTA (TAEDTA) and diammonium EDTA (DAEDTA) at two different temperatures is compared in Figure 4.9. The DAEDTA was as effective as TAEDTA at 100°F and lower temperatures.

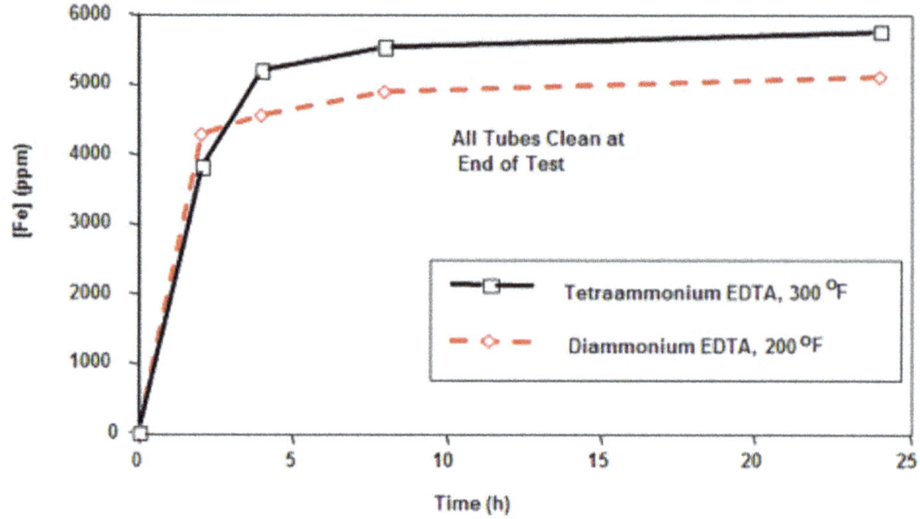

Figure 4.9 Cleaning of Boiler Tubes in Stirred Bombs (All Contain 0.25% Inhibitor)

The DAEDTA formulation has proven to be a very useful solvent since the low pH salt solution (available commercially as a ˜40% active solution) can be used to clean units at relatively low temperatures (<250°F). Then, when the iron is removed, the pH can be raised with ammonia and passivation and/or copper removal stages can be accomplished with only one fill of the unit. This solvent also has been used to perform preoperational cleaning of combined cycle electrical generating units[68] in which a mixture of a surfactant and DAEDTA provides a low-temperature cleaning operation for removing greases and mill scale.

A mixture of TAEDTA, buffered to pH 5 with citric acid, and a special corrosion inhibitor (Frenier and Settineri[69] and see Section 6.4.2) was used to clean iron oxides from a cooling water system online at a chemical plant. The details of this case history are in Section 4.6.2 as an example of best practices.[70]

Ammoniated HEDTA was tested as a solvent for iron oxides at a high pH,[71] and is about as effective as EDTA. However, HEDTA acid has much higher solubility in water than EDTA (6% at room temperature vs. 0.1% for EDTA). Therefore, low-pH chelating solvents with exceptional iron oxide dissolution abilities are possible.[72]

Frenier and Hoy[73] showed that formulations of HEDTA, HEDTA with formic acid, or HEDTA with H_2SO_4 clean iron oxides from boiler tubing much faster than other acidic solvents, such as hydroxyacetic/formic or diammonium EDTA. The increase in cleaning rate was attributed to the low pH (1.2–2.3) and the strong chelate bonds formed with Fe(III) even at a low pH. The removal rate of iron from a pipe scaled with red rust for two different solvents is compared in Figure 4.10. The low-pH solvent consisted of HEDTA/H_2SO_4, whereas the pH 5 solvent contained diammonium EDTA. Low-pH (pH 2) solvents based on sodium HEDTA and HCl are used in the oil and gas industry[11] for cleaning and well stimulation processes.

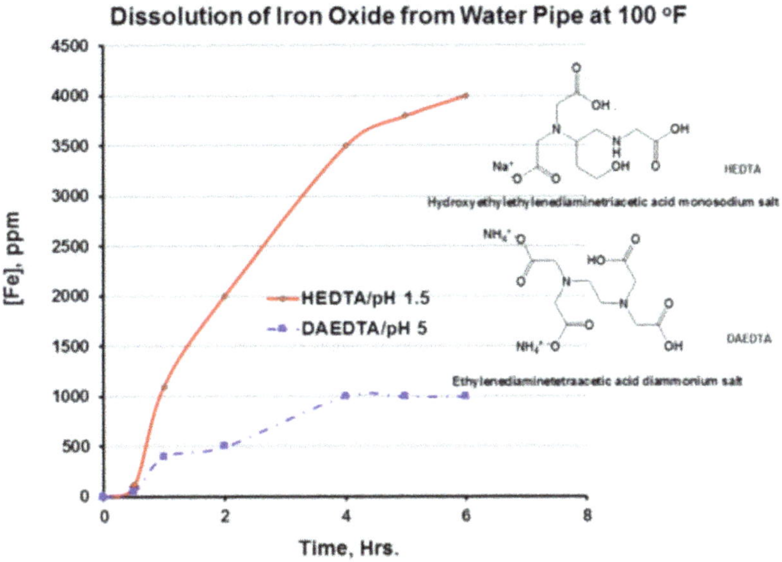

Figure 4.10 Cleaning of Water Piping at 100°F (38°C).

If the scale is mostly ferric oxide (FeOOH or red rust), a solvent[74] using HEDTA and the reducing agent erythorbic acid can be used to clean cooling water systems at temperatures at or below 150°F (65°C).

DTPA-containing solutions also could be used for chemical cleaning and have the highest stability constants of the alkylenepolyamine polyacetic acids (APAPAs) for most metals. Although the necessary high mass ratio (lb. chelate/lb. scale) makes this chemical too expensive for most uses, DTPA-based solvents have been proposed for dissolving barium sulfate scales. See Section 4.4.1.

Phosphonates. Phosphonates are analogues to carboxylic acids in which the carboxylate group is replaced with -PO_3H_2. They form chelate complexes that are like the amino carboxylic acids and were proposed as cleaning compositions. Frost et al.[75] described solutions for dissolving magnetite and copper that consist of 1-hydroxyethylidene-1,1-diphosphonic acid, N,N,N',N'-Ethylenediamine tetra-(methylenephosphonic acid), or nitrilotri-(methylenephosphonic acid) and are brought to a pH of 9 with ammonium hydroxide. The phosphonic acids are used at temperatures above 180°F (82°C) to dissolve magnetite from boiler tubes. See Figure 4.11 for several phosphonate structures.

The iron solutions formed from the dissolution of magnetite are used with oxidizing agents such as $NaNO_2$, $NaBrO_3$, or H_2O_2 to remove copper. Descaling formulations of aliphatic phosphonates and nitrates were described by Halsey.[76] The patent claims that a solution of aliphatic phosphonate and ammonium nitrate at a pH of 8.0 removed magnetite and copper oxide from a boiler tube at 208°F (98°C). Mixtures of aminopolycarboxylic acids and aminotrimethylenephosphonic acids (AMTP) are reported to be superior to either chelating agent used alone. Arrington and Bradley[77,78] tested mixtures of NTA and AMTP for dissolving magnetite from a boiler tube with added copper. The pH 9 formulations were heated to 275°F (135°C) to dissolve the iron oxide and then were cooled to 150°F (65°C) and oxidized with air and $NaNO_2$ to dissolve the copper. The data showed that the mixture removed the deposits better than NTA alone and did not produce any undesirable precipitates that may have been formed if the phosphonates were used alone.

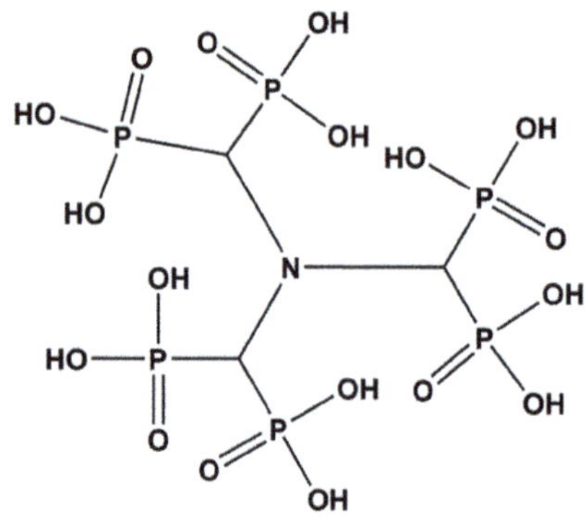

Nitrilotri-(methylenephosphonic acid) NTMP

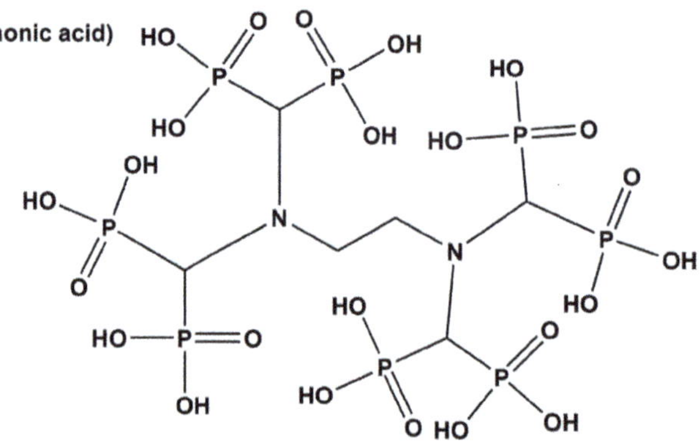

Ethylenediamine-N,N,N,N-tetrakis-(methylenephosphonic Acid) EDTMP

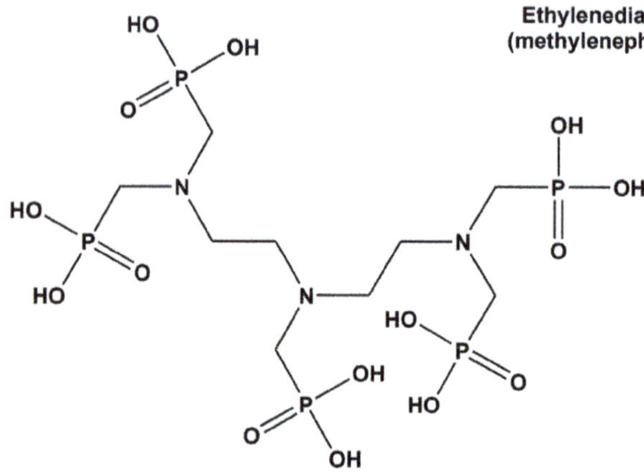

Diethylenetriaminepentamethylenephosphonic acid- DETPA

Figure 4.11 Phosphonate Chelating Agents

This section has identified the range of chemicals (mineral acids, organic acids, and chelating agents) that have been applied to remove iron oxides from iron surfaces. The actual choice of solvent will depend on the additional metal salts in the scale and the type (including metallurgies) and operating conditions of the unit to be cleaned. Detailed mechanistic details for both mineral acids and chelating agents are in Section 4.3.

For power-producing boilers or industrial units (and some nuclear plants), the fouling scale may contain copper. The copper may be plated on an iron surface or may exist as a copper oxide (see Section 3.4). The copper must be removed to prevent galvanic corrosive[79] attack on the iron-containing surfaces. The methods employed are described in Section 4.2.

4.2. Solvents for Removing Copper Deposits

Copper-containing deposits may be found in process equipment whenever the fluids meet copper-base alloys. Although most copper alloys are relatively resistant to corrosion, very small amounts of copper will dissolve and then may replate on contact with an iron-base alloy. Copper that is plated onto the steel will accelerate corrosion owing to the galvanic nature of the iron/copper couple. The copper acts as the cathode for the oxidizing agent, and the anodic iron corrodes. Many different techniques have been developed for removing copper deposits. Included here are discussions of alkaline oxidizers, HCl/complexers, and chelating agent solvents. Some of the removal methods require different stages of solvents, and others are one-stage (or one-fill) methods.

4.2.1. Copper Removal with Alkaline-Oxidizing Agents

Ammonia-oxidizing solutions are capable of dissolving copper from iron surfaces without corroding the surfaces. Nesbitt[80] noted in a review the use of alkaline inorganic oxidizers, including several ammoniacal solutions. Examples are the use of ammonium persulfate (1948) and ammonium bromate (1951). If the only fouling scale was copper oxides or plated copper, these types of solvents could be used as the only treatment. However, they also may be staged with HCl using a multi-solvent process that includes an acid for iron oxide dissolution, a rinse, and copper removal stages.

Alquist et al.[81] described several ammoniacal oxidizers for cleaning steam generating units. A bromate solvent was developed that consists of sodium bromate, ammonium hydroxide, and ammonium bicarbonate, and is used at 120–175°F (49–79°C). It is effective, but ammonium bromate can form a very explosive mixture under some circumstances. Hydrogen peroxide also causes rapid (explosive) reactions with concentrated organic materials, including EDTA.

The bromate reaction is shown in Equation (4.13), and the ammonium persulfate reaction is shown in Equation (4.14).

$$BrO_3^- + 3H_2O + 3Cu + 12NH_3 = Br^- + 6OH^- + 3[Cu(NH_3)_4^{2+}] \qquad \text{(Equation [4.13])}$$

$$S_2O_8^{2-} + Cu + 4NH_3 = 2SO_4^{2-} + [Cu(NH_3)_4^{2+}] \qquad \text{(Equation [4.14])}$$

An important improvement is the patented copper removal technology using ammonium carbonate and oxygen.[50] This process uses ammonium bicarbonate, aqua ammonia, and gaseous oxygen to dissolve complex copper and thus remove it from surfaces. The oxidation reaction is shown in Equation (4.15).

$$O_2 + 2H_2O + 2Cu + 8NH_3 = 4OH^- + 2[Cu(NH_3)_4^{2+}] \qquad \text{(Equation [4.15])}$$

Because the oxidizer can be continually replenished by introducing more oxygen, very large amounts of copper may be removed from boilers and other copper-fouled equipment. The spent solution also does not have to be reduced before waste treatment. This solvent is applied as a foam and is used to remove copper from turbine blades. See Section 8.4.

Another oxidizing agent described for use in removing copper is sodium perborate ($NaBO_3 \cdot 4H_2O$), which can be used with ammonia, amines, or EDTA.[77] The alkaline oxidizers containing ammonia passivate (see Section 4.2.4) the steel as the copper is dissolved. This passivation process prevents the copper from replating onto the steel. The redox reactions and their half-cell potentials are listed in Table 4.2.

Table 4.2 Redox Reactions[1]

Half-Cell Reaction	Standard Potential (V)
$Cu^{2+} + 2e^- = Cu$	0.34
$Cu^+ + e^- = Cu$	0.52
$2H^+ + 2e^- = H_2$	0.00
$BrO_3^- + 3H_2O + 6e^- = Br^- + 6OH^-$	0.61
$Fe^{3+} + e^- = Fe^{2+}$	0.77
$Fe(III)EDTA^- + e^- = Fe(II)EDTA^{2-}$	0.25
$Fe = Fe^{2+} + 2e^-$	0.45
$HO_2^- + H_2O + 2e^- = 3OH^-$	0.88
$O_2 + 2H_2O + 4e^- = 4OH^-$	0.40
$2NO_2^- + 3H_2O + 4e^- = N_2O + 2OH^-$	0.15
$NO_3^- + H_2O + 2e^- = NO_2^- + 2OH^-$	0.01

The alkaline oxidizers described above can remove copper deposits from power boilers but have to be staged with mineral acid (usually inhibited hydrochloric acid [HCl]) to remove iron oxide deposits that contain copper or copper oxides. In practice, this resulted in the use of acid/bromate stages or bromate/acid/bromate stages for very heavy copper-containing scales. Although these processes were effective, the overall time of the job was very long, and massive volumes of waste were generated. Additional research yielded the acid/thiourea cleaning process.[82]

4.2.2. HCl-Thiourea Process for Cleaning Boilers

This procedure (and a procedure using sulfuric acid and allylthiourea) allowed the cleaning industry to simultaneously remove iron oxides and copper deposits from utility power boilers. Solutions of 5%–10% HCl containing 0.5%–3% thiourea were used for many years. This dual process reduced the number of stages required to clean a boiler (and thus the amount of waste solvent). The major problem was that the necessary amounts of thiourea (per pound of copper) *decreases* as the concentration of copper increases. For boilers with a small amount of copper, a considerable excess concentration of thiourea is required to prevent precipitate formation, especially with localized deposits. It is important to note that the ferric ions generated during the dissolution of the magnetite oxidize the copper and allow the thiourea to complex it. As the amount of thiourea increases, the necessary acid strength also increases, and it may become more difficult to control corrosion to acceptable levels.

This process will not dissolve any plated copper that is not oxidized by the ferric ions released from the scale or that is in the scale as a copper oxide. See Equation (4.16) for the iron reaction.

$$Fe^{3+} + Cu + 4[CH_4N_2S] \rightarrow Fe^{2+} + [Cu(CH_4N_2S)_4^+] \quad \text{(Equation [4.16])}$$

In addition, McCoy[83] proposed that thiourea may reduce any 2+ oxidation state copper to the 1+ oxidation state (see Equation [3.15], which shows the formation of elemental sulfur and urea).

$$2\,Cu^{2+} + [CH_4N_2S] + H_2O \rightarrow 2Cu^+ + S^0 + CH_4N_2O \quad \text{(Equation [4.17])}$$

Improvements in the acid/thiourea process resulted from the discovery that mixing thiourea with other thiourea derivatives (hexahydropyrimidine-2-thione or N-[hydroxyethyl)-ethylthiourea) produced formulations[84,85] that were more resistant to precipitate formation than thiourea itself. These patents claim that the precipitate formed in the presence of the mixed complexer is more flocculent than the precipitate formed with thiourea and thus is easier to flush out of the boiler than other materials. See Figure 4.12 for some of the chemical structures.

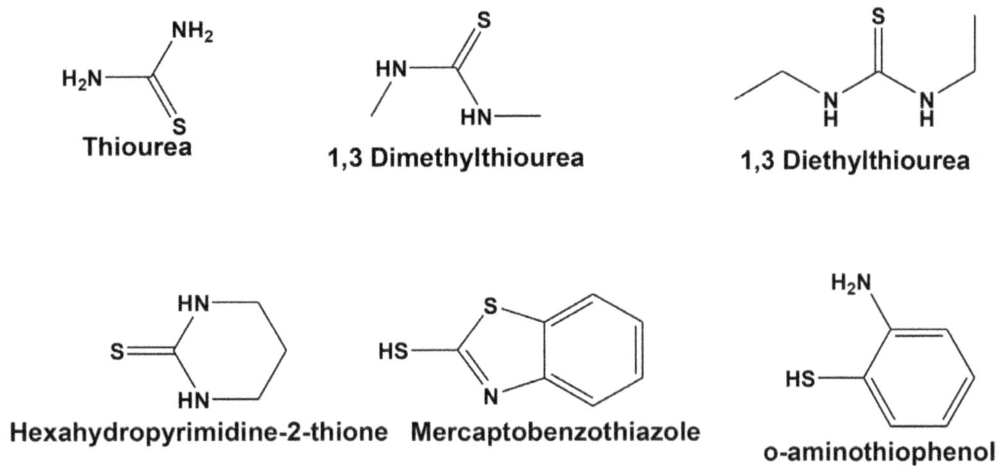

Figure 4.12 Thiourea and Derivatives

Frost[86] used polarography methods to determine the cumulative stability constants (β: four ligands would be represented as β_4) of substituted thiourea derivatives with Cu(I) in acid. Values for β_4 that were greater than 10^{15} were determined for thiourea, hexahydropyrimidine-2-thione, and several other complexes with Cu(I). These very high stability coefficients in acid explain the ability of these ligands to prevent copper from plating onto steel in HCl at acid concentrations >1M.

The discovery of thiourea and derivatives for Cu(I) complexes, and not Cu(II) complexes, was a major technical achievement. Significant (unpublished and unpublished) work was done to find other compounds that would stabilize Cu(I) and produce more effective one-stage Fe/Cu removal processes.

Thompson et al.[87] described an inhibitor mixture to prevent replating of copper from cleaning solutions containing HCl. The mixture included bis(hydroxyethyl) aminomethyloctyl sulfide and a polyethylene glycol surfactant.

Although these processes were effective for some boiler cleaning situations, safer methods that used less corrosive or toxic chemicals also were desired by the industry. In addition, a unit cleaned with HCl/thiourea still requires several rinses (and neutralizations) as well as a passivation stage (see Section 4.2.4). Shields[88] described the acid/thiourea (A/T) process as well as other solvents discussed in this book, but Shields did not comment on the current applications of A/T.

4.2.3. One-Fill Iron Oxide and Copper Processes Based on Organic Acids/Chelants

The applications of ammoniated citric acid solutions to dissolve iron oxide deposits and the use of the same fill of solvent to dissolve copper were developed by Alfano[43] and Bell.[44] This technology led to the development of the Alfano/Bell processes (a.k.a. Citrosolv). The chemistry involves the use of inhibited ammonium citrate (pH 3.5–5.0) to dissolve iron oxides from steel surfaces (see Section 4.1.3).

The spent iron-containing solvent is then cooled (from 200°F [93°C] to 150°F [66°C]), and additional ammonia is added to elevate the pH to 9–10. An oxidizer (air and/or sodium nitrite) is added to dissolve the copper, which is then complexed by ammonia as $(NH_3)_4Cu(II)$. During this oxidation process, the ferrous citrate generated during the iron oxide stage becomes oxidized to ferric citrate.

The original Alfano/Bell process was used at a temperature of 200°F (93°C) for the iron stage and at 150°F (66°C) for the copper stage. Improvements have included the use of ammonium difluoride to speed the rate of iron oxide removal and the use of higher temperatures (up to 300°F [149°C]). The increased temperatures require that the iron stage is conducted at a pH of 6.0 so that the inhibitors can adequately control the corrosion rate of the metal being cleaned. Additional variations include the use of triethanolamine (TEA–instead of ammonia) for primary pH control.[89] This variation improves the safety of the solvent (TEA is less toxic than ammonia) and also reduces the cracking incidents of copper-base metals in mixed metal systems (ferrous and copper).

Work by Lesinski[53] resulted in the development of an ammonium EDTA-based solvent (tetraammonium EDTA) for dissolving magnetite (Fe_3O_4) from utility boilers. However, copper dissolved (owing to the ferric ions in the scale) and then replated onto the boiler surfaces.

To remove the copper, researchers developed an air-blowing process to convert Fe(II)EDTA to FE(III)EDTA. The researchers also realized that a protective passive film would enable the copper to remain in solution after the ferric EDTA dissolved it. The ideas of passivation as well as copper removal were incorporated into the patents for this technology.[55, 90]

During copper removal procedures that use the iron solutions of EDTA or citric acid[51] as the reaction medium, operators noted that it is much easier to remove copper from plated steel in EDTA than in ammonium citrate. This observation was difficult to explain because the procedures are about the same: the iron solutions that remain in the boiler being cleaned are oxidized with air or air + $NaNO_2$. These chemicals are passed through the ammoniacal chelate solution in both cases. To understand the mechanisms of copper removal in chelate solutions, researchers studied the dissolution in chelating agents with and without iron. Two types of tests (described in Frenier and Kennedy[51]) were conducted:

Copper-plated steel coupons were cleaned in the electrochemical cell shown in Figure 4.13. All the test solutions were maintained at 150°F, and various oxidizers were added to the solutions of ammonium EDTA or ammonium citrate (pH 9.5), with or without 7000-ppm ferrous Fe(II) iron.

Iron and copper chelate solutions were analyzed using DC polarography (dropping Hg electrodes) using the classic methods of Kohlthoff and Lingane.[91]

The study demonstrated that during the reactions in ammonium citrate, the dissolved oxygen and sodium nitrite are the primary oxidizing agents for the plated copper. See Equation (4.18). These investigators then compared the ammonium citrate process with EDTA-based iron/copper solvents (discussed later in this section), in which the oxidized iron reacts with the copper, and the external oxidant acts to generate the ferric EDTA. Because the *concentration* of iron(III) EDTA is much higher than dissolved oxygen (7000 ppm vs. about 8 ppm), the EDTA/iron solution dissolved copper much faster than citric/iron solutions. See Equation (4.19).

$$4[\mathrm{Fe(II)Chel}] + 2H_2O + O_2 \rightarrow 4OH^- + 4\mathrm{Fe(III)Chel} \qquad \text{(Equation [4.18])}$$

$$4[\mathrm{Fe(III)Chel}] + 2\mathrm{Cu} + 2\mathrm{Chel} \rightarrow 2[\mathrm{Cu(II)Chel}] + 4\mathrm{Fe(II)Chel} \qquad \text{(Equation [4.19])}$$

Figure 4.14 shows the copper concentrations found during dissolution tests with four different 7000 ppm Fe(II) solutions: EDTA with air, EDTA with air + $NaNO_2$, citrate + air, and citrate with air + $NaNO_2$. The EDTA solutions removed more than 10 times as much copper as the comparable citrate solutions. The EDTA solutions with the additional oxidant (nitrite) removed more copper than the air-only solution.

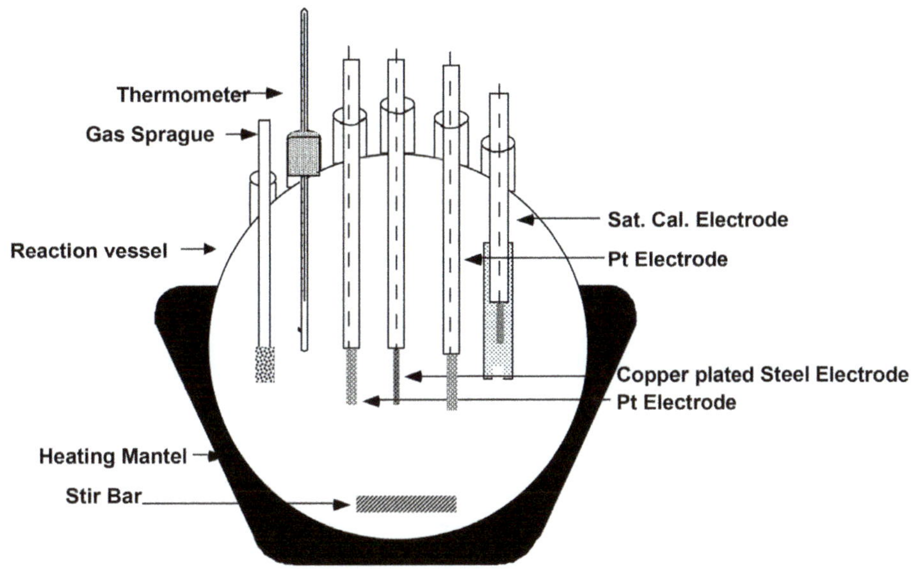

Figure 4.13 Copper Removal Cell

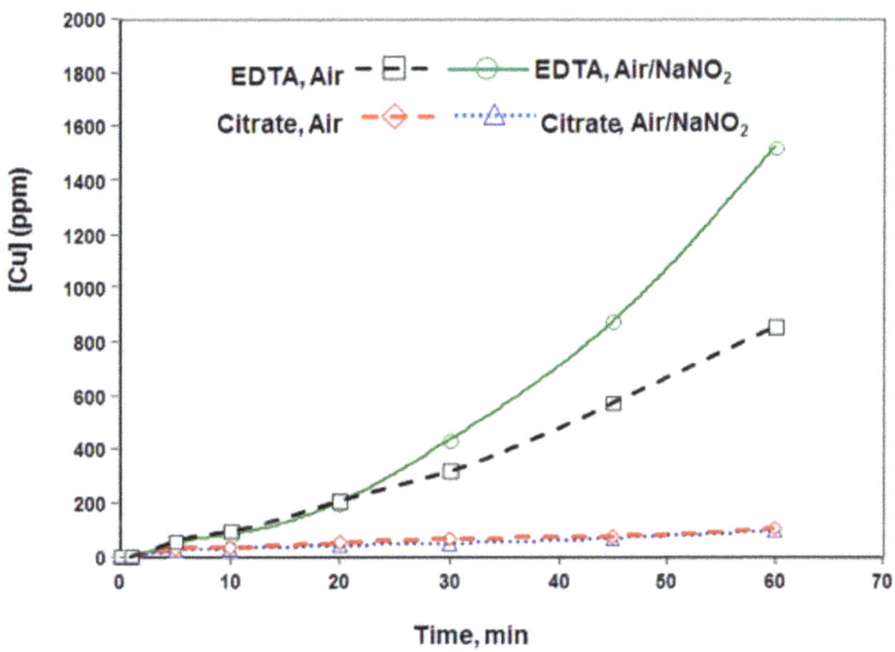

Figure 4.14 Dissolution of Copper from Plated Steel in 7000 ppm Fe(II) Chelate

In other tests,[51] citrate solutions actually removed more copper in the absence of iron than in the presence of iron. EDTA-iron solutions oxidized with H_2O_2 removed more than twice as much iron as the air-oxidized solutions.

Additional mechanistic studies of the copper removal processes in EDTA and citric acid are described by Frenier and Kennedy.[51]

According to Kohlthoff,[91] the half-wave potentials are approximately equal to the equilibrium potentials (E^o) for the reaction: oxidized species + e^- = reduced species. By comparing the measured half-wave potentials of the chelated species with the E^o value,

$$Fe^{2+} = Fe^{3+} + e^- \qquad \text{(Equation [4.20])}$$

The iron and copper chelate solutions exhibited classic oxidation and reduction waves that allowed the equilibrium (half-wave) potentials to be estimated. See Table 4.3.

Table 4.3 Polarography Data for 0.001m Metal Chelate Solutions

Metal	Chelate	$E_{1/2}$, measured mV	$E_{1/2}$, lit[92] mV	I_d mA	Reaction
Fe(II)	EDTA	-240	-250	4.0	Fe(II)→ Fe(III)
Fe(III)	EDTA	-260	-250	3.2	Fe(III)→ Fe(II)
Fe(II)	Citrate	-220 -465	-475	1.0 1.0	Fe(II)→ Fe(III) Fe(II)→Fe(III)
Fe(III)	Citrate	-475	-475	2.0	Fe(III)→ Fe(II)
Cu(II)	Citrate	-179 -340	-170 -340	2.0 2.2	C(II)→Cu(I) Cu(I)→ Cu°
Cu(II)	EDTA	-475	Not known	5.0	Cu(II)→Cu⁰

The value for the Fe(II)/Fe(III) couple is -525 mV. From this, an estimate of the relative ratios of the equilibrium constants for the ferrous and ferric chelates of citric acid and EDTA is calculated using Equation (4.21).

$$E_{1/2}(\text{complexed ion}) - E_{1/2}(\text{Standard}) = 0.059\log\left(\frac{K_{ox}}{K_{red}}\right) \qquad \text{(Equation [4.21])}$$

Here, K_{ox} and K_{rd} are the disassociation constant for the oxidized and reduced complexes, respectively. Using a E^o value of 0.525 mV for Fe(II) → Fe(III), the ratios of the equilibrium constants for EDTA is calculated to be 1.5×10^{-14} for (Fe(III)/Fe(II)) and 6.1×10^{-17} for the citrate couple (data from Martell and Smith[2]). This calculation shows that the citrate ion stabilizes the ferric species 10^3 times more than the EDTA ferric/ferrous couple.

The half-wave potentials for the copper citrate solutions are essentially identical to the literature values for the aquo-ammonium/copper species in solution, while the values for EDTA/Cu indicate a much more stable complex.

On the basis of these data, the following mechanisms are proposed:

$$\begin{aligned}
&\text{A. Oxidant} + \text{Fe(II)EDTA} \xrightarrow[k_1\ \text{fast}]{} \text{Fe(III)EDTA}\\
&\text{B. 2Fe(III)} + \text{Cu} + \text{EDTA} \xrightarrow[k_2\ \text{rate determining}]{} \text{2Fe(II)EDTA} + \text{Cu(II)EDTA}\\
&\text{C. Oxidant} + \text{Cu} + \text{EDTA} \xrightarrow{k_3\ \text{slow}} \text{Cu(II)EDTA}
\end{aligned} \qquad \text{(Equation [4.22])}$$

$$\begin{aligned}
&\text{A. Oxidant} + \text{Fe(II)Citrate} \xrightarrow[k_4\ \text{fast}]{} \text{Fe(III)Citrate}\\
&\text{B. 2Fe(III)} + \text{Cu} + 4NH_3 \xrightarrow[k_5\ \text{slow}]{} \text{2Fe(II)Citrate} + \text{Cu(II)}[NH_3]_4\\
&\text{C. Oxidant} + \text{Cu} + 4NH_3 \xrightarrow{k_6\ \text{rate determining}} \text{Cu(II)}[NH]_4
\end{aligned} \qquad \text{(Equation [4.23])}$$

Thus, the major oxidizing agent for the EDTA system is Fe(III)EDTA, and the external oxidizer (oxygen, nitrite) oxidizes the copper in the iron citrate system. The external oxidizer does oxidize the ferrous citrate to ferric citrate, but this species does not dissolve much of the copper but acts as a sink for the external oxidizer.

The only benefit of using the iron citrate solution is that an additional oxidizing fill of the unit is avoided. This analysis explains the data described in Figure 4.14.

Thus, in the commercially developed process, the ferrous EDTA generated during the dissolution of magnetite is oxidized by air (sometimes containing sodium nitrite). The oxidized chelating agent (ferric EDTA) passivates the steel and then oxidizes the plated copper. Passivation of the steel is a critical process and was monitored indirectly by measuring the EMF of the solution. Frenier and Kennedy[51] demonstrated that the solution potential reflects the ratio of Fe(II):Fe(III) and not the real passivation potential of the surface. However, when the EMF solution is in the range specified by the patents,[55] the surfaces usually have become passive. The passivation is caused by the strong oxidizer, Fe(III)EDTA, diffusing to the surface and raising the surface EMF into the passive region.

Additional improvements in the copper removal technology in EDTA included the use of H_2O_2 as the oxidant during the copper stage.[56] Use of this chemical can substantially increase the rate at which the ferric EDTA is generated. The time to achieve passivation is shortened, and copper removal is enhanced. In a similar manner, use of air/$NaNO_2$ and gaseous oxygen (O_2)[57] enhances passivation of the boiler metal, speeds conversion of ferrous EDTA to ferric EDTA, and thus increases the rate of copper dissolution. Using the oxygen or the hydrogen peroxide methods, copper removal can be accomplished in as little as 2 hours.

The oxygen process has been used when there is little external circulation. In these situations, the high concentration of hydrogen peroxide may cause the chelant to degrade, or the iron salts may decompose the peroxide. Frenier observed several successful boiler jobs using the H_2O_2 process. However, great care was taken, and the handling equipment for the peroxides was dedicated only for that purpose. On the basis of discussions with industry experts (not documented), Frenier concludes that the use of O_2 or O_2 mixed into air may be the safest and most economical process.

Several additional formulations with EDTA (or citric acid) and other materials have been proposed. Martin and Bradley[93] patented a mixture of hydrogen peroxide, oxalic acid, and an EDTA or citric acid-type complexer for removing copper deposits (iron oxides are removed using an additional stage). An unusual aspect of this process is that it is performed at a pH of 3.0–6.0. For passivation of the steel to be effective, oxidative copper removal processes usually are conducted at pH values >9.0.

A successful commercial process for removing copper from steel was developed by Jevec and Leedy[94] and Cleary and Gockley.[95] This process uses 5% EDTA, 2% H_2O_2, and ammonia to raise the pH to 7.0, then ethylenediamine to raise the pH to 10.0. This formulation is used after the EDTA/hydrazine iron oxide stage in the Electric Power Research Institute/Steam Generator Owners Group (EPRI/SGOG) process for removing iron oxide and copper from utility boilers. This procedure was successfully used to dissolve copper during cleaning of nuclear steam generators. Although it is an effective process, the large amount of steel, Inconel, and sludge causes much of the hydrogen peroxide to decompose by an auto redox reaction:

$$2H_2O_2 = 2H_2O + O_2^{\bullet} \qquad \text{(Equation [4.24])}$$

After the peroxide decomposes, the solvent's capacity for dissolving copper is greatly diminished. This reference noted above contains many citations of solvents that led to the final development of this process. The high concentrations of peroxide needed for this process also may cause the EDTA to be degraded. This reaction is known as the Fenton process, and the active oxidant probably is the hydroxyl radical, •OH. See the discussion of chelate degradation by Chen et al.[96] Chapter 5 has more details of cleaning nuclear-related plants in which much tighter specifications (but not necessarily different chemistries) are required.

A dual-stage (one-fill) process using polyphosphonic acids, such as nitrilotrimethylenephosphonic acid and ortho-hydroxyethylidene-1,1-diphosphonic acid,[75] used ammonia or amine salts to dissolve iron oxides. The same solutions are oxidized using hydrogen peroxide, sodium bromate, or sodium nitrite to dissolve copper.

The technology is similar in concept to the EDTA-based technology described above. Precipitation of ferrous/phosphate–containing solids is prevented by adding ammonium difluoride.[97]

Research into methods to remove iron oxide and copper deposits from power generating equipment constituted a large portion of the activities of laboratories involved in chemical cleaning operations. Except for the work on corrosion inhibitor mechanisms, there is more published information on the chemistry of iron and copper removal than other techniques. Section 4.3 reviews this information.

4.2.4. Chemical Cleaning Passivation: Basis for One-Fill Treatments

Passivation processes are needed after most chemical cleaning operations to prevent rapid flash rusting of unprotected surfaces. For some of the one-fill treatments, passivation is an integral part of the process. This section provides more information on passivation chemistries that are or were in use.

Uhlig[98] gave two definitions of passivation that are germane:

1) A metal active in the EMF series (or an alloy composed of such metals) is passive when its electrochemical behavior approaches that of an appreciably less active or noble metal.

2) A metal or alloy is passive if it substantially resists corrosion in environments in which, thermodynamically, there is a large free-energy decrease associated with its passage from the metallic state to the appropriate corrosion products.

The passivating film is thought to be composed of some type of iron oxide if iron is the active metal. The formation, composition, and breakdown of passivation is a very complex subject. At a symposium sponsored by the Electrochemical Society[99] in 1986, 21 papers were devoted to passivation theory and passivation breakdown. Formation of a passive film on carbon steel usually requires pH values of >8.5 and oxidizing conditions. Usually oxygen (or something that yields oxygen) must be present. The chemical composition of the passive film will vary depending on the aqueous fluids present but consists of various mixtures of iron oxides (magnetite, γ-Fe_2O_3, γ-FeOOH) or an amorphous oxide of iron. The thickness may range from 10Å to 30Å. Although the passive film is thought to consist of mostly iron oxides, varying amounts of ions from the passivating medium (such as chromium or phosphate) may also be present. Ligands that strongly complex with iron (EDTA, citrate, and chloride) will tend to disrupt the passive film or make it more difficult to form.

An effective method for studying passive film formation involves examining polarization curves.[51] Theoretical anodic and cathodic polarization curves are shown in Figure 4.15. This diagram depicts the behavior of iron in HCl. This system cannot achieve passivation because the chloride on the surface and the high acidity require unrealistic concentrations of an oxidizing agent. The chemical criteria are for the passive film (an iron oxide) to be stable in the passivating fluid. This usually requires an alkaline environment and an oxidizing agent.

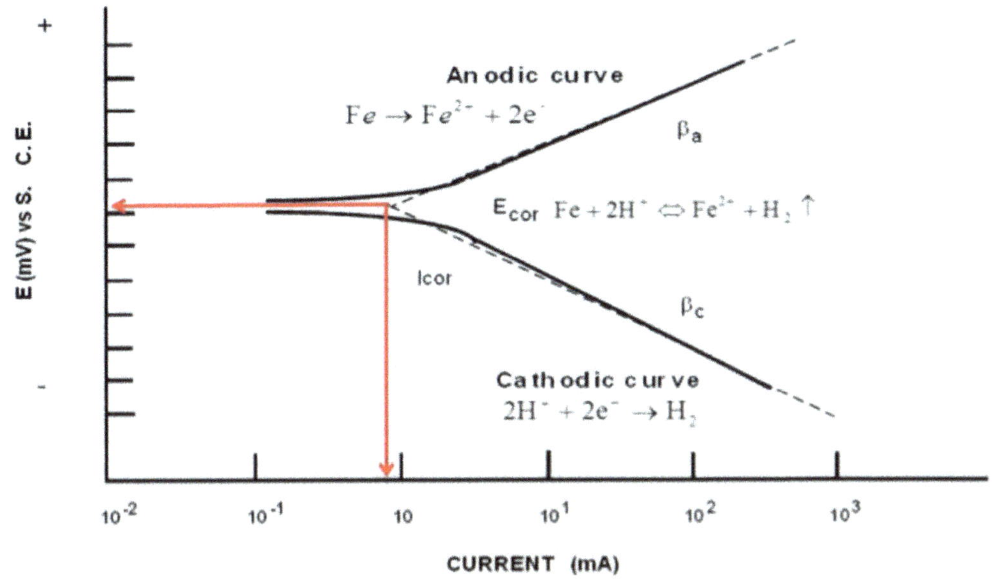

Figure 4.15 Theoretical Polarization Curve for Iron (Fe) in Acid

A theoretical polarization curve for an active passive system is shown in Figure 4.16. As the potential is shifted in the anodic (+) direction, the current (corrosion rate) increases. At a critical current density ($\mathbf{I}_{Crit}$), the current drops to a low value–indicating the onset of passivation. Using polarization curves, investigators can determine and describe the solution conditions controlling passivation. Polarization curve analysis and data are extensively used in the subsequent sections to describe passivation in chemical cleaning solutions.

Non-Chelating Agent Passivators

Three different common passivating systems used to passivate steel after a cleaning treatment (such as HCl or HCl/thiourea) were studied using an electrochemical cell as described in Figure 4.13.

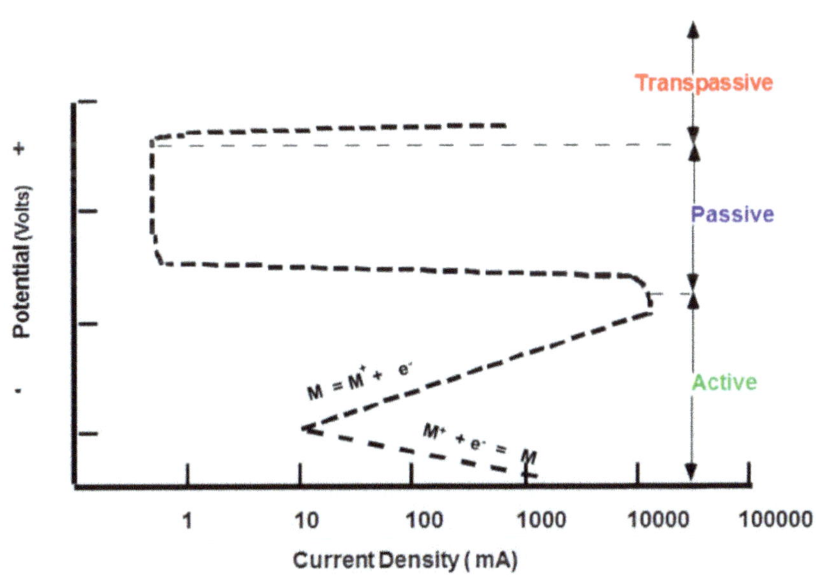

Figure 4.16 Theoretical Passivation Diagram

The anodic polarization curves were determined for mild steel in phosphate/nitrite, sodium carbonate/nitrite, and ammonia/hydrazine. These are the most commonly used treatments for acid-cleaned surfaces. The data listed below are from the paper by Martin and Banks.[100]

These formulations frequently are used after acid-cleaning treatments to leave the surfaces in a condition (passive) that resists corrosion from the ambient atmosphere. Note, however, that the metal is passive in an electrochemical sense only while it is in contact with the solvent. Upon exposure to the air in the plant, passivating conditions rapidly cease to exist.

Phosphate system. This system consisted of 0.25% NaH_2PO_4, 0.25% Na_2PO_4, and 0.5% $NaNO_2$. The system also contained dissolved air. Without air, the polarization curves did not show spontaneous passivation; however, this behavior is present with air and phosphate.

Sodium carbonate. This solution consists of 1% $NaCO_3$ with 0.5% $NaNO_2$. The polarization curves also show that this system achieves spontaneous passivation only when air is present.

Ammonia/hydrazine. Polarization curves did not indicate spontaneous passivation; however, one specimen did develop a dark gray coating in the presence of air. The results of the exposure tests are shown in Table 4.4.

Table 4.4 Passivation Tests[100]

Components	Temperature (°F)	Air Present	Satisfactory Passivation
Phosphate/Nitrite	150	Yes	Yes
Phosphate	150	Yes	No
Phosphate/Nitrite	150	No	Yes
Carbonate/Nitrite	200	Yes	Yes
Carbonate/Nitrite	200	No	No
Carbonate	200	Yes	Yes
Carbonate	200	No	No
Ammonia/Hydrazine	200	Yes	Yes
Ammonia/Hydrazine	200	No	No

In these tests, only the phosphate/nitrite (a mild oxidant) solutions achieved any passivation without air. All the systems benefited from introduction of the second oxidizer (air). Additional studies evaluated citrate/iron/nitrite (as after an Alfano/Bell treatment), phosphate/hydrazine, and carbonate/gluconate/phosphate. The citrate/iron/nitrite was a very effective passivator, but the other two solutions were not as effective. The conclusion is that more powerful oxidizers are more effective passivating solutions.

A passivation stage may be required after the use of the EDTA/hydrazine (EPRI) process (Welty et al.[67]). This consists of ammonia or an amine and H_2O_2. More details are in Section 5.3.

Passivation with Iron Chelates

Ferric EDTA [Fe(III)EDTA] is generated during the oxidation stage of the iron/copper removal treatments described in the previous section and acts as an in situ passivator. Note from Equations (4.22 and 4.23) that there are significant differences in the chemical behaviors of ferric EDTA compared with ferric citrate, which are generated during the oxidation stages of the two solvent systems. The analyses of the polarography tests indicate that the ferric EDTA species is less stable than ferric citrate and is able to accept an electron and to oxidize copper.

To understand the role of iron EDTA in passivation and copper removal processes, Frenier and Kennedy conducted extensive lab tests.[63] The laboratory system used a rotating disk electrode (RDE; see Figure 4.17) and the setup depicted in Figure 4.18.

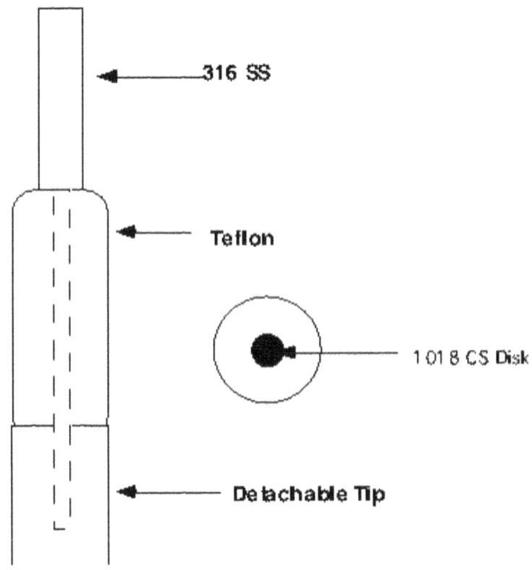

Figure 4.17 Rotating Disk Electrode

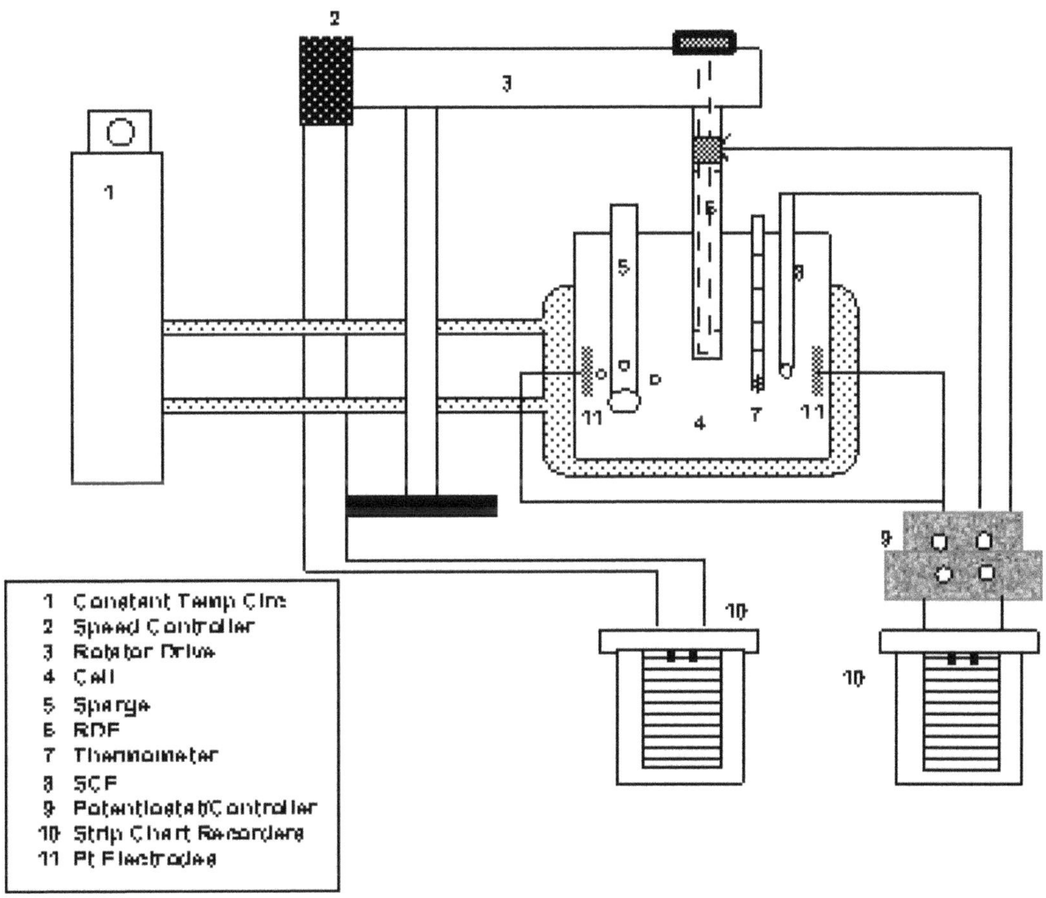

Figure 4.18 Passivation Setup with a Rotating Disk Electrode

The investigators[63] theorized that this chemical species may be diffusion-limited when acting as a passivator or corrodents because field data indicated that turbulence and shear of pumping hastened the copper removal efforts. Note the studies of surface reaction and diffusion control by Fredd and Fogler.[101]

An effective way to study the electrode reactions of a diffusion-limited species involves the use of an RDE. Eisenberg et al.[102] and Kuo and Landolt[103] studied Fe in acidic ferric solutions using an RDE. The limiting diffusion current with an RDE is given by

$$I_{Lim} = 59816 D^{2/3} \nu^{-1/6} \omega^{1/2} [Fe^{3+}] \quad \text{(Equation [4.25])}$$

Here, D is the diffusion coefficient, ν is the kinematic viscosity (Stokes), and ω is the angular velocity (radians [rad]/s). The diffusion coefficient can be found by varying the rotation rate and making a plot of I_{Lim} vs.$\omega^{1/2}$. Thus,

$$D = \left(\left(\frac{dI_{Lim}}{d\omega^{1/2}}\right) \frac{\nu^{1/6}}{59816[Fe^{3+}]} \right)^{3/2} \quad \text{(Equation [4.26])}$$

D and ν vary with temperature and must be experimentally determined. At passivation, I_{Lim} should be equal to or greater than the critical passivation current (I_{Crit}). The I_{Crit} values were measured from the polarization curves of steel RDEs in $(NH_4)_4$EDTA at various pH values, temperatures, and EDTA. Via actual passivation experiments, I_{Lim} at passivation was also estimated. The relationship between these current densities (CDs) should reveal important information concerning the passivation reaction.

Anodic polarization curves in $(NH_4)_4$ EDTA under various conditions of pH, temperature, and $(NH_4)_4$(EDTA) were run with an RDE. A theoretical example of an RDE run with ferric EDTA is shown in Figure 4.19.

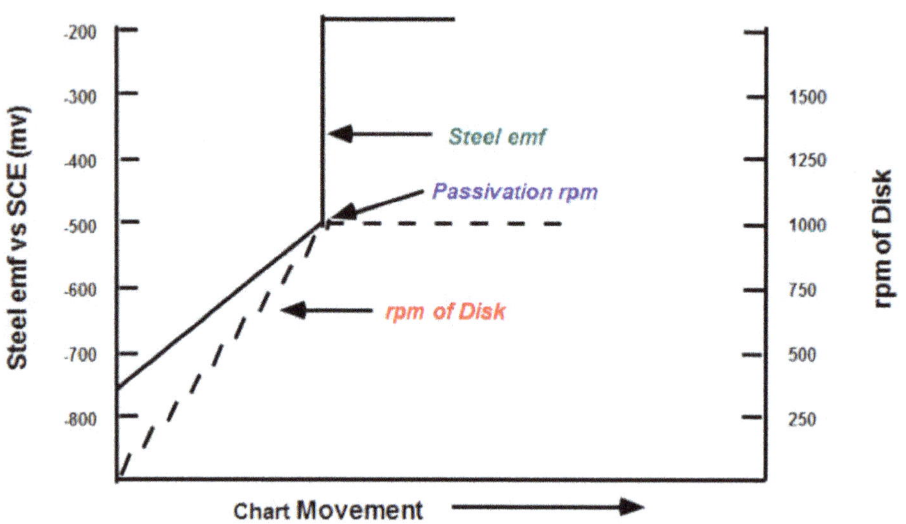

Figure 4.19 Steel EMF and Disk rpm Test

Iron was excluded from the tests, and the I_{Crit} value was determined. Various concentrations of Fe(III)EDTA were prepared, and cathodic polarization curves were determined so that *D* could be calculated. Finally, actual passivation tests were run to relate the passivation to pH, temperature, $(NH_4)_4$EDTA, flow, and Fe^{3+}. The I_{Lim} was calculated at passivation and compared to $\mathbf{I_{Crit}}$ determined from the anodic curves. In addition, the effects of various oxidizers used to generate ferric EDTA were evaluated in short- and long-term tests. For example,

$$\begin{aligned} &\text{A. Ox} + \text{Fe(II)EDTA} \rightarrow \text{Fe(III)EDTA} \\ &\text{B. Fe(III)EDTA} + \text{Fe} \rightarrow \text{Fe(Pass)} \end{aligned} \quad \text{(Equation [4.27])}$$

See Frenier[51,63] for the details of the tests. The basic equipment was an RDE made from 1018 carbon steel (Figure 4.17) in the electrochemical arrangement in Figure 4.18. The rotation rate could be varied from 20 to 10,000 rpm. A jacketed glass cell held ~600 mL of solution. The cell could be purged with gas. The saturated calomel electrode (SCE) was in a separate compartment connected to the cell with a KCl bridge and fritted glass. The potentials and currents were controlled with a potentiostat. Potentials vs. rpm were recorded on a two-pen, strip-chart recorder.

Polarization curves. The test solution was made at laboratory temperature, and the pH value was adjusted with NH_4OH or HCOOH. A total of 600 mL was thermostated in the closed cell. At the test temperature, a sample was removed and cooled, and the pH value was rechecked at laboratory temperature. Air was not excluded in these tests. The RDE steel tip was polished to 600 grit before all the runs and activated in boiling $(NH_4)_4$EDTA. The tip was put on the RDE, lowered into the solution, and the rotator activated at the test rpm. The potentiostat was placed on direct reading, and the open circuit potential (OCP) could equilibrate for ~10 min. Anodic or cathodic scans were then run from the OCP at 2 mV/s.

The $\mathbf{I_{Crit}}$ was determined from the anodic scans of the iron RDE in iron-free EDTA solutions. The $\mathbf{I_{Crit}}$ value reported is the average of runs at three to five different rpms. The cathodic curves of Fe(III)EDTA at various iron concentrations allowed a calculation of D at three temperatures. The kinematic viscosity was measured at three temperatures using a bulb viscometer.

Passivation rpm. Per the first definition of passivation, a metal is passive when it behaves like a noble metal. In these iron EDTA solutions, a noble metal such as platinum measures the Fe^{2+}/Fe^{3+} ratio of the solution.[4] From experience, the EMF measured (Pt vs. SCE) in solutions containing predominantly Fe(III)EDTA will be more positive than -100 mV. The active steel usually exhibits an OCP (Fe vs. SCE) value of about -800 mV. When the steel passivates, the OCP will increase by several hundred mV and will eventually reach about 100 mV. To conduct the experiment, the polished active RDE was lowered into the heated prepared Fe(III)EDTA solution. Beginning at 0 rpm, the rotation rate was slowly increased until the steel EMF suddenly rose to about -300 mV. The EMF and rpm were recorded on the two-pen recorder. The passivation rpm was determined from the chart recorder traces. A theoretical trace is shown in Figure 4.20. The x-axis reflects chart movement or time to passivation, but the time factor was not studied. The tests were repeated three to five times, and the rpm (pass) was averaged. Several graphical relationships between Fe^{3+} rpm, free EDTA, and temperature were derived from these data. In addition, using the values of D calculated with Equation (3.34), I_{Lim} at passivation was determined.

Anodic polarization curves in ammonium EDTA. Anodic curves were determined for two $(NH_4)_4$EDTA concentrations and three temperatures. Typical traces are shown in Figure 4.20.

Excellent active/passive behavior is seen, and I_{Crit} is easily determined.

The effect of pH was studied by lowering the pH value with formic acid. Three anodic curves are shown in Figure 4.20.

Not only is I_{Crit} increased as the pH value drops, but the passivation region is distorted. At a pH of 8.6, there is no passivation region in pure EDTA. As will be seen in later sections, other factors influence passivation at lower pH values.

The passivation potential is approximately -300 mV; Table 4.5 lists the I_{Crit} values at a pH of 9.3. As expected, the critical CD for passivation increased with temperature and free EDTA concentration. The active corrosion CDs before achieving passivation are quite high, so I_{Crit} must be exceeded as rapidly as possible.

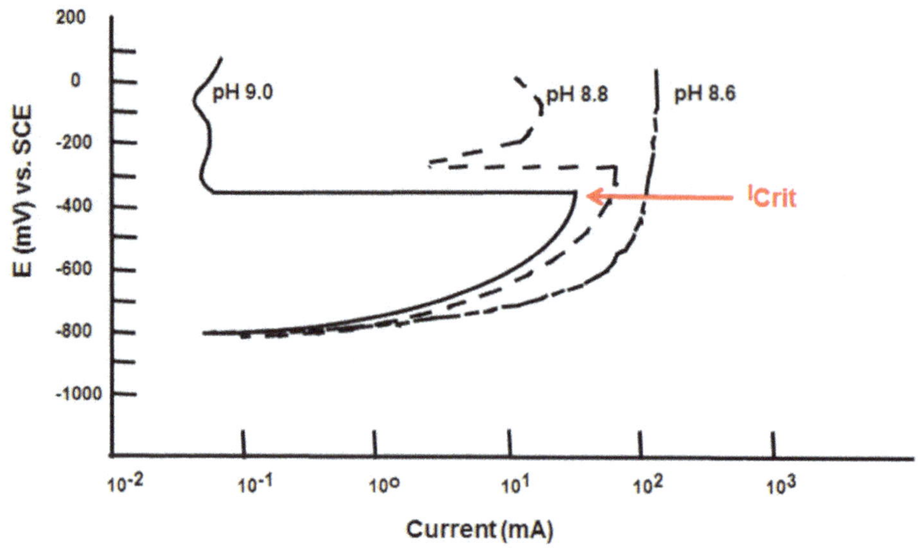

Figure 4.20 Anodic Polarization Curves in Ammonium EDTA at Three pH Values

Table 4.5 I_{Crit} Values in EDTA (pH of 9.3)

Temperature (°C)	I_{Crit}
0.8% $(NH_4)_4$EDTA	
54	8.0 mA/cm^2
65	13.2
77	20.3
1.6% $(NH_4)_4$EDTA	
54	16.2 mA/cm^2
65	24.8
77	45.3

Cathodic curves with Fe(III)EDTA. Cathodic curves in the presence of Fe(III)EDTA were run at five iron concentrations and three temperatures. The goal was to calculate D at the three temperatures. A sample of the cathodic curves is seen in Figure 4.21. The expected diffusion-controlled nature of the reduction [Fe(III) EDTA$^-$→Fe(II)EDTA^{2-}] is confirmed by these graphs. Eisenberg and Tobias[102] measured D for Fe^{3+} in H_2SO_4 to be 5.2 ± 0.3 × 10^{-6} cm^2/s at 77°F (25°C). Because the coefficient should increase as the temperature increases, the present values (7.5 ± 2.5×10^{-6} cm^2/s) are in reasonable agreement with the literature, and they should increase our confidence in the RDE technique.

Passivation tests. As described above, many tests were run to determine the rpm at which passivation occurs. From plots of $\omega^{1/2}$ or its equivalent, ($Re^{1/2}$) vs. [Fe^{3+}] should yield a straight line at a given temperature.

$$\mathrm{Re}(RDE) = r^2 \frac{\omega}{\upsilon} \quad \text{(Equation [4.28])}$$

Here:

Re = Reynolds number
R = Radius of the disk
ω = Angular velocity
ν = Kinematic viscosity

However, this is not a very convenient form of display for comparisons. Therefore, the graph of passivation Re vs. [Fe^{3+}] on a semi log plot is shown in Figure 4.22. These curves allow a better pictorial sense of the relationships between Re, Fe^{3+}, EDTA, and temperature to the passivation of steel. As noted earlier (Figure 4.20), pH has a pronounced effect on passivation. In uncontaminated EDTA, the passive region disappears at a pH of 8.6. The RDE passivation tests were run at a pH of 7.3, 8.3, 8.75, and 9.3 in solutions containing 0.8% free EDTA, and an EMF *change* indicating the onset of passivation was observed at a pH as low as 7.3. However, very high rotation rates were required. Therefore, although passivation may be possible at lower pH values, the very high critical CDs required probably limit the usefulness of the system to pH levels of ± 9.0.

The effect of pH was also studied in the presence of 3600 ppm Fe^{3+} and 0.8% free EDTA using full polarization curves. The electrode was not rotated to initiate the scan from the active state. In the presence of the added iron, a single cathodic loop corresponding to the reduction of Fe^{3+} to Fe^{2+} is seen at a pH of 9.3. The onset of passivation occurs at -300 mV, and the CD in the passive region is ˜20 μA/cm². At a pH of 8.0 (with Fe^{3+}), two cathodic loops are present, corresponding to the reduction of Fe^{3+} as well as H^+. Evidence for the onset of passivation (at -300 mV) exists because the anodic CD is about 10 times lower than the active (cathodic) CD. In the absence of ferric ions (Figure 4.20), the passive region does not exist below a pH of 8.6. At a pH of 7, the current resulting from H^+ reduction is increased, and little evidence of passivation exists (corrosion CD of ˜1000 μA/cm²).

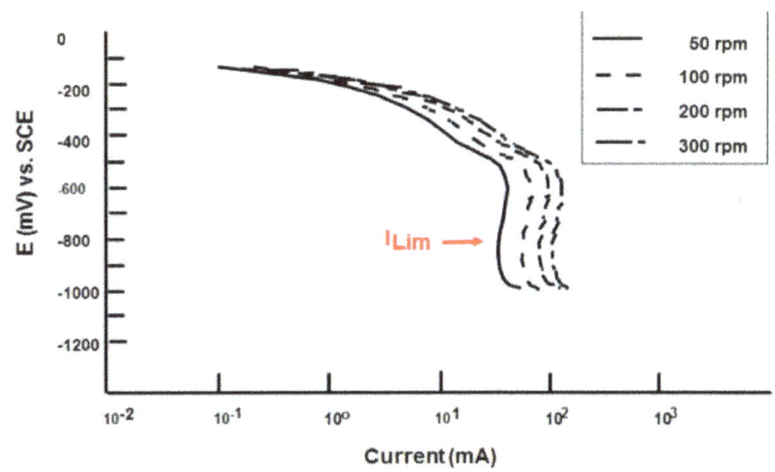

Figure 4.21 Cathodic Polarization Curves in Ferric EDTA

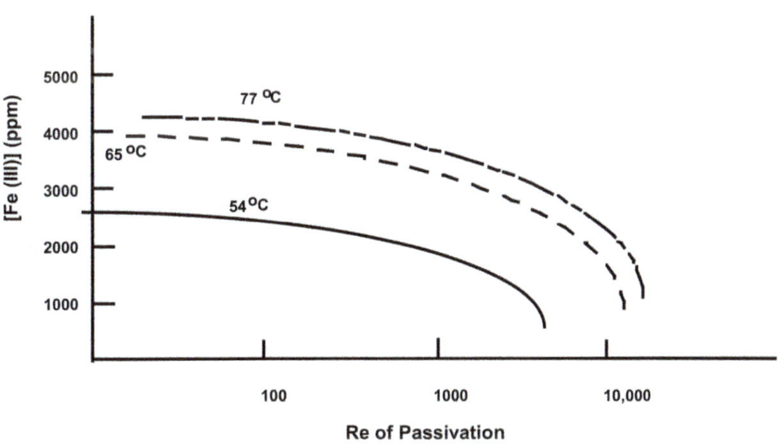

Figure 4.22 Reynolds Number (Re) of Passivation

The data demonstrate the significant relationships between Fe^{3+}, EDTA, temperature, flow rate, and pH. Passivation is most easily achieved at 1) high flow rates, 2) high iron concentrations, 3) low free EDTA concentrations, and 4) low temperatures. For a practical application of this technology to the cleaning of steel, ferric EDTA must be generated from the ferrous EDTA available at the end of the iron removal stage. A definitive kinetic study of the reaction, oxidant + Fe(II)EDTA $\rightarrow$ Fe(III)EDTA, has not been made. However, with constant aeration, the relative rates are H_2O_2 >air + $NaNO_2$ >air >$NaNO_2$. Chen et al.[96] determined the absolute half-life for the reaction between O_2 and Fe(II)EDTA to be 0.11 s at 77°F (25°C) and at a pH of 9.3.

Thus, mass transport effects control the effective rate of ferric generation for air. In situations in which the introduction and distribution of air are limited, the use of the more vigorous oxidants such as H_2O_2 or air + $NaNO_2$ or O_2 is strongly indicated so that the correct amount of ferric can be generated, and $\mathbf{I}_{\mathbf{Crit}}$ is surpassed in a minimum amount of time. Until $\mathbf{I}_{\mathbf{Crit}}$ is achieved, the equipment being cleaned will experience very high corrosion rates.

As stated earlier, I_{Lim} should be > I_{Crit} at passivation. The critical CDs have been measured from the anodic polarization curves and are shown in Table 4.6. I_{Lim} at passivation was calculated using Equation (2). The calculated I_{Lim} values along with the observed I_{Crit} values are listed in Table 4.6. At 0.8% $(NH_4)_4$EDTA, the observed limiting diffusion currents are significantly lower than I_{Crit} at the higher Fe^{3+}. A similar trend can also be seen at 1.6% $(NH_4)_4$EDTA, but the effect is not as pronounced as that at 0.8% $(NH_4)_4$EDTA. At 171°F (77°C) with 1.6% $(NH_4)_4$EDTA, the limiting currents are significantly higher than I_{Crit}. If the only role of ferric EDTA was as a cathodic depolarizer, supplying a source of CD to maintain passivation, I_{Lim} should not vary with iron concentration. This conclusion is required by Equation (3.23). Clearly, the iron EDTA also is changing the *anodic* reaction: Fe $\rightarrow$ Fe^{2+} + $2e^-$.

The treatment of the surfaces with ferric EDTA allows the solvent to dissolve any copper plated onto the steel surfaces and leaves the cleaned metal in a passive condition like those generated by the other alkaline oxidizers described above.

Table 4.6 Limiting Diffusion Current at Passivation and Critical Passivation Currents (mA/cm²)

0.8% $(NH_4)_4$EDTA						
Fe Conc.	129°F (54°C)		149°F (65°C)		171°F (77°C)	
(ppm)	I_{Crit}	I_{Lim} Pass	I_{Crit}	I_{Lim} Pass	I_{Crit}	I_{Lim} Pass
	8.0		13.2		20.3	
800		8.8		12.5		27.7
1680		7.0		11.9		35.2
2700		--		11.4		15.8
3600		--		9.5		9.2
4100		--		5.5		--
1.6% $(NH_4)_4$EDTA						
Fe Conc.	**129°F (54°C)**		**149°F (65°C)**		**171°F (77°C)**	
(ppm)	**I_{Crit}**	**I_{Lim} Pass**	**I_{Crit}**	**I_{Lim} Pass**	**I_{Crit}**	**I_{Lim} Pass**
	16.2		24.8		45.3	
1680		20.3		23.9		59.7
2700		18.1		29.0		68.0
3600		14.7		24.0		5.0
4100		15.0		25.0		59.0

4.3. Details of Iron Oxide Dissolution Mechanisms

The chemical dissolution of iron oxide deposits involves many processes that are designed to enhance the service life of steel industrial equipment (such as heat exchangers, boilers, and steam generators). Deposit removal improves heat transfer and reduces pitting corrosion.

Section 4.1 introduced the chemistries for removing iron oxides from industrial surfaces. In addition, the dissolution of copper (Section 4.2) must also be mentioned since some systems can remove copper in a single fill. This section (Section 4.3) provides more details of the complex iron oxide mechanism. This is a detailed review of the literature of iron oxide dissolution and can provide a starting point for further investigations. Much of the literature cited in this section is becoming increasingly difficult to obtain in the original form, so this summary preserves what is available. The references to the papers cited in this book have been cited with more information than was in the original NACE documents,[59, 60] where possible. The author of this book has attempted to access the originals online. However, some of these papers are no longer available. Because of the specialized nature of this information, some readers may choose to skip this section.

The purpose of this section is to review the literature on the mechanisms and kinetics of iron oxide dissolution processes. Emphasis is placed on pure iron oxides and iron oxides on steel surfaces. The limited scope of this review precludes complete coverage of the vast amount of work on passive film formation and dissolution. The details described in the following subsections include electrochemical effects on dissolution, dissolution of free magnetite, removal of iron oxide when it is on an iron surface, reactions in chelating agent solvents, and the effects of scale morphology. Most of the information in the next sections came from papers by Frenier[59] and Frenier and Growcock,[60] and more details of some of the experimental procedures are in these documents.

The stability and dissolution of iron oxides directly affect these processes. Diggle[104] reviewed the subject of iron oxide removal from oxide surfaces in 1973, with an emphasis on the electrochemistry of the processes involved. Over the last 35 years, removal of scale (oxide on a metal surface) began to receive scientific attention. In addition, the study of well-defined systems was perceived to be even more necessary for understanding the dissolution process; much of the work in this area focused on the dissolution of powdered magnetite since this chemical can be obtained in a pure form. Additional iron oxides such as hematite (Fe_2O_3) are not studied since in its pure (not hydrated form) it is very insoluble, even in HCl.

4.3.1. General Mechanisms of Iron Oxide Dissolution

This section has several parts:

- Electrochemistry effects on dissolution of magnetite
- Dissolution of free (not coupled to surfaces) magnetite
- Dissolution of magnetite on iron surfaces

Electrochemical Processes During Iron Oxide Dissolution

The studies of Vermilyea[105] provided some of the earliest theoretical bases for the dissolution of iron oxides. Vermilyea expressed the rate of oxide dissolution as

$$r = \frac{2rf}{\exp(-\frac{\alpha^- z^- F\eta}{RT}) + \exp(-\frac{\alpha^+ z^+ F\eta}{RT})} \qquad \text{(Equation [4.29])}$$

Here, r_f is the rate of the freely dissolving crystal, η is the applied overpotential, α designates the transfer coefficients of the ions, and z designates the ionic charges. The discussion is concerned only with changes in the overpotential at the double layer, and the author states that a doubling of the dissolution rate is the maximum rate change achievable, regardless of η. The driving force for oxide dissolution is the potential difference across the Helmholtz layer (see Figure 4.23, which applies to dissolution of magnetite [Fe_3O_4] from a metal [Fe] surface).

$$\phi_S - \phi_{OHP} = \frac{RT}{ZF}\ln(a/a_o) \quad \text{(Equation [4.30])}$$

ϕ_S and ϕ_{OHP} are the potentials at the surface and outer Helmholtz plane, *a* is the activity of the potential determining ion, and a_0 is its activity at the zero-charge potential (ZCP). In the absence of interfering ions, the proton (H^+) is the potential determining ion. Thus, pH controls the dissolution rate.

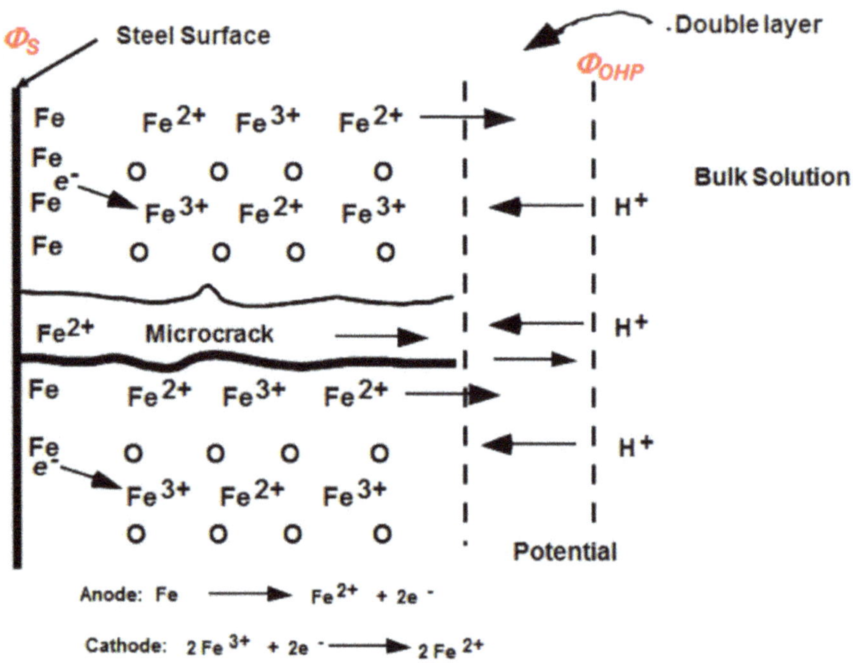

Figure 4.23 Iron Oxide Dissolution Mechanism

Other electrochemical phenomena also must be considered when the oxide is on an iron surface (see above). Most authors agree that iron is released from the oxide only as the ferrous (Fe^{2+}) ion. Hickling and Ives[106] stated that an Fe^{3+}/Fe^{2+} redox system (reductive dissolution) in the solid state is established and produces a *potential determining exchange current.* Therefore,

$$E = E' - 0.059pH - 0.059\log[aFe^{2+}(\text{soln})] \quad \text{(Equation [4.31])}$$

Haruyama and Masamura[107] stated that the reductive dissolution of magnetite occurs with 100% efficiency in the potential range from 0 to 900 mV (vs. SHE), as follows in acids:

$$Fe_3O_4 + 8H^+ + 2e^- = 3Fe^{2+} + 4H_2O \quad \text{(Equation [4.32])}$$

Although there is some disagreement about the effect of potential, these authors[106,107] agreed that the cathodic current is essential and that the rate-determining step is the transfer of the Fe^{2+} and O^{2-} (or OH^-) ions across the interface, i.e.,

$$Fe^{2+}_{solid} \rightarrow Fe^{2+}_{solution}$$

and

$$O^{2-}_{solid} \rightarrow O^{2-}_{solution}$$

or (Equation [4.33])

$$O^{2-}_{solid} + 2H^{+} \rightarrow H_2O$$

or

$$OH^{-}_{solid} \rightarrow HO^{-}_{solution}$$

Haruyama and Masamura[107] concluded that the overpotential changes (η) the magnetite by a transfer of ions across the interface until the chemical potential of the potential determining ions (Fe^{2+}) balances the overpotential:

$$\eta = E = E_0^{'} - 0.059\log[\frac{Fe^{2+}}{Fe^{3+}}] \qquad \text{(Equation [4.34])}$$

Equations (4.31 and 4.34) are useful in analyzing other systems including passivation (Section 4.2.4). An interesting observation of Hickling and Ives[106] is that the redox potential of dissolving magnetite and the Flade potential (the potential at which the passive film dissolves) in a similar solution are almost identical, implying that passivation and dissolution of oxides are simply reversible reactions of the same type.

Mancey et al.[108] extensively studied oxide-covered iron electrodes in EDTA plus oxalic acid and oxalate salts (see structures in Figure 4.1). Distinct potential regions were observed.

- In Region I (E >-100 mV vs. SCE), little Fe^{2+} was released, and there was only minor oxide dissolution. This is considered the induction period for pore formation.
- In Region II (-450 mV < E < -100 mV), potential values were between those of magnetite reduction (reductive dissolution) and metal dissolution, suggesting auto reduction. Thus, Equation (4.33) was proceeding to the right. The rate of Fe^{2+} release increased with negative potential changes in agreement with Equations (1 and 2).
- At potential more negative than -450 mV, H_2 evolution predominated, and oxide dissolution ceased.

They also claimed that for oxalate concentrations greater than about 2×10^{-3} mol dm^{-3}, magnetite dissolves by direct chemical dissolution. The mechanism appears to involve adsorption of oxalate ions at ferric ion sites in the oxide lattice, followed by proton attack and desorption of cationic species. Once metal dissolution starts, β-ferrous oxalate dihydrate is precipitated on the electrode, leading to erratic fluctuations in the electrode potential and eventually to inhibition of metal dissolution. The use of oxalic acid for dissolving nuclear sludge and deposits in steam generators is described in Chapter 5. The work of Panias et al.[25] and Al-Mobarak[109] is also reviewed in Chapter 5.

Two electrochemical studies of iron oxide in EDTA solutions by Brunet and Turluer[110] confirmed reductive dissolution of Fe_3O_4 from steel surfaces. Ogura and Sato[111] showed that cathodic reduction of iron oxides on iron is drastically affected by chelating agents such as glycine, NTA, EDTA, and iminodiacetic acid; this probably is the result of a shift in $\alpha Fe^{2+}/\alpha Fe^{3+}$ caused by the preferential formation of Fe(ll) or Fe(III) complexes.

Both magnetite dissolution and dissolution of the base metal contribute Fe^{2+} to the solution. Separating the two contributions is at best a difficult procedure. Most of that effort focused on the dissolution rate of the base metal through a passive film. The rate of that process appears to depend on the ratio of film thickness, L, to the metal/solution potential difference, ϕ_M. The dissolution process implies transfer of metal ions to the oxide phase, followed by ion diffusion through the oxide, transfer through the oxide solution interface, and finally hydration of the ions. Thus, the potential difference at the Helmholtz layer is given by

$$\phi_H = \frac{\phi_M}{1 + \frac{L_{\varepsilon H}}{S_{\varepsilon F}}}$$ (Equation [4.35])

S is the thickness of the Helmholtz layer, and ε_H and ε_F are the dielectric constants of the Helmholtz and oxide layers, respectively.

Dissolution of Free Magnetite Mechanism in Acidic Media

This section and the next one provide more details of the mechanisms and rate of reaction of iron oxide dissolution chemistries. Work on dissolution of powdered magnetite led to an increased understanding of environmental factors affecting its rate of dissolution in aqueous media.

A significant conclusion[112-115] is that additives, which shift the solution potential in the negative direction, accelerate the dissolution of magnetite. In mineral acid, formation of Fe^{2+} increases the rate, whereas Fe^{3+} decreases it. In aqueous Na_2EDTA, conditions that favor formation of HFe(II)EDTA accelerate the dissolution. The reaction order with respect to H^+ is 0.5 in the mineral acids; in chelants, however, the pH dependence of the rate is not simple. In chelants, complexes can involve H^+ and thereby affect the dissolution rate are formed. Furthermore, in Na_2EDTA, the rate appears to be independent of Na_2EDTA. Note that these electrochemical effects were also described in the previous section.

The adsorption effect of chelating agents onto the oxide appears to be very complex and probably is controlled by the pH, ZCP of the oxide, and the temperature. The action of the low oxidation state metal ions (LOMI) reagents (including ferrous ions) helps to explain the "S" shape and complex kinetics of many dissolution processes. An example is shown in Figure 4.8.

In the absence of an applied potential, the dissolution of free magnetite in mineral acids proceeds according to the following proposed mechanism (mass and charge balances are not attempted).[116]

$$\begin{aligned}
&(a)\ Fe_3O_4 + 2H^+ \rightarrow Fe_2O_3 + Fe^{2+} + H_2O \\
&(b)\ Fe_3O_4 + H^+ \rightarrow FeO_x + Fe^{2+} + H_2O \\
&(c)\ FeO_x + Fe^{2+} + An^{y-} \rightarrow (FeO_{x-1})An \\
&(d)\ (FeO_{x-1})An \rightarrow (FeO_{x-1})^{(x-1)+} + An^{y-} \\
&(e)\ (FeO_{x-1})^{(x-1)+} + H^+ \rightarrow Fe^{3+} + H_2O
\end{aligned}$$ (Equation [4.36])

Data from Gorichev[116] suggest that step (c) is the slowest step of the reaction. Development of the surface area of the FeOx phase during the early stages of dissolution might be manifested in the form of an induction (incubation) period. Experimental evidence confirms the presence of an induction period. Consistent with this picture are the good fits of the data in all media (acids with or without chelants) to Equation (4.36).

However, the above mechanism does not properly account for the lack of an induction period in the case of Fe_2O_3 dissolution, i.e., dissolution of either Fe_3O_4 or Fe_2O_3 should, according to the mechanism, lead to formation of FeOx, whose decomposition is presumed to be the slow step. This author[116] also says that this criticism could be eliminated, however, if the initial disproportionation reaction, step (a), is even slower than step (c). This is indeed a strong possibility because step (a) involves two rather slow processes: lattice restructuring and solid-state diffusion of O^{2-}.

Additional experiments have provided kinetic information on the dissolution of free magnetite. Tests with dissolution of powdered magnetite of a particle size < 100 μm in HCl, H_2SO_4, H_3PO_4, and Na_2EDTA at 68–176°F (20–80°C) were investigated by Gorichev et al.[112] Reaction rate curves—fractional conversion, α, vs. time—were S-shaped (sigmoidal) in all cases. The rate curves for hematite (Fe_2O_3) and wustite (FeO), on the other hand, lacked the initial induction period observed with magnetite. Sigmoidal curves are typical of second-order autocatalytic reactions according to Pearson and Moore[117] of the type

$$A \longrightarrow B + C + \ldots\ldots\ldots$$

where the rate of loss of A (α is the activity of A) can be expressed by

$$d\alpha / dt = k\alpha(1-\alpha) \qquad \text{(Equation [4.37])}$$

Here, κ is the reaction rate coefficient. Reactions that are catalyzed by a product can often be treated with Equation (4.37). Many growth[118] and decay[119] processes also give sigmoidal curves. Higher order autocatalytic reactions may likewise be fitted into this scheme. For reactions of the type

$$A + B \longrightarrow C + D,$$

where C catalyzes the reaction, the rate at which A is lost can be expressed by

$$d\alpha / dt = k\alpha^{p}(1-\alpha)^{q} a^{r} B \qquad \text{(Equation [4.38])}$$

p, *q*, and *r* depend on the elementary steps in the mechanism, and α is the activity of reactant B. The rate curves for hematite and wustite can be treated by the simple "mth" order rate expression:

$$d\alpha / dt = k\alpha^{m} \qquad \text{(Equation [4.39])}$$

where m ˜ 1.

Gorichev et al.[112] found that their dissolution data for Fe_3O_4 fit the following for localized corrosion, or etching:

$$-\ln(1-\alpha) = a\sinh(kt) \qquad \text{(Equation [4.40])}$$

The model assumes that the rate is controlled by the formation (nucleation) and growth in one dimension (filaments) of a solid intermediate. The total rate of filament growth is a function of the rate of nuclei formation. The rate expression from which Equation (4.40) was obtained can be approximated by

$$d\alpha / dt = k\alpha^{m}(1-\alpha) \qquad \text{(Equation [4.41])}$$

Note that Equation (4.41) is a special version of Equation (4.38).

Various rate laws in addition to Equation (4.40) and its integrated form Equation (4.41) were examined, but these were less successful at fitting the data for magnetite dissolution. Most notable of these is the Erofeev relation,[120] which like the Gorichev et al.[112] expression, was developed to treat solid-state decomposition reactions in which a solid product phase is formed. This has the form

$$-\ln(1-\alpha) = kt^{m} \qquad \text{(Equation [4.42])}$$

In one scenario,[114] the dissolution of magnetite entails formation of an unstable iron-deficient oxide phase produced by the more rapid migration of iron ions out of the oxide matrix; the reaction rate is governed by decomposition of this peroxide-like species. Gorichev et al.[114] determined a value for *m* of 2.2 ± 0.2 in H_2SO_4, but it was apparent that the fit of the data was poor. The Erofeev reaction does not appear to be suitable for the treatment of data obtained before the induction period.

Azuma and Kametani[121] performed a series of experiments on the dissolution of ferric oxides in mineral acids. S-shaped curves similar to those of Gorichev et al.[114] were found. They treated their data with the expression

$$\log W = \log C + m \log t \qquad \text{(Equation [4.43])}$$

Here, W is the wt.% oxide dissolved at time t, and C is a rate constant. This rate law is a rearrangement of the Erofeev relation (Equation [4.42]).

In a different study, Gorichev et al.[113] found that the dissolution rates of wustite, hematite, and magnetite in EDTA all peaked at a value of 0.5 to 0.6 (Equation [4.42]). In the mineral acids, magnetite gave a constant value of A = 0.09, whereas in H_3PO_4, hematite gave a ˜ 10.

The reaction rate coefficients, k, decreased in the order H_3PO_4 > HCl > H_2SO_4. However, the reaction orders, n, with respect to molar acid concentration were 1.5, 2.3, and 1.25 for H_3PO_4, HCl, and H_2SO_4, respectively, at concentrations up to 6M. Azuma and Kametani[121] observed reaction orders of 1.2 to 1.8 for the same acids. The latter also measured the temperature dependence of the dissolution process: 18.2 to 21.9 kcal/mol. This relatively low value is approximately one-half that obtained by Gorichev et al.,[113] which suggests that the particle size used by Azuma and Kametani[121] was large enough that diffusion through the material limited the reaction rate. This phenomenon may be expected to lead to an (m + 1)/2 reaction order with respect to acid concentration, rather than m. Thus, if the *intrinsic* reaction rate in HCl displays a reaction order of 2.3, the Azuma and Kametani[121] data may be expected to yield n = 1.65; this is quite close to their observed value of 1.8.

Azuma and Kametani[121] also explored the effect of anion complexer strength. A direct relation between the stability constant of the Fe^{3+}/anion complex and dissolution rate was found. Of all the acids examined, HF, which forms the most stable Fe^{3+} complex, was the most vigorous solvent. Gorichev[115] also investigated the effect of the anion of the acid. He found that data for all the acids at various concentrations could be represented as one plot of α vs. $t/t_{0.5}$, where $t_{0.5}$ is the half-life of the oxide. This has been interpreted to mean that the rate-determining step is the product of two independent functions and may represent the same for all of the acids.[122] Thus, the rate coefficient of Equation (4.42) may be expressed by

$$k = f(T, E, H^+) \qquad \text{(Equation [4.44])}$$

Here, E is the potential for $Fe^{3+} + e^- = Fe^{2+}$, and T is the temperature. The rate of dissolution has been shown to be related to the transfer of protons from the acid to the iron oxide. The dependence of the specific rate of dissolution (Equation [4.44]) on the potential was shown by Gorichev et al.[112] to translate into a dependence on the concentrations of Fe^{2+}, Fe^{3+}, and the anion of the acid. This may be expressed by

$$k = k_j (aH^+)^{0.5} (aFe^{2+})^{0.5} (aFe^{3+})^{-0.5} (aAn^{y-})^u \qquad \text{(Equation [4.45])}$$

Here, a_j is the activity of species j, and ($a_{An}{}^{y}$-) is the activity of the anion. Gorichev et al.[113] found u = 0.5 for $SO_4{}^{4-}$ and $PO_4{}^{3-}$ – while u = 0.7 for Cl^-. Thus,

$$d\alpha / dt = k_j (aH^+)^{0.5} (aAn^{y-})^u \left(\frac{aFe^{2+}}{aFe^{3+}}\right)^{0.5} \alpha^m (1-\alpha) \qquad \text{(Equation [4.46])}$$

A significant conclusion of the Gorichev et al.[112,113,114,116,123] work may be summarized as follows: additives, which shift the potential in the negative direction, accelerate the dissolution of iron oxides.

Whenever acids are used to remove iron oxide from steel or iron, organic CIs are frequently used. While accelerative and inhibitive effects on the dissolution kinetics of the oxide were observed,[124, 125] few authors have tried to rationalize the results. Cumper et al.[126] looked at the effect of heterocyclic nitrogen compounds (indole and pyrrole) on the dissolution rate of magnetite in 2N HCl. When excess acid was used, the dissolution rates obeyed first-order kinetics for the first half of the dissolution process.

Pyrrole, indole, and 2,5-dimethylpyrrole all enhanced the dissolution rate, and pyrrole was the most powerful enhancer. The enhancement is explained by adsorption of the ring compound onto magnetite-pyrrole through its lone pair of electrons, and indole via its lone pair and its system.

Equation (4.42) contains only information about the development of an active surface area but does not account for the dissolution medium. For mineral acids[113] and Na_2EDTA, Gorichev[115] developed Equations (4.46 and 4.47), respectively, which contain explicit information about the dependence of the rate on the activities of H^+, An^{y-}, free Fe^{2+} and Fe^{3+}, and chelated Fe^2+ and Fe^3+. The observed reaction order, n, with respect to acid concentration is significantly greater than that given by Equation (4.45) or Equation (4.46).

$$d\alpha / dt = k_j (aH^+)^{0.5} (aAn^{y-})^u (\frac{\gamma \mathrm{Re}\, d}{\gamma Ox})^{0.5} (\frac{C_{II}}{C_{III}})^{0.5} \alpha^m (1-\alpha) \qquad \text{(Equation [4.47])}$$

The reaction order with respect to H^+ was assumed to be 0.5. Here, C_{II} and C_{III} are the overall concentrations of Fe(II) and Fe(III) containing ions, and γ_{RED} and γ_{OX} are the fractions of the free ions (i.e., Fe^{2+} and Fe^{3+}). The composition of the Fe(II) and Fe(III) complexes depends on the pH. The increase in dissolution rate brought on by increasing a_H^+ results in formation of the complex H[Fe(II)EDTA]. Reducing the pH to less than 2.3 causes a decrease in the dissolution rate, which is brought about by decomposition of that complex.

The different reaction orders may be reconciled by considering that at a high concentration, the activity of a solute does not linearly vary with molar concentration. For example, increasing the molar concentration of HCl from 1M to 4M increases the activities of H^+ and Cl^- by a factor of 8, rather than 4. Hence, Equation (4.47) may be expressed by

$$d\alpha / dt = k_j [H^+]^{1.0} (Cl^-]^{1.4} \left(\frac{[Fe^{2+}]}{[Fe^{3+}]} \right)^{0.5} \alpha^m (1-\alpha) \qquad \text{(Equation [4.48])}$$

in the concentration range of 1.0M to 4.0M. (This expression would be satisfactory down to 0.05M.) The sum of the reaction orders for H^+ and Cl^- is 2.4, which is quite close to the observed value of 2.3.[114] Equations (4.47 and 4.48) are difficult to work with because values of $\alpha Fe^{2+}/\alpha Fe^{3+}$ or Fe^{2+} Fe^{3+} are required. These are generally not measured. Hence, further development of kinetics treatments, with an eye toward eliminating the dependence of the rate on $\alpha Fe^{2+}/\alpha Fe^{3+}$, appears to be in order.

The work of Gorichev[116] on the dissolution of magnetite in Na_2EDTA closely paralleled his studies in mineral acids. A measured activation energy of 39 kcal/mol demonstrated the lack of mass transport limitations on the rate. It should be noted that the pH dependence of the rate was found to be more complex than in the case of the acids. A maximum in the rate at 140–176°F (60–80°C) occurred at a pH (controlled by H_2SO_4) of 2.3 and at a = 0.6. The rate coefficient, k, was found to be independent of EDTA concentration (zero order). The rate of dissolution was found to depend both on H^+ and on the oxidation/reduction potential of the solution, just as in the case of the mineral acids.

Rubio and Matijevic[127] looked at the dissolution of FeOOH in EDTA. The maximum rate occurred at a pH of 9.7. This was interpreted to mean that dissolution of FeOOH involves adsorption of EDTA and stabilization of the ferric EDTA species. Thus, different oxidation states of iron in the scale may lead to completely different dissolution mechanisms.

Two papers[60,127] expanded the knowledge of the adsorptive behavior of chelating agents and the role of surface interactions in the dissolution mechanism. These studies were focused on deposit removal from nuclear steam generators. Chang et al.[128] showed that the adsorption of EDTA and other polyamine polycarboxylate acid chelants onto hematite is strongly pH dependent and becomes very weak above a pH of 9.0. Although the absolute amounts of chelant absorbed vary with chelant structure, they all follow the same master curve when normalized for molecular size. When the dissolution of hematite was investigated by Chang and Matijevic,[129] a very complex function of chelant structure, pH, and temperature emerged. Two mechanisms were proposed to account for the observed behavior.

$$\begin{aligned}&a)\ \alpha Fe_2O_3 + 3H_2O \rightarrow 2Fe^{3+}(aq) + 6OH^- \\ &b)\ Fe^{3+}(aq) + \sum H_nL^{n-1} \rightarrow \sum FeH_nL^{3+n-1} + \sum Fe(OH)_mL^{3-m-1} \\ &c)\ Fe^{3+}(aq) \rightarrow Fe_x(OH)_y^{(3x-y)+}\end{aligned} \qquad \text{(Equation [4.49])}$$

$$\text{Solid} - Fe(OH) + H^+ + H_nL^{n-1} = \text{Solid} - FeH_nL^{n+l-i} + H_2O \rightarrow FeH_nL^{n+l-i}(aq) \qquad \text{(Equation [4.50])}$$

The mechanism of Equation (4.49) dominates at a pH of 9.0, whereas the mechanism of Equation (4.50) dominates at lower pH values, where adsorption is maximized. No correlation was found between the dissolution rate and the thermodynamic properties of the closely related chelants. This observation contrasts with that of Frenier[59] who claimed that a relationship exists between dissolution rate and thermodynamic properties at a pH of 9.2, when ammoniated chelants are used to remove iron oxide from steel surfaces.

The dissolution of magnetite in EDTA was described by Blesa and Maroto,[130] but the pH vs. reaction rate data were significantly different from those obtained by Chang and Matijevic[129] for hematite. Both authors found decreased dissolution at a very low pH (4.5) compared with moderate pH values (6); however, Blesa and Maroto found even lower rates at a pH of 8, whereas Chang and Matijevic found the opposite effect. Their studies were conducted at 77°F (25°C). As the temperature increased, the chelants in both cases became much more effective at low rather than at high pH values. Blesa and Maroto[130] claimed that the adsorbed EDTA blocks the surface and thus retards dissolution. As the temperature increased, this blocking action presumably becomes less effective, and the rate of oxide bond scission exceeds that of chelant/oxide formation.

As mentioned earlier, many iron oxide dissolution curves have a pronounced induction period that is characteristic of an autocatalytic process. In describing the dissolution of magnetite by oxalic acid, Baumgartner et al.[131] proposed a mechanism in which the ferrous oxalate ion catalyzes the dissolution through a fast electron transfer reaction (reductive dissolution). They proved this thesis by adding more ferrous ions and noting the disappearance of the induction period. An important insight is that chelated ferrous ions can act as reducing agents for ferric ions in the scale if the chelant preferentially stabilizes the oxidized state with respect to the reduced state. Presumably, this is true for oxalic acid. It is true for EDTA and citric acid at a pH of 9.[51] Baumgartner et al.[131] showed that the standard cell EMF (E_o) is changed by 790 mV (in the negative direction) for iron EDTA and 990 mV for iron citrate compared with the aquo couple. Thus, the mechanism proposed by Baumgartner et al.[131] is probably active in other chelant systems.

Other work on the mechanism of magnetite dissolution in chelants was recently reported. Hausler[132] looked at magnetite dissolution in ammonium EDTA at pH values from 4.2 to 7.0 and found that hydrazine accelerated the dissolution rate at a pH of 7 but not at 4.2. Instead of invoking the accepted reductive dissolution mechanism, he proposed an unusual N_2H_4-Fe(III)EDTA complex to explain his results.

In the absence of a steel surface, reducing agents are frequently added to chelant solvents to increase the reaction rate of magnetite dissolution. Mixtures of ammonium EDTA and N_2H_4 were developed to remove deposits from nuclear steam generators. A great deal of work was done to develop practical solvents for cleaning the secondary side of steam generators, but the five references listed[42-46] are illustrative of the process. These authors apparently assume that N_2H_4 fosters the reductive dissolution mechanism, but the details are not addressed. The one exception is Hausler,[132] who favors an unusual iron hydrazine complex. Because the deposits in nuclear steam generators frequently contain mixed oxides, such as iron/chromium and iron/nickel species, other reducing metal complexes were suggested. These are called low oxidation state metal ion (LOMI) reagents and include Fe(II)EDTA, Cr(III), and V(II) picolinate. These materials have been used extensively to clean the primary side (usually called decontamination). Presumably, they catalyze scale dissolution by the Baumgartner[131] mechanism. More details of nuclear deposit removal are described in Chapter 5.

Dissolution Mechanism of Iron Oxide from Steel Surfaces

Compared with the large amount of work on the mechanism of free iron oxide dissolution, there is relatively little information available on the dissolution of oxides from steel surfaces. See Brunet et al.[133] and Hausler.[132] A major contribution of these studies is confirmation that the reductive dissolution mechanism is operative, thus leading to oxidation of the base metal at the expense of ferric reduction in the scale. Because operational iron oxide scales undoubtedly contain numerous microcracks,[134] the anodic reactions probably take place at these cracks.

Figure 4.23 shows Frenier's schematic representation of magnetite dissolution from a ferrous metal surface. The iron is oxidized at the microcracks, releasing electrons that reduce Fe^{3+} in the scale. The Fe^{2+} ions can pass across the Helmholtz layer into the bulk solution. Protons also cross the double layer to react with the oxide ions initially forming OH^- ions. The passage of the products across the interphase is considered the rate-determining step. Properties of the solution that are known to influence the reaction rate include the presence of strong complexers such as chelants (EDTA) and inorganic reactants, e.g., F^- ions, and the pH. The complexers affect the rate by removing Fe^{2+} ions from the region around the double layer or by changing the surface charge, thereby changing the potential.[1] Although the rate may depend on the strength of the complexer,[60, 111, 121] it may not depend on the concentration of a free complexer.[113] However, the pH has a marked effect on the rate. Most authors suggest a half order dependence on H^+ in the absence of complexing agents, but the pH dependence in complexer solutions is far from straightforward.[115]

Two overall equations can be written for magnetite dissolution from steel surfaces. Equation (4.32) summarizes the species that are involved when no chelant is present.

If an *ammonium* chelant is present, Equation (4.32) becomes Equation (4.12). If an acidic chelant is present, the equation is

$$Fe_3O_4 + Fe + 8H^+ + 4Chel \rightarrow 4FeChel^{2+} + 4H_2O \qquad \text{(Equation [4.51])}$$

Although the general aspects of the mechanism seem to be worked out, many details of the kinetics still need to be established. The dissolution of bulk Fe_3O_4 on steel needs to be examined via detailed studies such as those of Gorichev et al.,[112,113,114,116,123] which will, out of necessity, entail flow experiments, e.g., in a loop or with rotating disk specimens.

Electrochemical studies of such composites also would be useful. A major problem will continue to cloud experiments in which the concentration of Fe in solution is used as the primary kinetic marker; the corrosion of the base metal by the solution (with or without inhibitors) must be subtracted from the oxide dissolution rate. Unfortunately, no information exists on the effect of the oxide cover (other than passive layers) on the corrosion rate of the base metal. Finally, the scale morphology may play a major role in the application of these theories to descaling of real heat transfer equipment. Much of following description of reduction dissolution is taken from an article by Frenier and Growcock.[60] Several experiments are noted in the next paragraph and in Section 4.3.2.

Reductive Dissolution Tests

The reductive dissolution of magnetite is well established in acidic solutions.[105,107] Also see the electrochemical papers noted earlier in Section 4.3.1. Per this mechanism, Fe_3O_4 dissolves and directly releases Fe^{2+} ions, and the base metal or some other reducing agent is consumed. This mechanism may also be active in chelate solutions. Magnetite electrically coupled to steel coupon dissolves at a faster rate than isolated scale (Table 4.7). Pocock and Leedy[66] compared several solvents for the removal of Fe_3O_4 from nuclear steam generators. Because the generators contain little carbon steel, chemical reducing agents were used. A solution of hydrazine and ammonium EDTA was the most effective solvent. In these tests, Fe_3O_4 was stirred into an autoclave with ammonium EDTA at a pH of 7.5 and at 203°F (95°C). The chemical reducing agent accelerated the dissolution rate (Table 4.8).

Table 4.7 Reductive Dissolution Test, 0.13M $(NH_4)_2EDTA$, 203°F (95°C)

Scale Condition	Rate (g/cm²/h)
Isolated	1.3×10^{-2}
Electrically coupled	7.1×10^{-2}

Table 4.8 Hydrazine Concentration Test,
$(NH_4)_2EDTA$, pH of 7.5, 203°F (95°C)

	% Fe_3O_4 Dissolved			
% N_2H_4	2 h	4 h	6 h	8 h
0.25	21	25	28	26
1.0	67	84	86	90

4.3.2. Details of Magnetite Removal from Steel Surfaces in Chelant Solutions

Because of the importance of chelating agents to boiler cleaning, several lab tests were performed. See Frenier[60] for more details.

Reductive Dissolution Tests in Chelating Agent Solutions. A test for the ratio of Fe(II)/Fe(III) in solution during a cleaning test was devised using an electrochemical method.

Reductive dissolution was studied by determining the oxidation state (2+ or 3+) of the iron during simulated, dynamic scale dissolution tests. The ratio of $Fe(II)EDTA^{2-}/Fe(III)EDTA^{-}$ can be determined by measuring the redox potential (EMF) of the solution. For a one electron change at 77°F (25°C), the Nernst equation predicts that

$$E = E_0 - 0.059 \log \frac{[Fe(II)EDTA^{2-}]}{[Fe(III)EDTA^{-}]} \qquad \text{(Equation [4.52])}$$

This equation assumes concentration equals activity. Here, E is the redox potential of the cell, and E_0 is the standard redox potential at unit activity. Thus, a plot of EMF vs. log[Fe(II)EDTA/Fe(III)EDTA] should give a straight line with a slope of -0.059 mV. For the iron/EDTA system, this was measured (with the cell described in Figure 4.13) by following EMF changes during the air oxidation of $Fe(II)EDTA^{2-}$. Note that $Fe(II)EDTA^{2-}$ is colorless in the visible spectrum, but at a pH of 9.2, $Fe(III)EDTA^{-}$ has a strong absorbance at 345 nm as measured on a Beckman DB spectrophotometer. From a Beer's law plot of absorbance vs. concentration, $[Fe(III)EDTA^{-}]$ could be determined and $[Fe(II)EDTA^{2-}]$ calculated from knowledge of total Fe in solution.

A stock solution of $Fe(II)EDTA^{2-}$ was prepared by refluxing iron powder with 10% tetraammonium EDTA for 36 hours, followed by filtration to remove unreacted iron. The solution was reduced to $Fe(II)EDTA^{2-}$ by stirring with excess 64% N_2H_4 overnight. The total Fe concentration was then determined by atomic absorption. To determine the EMF relationships, 750 mL of 4% tetraammonium EDTA was thermostated at a desired temperature, and prepurified N_2 was bubbled through for 1.5 hours. Enough stock $Fe(II)EDTA^{2-}$ was introduced to give a desired total Fe concentration (usually 0.8%), and the pH was carefully adjusted by passing NH_3 through the solution, or by adding glacial acetic acid. The potential (EMF) of the solution was measured between a SCE and the platinum electrode using an NLS X-2 digital voltmeter. Air was introduced with a gas sparge; at various intervals, the EMF was recorded, and the $Fe(II)EDTA^{2-}/Fe(III)EDTA$ ratio was spectrophotometrically determined on aliquots. At all times, the pH was maintained at a ±0.05 pH unit. By using various pH values, plots of EMF vs. [Fe(II)]/[Fe(III)] were developed. As a check, runs were made by adding known amounts of ferrous and ferric sulfate to ammonium EDTA. These points fell on the same line as those determined with $Fe(II)EDTA^{2-}$ made from digested Fe. Therefore, if the pH and temperature are known, the ionic ratio can be determined from the EMF solution. Note that the slopes are somewhat different from those predicted by the Nernst equation, and there is a strong pH dependence.

A dynamic test loop (Figure 2.22) was used to simulate cleaning environments to test reductive dissolution mechanisms. The loop was a completely PTFE coated circulation system capable of maintaining temperatures above 320°F (160°C) and pressures above 725 kPa. The volume was about 34 L, and flow could be regulated from 0 to 270 L/min. For this test, a 3-ft (91-cm) section of 6.35-cm ID boiler tubing taken from a Midwestern U.S. drum boiler was used. An X-ray diffraction analysis of the scale is seen in Table 4.9.

Table 4.9 Analysis of Scale Used in Loop Test

Material	%
Fe_3O_4	35-45
Fe_2O_3	30-40
Cu	15-20
Cu_2O	5-10

The loop was filled with 6% tetraammonium EDTA at a pH of 9.2, purged with N_2 for approximately 1 hour, and brought to an operating temperature of 320°F (160°C) with a flow velocity of about 300 cm/s. During the first hour, samples were taken every 10–15 minutes, and [Fe] was determined by atomic absorption (AA) spectrophotometry. After the first hour, samples were taken every 30 minutes. The EMF was determined by running a sample into a 100-mL cell containing a Pt and a SCE. Before taking the sample, the cell was purged with N_2 for 5 minutes. From the EMF solution and the Fe, ionic concentrations were calculated.

Figure 4.24 shows the results of the test to determine the oxidation state of Fe in solution. Initially Fe(III) was the dominant oxidation state; however, as time progressed, the concentration of Fe(II) became the dominant state, and the rate showed a steady increase until a plateau was reached after about 3.5 hours. The tube was free of magnetite at the end of the test. See Frenier[60] for more details.

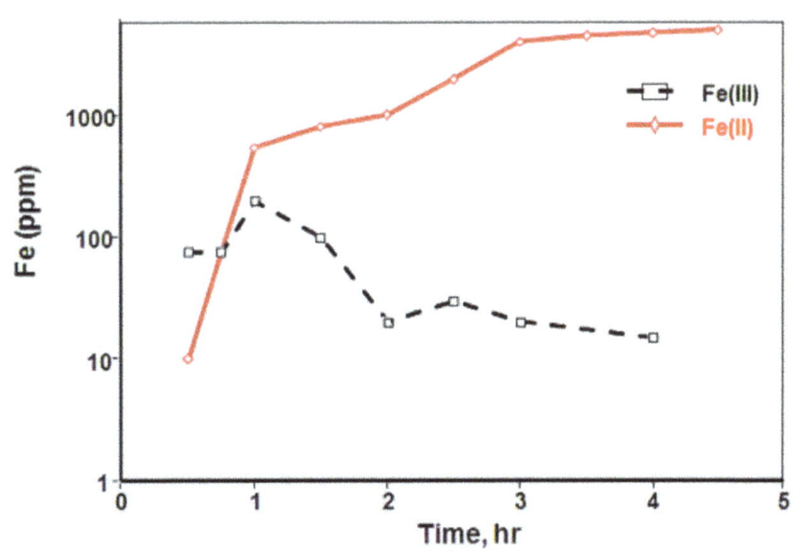

Figure 4.24 Concentrations of Fe(II) and Fe(III) during Tube Cleaning at 320°F (160°C)

Reaction Kinetics of Iron Oxide in Chelants

Frenier and Growcock[60] investigated magnetite dissolution from steel surfaces. They demonstrated that the protons could be supplied by ammonium ions (NH_4^+) at a pH of 9. Also, the reaction rate depended on the strength of the iron-chelating agent [the Fe(II) species presumably is stabilized]. Thus, the reaction rates were found to decrease in the order EDTA > NTA > citric acid. Because of the importance of EDTA-type chelating agents to the practice of modern chemical cleaning of steel surfaces, this subject is considered in more detail in the next chapter.

The data from these three experiments indicate that reductive dissolution occurs in ammonium EDTA solutions. Although some magnetite may directly dissolve, releasing ferric ions that subsequently reduce on the bare steel, reduction of ferric iron in the scale by base metal or N_2H_4 appears to predominate. The accelerative effect of reductive dissolution in acidic solution is frequently attributed to potential changes across the Helmholtz layer.

The Erofeev empirical equation[120] (Equation [4.42]) was used to analyze the iron dissolution data. In most cases, the boiler tubes were 100% clean after 6 hours, and the fraction dissolved (α) was calculated from the final concentration of Fe in solution at the end of the tests. In those cases, where the tubes were not clean, a final Fe of 3000 ppm, based on the average data from the other tests, was used to estimate α. The reaction order, n, and the dissolution rate constant, k, were calculated from a log-log plot of ln $(1/1\text{-}\alpha)$ vs. t.

The rate constants and corrosion rates of the 1010 (carbon steel) coupons for five ammonium chelants tested at 320°F (160°C). The dissolution curves are seen in Figure 4.8, and the data are summarized in Table 4.10.

An EDTA solvent also was tested with two different cations—Na^+ and NH_4^+. The cation effect is shown in Table 4.11 and Figure 4.25. The effect of temperature is described in Figure 2.20 and Table 4.12. The corrosion rates of the 1010 CS coupons were very low, indicating that the iron detected was associated mostly with scale dissolution, not general solvent corrosion. The data helped to quantify the effects of chelate type (stability constant for Fe), temperature, and cation.

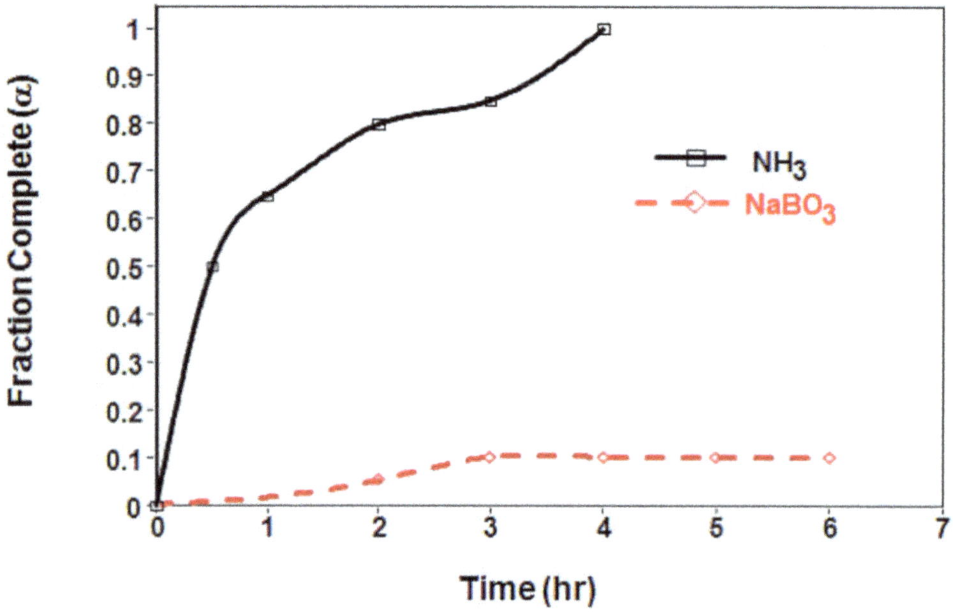

Figure 4.25 Iron Oxide Dissolution Tests at a pH of 9.2 with EDTA Test Temperature at 320°F (160°C)

Table 4.10 Rate Constants for Iron Oxide Dissolution in Ammonium Chelant Solutions

Iron Oxide Dissolution at a pH of 9.2; adjusted with NH_4OH, 0.12M Chelant, 320°F (160°C)			
Chelant	**k(h^{-1})**	**n**	**Rate, 1018 (lb./ft^2/day)**
Citric acid	0.08	0.65	0.005
NTA	0.12	0.20	0.004
EDTA	0.28	0.70	0.005
HEDTA	0.28	0.96	0.003
DTPA	0.35	0.65	0.004

Table 4.11 Rate Constants for Iron Oxide Dissolution in Na^+ and NH_4^+ EDTA

Iron Oxide Dissolution at a pH of 9.2 With 0.12M EDTA, 320°F (160°C)			
Buffer	**k(h-^{1})**	**N**	**Rate, 1018 (lb./ft^2/day)**
$NaBO_3$	0.003	-	0.002
NH_4OH	0.28	0.70	0.005

Table 4.12 Rate Constants for Iron Oxide Dissolution Temperature Effects in $(NH_4)_4$EDTA

Iron Oxide Dissolution at a pH of 9.2, 0.12M $(NH_4)_4$EDTA			
Temp, °C	**k (h^{-1})**	**N**	**Rate, 1018 (lb./ft^2/day)**
160	0.28	0.70	0.005
130	0.28	0.63	0.005
120	0.12	0.72	0.003

In the present case, in which the magnetite is electrically coupled to steel and reductive dissolution is invoked, a mass action mechanism can be proposed wherein the chelant removes Fe^{2+} (and some Fe^{3+}) ions from the interface. The rate coefficients (k, Table 4.12) are in the same order as the Fe^{2+} stability constants (Table 1.2) of the ligands tested (Citric < NTA < HEDTA = EDTA < DTPA), but the differences in the rates are large only when comparing the weak chelants with the strong ones. At a pH of 9.2, citric acid is such a poor complexer for Fe^{2+} that the reaction rate is almost zero. This observation of the effect of complexer strength is similar to that of Azuma and Kametani,[121] who reported that mineral acids with strong complexing ions (e.g., HF) were more active solvents than those (H_2SO_4) with weaker complexers. A total reaction order of <1.0 was observed in all the tests. Fractional reaction orders indicate complex mechanisms and are observed when the rate-controlling step is limited by diffusion. This total reaction order is in the same range as reported by Brunet.[135]

When sodium is substituted for ammonium in the chelant formulation, the reaction rate becomes very low at alkaline pH values. Both Gorichev[115] and Lesinski [54] made similar observations. Apparently, the ammonium ion provides the protons necessary to support the scale dissolution reaction. Note that Equation (4.53 B) is essentially the same as Equation (4.51).

$$\text{A) } NH_4^+ = H^+ + NH_3 \qquad \text{(Equation [4.53])}$$
$$\text{B) } Fe_3O_4 + Fe + 8H^+ + 4Chel \rightarrow 4Fe(Chel)^{2+} + 4H_2O$$

Because ammonia is at its maximum buffer point (pH of 9.2), the pH remains almost constant. The results described in Table 4.11 and Figure 4.25 confirm the effect of the proton donor on the scale dissolution reaction. $NaBO_3$ was chosen as the basic component instead of NaOH so that a buffer comparable to NH_4OH would be present. Thus, large pH changes should not have obscured the cation effect.

The effect of temperature change on the rate of magnetic dissolution in $(NH_4)_4$ EDTA is shown in Table 4.12. Although the data are insufficient to calculate an activation energy with precision, the temperature dependence is small (~15 kcal/mole) in the range tested. This small number suggests a diffusion-limited, rate-controlling step.

From the data gathered in these experiments, the following mechanism is proposed for the dissolution of magnetite from a steel surface using an alkaline chelate. See Figure 4.23.

$$\begin{aligned}
&\text{A) } Fe \rightarrow 2e^- + Fe^{2+} \text{(anode – microcracks in scale)}\\
&\text{B) } Fe^{3+} + e^- \rightarrow Fe^{2+} \text{(cathode)}\\
&\text{C) } Fe^{2+}\text{(solid)} \rightarrow Fe^{2+}\text{(soln – RDS)}\\
&\text{D) } Fe^{2+} + \text{Chel} = \text{Fe(II)Chel}\\
&\text{E) } H^+ + O^{2-} \rightarrow OH^-\\
&\text{F) } OH^-\text{(solid)} \rightarrow OH^-\text{(soln)}\\
&\text{the overall rxn. is:}\\
&\text{G) } Fe_3O_4 + Fe + 8H^+ + 4\text{Chel} \rightarrow 4\text{Fe(Chel)}^{2+} + 4H_2O\\
&\text{If ammonia is present the overall rxn. is:}\\
&\text{H) } Fe_3O_4 + Fe + 8NH_4^+ + 4\text{Chel} \rightarrow 4\text{Fe(Chel)}^{2+} + 4H_2O + 8NH_3
\end{aligned} \quad \text{(Equation [4.54])}$$

All relevant information indicates that step D (diffusion of Fe^{2+} across the Helmholz layer) is rate determining when magnetite is in physical contact with iron or another electron source. This conclusion is most strongly supported by the small heat of activation, the fractional reaction order, and the relatively mild effect of chelating agent strength on the rate. A significant effect is noted only when very weak chelants (citrate and NTA) are used.

4.3.3. Effects of Morphology of Oxide Deposits

Experience in the chemical cleaning industry showed that no two iron oxides are alike. The heat history, density, and impurities may have a very large effect on the actual rate of deposit removal. Frenier[136] has shown examples of highly mixed deposits of iron oxide, iron sulfide, and possibly alkaline earth salts in tubing scales. Note Figure 3.17 and Figure 3.18.

Azuma and Kametani[121] made systematic studies of the effect of sample preparation on oxide dissolution. The formation temperature of the oxide markedly influenced the dissolution rate. The physical properties apparently controlling the reaction were the surface area (calculated by a permeability method) and a normalizing factor that is dependent on the aggregate crystal structure. Fields[134] also demonstrated the importance of the scale structure (layers and microcracks). McPherson and Martin[137] showed that the presence of chromium-containing spinels has a marked influence on deposits found in the super-heat or reheat sections of power boilers.

4.4. Solvents for Deposits Containing Calcium, Barium, or Silicon

Except for halides and carbonates, deposits containing these elements that form salts with other anions are not soluble in HCl and thus can be very difficult to remove from many types of industrial units. However, these types of deposits can form in many aqueous solutions and foul surfaces. See Sections 3.5 and 3.6. Although these solids can constitute most of a fouling scale, they may be mixed with acid-soluble minerals that can require complex dissolution processes.

4.4.1. Solvents for Alkaline Earth Salts

Carbonates. Calcium carbonate as well as iron carbonate are soluble in HCl and are easily removed from industrial equipment using this solvent or organic acids and chelating agents. The dissolution of alkaline earth carbonates by acids is one of the most investigated surface reactions, especially as it is applied to acidizing and permeability stimulation by worm-hole formation. The details are beyond the scope of this book. See the reviews by Hill and Schechter,[138] Fredd,[139] and Fredd and Fogler.[140] Additional studies that are applicable for calcite scale are described below.

If the solvent has available protons, the reaction with the carbonate to produce CO_2 also acts to help break up the scale lattice because the structure of the crystal is destroyed. Since the surface reaction rate is not a major factor (especially at elevated temperatures) in the overall process, the key consideration is the capacity of the solvent, and this is controlled by the solubility of the calcium salt formed by this reaction, as well as by the molecular weight of the anions. Calcium chloride formed from HCl is very soluble.

Common organic acids such as formic acid and acetic acid also readily dissolve calcium carbonate but form insoluble precipitates at high enough concentrations of calcium to limit the capacity of organic acid solvents. The molecular capacity of the solvents also is important. HCl has the lowest ratio of molecular weight to molecular weight of calcite of any practical solvent, so the theoretical capacity is the highest for any solvent. Table 4.13 shows laboratory data for dissolution tests performed at atmospheric pressure that compares capacities of several possible solvents. They reflect realistic solvent concentrations. These tests (described in Frenier et al.[141] and LePage et al.[142]) were run at 150°F in a loosely closed reactor for 24 hours that contained an excess of calcite.

Table 4.13—Calcite Dissolution Capacity of Various Dissolvers (Excess Chelant and Calcite)

Dissolver	$CaCO_3$ dissolved (lb./1000 gal)
20% Trisodium HEDTA (pH 2.5)	738@
20% Tetrasodium EDTA (pH 12)	520@
21% Diammonium EDTA (pH 4.5)	735@
20% Disodium HEIDA (pH 2.5)	250@
20% Tetrasodium GLDA (pH 1.5)	1600#
15% HCl	1833@
10% Formic acid	845@
5% Citric acid	5@
20% Trisodium NTA (pH 4)	5@
10% Acetic acid	422@

@ Frenier (2001a) # LePage et al.[142]

The reaction equation for acetic acid with calcite is

$$CaCO_3 + 2(CH_3\text{-}COOH) = Ca(CH_3\text{-}COO^{-}) + CO_2 + H_2O \qquad \text{(Equation [4.55])}$$

Sulfates. The three major sulfate minerals that could foul industrial equipment are gypsum, anhydrate, and barite. Unless an x-ray diffraction test has been performed, the deposit will be described only as barium or a calcium sulfate. The state of hydration may dramatically affect the dissolution rate of calcium sulfate, but this will be a morphology issue unless the exact crystalline form is known. It is also possible for calcium and barium sulfate crystals to be mixed into a composite scale. In addition, calcite can mix with the sulfate. Some fields also have reported scales that contain strontium sulfate, and some sulfate deposits can contain radium or radon. These mildly radioactive deposits are called naturally occurring radioactive materials (NORM). These different solids may also contaminate pigs that have cleaned lines that are associated with some oil/gas fluids and may require cleaning before reuse or disposal.

Sulfate deposits are difficult to dissolve because they are not soluble in acids. This is the case since there is no way to effectively remove the sulfate anion from the system without removing the anion from the solvent. See the equilibrium Equations (3.12–3.16) in Section 3.1.

In addition, oxides or carbonate scales become opened-up physically during the dissolution process because the formation of the CO_2 and H_2O reaction products can produce microcracks. This process can improve penetration of the scale by the fresh solvent. Therefore, the effective rate of dissolution of a carbonate or oxide increases dramatically during the process as the available surface area increases. The dissolution of a sulfate scale takes place on the surface of the crystal until natural inhomogeneities cause the mass to fragment. This fragmentation is important to the overall dissolution rate but is not a predictable phenomenon. In addition, there are large differences in the solubility of calcium sulfate and barium sulfate (Table 3.2). As described below, barium salt is especially difficult to remove. Since both sulfates are products of the mixing of production water streams, analyses of these water sources will give an indication of either barium or calcium sulfates. However, only a chemical examination of the deposit found in a line (Section Section 2.6.1) can distinguish between these two deposit types to select a specific solvent.

Before the introduction of chelating agents to dissolve sulfate deposits, the only method for removal of these scales (usually gypsum deposits in surface equipment) consisted of stages of alkaline agents such as sodium carbonate (soda ash), which were soaked into the deposit to convert the surface to calcium carbonate.[143] The converted deposit is then soluble in HCl. Multiple stages of soda ash followed by HCl were often needed to remove the deposit. This was a very slow process and required large volumes of chemicals.

Cowan and Weintritt[18] have a lengthy section on methods to disintegrate gypsum. They note that KOH, NaOH, and sodium carbonate (soda ash) are used to convert the gypsum to a more soluble salt, and then HCl is the used to remove these salts. They also claim that low mol weight aminomethylenephosphonates (AMP) can remove scale deposits, and the products based on AMP derivatives have been used to remove calcium sulfate scale in oilfield applications. These are chelating agents that act in a mechanism like EDTA.

Solutions of sodium EDTA can be used to dissolve deposits containing calcium sulfate or calcium carbonate. In fact, the removal of calcium was one of the first procedures performed using EDTA formulations. Bersworth[144] claimed the use of alkali metal (sodium or potassium) salts of EDTA for removing hardness deposits from boilers. The pH was >8, and the temperature was at about the boiling point of water. He even developed a method for regenerating the EDTA used to dissolve calcium salts by adding sulfuric acid to precipitate calcium sulfate. The pH is then elevated to the use level (12) with sodium hydroxide. Hayashi and Kawazoe[145] cleaned well tubing at a geothermal plant using a pH 6 solution of potassium EDTA. The treatment temperature was up to 400°F (200°C), but this solvent was a factor of 10 less corrosive than inhibited HCl to the well tubing (steel). Also, Vinso[146] used sodium EDTA at pH 8–11 to remove hardness, then lowered the pH to 4–5 with acid, and used the same solution to dissolve iron oxides.

Lawson[147] described a hydroxamic acid-based solution, preferably containing a potassium salt, that will disintegrate and dissolve gypsum. The most preferred material is potassium methylhydroxamic acid (Potassium methylhydroxamic acid 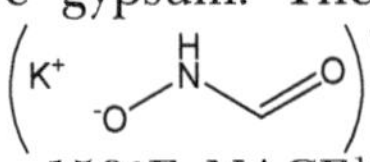

). The formulation used was very alkaline, and the preferred reaction temperature was 150°F. NACE[148] has reviewed test methods for comparing gypsum solvents, and the report reviewed data from various testing labs. The test methods were similar to the low-temperature and high-temperature processes described in Section 2.6.4. Potassium salts of various chelating agents have also been proposed for dissolving barite. See the next paragraph.

Barium-containing scales are more difficult to remove than calcium-containing scales because the kinetics of the dissolution reactions are much slower than in the calcium scales. There has been considerable interest in removing $BaSO_4$ scales from the tubulars and the near wellbore region. Conventional chelant solvents have slow reaction rates for these scales. An additional concern with the barium sulfate deposits is the occurrence of NORM in conjunction with the sulfate scales. Paul and Morris[149] proposed several solvents for the sulfate scales that are based on the use of a scale converter or dissolution catalysts with a chelating agent. The chelate can be EDTA based, but DTPA is claimed to be more effective in these formulations. Oxalic acid, fluoride, and dithionate are possible catalysts. The basic idea is that the converter or catalyst reacts at the surface of the scale and converts the deposit to a more soluble form that the chelate can then complex.

Chelating agents may remove various deposits in one treatment (if there is enough solvent capacity). The reactions of chelating agents with calcite are shown in Equations (4.56–4.58).

$$CaCO_3 + Na_4EDTA = CaEDTA^{2-} + 4Na^+ + CO_3^{2-} \qquad \text{(Equation [4.56])}$$

$$CaSO_4 + Na_3HEDTA = CaHEDTA^- + 3Na^+ + SO_4^{2-} \qquad \text{(Equation [4.57])}$$

$$BaSO_4 + K_4EDTA = BaEDTA^{2-} + 4K^+ + SO_4^{2-} \qquad \text{(Equation [4.58])}$$

For a chelating type solvent to be effective, the ratio of the solubility product to the chelate equilibrium constant (K_{sp}/K_{eq}) must be greater than 1. See the discussion of solubility in Section 3.1. Note that EDTA and DTPA could theoretically dissolve $CaSO_4$ because the ratio is > 1, whereas none of the chelating agents have a ratio > 1 for $BaSO_4$. This makes dissolution of barite much more difficult (slow) compared with gypsum dissolution. The use of chelates are effective for dissolving gypsum at moderate temperatures (see Figure 4.26). Here, HEDTA and EDTA at pH 12 were effective, whereas the pH 2.5 solvent dissolved much less scale.

Solvents for barite are effective only at high pH (12) and high temperatures (>150°F). Figure 4.27 shows examples of reaction rates.

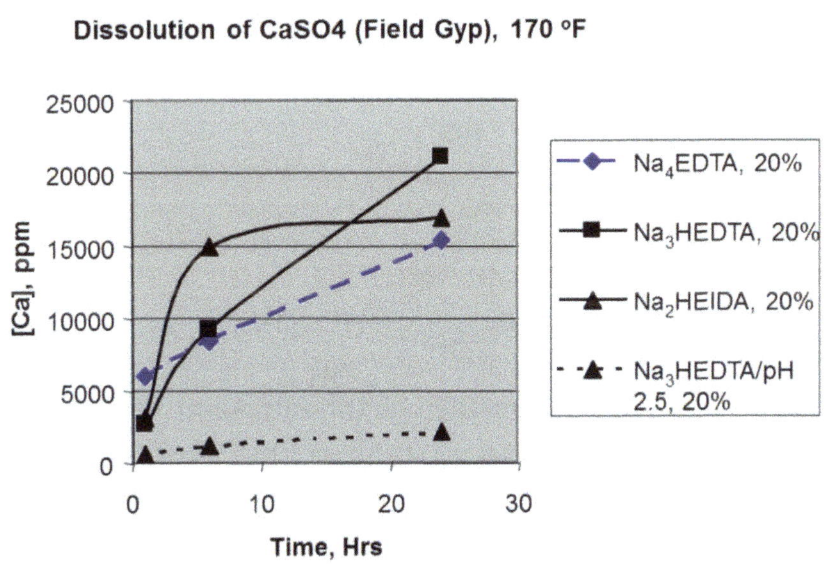

Figure 4.26 Dissolution of Gypsum

Note that several cheating agents at high pH (12) will readily dissolve gypsum; barite is very difficult to dissolve, and after 50 hours (Figure 4.27) less Ba is in solution than the gypsum (Ca) after only 5 hours. In addition, only K_4EDTA with an "X" additive was effective at the lower level. There is a large number of "additive" kinetic accelerators for dissolving barite that is reviewed in Chapter 5 of Frenier and Ziauddin.[4] Also see the patent by Paul and Morris.[149]

A patent by Zaid and Wolf[150] lists a solvent for alkaline earth sulfates that contains a combination of ingredients that include EDTA, a bicarbonate, ammonia, and a phosphonate chelating agent such as nitrilotri(methylphosphonic acid) (See Figure 4.28). They placed 25 mL of this solution in a beaker and heated it to about 160°F in a hot water bath. Per this patent, barium sulfate was added to the heated solution. The barium sulfate was completely dissolved within 15 minutes. However, the commercial application of this solvent has not been confirmed.

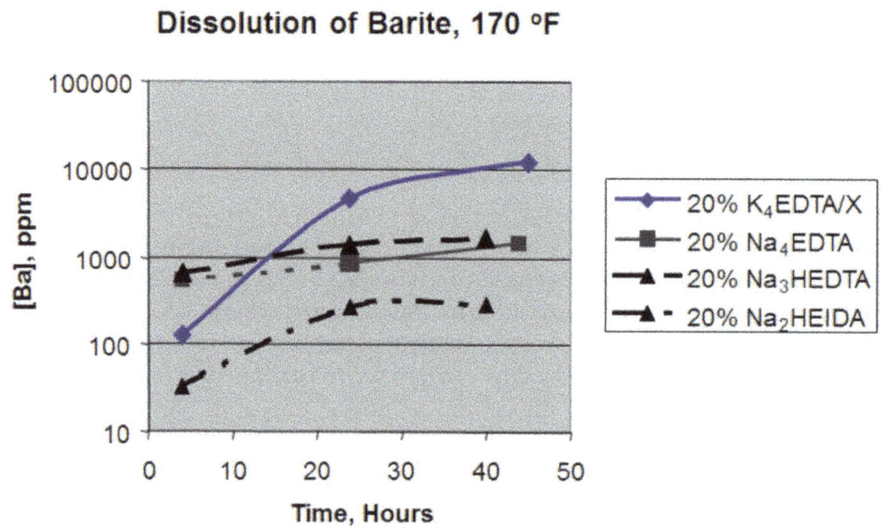

Figure 4.27 Dissolution of Barite

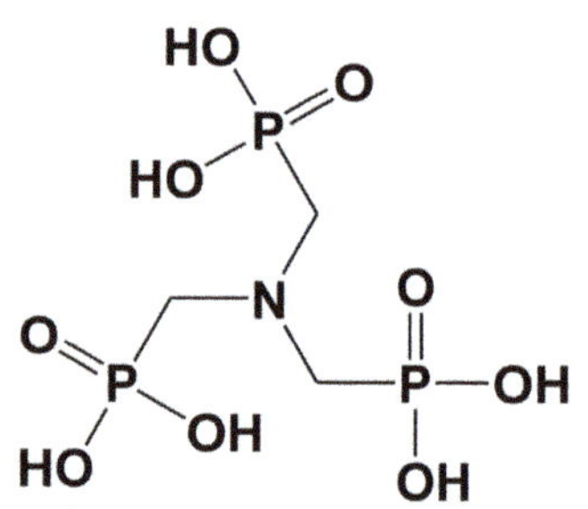

Figure 4.28 Component of a Barium Sulfate Solvent

4.4.2. Solvents for Silica and Silicate Deposits

As noted in Section 3.6, silica or metal silicates are soluble only in HF-based solvents or in high-pH alkalis. Section 4.1.1 and Equation (4.7) describe HCl/HF-containing fluids that will dissolve silica and some metal silicates, but the pH and other ions must be carefully controlled (Frenier and Ziauddin[11]), or various precipitates may form. These include CaF_2 as well as hydrated (amorphous) silica. Additional problems with the use of HCl/HF solvents is the toxicity of HF. Also, the addition of fluoride to the HCl solution can make it more difficult for the CIs (Section 6.2) to protect a steel surface during the cleaning procedure.

This section highlights developments for using non-HF solvent for dissolving silicate and silica-containing deposits.

Silica (SiO_2) or hydrated silica ($H_{2x}SiO_{2+x}$) is soluble in NaOH (or KOH). The two-stage reaction is

$$\begin{aligned} &1)\ H_{2x}SiO_{2+x} + 2xNaOH \rightarrow Na_{2x}SiO_{2+x} + 2xH_2O \\ &2)\ Na_{2x}SiO_{2+x} + (2-2x)NaOH + xH_2O \rightarrow 2Na^+ + H_2SiO_4^{2-} \end{aligned} \quad \text{(Equation [4.59])}$$

Thus, strong caustic fluids can remove silica or polysilica from surfaces, but the reactivity with the specific silicon-containing solid must be evaluated. Several alkaline/additive solvents are described.

According to Gahraibeh et al.,[151] a silica solvent that is based on NaOH solutions can dissolve different silicate minerals. These authors note that although strong NaOH may dissolve silica, the dissolution of common *silicates* may require the addition of chelating agents and/or surfactants to deal with the other cations in a field scale. The report details the development of a solvent that was tested using three types of silicate scales. See Table 4.14.

Table 4.14 Powder X-Ray Diffraction Analysis Results for Samples I, II, and III

Sample	Compound	Chemical Formula	Color in Diffractogram	Content%
I	Kanemite	$Si_2NaO_8H_8$	Black	83
	Amorphous part			17
II	Kanemite	$Si_2NaO_8H_8$	Black	80
	Amorphous part			20
III	Wairakite	$NaCaSi_4Al_2O_{14}H_4$	Black	34
	Wairakite	$CaSi_2AlO_7H_2$	Grey	40
	Amorphous part			27

The investigators tested a series of mostly NaOH-based solvents with different types of chelating agents and surfactants. With the exception of a polyglucoside surfactant, the exact compositions were not revealed. One solution (a mix of NaOH, different chelants, and a surfactant, called the best solution [BS]) was shown to dissolve almost all of samples I and II and about 80% of scale sample III. Three photographs of different dissolution tests with different methods are shown in Figure 4.29.

Figure 4.29 A Photograph of the Dissolution Effect of 7% BS + 15% NaOH on Sample I. (a) 7% BS + 15% NaOH after Mixing, (b) 7% BS+ 15% NaOH with ~1 g of Sample I Immersed in the Solution, and (c) after Sitting in the Water Bath at 70°C for 24 hours[148]

Because essentially all dissolution processes (except HCl with calcite) exhibit surface reaction control kinetic behavior, the effect of stirring (a vs b), should be expected. The results[151] show that the sodium silicates are much more soluble in these fluids than alkaline earth (such as Ca)-based silicates.

Some solvents with similar compositions have been investigated by Varrin et al.[152] This patent describes a process that uses tetramethylammonium hydroxide solutions that also may contain chelating agents to dissolve aluminum oxides and some silica. A formulation of trimethylammonium hydroxide and EDTA could dissolve significant amounts of AlOOH and smaller amounts of silicon dioxide. Here, the application was the dissolution of various metal salts from deposits found in nuclear steam generators. More details of the testing of these types of solvents are in Section 5.3.2.

Demadis and Mavredaki[153] described tests of several chelating agents that were added to an alkaline solution (pH 10) to increase the dissolution of amorphous silica. The authors found that EDTA and phosphonobutane-1,2,4-tricarboxylate (PBTC) increased the dissolution rate more than the other additives. These solutions were almost as effective as pH 4 NH_4-HF_2 solution for dissolving the silica in these tests (presumably at ambient temperature). A mechanism of dissolution proposed by Demadis and Mavredaki[153] includes several steps:

- Initial dissolution is catalyzed by hydroxyl ions; therefore, it may be reasonable to assume that OH^- ions attack the surface tetrahedral silicon centers belonging to deprotonated silanol groups ($\equiv Si{-}O^-$).
- Inner silicon centers are unreactive because they are well embedded within the silica particle core.
- Once OH^- forms a Si-OH bond with surface silicon, the Si-O network that connects surface silicon atoms with internal silicon centers starts to collapse, thus exposing additional silicon sites that become susceptible to attack.
- Additives containing chemical groups that are strongly anionic, such as $-COO^-$ and $-PO_3^{2-}$, may react with silicon centers in the amorphous network of SiO_2 in a similar fashion as OH^-.
- The formation of silicate esters also was proposed.

A subsequent paper by Demadis et al.[154] claims that DETPA is also an effective silica dissolver in alkaline solutions. The behavior of DETPA is similar to that of PBTC, and its best concentration is 7500 ppm, which yields dissolution of 322 ppm silica (after 24 hours), 340 ppm (after 48 hours), and 333 ppm (after 72 hours). Synthetic lab produced (SSD) silica is a very recalcitrant deposit showing resistance to dissolution even by the most effective additives, PBTC and DETPA.

Other investigators (Gharaibeh et al.[151]) have also observed that the solubility of a silica/silicate deposit is dependent on morphology and specific chemical structure.

4.5. Additional Chemicals Required for Cleaning Formulations

CIs are the most important additives in an aqueous chemical cleaning formulation. These chemicals are described in Chapter 6. Solvents for removing iron oxide and copper from power generating equipment may be complex. The pH values may be adjusted using ammonia (to produce ammonium salts) or with sodium or potassium hydroxide as required by the solvent tests in Section 2.6.

Also, a chelating agent may be formulated with a reducing agent to improve dissolution rates, and the pH may be adjusted to a set value. Different reducing agents including hydrazine and ascorbic acid have been added to several different formulations to improve dissolution of iron oxides. The function of these agents have been described in previous sections.

Fluorides have been added to organic acid solvents to increase the rate of dissolution of iron oxides from supercritical boilers. Fluorides are also claimed to reduce corrosion. A degreasing solvent for pre-operation cleaning may contain NaOH, a phosphate salt, and a silicate salt along with a surfactant to remove loose scale dirt and light greases. Surfactants also have been added to chelants/acids to remove mill scale and light greases. These applications have been referenced in earlier sections of Chapter 4.

Solvents for mixed deposits (inorganic/organic solids) may include acids, chelating agents, and surfactants. These formulations are described in Section 7.5.5.

4.6. Summary of Chemical Removal of Inorganic Scales and Best Practices

This section summarizes the information in this lengthy chapter and provides practical advice from several authors.

4.6.1. Summary

This chapter reviews the chemistries for dissolving inorganic deposits form a wide variety of industrial equipment. When the original book[1] was published in 2001, cleaning of power boilers was a primary focus of many operators and cleaning vendors. Thus, most of the chemistry discussions focused on iron oxide and copper removal. On the basis of the author's observations and Hansen,[155] the cleaning chemistries have not changed significantly since 2001. Therefore, many of the processes and chemicals described in Sections 4.1 and 4.2 are still in use.

However, according to Hansen,[155] there have been changes in the types of deposits and the amounts of deposits found in utility boilers owing to changes in feed water treatments and other engineering improvements. Table 4.15 includes information on typical cleaning intervals proposed by Hansen.

Table 4.15 Current Boiler Cleaning Intervals (2012[155])

Boiler Type	Cleaning Interval	Comments
High-pressure drum type	4–7 years	Some approaching 10–12 years between cleaning
High-pressure drum type, post 1980s construction	Longer intervals then A	The use of condensate polishing and/or stainless steel feedwater heaters has had a positive impact
High-pressure drum type with feedwater heaters with Cu alloys	Shorter intervals than A or B	
Once-through boilers	1–5 years	Units on All Volatile Treatments (ATV)
Once-through boilers	Longer than D	Those units converted to Oxygenated Water Treatment have extended their intervals between cleanings.

This chapter also reviews the information on the known mechanisms of iron oxide and copper dissolution, and this knowledge has resulted in more precise cleaning techniques for power boilers. The information on the dissolution of calcium carbonates as well as various sulfates was produced in the oil and gas industries and can be applied to industrial equipment fouled with these solids.

Note that several improved solvents for sulfate scales and silicates that are based on new chelating agent technologies have been developed and commercialized.

4.6.2. Best Practices

Several authors have proposed best practices for cleaning different types of boilers and combined cycle generating units with heat recovery steam generators (HRSGs) and cooling water systems.

Cowan and Weintritt[18] describe the removal of silica deposits from the blades of steam turbines (Figure 1.5). The authors claim that if the deposit is extremely heavy, mechanical removal is generally favored over chemical removal. If not, however, a removal method involving chemical washing can be used. Usually, the deposits are washed with water to remove any soluble deposits. Washing can be done online under light load. Following the water wash, a solution of 10%–20% NaOH is added, and washing is continued until the deposit is removed. Also see Gahraibeh et al.[151]

Venturini and Duncan[156] describe boiler cleaning operations in Australia. The authors discussed the reasons for chemically cleaning a boiler at both the pre-operation (pre-commission) stage and the post-operation (post-commission) stage. The most common type of pre-operation cleaning conducted in Australia is a two-stage process involving alkaline detergent cleaning to remove oil and grease and ammoniated citric acid cleaning to remove mill scale and corrosion products. The most common type of post-operation cleaning conducted in Australia is hydrochloric acid cleaning.

HRSG plants have a gas turbine at the beginning of the cycle to produce electricity very rapidly, and the hot gases are captured to make steam in a recovery boiler. See Figure 1.4. Shields[88] has reviewed the cleaning of these types of units and claims that ammonium EDTA as well as sodium or ammonium citrate have been used to preoperationally clean them. This paper gives many details of the planning process. Note that a pre-stage of NaOH with surfactants may precede the acid stage, which can be accomplished by adding sufficient citric acid to lower the pH.

McGraw and Fawcett[157] also provide details about the planning process for cleaning HRSG units (and other types of plants).

The conclusions are below with notations of references in this book:

- Determine the condition of the unit (Section 2.6)
- Determine the type of cleaning needed (chemical, mechanical, or both; Section 2.6)
- Plan the site layout of equipment and utilities (Section 8.1)
- Design the cleaning for both chemistry and equipment
- Engineer the procedure
- Review all plans
- Plan for communication between the contractor and the operator
- Plan and execute a safety process (Section 10.2)
- Prearrange waste disposal (Chapter 9)
- Plan for a final inspection and provide reports to the operator (some data from chemical tests; Section 2.6.6).

MacDonald[158] described the planning and execution of pre-operational cleanings of HRSG units that includes details of various stages (degreasing and inorganic scale removal). The paper also describes useful process details.

Majnouni and Jaffer[159] have reviewed the cleaning processes used in industrial boilers. As noted in Section 1.3.1, there are a very large number of industrial boilers of many designs. Note that scales have many of the compounds listed in Table 1.2 of this book, but because of possibly less water treatments compared with utility boilers, copper and the alkaline earth scale accompany the iron oxides. The amount of deposit may exceed that of utility units and cleaning is recommended within 1 year if the deposit weight exceeds 23 g/

ft^2. This paper recommends specific chemicals for degreasing, sulfate removal, copper dissolution, and iron oxide dissolution. This paper includes several tables with specific procedures and is recommended as a first source in planning for cleaning.

Therkildsen[160] provides information on the cleaning of a superheater in a 23-year-old plant. The operators chose inhibited HCl containing HF at about 62°C. Although this solvent cleaned the tubes, some pitting was found. This paper illustrates the necessity for examining the use of solvents and the possible postcleaning consequences. Because of the postcleaning corrosion issues, the sensitive alloy tubes in SuperHeat/ReHeat sections of the power plant also can be cleaned with EDTA or citric acid-based solvents.

Leininger and Wood[161] describe on-stream cleaning to remove iron oxide in an open cooling system. This case history demonstrated that an ammonium EDTA/citric acid fluid containing a special CI[69] can be used without having excess ferric ion corrosion. See more details of the inhibitor in Section 6.4.2. The cooling tower basin was drawn down enough to accommodate the chemicals. The cleaning solution was pumped into the system near the intake to the system circulating pumps. During the cleaning, all blowdown was terminated. Analytical equipment was provided by a contractor in a mobile laboratory. The laboratory contained instrumentation used to determine iron concentration, residual chelant, pH, conductivity, and a means to measure the effectiveness of the inhibitor to protect the system from ferric corrosion. Specimen coupons were placed at different points in the system.

Corrosion coupons of metals in the system were exposed to the cleaning solution during the entire cleaning period. These corrosion coupons were reweighed after the cleaning, and corrosion rates were calculated. All data on corrosion rates were well within expected values. At several points in the system, bypass loops were constructed. With these loops, it was possible to install pieces of heavily scaled piping. At various times during the job, these samples could be visually inspected.

The results of the job were that a total of almost 16,000 lbs. of deposit expressed as iron oxide was removed from the system. Virtually all the piping in the system was cleaned—close to 100%. Flow improvement was almost immediate. Water flow in the soaking pit area increased to such a point that after the cleaning, the flow had to be cut back to 50% because the gravity return could not carry the additional flow.

While the chemistries for dissolving iron oxide and copper have not changed in the non-nuclear industries, there have been some innovations for dissolving calcium and barium deposits that come from the oil and gas stimulation industries. These are described in Section 4.4. Subsequent chapters describe innovations in cleaning nuclear plant equipment (Chapter 5), improvements in inhibitors for cleaning solvents (Chapter 6), removing organic solids (Chapter 7), new equipment and mechanical methods (Chapter 8), disposal of wastes (Chapter 9), and general job design requirements (Section 11.2).

References

[1]W.W. Frenier, *Technology for Chemical Cleaning of Industrial Equipment* (Houston, TX: NACE International, 2001).

[2]A.E. Martell, R.S. Smith, *Critical Stability Constants* (New York, NY: Plenum Press, 1978).

[3]P.W. Jones, D.R. Williams, "Chemical Speciation Used to Assess [S,S']-Ethylenediaminedisuccinic Acid (EDDS) as a Readily-Biodegradable Replacement for EDTA in Radiochemical Decontamination Formulations," *Applied Radiation and Isotopes* 54, 4 (2001): 587-593.

[4]W.W. Frenier, M. Ziauddin, *Formation, Removal and Inhibition of Inorganic Scale in the Oilfield Environment* (Richardson, TX: Society of Petroleum Engineers, 2008).

[5]W.W. Frenier, *Chemical Methods for Controlling Corrosion During Oil and Gas Activities* (Houston, TX: NACE International, 2017).

[6]J. Grebe, "Treatment of Deep Wells," U.S. Patent 1877504, 1932.
[7]P.C. Samuel, *Steam Fuel Users J* (1981).
[8]G. Trabanelli, F. Zucchi, A. Frignani, M. Zucchini, V. Regis, G. Rocchini, "A Contribution to the Acid Cleaning of Boilers," *Werk. u. Korr.* 30, 6 (1979): 426-432.
[9]R.D. Ruiz, J.G. Frost, "Chemical Cleaning-an Important Tool for Maintaining Boiler Efficiency and Integrity" (Duncan, OK: Halliburton Industrial Services Division, 1985).
[10]J. Rusch, "Paper Mill Equipment Chemical Cleaning," CORROSION/79, NACE paper no. 79006 (Houston, TX: NACE, 1979).
[11]W.W. Frenier, M. Ziauddin, *Chemistry for Enhancing the Production of Oil and Gas* (Richardson, TX: Society of Petroleum Engineers, 2014).
[12]G.W. Bradley, J.A. Smith, "Removal of Iron Oxide Deposits without Ferric Ion Corrosion," *MP* 12 (1973): 48-52.
[13]M. Alfandry et al. Ferrara Univ. Annali, Sez. 5 (1971): 137-146.
[14]G. Poff, "Effect of Ammonium Bifluoride on the Corrosion Rates of Some Common Chemical Cleaning Solvents," *MP* 17, 10 (1978): 24-26.
[15]F. Spillner, "Die Reaktionsbeschleunigende Wirkung Von Fluoriden Beim Beizen Verzunderter Eisenteile" *Werk. u Korr.* 18, 9 (1967): 784-798.
[16]U. Bieller, H.P. Brochardt, "Use of Hydrofluoric Acid in Chemical Cleaning of Power Plants" *VGB Kraftwerkstech* 58, 12 (1978): 927-930.
[17]J.H. Rusch, "Method for Treating Ferrous Metal Surfaces," U.S. Patent 3460989, 1960.
[18]J.C. Cowan, D.J. Weintritt, *Mineral & Organic Scale Deposits* (Lafayette, LA: Cowan & Weintritt Publishing Company, 2004).
[19]S.M. Garcia, "Aluminum Cleaning and Brightening Composition and Method of Manufacture Thereof," U.S. Patent 4970014, 1990.
[20]S.M. Garcia, "Composition for Copper and Copper Alloys and Method of Manufacture Thereof," U.S. Patent 5019288, 1991.
[21]W.W. Frenier, M.D. Coffey, J.D. Huffines, D.C. Smith, "Composition and Method for Removing Sulfide-Containing Scale from Metal Surfaces," U.S. Patent 4220550, 1980.
[22]W.W. Frenier, "Method and Composition for Removing Sulfide-Containing Scale from Metal Surfaces," U.S. Patent 4310435, 1982.
[23]J.N. Tanis, A.D.A. Systems, "Standard Practice for Descaling Stainless Steel Parts," in *Procedures of Industrial Water Treatment* (Ridgefield, CT: Ltan, 1987).
[24]J.N. Tanis, Chapters 14-15, in *Procedures of Industrial Water Treatment.* (Ridgefield, CT: Ltan, 1987).
[25]D. Panias, M. Taxiarchou, I. Paspaliaris, A. Kontopoulos, "Mechanisms of Dissolution of Iron Oxides in Aqueous Oxalic Acid Solutions," *Hydrometallurgy* 42 (1996): 257-265.
[26]J.F. Eberhard, R.B. Rosene, "Removal of Scale Deposits," U.S. Patent 3003899, 1961.
[27]C.F. Reich, „Scale Removal," U.S. Patent, 3003898, 1961.
[28]L.G. McLaughlin, "Improved Acid Solution for Boilers Removes Oxides without Precipitate," *MP* 2, 8 (1961): 52-56.
[29]J.M. Sullivan, J.W. Horvath, "Chemical Cleaning of Utility Once through Generating Units That Have Been on Long Term Oxygenated Treatment," CORROSION/2003, paper no. NACE 03583 (Houston, TX: NACE, 2003).
[30]R.B. Dooley, K. Shields, A. Aschoff, M. Ball, A. Bursik, "Cycle Chemistry Guidelines for Fossil Plants: Oxygenated Treatment" (Palo Alto, CA: Electric Power Research Institute, 1994).
[31]W.W. Frenier, D. Larson, "Development and Testing of a New Solvent for Cleaning Once-through Boilers," Paper IWC 94-74, 55th International Water Conference, Pittsburgh, PA, Oct. 31–Nov. 3, 1994.
[32]W.W. Frenier, "Formic-Carboxylic Acid Mixtures for Removing Iron Oxide Scale from Steel Surfaces," U.S. Patent 5529637, 1995.
[33]C.A. Poulos, "A Comparative Study on the Dissolution of Magnetite by Organic Acids," *MP* 23, 8 (1984): 19-21.
[34]J.P. Veysset, P. Saurin. "Experience with EDF Chemical Cleaning Process in Nuclear Power Plants," CORROSION/93, paper no. NACE 93367 (Houston, TX: NACE, 1993).
[35]D.C. Horner, "Citric, Tartaric and Gluconic Acids in Metal Finishing," *Elec. Metal Finish* 21, 3 (1968): 75-80.
[36]B.J. Wiersma, "Hydrogen Generation During the Corrosion of Carbon Steel in Oxalic Acid," (Aiken, SC:

Westinghouse Savannah River Company, 2004).
[37]M.A. Blesa, H.A. Marinovich, E.C. Baumgartner, A.J.G. Maroto, "Mechanism of Dissolution of Magnetite by Oxalic Acid-Ferrous Ion Solutions," *Inorganic Chemistry* 26, 22 (1987): 3713-3717.
[38]C.M. Loucks, E.B. Morris, E.A. Pirsh, "Cleaning of Heat-Exchange Equipment," Annual Meeting of ASME, New York, NY, 1958.
[39]S.T. Arrington, G.W. Bradley, EPRI Service Water System Reliability Improvement Seminar, Charlotte, NC (1989).
[40]E.B. Morris, "Precautions in the Use of Citric Acid for Chemical Cleaning," *J. Engineering for Power* 10 (1961): 367-370.
[41]R.E. Hamm, C. Shull Jr., D. Grant, "Citrate Complexes with Iron (II) and Iron (III)," *J. American Chem. Soc.* 76 (1954): 2111-2114.
[42]M. Bobtelsky, J. Jordan, "The Structure and Behavior of Ferric Tartrate and Citrate Complexes in Dilute Solutions," *J. American Chem. Soc.* 69 (1947): 2286-2290.
[43]S. Alfano, "Process for Removing Copper-Containing Iron Oxide from Metal Surfaces," U.S. Patent 3072502, 1963.
[44]W. Bell, "Scale Removal," U.S. Patent 3248269, 1966.
[45]E. Wackenhut, L.W. Lamb, J.P. Engel, "Use and Disposal of Boiler Cleaning Solvent," *Power Engineering* 77, 11 (1972): 68.
[46]A. Klein, K.L. Atwood, American Power Conference, Chicago, IL (1964).
[47]W.E. Bell, *Combustion* 5 (1964): 25-28.
[48]W.J. Blume, "Citric Acid-Based Chemical Cleaning Processes," CORROSION/76, paper no. NACE 76128 (Houston, TX: NACE, 1976).
[49]G.T. Blair, P.W. Stal, M.F. Zienty, "Citric Acid in Creative Chemical Cleaning" (South Bend, IN: Miles Laboratories, 1980).
[50]S.T. Arrington, G. Bradley, "Process for Removing Copper and Copper Oxide Deposits from Surfaces," U.S. Patent 5009714, 1991.
[51]W.W. Frenier, W.C. Kennedy, "Mechanism of Copper Dissolution in Chelant Solvents," CORROSION/83, paper no. NACE 222 (Houston, TX: NACE, 1983).
[52]D. Daniels, "Boiler Chemical Cleaning: Doing It Correctly," *Power Magazine* Sept. 1, 2014, http://www.powermag.com/boiler-chemical-cleaning-doing-it-correctly/?pagenum=1 (2014).
[53]C.A. Lesinski, "Scale Removal, Ferrous Metal Passivation, and Compositions Therefor," U.S. Patent 3308065, 1967.
[54]C.A. Lesinski, "Scale Removal, Ferrous Metal Passivation, and Composition," U.S. Re. 30796, Reissue of U.S. Patent 3308065, 1981.
[55]L.W. Harriman, P.E. Muelberg, F.N. Teumac, "Removal of Copper Containing Incrustations Form Ferrous Surfaces," U.S. Re. 30714, Reissue of U.S. Patent 3438811, 1981.
[56]T.G. Cook, "Process for Removing Copper and Copper Oxide Encrustations from Ferrous Surfaces," U.S. Patent 4443268, 1984.
[57]S.T. Arrington, G.W. Bradley, "Removal Of Copper From Utility Boilers Utilizing An Environmentally Acceptable Oxidant," 50th International Water Conference, Pittsburgh, PA (1985).
[58]J.L. Slicker, "Alkaline Copper Removal," 28th International Water Conference, Pittsburgh, PA, 1967.
[59]W.W. Frenier, "The Mechanism of Magnetite Dissolution in Chelant Solutions," *Corrosion* 40, 4 (1984): 176-180.
[60]W.W. Frenier, F.B. Growcock, "Mechanism of Iron Oxide Dissolution," *Corrosion* 40, 12 (1984): 663-668.
[61]Dow, "Keys to Chelation with Versene[tm], Chelating Agents," (Midland, MI: The Dow Chemical Company, 1990).
[62]J.D. Vinso, "In-Service Cleaning of Cooling Water Systems," U.S. Patent 3527609, 1970.
[63]W.W. Frenier, W.C. Kennedy, "Passivation of Steel in Ammonium EDTA Solutions," *Corrosion* 42, 10 (1986): 613-522.
[64]B.O. Shortouse, "Boiler Circulation During Chemical Cleaning," *Chemical Engineering* 90, 17 (1983): 75-79.
[65]E.C. Pitzer, "Neutral Metal Cleaning Compositions Containing Hydrazine and a Polycarboxylamino Acid," U.S. Patent 3297580, 1967.
[66]F.J. Pocock, W.S. Leedy, "Chemical Cleaning Research for Steam Generators," Proc. 32nd International Water Conference, Pittsburgh, PA, Nov. 2-4, 1971.

[67]C.S. Welty, J.M. Jevec, W.S. Leedy, "Laboratory Studies for Removing Sludge from Secondary Side of Nuclear Steam Generators," CORROSION/82, paper no. NACE 82034 (Houston, TX: NACE, 1982).

[68]Siemens, Efficient Power Generation Using Fossil Fuels, 2016, Siemens, Munich Germany, http://www.energy.siemens.com/us/en/sustainable-energy/power-generation.htm?stc=usccc023878&s_kwcid=AL!462!10!10539230002!41136103822&ef_id=Vb9LnQAABcZiI3d8:20160209192206:s

[69]W.W. Frenier, W.J. Settineri, "Sulfonium Compound as a Corrosion Inhibitor in Aqueous Acid," U.S. Patent 3764543, 1973.

[70]W.W. Frenier, S.J. Barber, "Choose the Best Heat Exchanger Cleaning Method," *Chemical Engineering Progress* 94, 7 (1998): 37-45.

[71]W.W. Frenier, D.A. Wilson, "Aqueous Composition and Method of Use," U.S. Patent 4430128, 1984.

[72]W.W. Frenier, D. Wilson, "Aqueous Acid Composition and Method of Use," U.S. Patent 4636327, 1987.

[73]W.W. Frenier, E.F. Hoy, "HEDTA - a Versatile Chemical for Cleaning Steel Surfaces," *MP* 25, 4 (1986): 18-24.

[74]E.F. Hoy, "A Safe-Low Temperature-Ferric Oxide Solvent System," paper 87-43, 47th International Water Conference, Pittsburgh, PA, Nov. 2-4, 1987.

[75]J.G. Frost, L.D. Martin, W.P. Banks, "Method for Removing Magnetite," U.S. Patent 3854996, 1974.

[76]G. Halsey, GB, U.S. Patent 1397451, 1975.

[77]G.W. Bradley, S.T. Arrington, "Methods and Compositions for Removing Copper and Copper Oxides from Surfaces," U.S. Patent, 4586961, 1986.

[78]S.T. Arrington, G.W. Bradley, "Method of Removing Iron and Copper-Containing Scale from a Metal Surface," U.S. Patent 4666528, 1987.

[79]M.G. Fontana, *Corrosion Engineering*, 3rd ed. (Boston, MA: McGraw-Hill, 1986).

[80]A.D. Nesbitt, "A History of the Chemical Cleaning Business," CORROSION/93, paper no. NACE 93362 (Houston, TX: NACE, 1993).

[81]F.N. Alquist, J.L. Wasco, C.R. Irons, "Removing Copper-Containing Incrustations from Steam Generators," U.S. Patent 2567835, 1951.

[82]R.C. Martin, W.T. Abel, "Copper and Iron Containing Scale Removal from Ferrous Metal," U.S. Patent 2959555, 1960.

[83]J.W. McCoy, *Industrial Chemical Cleaning* (New York, NY: Chemical Publishing Co., 1984).

[84]B.B. Arnold, J.G. Frost, "Method of Removing Copper Containing Iron Oxide Scales from Iron," U.S. Patent 3547697, 1970.

[85]J.A. Knox, J.A. Smith, R.F. Stout, R.E. Dixon, U.S. Patent 3730901, 1973.

[86]J.G. Frost, "Nature and Cumulative Stability Constants of Copper(I) Complexes of Some Substituted Thioureas," *Inorganic Chemistry* 15 (1976): 940-943.

[87]N.E.S. Thompson, D. Redmore, B.A. Oude-Alink, B.T. Outlaw, "Processes for Inhibiting Corrosion Using Compounds Containing Sulfur and Amino Groups," U.S. Patent 4450137, 1984.

[88]K.J. Shields, "Chemical Cleaning of Fossil Power Station Steam Generators; Past, Present and Future," 14th International Conference on the Properties of Water and Steam in Kyoto, Japan, Aug. 29–Sept. 3, 2004.

[89]J.D.D. Uhr. "Citric Acid-Based Cleaning of Mixed Metal Systems," CORROSION/80, paper no. NACE 80217 (Houston, TX: NACE, 1980).

[90]F.N. Teumac, J.S. Scruggs, "Cleaning of Ferrous Metal Surfaces," U.S. Patent 3627687, 1971.

[91]I.M. Kohlthoff, J.J. Lingane, *Polarography* (New York, NY: Interscience, 1952).

[92]S. Baumgarten, R.E. Cover, H. Hofsass, S. Karp, P.B. Pinches, L. Meites, "Half-Wave Potentials of Metal Ions in Organic Hydroxyacid Supporting Electrolytes," *Analytica Chimica Acta* 20 (1959): 397-404.

[93]L.D. Martin, G.W. Bradley, "Method of Removing Copper and Copper Oxide from a Ferrous Metal Surface," U.S. Patent 4452643, 1984.

[94]J.M. Jevec, W.S. Leedy, "Chemical Cleaning Solvent and Process Testing" (Palo Alto, CA: Electric Power Research Institute, 1983).

[95]W.F. Cleary, G.B. Gockley, "Chemical-Cleaning Process Evaluation: Westinghouse Steam Generators. Final Report" (Palo Alto, CA: Electric Power Research Institute, 1983).

[96]D. Chen, A.E. Martell, D. McManus, "Studies on the Mechanism of Chelate Degradation in Iron-Based, Liquid Redox H_2S Removal Processes," *Canadian Journal of Chemistry* 73, 2 (1995): 264-274.

[97]J.G. Frost, L.D. Martin, W.P. Banks, "Method for Removing Scale from Metallic Substrates," U.S. Patent 3996062 1976.

[98]H.H. Uhlig, *Corrosion and Corrosion Control* (New York, NY: John Wiley and Sons, 1971).
[99]E. McCafferty, R.J. Brodd, "Surfaces, Inhibition, and Passivation" (Pennington, NJ: The Electrochemical Society, 1986).
[100]L.D. Martin, W.P. Banks, "Electrochemical Investigation of Passivating Mild Steel Surfaces," *MP* 14, 9 (1975): 33-36.
[101]C.N. Fredd, H.S. Fogler, "Influence of Transport and Reaction on Wormhole Formation in Porous Media," *AIChE Journal* 44, 9 (1998): 1933-1949.
[102]M. Eisenberg, R. Tobias, C.R. Wilke, "Ionic Mass Transfer and Concentration Polarization at Rotating Electrodes," *J. Electrochem. Soc.* 101, 6 (1954): 306-320.
[103]H.C. Kuo, D. Landolt, "Rotating Disc Electrode Study of Anodic Dissolution or Iron in Concentrated Chloride Media," *Electrochimica Acta* 20, 5 (1975): 393-399.
[104]J.W. Diggle, *Dissolution of Oxide Phases from Oxides and Oxide Films* (New York, NY: Marcel Dekker, 1973).
[105]D.A. Vermilyea, "Dissolution of Ionic Compounds in Aqueous Media," *J. Electrochem. Soc.* 113, 10 (1966): 1067.
[106]A. Hickling, D.J.G. Ives, "The Electrochemical Behavior of Iron Oxides in Dilute Sulfuric Acid and Interpretation of the Flade Potential of Iron," *Electrochimica ACTA* 20, 1 (1975): 63-69.
[107]S. Haruyama, K. Masamura, "Electrochemical Methods in Passivity," *Electrochim. Acta* 20 (1975).
[108]D.S. Mancey, D.W. Shoesmith, J. Lipkowski, A.C. McBride, J. Noël, "An Electrochemical Investigation of the Dissolution of Magnetite in Acidic Electrolytes," *Electrochem. Soc.* 140, 3 (1992): 637-642.
[109]N.A. Al-Mobarak, "Effect of Oxalic Acid on the Dissolution of Magnetite Coupled with Iron of Various Surface Area," *Int. J. Electrochem. Sci.* 3 (2008): 666-675.
[110]S. Brunet, G. Turluer, "Ferrous Metals," 5th European Symposium Corrosion Inhibitors, Ferrara, Italy 513-538 (1980).
[111]K. Ogura, K. Sato, "Cathodic Dissolution of the Passive Film on Iron—I. Kinetics and Mechanism," *Electrochim. Acta* 25, 6 (1980): 857-862.
[112]I.G. Gorichev, S. Ashkharua, S. Vainman, "The Applicability of the Topochemical Model to the Dissolution of Certain Oxides in Acids," *Russ. J. Phys. Chem. Trans. from Zh. Fiz. Khim.* 50 (1976): 1853.
[113]I.G. Gorichev, L. Malov, V. Dukhanin, "Ratio of the Constants for the Formation and Dissolution of Active Centers of Magnetite and Hematite," *Russ. J. Phys. Chem., Trans. from Zh. Fiz. Khim.* 52 (1978): 681.
[114]I.G. Gorichev, N.A. Kipriyanov, "Influence of Iron (II) Ions on the Kinetics of Magnetite Dissolution in Hydrochloric-Acid," *Russ. J. Phys. Chem., Trans. from Zh. Fiz. Khim.* 53 (1979): 1293.
[115]I.G. Gorichev, "The Influence of the Ph of the Medium on the Kinetics of the Dissolution of Magnetite in a Solution of Sodium Ethylenediaminetetracetate," *Russian J. Physical Chemistry* 54 (1980): 744.
[116]I. Gorichev, "Tables for Calculations from Topochemical Equation," in *Tables for Calculations from Topochemical Equation* (Minsk, Belarus SSR: IZD, 1962).
[117]H.G. Pearson, J.W. Moore, *Kinetics and Mechanism,* 3rd ed. (New York, NY: John Wiley and Sons, 1981).
[118]H. Schmalzried, *Solid State Reactions,* translated by A.D. Pelton (New York, NY: Academic Press, 1974).
[119]D.A. Young, *Decomposition of Solids* (Oxford, U.K.: Pergamon, 1966).
[120]B.V. Erofeev, *Tables for Calculations from Topochemical Equation* (Minsk, Belarus SSR: IZD, 1962).
[121]K. Azuma, H. Kametani, "Kinetics of Dissolution of Ferric Oxide," *Trans. AIME* 230 (1964): 853-861.
[122]P. Barret, *Heterogeneous Kinetics* (Paris: Gauthier-Villars, 1973).
[123]I.G. Gorichev, I.V. Artamonova, E.E. Nifant'ev, "Comparative Evaluation of Magnetite Dissolution Efficiencies of Edta and Oedp Aqueous Solutions," *Russian Journal of Inorganic Chemistry* 54, 5 (2009): 808–818.
[124]O.L. Riggs, H.M. Hurd, "Effect of Inhibitors on Scale Removal in HCl Pickling," *Corrosion* 24, 2 (1968): 45.
[125]R. Roberge, "Inhibition of Carbon Steel Corrosion in Chemical Cleaning Solutions Containing Solid Magnetites," *Corrosion* 39, 12 (1983): 496.
[126]C.W.N. Cumper, R. Grzeskowiak, P. Newton, "Effect of Pyrrole, Indole and Their Dimethyl Derivatives on the Dissolution of Magnetite," *Corros. Sci.* 22 (1982): 551.
[127]J. Rubio, E. Matijevic, "Interactions of Metal Hydrous Oxides with Chelating Agents. I. Ð-Feooh—EDTA," *J. Colloid and Interface Science* 68, 3 (1984): 408-421.
[128]H.C. Chang, T.W. Healy, E. Matijevic, "Interactions of Metal Hydrous Oxides with Chelating Agents: III. Adsorption on Spherical Colloidal Hematite Particles," *Colloid Interface Sci.* 92, 2 (1983): 469-478.
[129]H.C. Chang, E. Matijevic, "Interactions of Metal Hydrous Oxides with Chelating Agents: IV. Dissolution of Hematite" *J. Colloid Interface Sci.* 92, 2 (1983): 479-488.

[130]M.A. Blesa, A.J.A. Maroto, "Mechanisms of Oxide Dissolution by Acid Chelating Agents," Conference on Decontamination of Nuclear Facilities, LeGrange Park, IL, 1982.
[131]E. Baumgartner, M. Blesa, H. Marinovich, "Heterogeneous Electron Transfer as a Pathway in the Dissolution of Magnetite in Oxalic Acid Solutions," *Inorganic Chemistry* 22, 16 (1983): 2224-2226.
[132]R.H. Hausler, "Role of Hydrazine in EDTA Containing Chemical Solvents," CORROSION/83, paper no. NACE 83030 (Houston, TX: NACE, 1982).
[133]S. Brunet, G. Pinard-Legry, G. Turluer, "Electrochemical Aspects of Chemical Cleaning of Steam Generators," 8th International Congress on Metallic Corrosion, Frankfurt, Germany, 1981: p. 1634.
[134]W.M. Fields, "The Growth, Structure and Breakdown of Magnetic Films on Mild Steel," Proc. 2nd Int. Cong. Met. Corrs. (Houston, TX: NACE International, 1963).
[135]S. Brunet et al., "Ferrous Metals," 5th European Symposium on Corrosion Inhibitors, Ferrara, Italy, 1980.
[136]W.W. Frenier, *Chemical and Mechanical Methods of Pipeline Integrity* (Richardson, TX: Society of Petroleum Engineers, 2017).
[137]W.G. McPherson, J.D. Martin, "Examination of Superheater Oxidation Products by SEM," 2nd International Congress on Metallic Corrosion, Houston, TX (1977).
[138]A.D. Hill, R.S. Schechter, "Fundimentals of Acid Stimulation," in *Reservoir Stimulation*, 3rd ed., eds. M. J. Economides, K. G. Nolte (New York, NY: John Wiley and Sons, 2000).
[139]C.N. Fredd, "Advances in Understanding and Predicting Wormhole Formation," Appendix in *Reservoir Stimulation*, 3rd ed., eds. M. J. Economides, K. G. Nolte (New York, NY: John Wiley and Sons, 2000).
[140]C.N. Fredd, H.S. Fogler, "The Kinetics of Calcite Dissolution in Acetic Acid Solutions," *Chemical Engineering Science* 53, 2 (1998): 3863-3874.
[141]W.W. Frenier, D. Wilson, D. Crump, L. Jones, "Use of Highly Acid-Soluble Chelating Agents in Well Stimulation Services," Paper SPE 63242, 2000 SPE Annual Technical Conference and Exhibition, Dallas, TX, Oct. 1-4, 2000.
[142]J.N. LePage, C.A. DeWolf, J.H. Bemelaar, H.A. Nasr-El-Din, "An Environmentally Friendly Stimulation Fluid for High Temperature Applications," paper SPE 121709, 2009 SPE International Symposium on Oilfield Chemistry, The Woodlands, TX, Feb. 20-22, 2009).
[143]R.L. Rybacki, "Scale Converters," U.S. Patent 4147647, 1973.
[144]F.C. Bersworth, "Metal Ion Control," U.S. Patent 2961311, 1960.
[145]Y. Hayashi, S. Kawazoe, Japan Patent 63028890, 1980.
[146] J.D. Vinso, "In-Service Cleaning of Cooling Water Systems," U.S. Patent 3527609, 1970.
[147]M.B. Lawson, "Formulation for the Dissolution of Gypsum," U.S. Patent 4096869, 1978.
[148]NACE, "A Review of Screening Tests for Gypsum Scale Removal Chemicals" (Houston, TX: NACE International, 2003).
[149]J.M. Paul, R.L. Morris, "Composition for Removing an Alkaline Earth Metal Sulfate Scale," U.S. Patent 5282995, 1994.
[150]G.H. Zaid, B.A. Wolf, "Method of Removing Metal Sulfates from Surfaces," U.S. Patent 6494218, 2002.
[151]S.A. Gharaibeh, T.R. McCartney, R.A. Shank, "A Novel Non-Toxic Method for the Decontamination of Silicate Scales: A Case Study," CORROSION/2017, paper no. NACE C2017-9087 (Houston, TX: NACE, 2017).
[152]R.D. Varrin Jr., M.J. Little, C.E. Anderson, "Method and Compositions for Removing Deposits," U.S. Patent 8728249, 2014.
[153]K.D. Demadis, E. Mavredaki, "Green Additives to Enhance Silica Dissolution During Water Treatment," *Environ Chem Lett* 3 (2005): 127-131.
[154]K.D. Demadis, E. Mavredaki, M. Somara, "Additive-Driven Dissolution Enhancement of Colloidal Silica. 1. Basic Principles and Relevance to Water Treatment," *Industrial & Engineering Chemistry Research* 50, 22 (2011): 12587-12595.
[155]K. Hansen, "Chemical Cleaning of Boilers," Paper IWC 12-67, 73rd Annual IWC, San Antonio, TX, Nov. 4-8, 2012.
[156]T. Venturini, N. Duncan, "The Chemical Cleaning of Boilers in Australia," CORROSION/2003, paper no. NACE-03589, (Houston, TX: NACE, 2003).
[157]J. McGraw, D. Fawcett, "Chemical Cleaning HRSGs, Top 10 Lessons Learned," CORROSION/2003, paper no. NACE-03580 (Houston, TX: NACE, 2003).

[158]S. MacDonald, "Preoperational Chemical Cleaning of Heat Recovery Steam Generators," CORROSION/2003, paper no. NACE-03582 (Houston, TX: NACE, 2003).
[159]M.D. Majnouni, A.E. Jaffer, "Chemical Cleaning of an Industrial Boiler: An Overview," 64th Annual Water International Conference, Pittsburgh, PA, Oct. 19, 2003.
[160]S.-E. Therkildsen, "Experience from Chemical Cleaning of a 23 Years Old Intermediate Pressure Superheater in Unit 5 at Asnaes Power Station (Asv5)," 14th International Conference on the Properties of Water and Steam, Kyoto, Japan, Aug. 29, 2004.
[161]J.W. Leininger, D.N. Wood, "The On-Stream Cleaning of an Entire Cooling Water System: A Case History," NACE Regional Meeting, McAfee, NJ, Oct. 8, 1976.

5 Cleaning Nuclear Plant Equipment

Processes to clean the water-contacted parts of nuclear power plants (Section 1.3.2) can be specialized applications of the iron oxide and copper dissolution methods described in Chapter 4. However, because of the radioactivity usually present in deposits that foul nuclear generating equipment, specialized procedures have been developed to clean some types of nuclear units. Three different types of processes are conducted and described in this chapter.

1) Chemical treatments are used to decontaminate (Section 5.2) parts of the units to allow maintenance or before decommissioning the plant (these are health physics issues). This can include parts of boiling water reactors (BWR) and units of CANDU and other pressurized water reactors (PWRs) as well as waste tanks at different facilities.

2) The secondary sides of PWR steam generators are cleaned (Section 5.3) to remove scale and crud (loose solids or debris) and to slow the formation of dents that restrict flow in the steam generator and could crush the tubes. This latter program is called *crevice cleaning* and may require a separate solvent stage.

3) Various service water–containing systems in nuclear plants, including heat exchangers, are cleaned for all the reasons described in Chapter 1.

5.1 Consequences of Formation of Deposits in Nuclear Equipment

Details of general scale formation are in Sections 3.2, 3.3, and 3.8. Most fouling and radioactive deposits are the result of the deposition of iron oxides and other transition metal oxides (containing chromium or nickel) that form due to the slow corrosion of the alloys in the aqueous fluids. If the water circuit has any copper-based alloys, copper metal and copper oxide also may be part of a fouling scale.

Hudson et al.[1] notes that corrosion products depositing in the secondary side of steam generators are detrimental to both the steam generator and steam generator tubes because these deposits can cause tube denting, pitting, and other corrosion mechanisms; loss of heat transfer within the steam generator; tube withdrawal; and overall steam generator degradation. Tube denting occurs when forces from corrosion deposit buildup between the tube and tube support structure (see Figure 1.8) place additional stress on the steam generator tube wall, causing localized constrictions. These dents can cause tube withdrawal, tube cracking, and increased steam generator tube inspections. However, dented tubes may necessitate plugging if inspections cannot be performed. Steam generator tube inspections are typically performed every 18 to 24 months.

These authors claim that as utilities seek extensions in their operation cycles, these inspections will be performed with less frequency; however, deposits will continue to accumulate on the secondary side. For example, a feedwater iron concentration of 1 ppb Fe will result in 100–200 lbs. (45–91 kg) of deposit per cycle, depending upon the secondary system design and cycle duration. For those utilities that have exhibited increased deposit loading, the frequency of planned inspections may increase as the unit ages. This cumulative sludge loading may damage the tubes; however, there are various options to avoid these possible outcomes of corrosion and deposits. These options include steam generator replacement, improving or altering chemistry control, sludge lancing or other mechanical deposit removal techniques, chemical cleaning, or continual plant operation and scheduled maintenance until decommissioning. The action performed to mitigate corrosion and remove deposits depends on the utility, the severity of the corrosion degradation and deposit loading, the system design, and operation objectives.

Marks et al.[2] claim that the buildup of scale also affects heat transfer in single-phase flow environments. However, in nuclear steam generators, tube deposits form in the presence of two-phase conditions that involve boiling at the tube surfaces. Thus, the authors claim that the structure of steam generator tube deposits are marked by the presence of porous voids within the deposit thickness, particularly in the outer portions of the layer thickness adjacent to the bulk secondary fluid flow (see Figure 3.16).

They also claim that this type of porous structure results in a net thermal resistance, which is close to zero or even negative–that is, the heat transfer efficiency in the presence of the porous deposit layer is often equivalent to or even more favorable than that realized with clean tubing because of an increased number of boiling nucleation sites and wick boiling. However, as time passes, this improvement in heat transfer is reduced as more scale forms. The desire to return the heating surfaces to this improved heat transfer state is one of the goals of some of the steam generator cleaning methods described in Section 5.3.2.

An unusual type of crevice corrosion can affect steam generators during chemical cleaning operations, which are described in Section 5.3. Marks et al.[3] show that the corrosion coupons that are placed in the steam generator can form a pit in the vicinity and partially under the bolt that attaches the coupon. See Figure 5.1(a), which depicts the authors' explanation for this type of corrosion pitting.

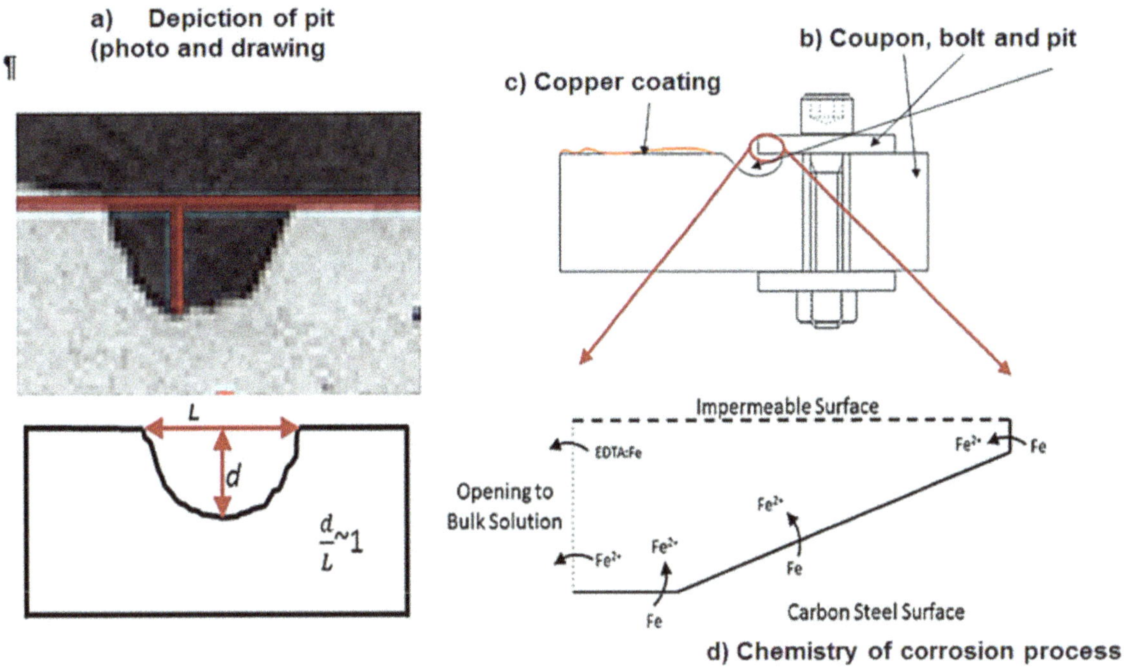

Figure 5.1 Chemistry of Pit Formation on Coupon During Chemical Cleaning

Their detailed explanation of this process is that following the ASCA process (a soft or preventative cleaning process with lower deposit dissolution capacity; see Section 5.3.2), slight non-uniform corrosion has been observed at coupon mounting hardware and at some free span locations (i.e., the coupon surfaces away from the mounting hardware).

In this case, they claim that non-uniform corrosion on the order of 50–75 μm would be considered typical with general corrosion on the order of 25 μm. Thus, non-uniform corrosion is two to three times the general corrosion. Note that the advanced scale conditioning applications process does not include a chemical corrosion inhibitor. Also, more non-uniformity in the corrosion is seen at plants with higher copper loadings in the deposits, owing to

1) The copper plate-out exfoliation or breaches. This copper plate forms during the "copper stage" of an EDTA solvent-based cleaning (see Sections 3.4 and 4.2.3).
2) The cathodic/galvanic corrosion that can occur as surfaces become copper coated.

Overall,[3] non-uniform corrosion may represent about 2% of the total corrosion. As with hard steam generator chemical cleaning processes, the morphology is one that does not support a significant stress riser or subsequent chemical hideout like that seen in tube to tube support crevices.

Note that the copper plate can subsequently be removed with a copper stage, but the pits will remain and must be analyzed. The remainder of this paper analyzes the details of the pitting process and the amount of corrosion loss that is measured on the coupons.

5.2. Decontamination Solvents and Processes

This section describes decontamination (decon: a reduction in radioactivity) in parts of nuclear electrical generation plants as well as in waste storage facilities.

5.2.1. Chemical Processes for Decontamination of Surfaces

The buildup of radioactive contamination on out-of-core surfaces leads to occupational radiation exposure when these surfaces are manually maintained or serviced. Decontamination is the process by which a *portion* of the radioactive species are removed to allow servicing of the unit or if required before unit decommissioning. These processes are also performed to decontaminate waste storage tanks as well as BWR surfaces and CANDU units and some sections (primary) of light water PWRs. Physical techniques, such as abrasive cleaning, have very restricted use in system decontamination since they are limited to surfaces that are relatively accessible. Chemical methods can be applied to an entire loop or an entire reactor unit. A very comprehensive review of decon chemicals and processes is in Wood.[4]

The metric for decontamination of a unit is the decontamination factor (DF), defined as the ratio of radioactivity *before* and *after* the decontamination of radioactively contaminated objects, waste water, air, or any other fluid or surface.

A report by Boing[5] states that there are several reasons to decontaminate a nuclear unit:

- Remove the contamination from components
- Reduce dose level in the installation (reduce dose during maintenance or dismantling)
- Minimize the potential for spreading contamination during further dismantling
- Reduce the contamination of components to such levels that may be
    - Disposed of in a lower category (lower cost)
    - Recycled or reused in the conventional industry (clearance of material)

Boing[5] describes some possible required DF values that depend on the conditions in a unit as well as the decontamination goals. See Figure 5.2, which is based on this report.

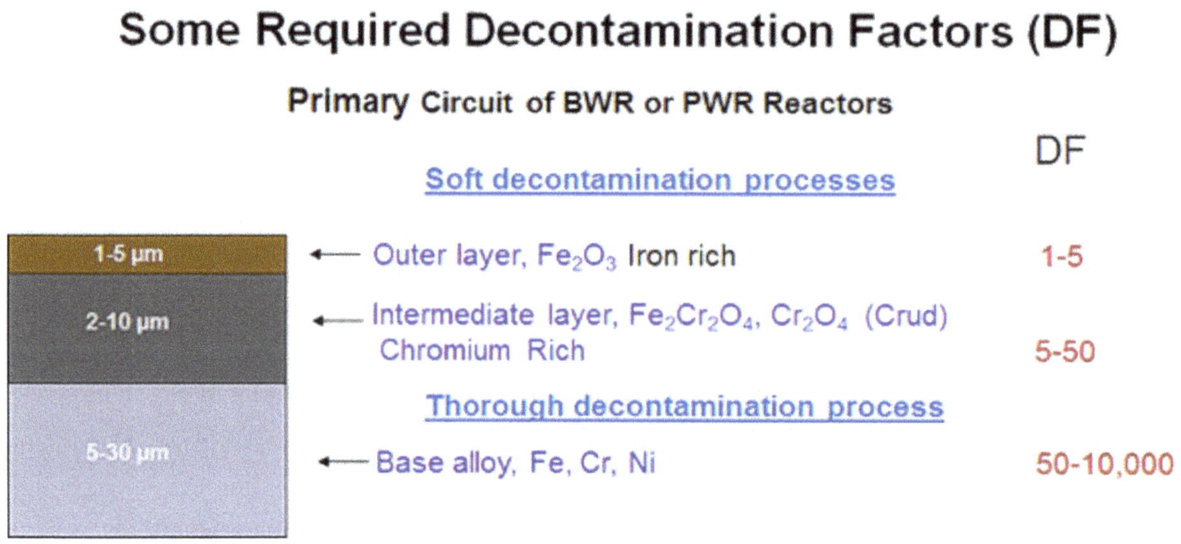

Figure 5.2 Possible DF Values for Different Requirements

Remark[6] claims that decontamination technologies can be classified into three categories of equipment in the reactor that may require cleaning:

1) Miscellaneous tools and equipment
2) Removable components
3) Isolable components

The authors state that within these categories decontamination techniques can be further divided into methods for removing loosely adherent contamination, tightly adherent contamination, or both types of contamination.

The authors also assert that the type of corrosion film that constitutes the contamination is critical in engineering the optimum technique. Various chemical processes for reducing radioactivity values are described in the next subsection.

Decontamination Solvents and Processes. Chapter 4 provides a general review of many of the chemicals used in cleaning fluids, and the structures are depicted in Figure 4.1. The chemical characteristics are listed in Table 4.1. Specific decon solvents and methods are reviewed in this subsection. Chemical as well as physical methods have been applied in different plants. Wood[4] notes that both complete ("hard," i.e., strong solvents) and dilute ("soft") solvent processes have been performed to decontaminate nuclear units. However, similar chemicals can be employed regardless of the extent of cleaning. Depending on the goal of the treatment, different masses of scale removal and activity reduction may be achieved.

Solvents to remove radioactivity from stainless steel or other similar alloys have included stages of

- Alkaline permanganate (AP) with ammonium citrate (AC)
- Ammonium citrate/ammonium oxalate (CITROX)
- Regeneration of weak chelate solutions with the CANDECON and CANDEREM processes (used on CANDU units)
- Oxalic acid (see King[7] and Wiersma[8]) also has been used to decontaminate storage tanks holding radioactive waste and for some steam generator cleaning
- In addition, the powerful reducing agents contained in low oxidation state metal ion (LOMI) technologies will be described later in this chapter.

A symposium chaired by Staehle[9] held at Ohio State University (OSU) in 1975 was sponsored by the Electric Power Research Institute (EPRI), the American Society of Mechanical Engineers (ASME) and OSU to summarize the state of the art at that time. Significant technologies described included the use of chelants such as EDTA and citric acid as well as regenerative processes such as CANDECON. Many of these processes have been used to decontaminate various parts of operating nuclear plants. Wood[4] reviewed advances in the field up to 1991 and more recently[10] from 1991 to 1994. Per this author, there had been more than 40 decontamination procedures performed before 1995. The World Nuclear Association[11] estimates that there have been more than 150 reactors decommissioned (and some required decontamination). As more units are decommissioned, this type of service will be required.

The technologies used include all the methods described above. Issues of concern include waste processing, corrosion, and the size of the DF achieved. The following sections will describe the development of some of the major processes currently used in commercial applications.

The AP/CITROX process uses alternating oxidizing and reducing stages:

- 12% alkaline permanganate ($KMnO_4$) as an oxidant to convert Cr(III) to Cr(VI)
- Reducing stages of 6% ammonium citrate/ammonium oxalate are used to dissolve iron oxides

These processes can produce DF values of 10 to 100. Although these treatments have been used since 1970,[12] these chemicals produce very large volumes of waste and are very aggressive and corrosive. Wood[4] also claims that in some cases, the addition of an oxidizing step, such as nitric acid/potassium permanganate (NP) has given only a small further reduction in residual fields. However, for some components, such as the recirculation pumps, use of NP has proved beneficial.

Several patents provide details of the reasons for using the AP/CITROX type of materials. Ishida et al.[13] note that the radionuclides are incorporated into oxides, which are present on surfaces of the component and piping. Examples are oxides containing large amounts of iron (iron-based oxides), e.g., hematite (a-Fe_2O_3), nickel ferrite ($NiFe_2O_4$) and magnetite (Fe_3O_4). In addition, oxides containing more chromium (hereafter referred to as chromium-based oxides), e.g., chromium oxide (Cr_2O_3) and iron chromite ($FeCr_2O_4$), also may be present. Ishida et al.[13] claim that the iron-based oxides are readily soluble in acids and reducing agents, and the chromium-based oxides are readily soluble in oxidizing agents. The patented method alternately applies an oxidation decontaminating agent and a reduction decontaminating agent for chemical decontamination to remove the iron-based oxides and the chromium-based oxides that are present on the surfaces of piping and components.

$KMnO_4$ is a powerful oxidizing agent in basic solution with E° = 1.23V. The basic fluid half-reaction is

$$MnO_4^- + 2H_2O + 3e^- = MnO_2 + 4OH^- \qquad \text{(Equation [5.1])}$$

Depending on the compounds being oxidized, the overall reaction can be complex. A detailed study of the oxidative dissolution of chromium(III) oxide (Cr_2O_3) to Cr(VI) in potassium permanganate is reported by Segal and Williams.[14] The kinetics of the reaction follow an inverse-cubic rate law under both acid and alkaline conditions. Also, no inhibition by MnO_2 product was observed. In alkaline solution, at temperatures ranging from 43°C to 140°C, they also found that the rate is dependent on permanganate and hydroxide concentrations, which are explained in terms of a Langmuir adsorption mechanism. The rate of dissolution is controlled by electron transfer within a surface complex in which both OH^- and MnO_4^- are bound. They claim that the thermodynamic parameters imply chemical bond formation between permanganate and ions in the oxide surface.

A patent by Ishida et al.[15] claims that if decontamination using an oxidation agent solution is conducted first, magnetite (Fe_3O_4) is formed on a surface of the structural member and changes into hematite, which is difficult to dissolve by a reduction decontaminating solution (see details of reductive dissolution processes in Sections 4.1 and 4.3). These authors suggest that the reductive solvent should contain oxalic acid as well as hydrazine, and the pH should be adjusted to about 2.5 to dissolve the iron oxides.

Bengel[16] describes the AP/oxalic acid (APO) process as using AP for chemically decontaminating oxidized metal surfaces wetted by primary water circulating in cooling loops of nuclear reactors, which maintains permanganate-containing primary water. It is applied at a temperature of about 90°C and at a pH of 9–12 to oxidize the wetted surfaces. Oxalic acid is then added to the water, and the oxalic acid–containing primary water is maintained at a temperature of at least 90°C and at a pH of 5–7 while it is circulated to destroy the permanganate ions and manganese dioxide. Note that other oxidizing agents, such as nitric acid, have been proposed as the first agent in this type of process, but only very resistant metals would withstand this acid.

The LOMI technologies were developed to provide less aggressive reagents that would still reduce radioactivity. Loose crud found on PWR primary side surfaces generally is composed of variations of nickel ferrite ($NiFe_2O_4$), whereas deposits on BWR surfaces include iron oxides, such as Fe_2O_3 or FeOOH. Fe_3O_4-containing deposits also may be present. The LOMI reagents provide reducing environments that cause some change in the oxidation state of the iron ions. See detailed discussions of reductive dissolution in Section 4.3.

Bradbury[17] traces the development of some LOMI reagents. He claims that the rates of oxide dissolution are highest for one-electron reducing agents such as V^{2+}, Cr^{2+}, $HSCH_2COOH$, and $Na_2S_2O_4$. He concludes that the reduction process of Fe^{3+}—>Fe^{2+} requires only one electron per site. Panias[18] contends that when using oxalic acid to dissolve iron oxides that are not on iron, Fe^{2+} ions can act as the reducing agent and catalyze scale dissolution.

Bradbury[17] claims that the most effective reducing agents are V(II) bipyridyl and V(II) picolinate. The bipyridyl reagent was not stable with the ionizing radiation in the reactor. In the presence of excess formic acid, the V(II) ion can be protected from oxidation by the hydroxyl radicals produced from water radiolysis:

$$\bullet OH + HCOO^- \rightarrow CO_2 + H_2O \qquad \text{(Equation [5.2])}$$

The carbonyl ion will reduce any V(III) produced from oxidation by the hydroxyl radical or the ferric iron in the deposit:

$$V(III) + CO_2^- \rightarrow V(II) + CO_2 \qquad \text{(Equation [5.3])}$$

Figure 5.3 shows the chemical structures of two ligands (a and b) and a flat depiction of V(II) complexes of these ligands (c and d). Note that the coordination number is 6, and their orientation will be roughly octahedral, depending on steric effects. Some of the complexes are considered to be kinetically inert in aqueous solutions.[19] Parts (e) and (f) of this figure also show additional reducing agents used in nuclear cleaning solvents.

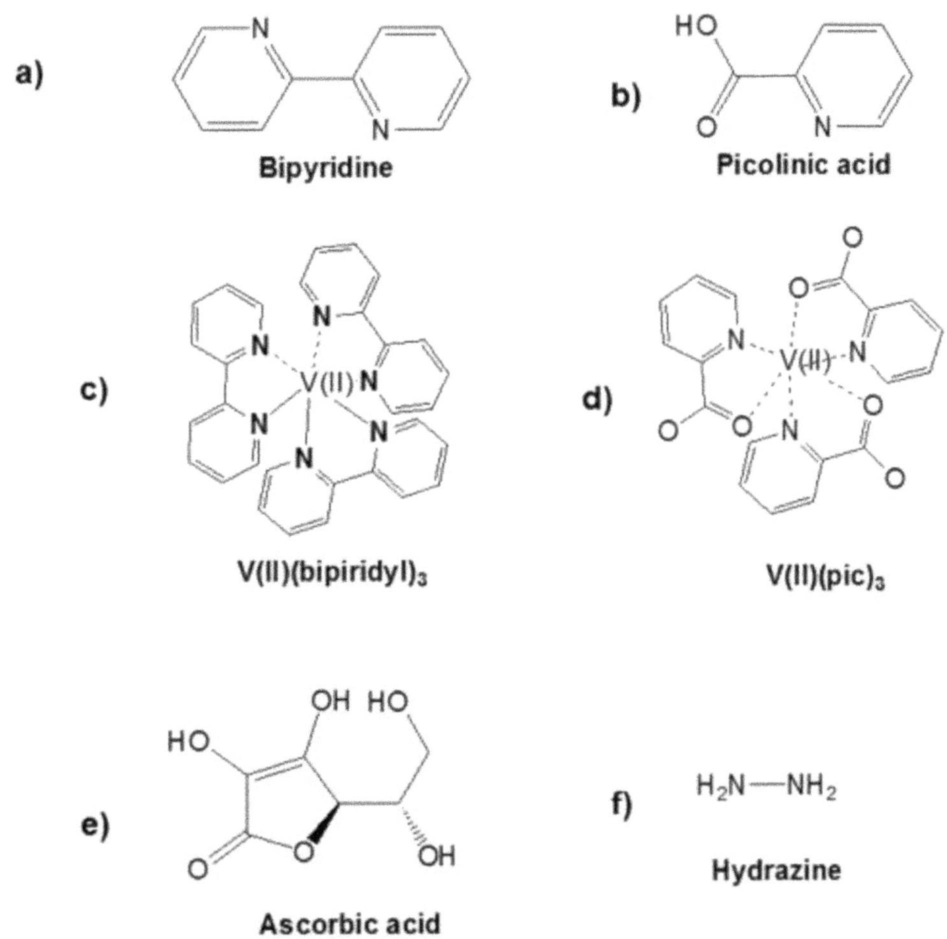

Figure 5.3 LOMI Ligands, Vanadium (II) Complexes, and Reducing Chemicals

For example, low concentrations (0.1%) of the vanadium picolinate $V(pic)_3^-$ were dissolved in formic acid (pH, ~4). These solutions have been used to remove the deposited oxides that were found in BWR systems. The circuit is filled with a degassed (the reagent is sensitive to air) solution of picolinic acid and an alkali to bring the pH to about 4 and is circulated at about 80°C. A concentrated vanadium formate solution is injected. The amount of vanadium should be equal (molar amount) to the iron in the deposit. After circulation for a time sufficient to achieve decontamination, the entire solvent is diverted through a mixed-bed ion-exchange bed to remove all the reagents. The Bradbury report[17] includes detailed investigations of the vanadium reagents and describes uses of other potential LOMI reagents, including Cr(II)bipyridyl and several based on EDTA.

Wood[20] (Table 5.1) shows results from the use of dilute reagents (LOMI as well as citrox) to clean different sections of primary BWR systems. The DFs were significant as well as the removal of radioactive material.

Table 5.1 Dilute Reagent Decontamination Cleanings of BWR Systems

Plant and System	Solvent Steps	Decontamination Factor (DF)	Radiation Removed (Curies)
Oyster Creek: RRS	LOMI/NP/LOMI	10.5	55.3
Quad Cities 2: RRS RWCU	 LOMI/NP/LOMI (LOMI on Suction) LOMI/LOMI	 4.4 2.5	 97 13.0
Dresden 2: RRS RWCU	 LOMI/NP/LOMI (LOMI on Suction) LOMI/LOMI	 9.8 2.5	 56.3 13/0
Brunswick 1 RRS- Loop A RRS- Loop B	 Citrox/AP/Citrox Citrox/AP/Citrox	 3.6 8.4	 23 Total
Peach Bottom 3: RRS B Pump RWCU RRS	 LOMI/NP/LOMI LOMI/NP/LOMI LOMI/NP/LOMI	 5.2 6.7 15.2	 7.6 13.1 26.8
J.A. FitzPatrick RHR RRS	 LOMI/NP/LOMI LOMI/NP/LOMI	 3.9 42.4	 0.2 61.4

Abbreviations: RRS, Reactor (water) recirculation system; RWCU, Reactor water clean-up systems; RHR, Residual Heat Removal system.

Wood[20] describes a regenerative LOMI process in more detail. This method incorporates an initial injection of a dilute LOMI solution (vanadium formate, picolinic acid, and sodium hydroxide) into a decontamination circuit followed by operation of a small cluster of cation exchange columns during the decontamination process. The cation exchange resin is used to remove metals in the same manner as in prior decontamination processes, but operation of the cation exchange resin is continued to allow picolinic acid initially bound to the cation exchange resin to be released and recycled to the LOMI solution.

Operation of the cation exchange columns ceases after the picolinic acid has been released but before the metals (e.g. sodium, iron, and vanadium) are released back to the LOMI solution.

The patent states that a cluster of cation exchange columns is operated in a sequence in which one column releases picolinic acid while another binds picolinic acid. The method includes continuous additions of vanadium formate and sodium hydroxide. Cleanup at the end of the method proceeds in the normal manner wherein larger cation and anion exchange columns are utilized. Because the concentration of the components is much lower than conventional LOMI processes, however, the amount of cation exchange resin required at this stage is greatly reduced. A process diagram is shown in Figure 5.4.

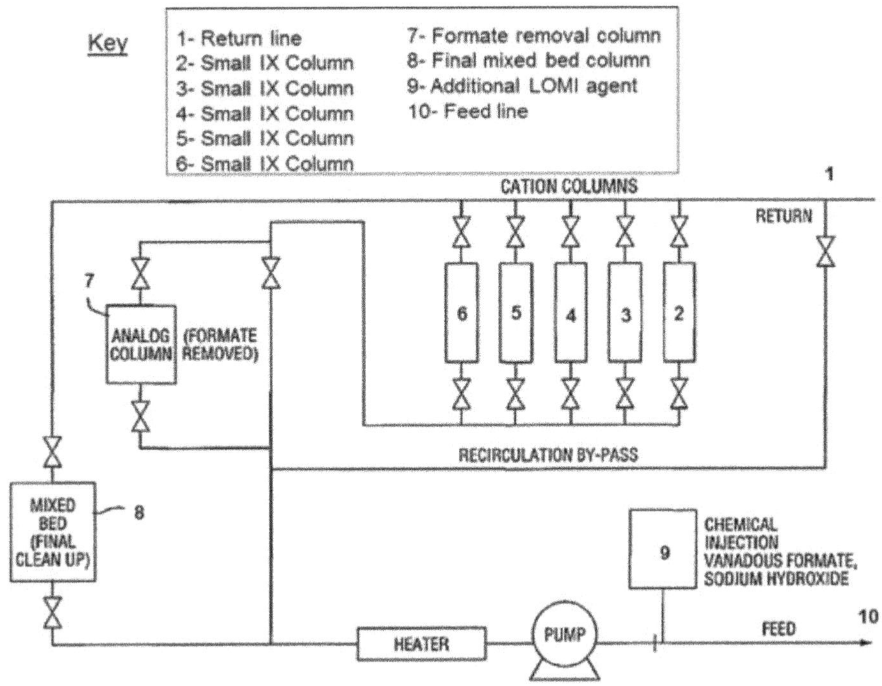

Figure 5.4 LOMI Regeneration Process Diagram[20]

A much more concentrated solution has been used to remove uranium dioxide (UO_2) contaminates from the Three Mile Island-2 (TMI-2) reactor.[21] The OPG solvent contained oxalic acid, sodium gluconate, gluconic acid, 8-hydroxyquinoline, and hydrogen peroxide.

Murray[22] describes a method for decontaminating metal surfaces with an oxide coating containing radioactive substances, such as the primary system surfaces of a pressurized water nuclear reactor. The process uses an aqueous solution of weak chelants and Fe(II). The weak (dilute) chelant maintains the dissolved metals in solution and prevents precipitation, while the ferrous iron improves the dissolution rate and minimizes base metal corrosion. The chelant can form multi-ligand complexes with the metals of the oxide coating (between 0.1% and 2.0% by weight based on the weight of the solution). The report claims that the dilute chelants should be complexing agents generally with an equilibrium constant for metal ions, such as ferric ions of 10^{12} to 10^{19}. Examples of such chelants are nitrilotriacetic acid (NTA), hydroxyethylenediamine tetraacetic acid (HEDTA), citric acid, and iminodiacetic acid (IDA), with NTA being preferred. Note the information in Table 2.3.

Murray[22] also claims that the solutions can be regenerated using ion exchange methods. The patent also has a useful review of other decon solvents and processes. Dilute solutions of organic acids that are *regenerated* using a cation exchange resin form the basis for the CAN-DECON[23] process. The organic acids (citric and oxalic acids) are added to the cooling water at about 0.1% by weight, and the solution is then passed through a cation exchange resin. This resin removes the metal ions and regenerates the acid in the solvent.

Speranzini[24–26] has investigated the corrosivity and decontamination effectiveness of mixtures of EDTA, citric acid, and oxalic acid. The authors found that mixtures of EDTA with citric acid were more effective than citric acid or citric/oxalic acids as solvents for BWR oxides. However, at 90°C, the rate constant for dissolution of oxide in citric acid/EDTA-solution is a factor of *two less* than the constant for dissolution in the citric acid/oxalic acid/EDTA-solution.

Speranzini[27] postulated that oxalic acid adsorbs at ferric ion sites in the oxide lattice and assists in dissolution of the oxide, and EDTA assists in the removal of ferrous and ferric ions from oxide surfaces by ligand exchange and complexation in solution:

$$\begin{aligned} &A)\ Fe(II)(C_2O_4) + H_nEDTA^{(4-n)} \Leftrightarrow FeEDTA^{2-} + C_2O_4^{2-} \\ &B)\ Fe(III)(C_2O_4)^{+} + H_nEDTA^{(4-n)} \Leftrightarrow FeEDTA^{-} + C_2O_4^{2-} \end{aligned} \qquad \text{(Equation [5.4])}$$

Their theory is that the rate-limiting steps imposed by high concentrations of $Fe(III)(C_2O_4)^+$ and $Fe(II)(C_2O_4)$ at oxide surfaces are therefore replaced by other limiting reactions, perhaps ligand exchange.

Panias[18] and Al-Modarak[28] have contributed to understanding of the use of oxalic acid to dissolve magnetite. They describe a series of steps:

1) Adsorption of the oxalate molecule on the iron oxide surface, where the active species depends on pH (see Figure 5.5)
2) Non-reductive dissolution, which is slow
3) Reductive dissolution that includes an induction period while Fe^{2+} ions build up and then an autocatalysis stage in which the ferrous oxalate catalyzes the reaction

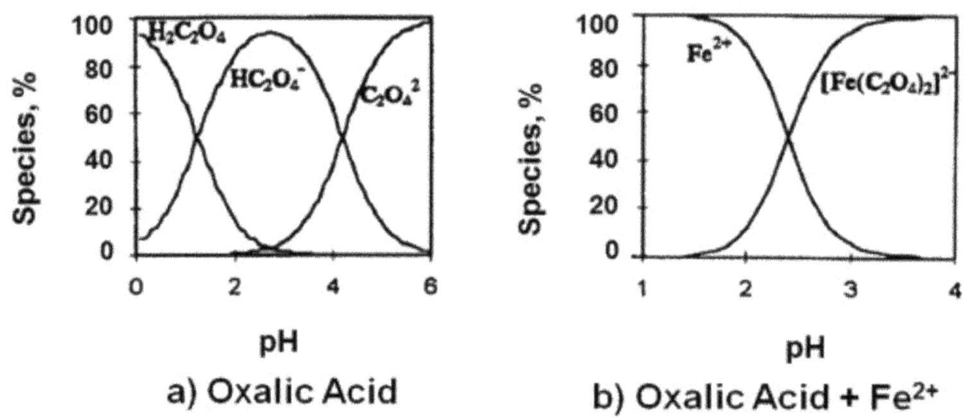

Figure 5.5 Speciation Plots for Oxalic Acid as Function of pH. (a) Without Fe^{2+} and (b) with Fe^{2+}

Note that some of the tests were run at pH 4, and the active species is the oxalate ion. The catalyzing ferrous ions can come from a magnetite scale, but in the case of hematite, they must be provided unless the scale is on an iron surface. The electron-transfer reaction is

$$[> Fe^{III} - C_2O_4^{\ 2-}] \leftrightarrow [> Fe^{II} - C_2O_4^{\ -}] \qquad \text{(Equation [5.5])}$$

Oxalic acid improves decontamination in some situations but also has contributed to increase in inter-granular attack of sensitized 304 SS.[27] In these studies, oxalic acid was removed from the CAN-DECON solutions, and EDTA was added. A modified solution system was named CAN-DEREM[27] and was used with stages of AP to decontaminate steam generator channel heads at Beaver Valley-I station. Decon factors of 4.8 to 6.5 were achieved with this process. Dilute chemical decontamination processes were described by Murray.[29] These processes use dilute solutions (about 0.1%) of organic acids and chelating agents[30] that are circulated for 2–48 hours and then are disposed of using ion exchange resins. These processes must be used more frequently than more concentrated solvents, however. Decon factors as high as 50 can be obtained using these reagents.

Oxalic acid–based solvents have been tested for decontaminating storage tanks containing nuclear wastes. King[7] claims that high level waste storage tanks will gradually be emptied of the bulk waste volume, leaving heel materials that require removal before tank closure. At Savannah River, some of the sludge contains Al and Fe hydroxides. Hematite was dissolved by oxalic acid.[31]

5.2.2. Examples of Decontamination Applications

Dresden-1. Dresden-1 was one of the first commercial nuclear reactors. Put into service in July 1960, this 200 mWe BWR, located outside Morris, IL, was owned and operated by Commonwealth Edison. The unit was taken out of service in October 1978, and decontamination procedures were need during the decommissioning.

A low pH mixed chelant system containing proprietary corrosion inhibitor and additives was selected to perform the cleaning task (called NS-1[32]) at the Dresden-1 plant. Hammer[32] states that extensive metallurgical verification, engineering reviews, and corrosion tests were performed and double-checked to qualify the process. The process called for the cleaning solution to be applied at 200°F and to be circulated throughout the primary circuitry, reactor vessel, steam drum, and secondary generators for a minimum of 100 hours and a maximum of 200 hours. Laboratory tests indicated that decontamination factors as high as 1000 could be achieved. During the actual cleaning, comprehensive chemical and isotopic analyses monitored the progress of the cleaning and provided data to determine if termination criteria had been met. Wood[4] states that this process showed decontamination factors that exceeded 11.3, and over 750 curies of activity were removed. He states that this was the only "hard-complete" decon of a BWR before 1990. The results shown in Table 5.1 described the "soft-dilute" cleaning processes.

Evaluation of the results showed that the solvent system performed as designed, and the goals of the decontamination were achieved.

Waste Tanks at Savannah River (SRS). King[7] states that oxalic acid is an industry standard chemical used for the cleaning and maintenance of nuclear power plants, although these operations often involve the removal of relatively small volumes of chemical scale materials with dilute acid followed by regeneration of the acid for reuse. They also cite information that the baseline chemical (oxalic acid) cleaning method was recently used for removing deposits in the "heel" area of two SRS waste tanks, and lower than expected amounts of solids were removed from the tanks.[33] They claim that the baseline method has not been optimized to minimize the use of oxalic acid, and the results indicate that better understanding of sludge dissolution chemistry in oxalic acid is needed to achieve more consistent and effective results.

The King[7] report presented the following evidence-based conclusions:

- Greater than 90% hematite dissolution occurred using oxalic/nitric acid mixtures within 1 week for two hematite sources and within 3 weeks for a third hematite sample with a larger average particle size. This dissolution rate appears acceptable for waste tank cleaning applications.
- Stoichiometric dissolution of iron phases in oxalic acid (based on the oxalate concentration) and the formation of the preferred 1:1 Fe to oxalate complex is possible with the addition of a supplemental hydrogen ion source (HNO_3 or H_2SO_4) and pH control.
- Sulfuric acid is nearly twice as effective as nitric acid (on a molar basis) at promoting hematite dissolution in oxalic acid solutions.

Wiersma[34] conducted corrosion tests in the proposed oxalic solvents noted in the previous paragraphs. Electrochemical and immersion (coupon) studies were conducted to investigate the corrosion behavior of carbon steel in simulated chemical cleaning environments. The effects of temperature, agitation, and the presence of sludge solids in the oxalic acid on the corrosion rate and the likelihood of hydrogen evolution were determined. The results showed that the corrosion rates decreased significantly in the presence of the sludge solids. However, agitation increased the corrosion rates, and the coupon tests were more predictive than the electrochemical tests.

5.2.3. Details of Oxalic Acid-Cleaning Chemistries

Oxalic acid has been used to clean steam generators as well as waste storage tanks. A review of some investigations of the chemistries of iron oxide scale dissolution follows.

Blesa[31] and Al-Mobarak[28] suggest that the dissolution of magnetite is similar to the processes described in Equations (4.9 and 4.10). However, they also claim that the dissolution of magnetite is an autocatalytic process. Thus, the mechanism involves an interfacial electron transfer between ferrous-oxalate complex and surface ferric ions, and the main reaction does not involve the oxidation of oxalate.

A significant observation of the use of oxalic acid is the solubility of the iron-oxalate complexes.[35] The Fe(II) mono complex is yellow-green with a solubility of ~1 g/L, and the Fe(III) complex solid is emerald-green and is slightly soluble. Electrochemical studies by Wiersma[8] of steel in the presence of oxalic acid has suggested the following corrosion reaction:

$$\mathrm{Fe}^0 + \mathrm{H_2C_2O_4} \rightarrow \mathrm{Fe}^{\mathrm{II}}(\mathrm{C_2O_4}) + H_2 \uparrow \qquad \text{(Equation [5.6])}$$

Wiersma also performed anodic polarization tests in 0.1 M oxalic acid on steel and interpreted the first peak, which occurs at approximately -0.45 V vs. SCE, as indicative of iron dissolution and the first minimum region (-0.3 to +0.1 V vs. SCE) as indicative of passivation by the ferrous oxalate precipitate. However, Wiersma also observed that the current density gradually rises with potential, indicating that the ferrous oxalate may provide an unstable passive film.

5.3. Cleaning Steam Generators

The steam generators of PWR nuclear reactors are very large heat exchangers in which the water from the core produces steam that is used to drive the turbines of the power plant. The water (and thus the deposits) on the shell (secondary) side (See Figures 1.7 and 1.8) of the steam generators should be non-radioactive, but owing to leaks, low levels of radioactivity may be present. The tubes usually are constructed of a nickel-chromium-iron alloy, whereas the shell and the plates that support the tubes are constructed from a carbon steel alloy (in early units). In units built after corrosion was recognized to be a problem, the plate is constructed of ferritic stainless steel. The tubes may be arranged as once-through or in a "U" tube design (see Table 1.1 and Figure 1.9).

The deposits that form in the generators usually contain magnetite as well as varying amounts of copper. The deposits may form a sludge (i.e., crud) that collects on the support plates and in a pile at the bottom of the generator. A more insidious linear fast-grown magnetite also can form in the annulus between the tubes and the support plates. This material can dent the tubes, can crack the support plates, and may cause the generator to lose capacity and thus power rating.

A lengthy report by IAEA[36] outlines several mechanical and chemical processes that may be applied to clean various parts of steam generators.

Mechanical methods suggested can include

- Standard tube sheet lancing (1 step)
- Enhanced sludge lancing (multiple steps)
- Inner bundle lancing
- Pressure pulse and water slap methods
- Upper bundle hydraulic cleaning

More details of some of the mechanical methods are described in Section 5.3.3.

The IAEA document also recommends several chemical cleaning methods that are described in Sections 5.3.1 and 5.3.2. These include both high-temperature and low-temperature hard cleaning procedures as well as soft cleaning processes.

Note that because access to the shell side is limited and flow is difficult to achieve, cleaning of steam generators is one of the greatest challenges the chemical cleaning industry faces. Several stages of cleaning may include iron oxide dissolution, copper removal, and crevice cleaning. These stages usually require different solvent packages. Depending on the technique used, a corrosion inhibitor (Chapter 6) may or may not be used.

5.3.1. Cleaning Processes to Remove the Bulk of Scale and Sludge (Hard Cleaning)

The cleaning of utility boilers and other process equipment usually is designed to remove all (or most) of the fouling deposits and return the heat transfer surfaces to original conditions. Since the development of steam generator cleaning in the 1980s, this was the original goal. This section documents the chemical technologies that have been developed to accomplish what has been called hard cleaning (see Little[37]). Processes using less concentrated chemicals are termed "soft" and are described in Section 5.3.2.

Since a portion of the magnetite deposit is not in electrical contact with a reducible substrate, an external reducing agent usually is a part of the solvent mixture (see Sections 4.3.1 and 4.3.2). Pitzer[38] proposed a mixture of ammonium EDTA with 0.1–0.5 molar concentration of hydrazine, used at pH 6.5–9.5. This solvent dissolved most of the magnetite pellets in his test matrix and a mixed (iron, nickel, cobalt, chromium) oxide scale from a reactor. The treatment temperature usually was about 95°C. Pocock[39] also recognized the need for a reducing stage when he tested several different solvents for dissolving iron oxide, nickel oxide, and chromium oxide.

Various EDTA and citric acid formulations have been used with reducing agents, including ascorbic acid, glucose, thioglycolic acid, and hydrazine. The solvents also often contained a commercial corrosion inhibitor. On the basis of static and dynamic tests, the most effective solvent was determined to be ammonium EDTA with 1% hydrazine (pH 7). This solvent was very effective in dissolving the iron oxide and somewhat effective for dissolving the nickel but was not very effective for the chromium oxides. The corrosion of the tube and support plate material was significant, and this concern dominates many of the subsequent studies of solvents for steam generator cleaning processes. The reaction is

$$2Fe_3O_4 + N_2H_4 + 6[H(NH_4)_3EDTA] \rightarrow 6[Fe(II)EDTA]^{2-} + N_2 + 8H_2O + 12NH_4^+ + 6NH_3 \qquad \text{(Equation [5.7])}$$

Pettit[40] described tests to develop a solvent for PWR deposits, especially to remove the denting deposits. A total of 14 solvent combinations were tested on a synthetic sludge that contained iron oxide, copper oxide, nickel oxide, chromium oxide, and various minor contaminants. The most promising solvents contained EDTA and citric acid or citric acid and ascorbic acid, as well as corrosion inhibitors (such as diethylthiourea). It was also determined that an oxidative copper stage would be required during the process.

Frost,[41] Blakrishnan,[42] and McSweeney[43] developed processes that use a copper removal pre-stage consisting of ammonia and air to remove the bulk of the copper. Iron removal stages follow. These then employ ammonium EDTA containing citric acid, hydrazine, and a corrosion inhibitor at 90–95°C for iron oxide removal.

This cleaning program was a multistage process consisting of sequences of copper removal stages and iron removal stages:

1) A copper removal stage using 10% ammonia solution at 25–35°C with air sparging to provide oxidizing conditions and assist solvent circulation.
2) An iron removal stage with a solution containing 8% EDTA, 2% citric acid, 0.4% hydrazine, and 0.2% inhibitor at pH 6.0 at 93°C. (Nitrogen sparging was used in this stage to maintain deaerated conditions and to circulate the solvent.)
3) A final copper stage using ammonia and air completed the process.

An anti-foam agent was used to minimize foaming in all cleaning stages. The actual cleaning process consisted of a total of six copper removal stages and four iron removal stages followed by flushing with demineralized water and by passivation of carbon steel surfaces.

The chemical cleaning resulted in the dissolution of 900 kg of deposits (magnetite, copper, and other metal and anionic species). This value was based on the mass of the metals dissolved determined by analyses of the spent solvents (See Section 2.6.6).

An unknown quantity of suspended solids was also removed by in-line filters and in the waste solvent. Corrosion of boiler materials during the cleaning were measured at 10 μm for carbon steel and 0.3 μm for Inconel 600. The procedure also was used to remove 500 kg of magnetite, 200 kg of copper, and various other deposits from the Nuclear Power Demonstration Nuclear Generating Station of Ontario Hydro in Canada.

In 1977, the Electric Power Research Institute (EPRI), in conjunction with the Steam Generators Owners Group (SGOG), sponsored work to develop a generic, non-proprietary system[44] for cleaning steam generators. This effort produced the EPRI/SGOG solvent system, as well as a bookshelf of reports. This section will review the most significant reports.

Welty[45] described the overall process. A total of 48 iron oxide solvents were tested to determine solvency as well as corrosivity toward A-36 carbon steel electrodes. The solvents included EDTA, HEDTA, NTA, DTPA, and citric acid salts (usually ammonium). Reducing agents included ascorbic acid and hydrazine (Figure 5.3). Two solvents (EDTA plus hydrazine at pH 6.0 and EDTA plus citric acid and hydrazine at pH 6.0) were tested in a "Mock-Up" apparatus containing several different metals. The inhibitor was the non-proprietary CCI-801 (see discussion below). The solvent without citric acid produced significantly less corrosion than the other material and was chosen for further development.

Two copper solvent approaches were tested: 10% ammonia with an air sparge or EDTA/ammonia/ethylenediamine (EDA)/H_2O_2. The latter system produced much less corrosion to exposed steel than the ammonia/air mixture and also removed large amounts of copper. The final solvent system was

1) Copper stage: 5% EDTA, 2% H_2O_2, pH 7.0 w/NH_3, pH 10 with EDA.

2) Iron stage: 10% EDTA, 1% N_2H_4, 0.5% CCI-801, pH 7.0/NH_3. With the exception of the amount and type of inhibitor used, the iron solvent is identical to the formulation recommended by Pocock[39] in 1971.

The development and testing of a non-proprietary corrosion inhibitor for the EPRI solvent was a very important part of the solvent development process. Hausler[46] tested more than 200 compounds from 15 different chemical categories in a solution of 10% EDTA, 1% N_2H_4, pH 7.0 with NH_3. The solution also contained 17.5 g/L Fe_3O_4 (to simulate dissolving scale).

The chemical categories tested included thiourea, several categories of aliphatic and aromatic amines, acetylenic alcohols, alkylpyridines, and pyrimidines. Only sulfur-containing compounds such as dibutylthiourea, TIA (tetraimidazoline derivative), and alkylpyridines provided more than 90% protection in the initial screen. The sulfur compounds and the alkylpyridines were rejected owing to solubility problems.

The paper[46] reported that a new class of compounds, imino-amido condensation products containing thioether linkages, were synthesized. The formulation provided >90% protection to alloy steel samples at 160°F (71°C) and 250°F (121°C). The final experimental formulation did not contain TIA because it would not protect SA-533-A steel at 250°F (121°C). The experimental formulation was then used in pilot-sized pot boilers and model boilers with EDTA/hydrazine to clean sludge and remove magnetite from simulated crevices.[45] Some concern was expressed about the sulfur content (~1.25%) in Hausler's[46] experimental formulation because the residual sulfur might cause cracking of the Alloy 600 tubes found in nuclear steam generators. Note the other corrosion inhibitors in Section 6.3 of this book.

Jevec[47] reviewed the work of developing the iron sludge solvent, including the tests to determine if citric acid was a desirable additive, the proper concentration of hydrazine, and the optimum pH. Three types of copper solvents were reviewed: EDTA/sodium nitrite, ammonia/air, and EDTA/peroxide/EDA. Development of analytical and monitoring techniques were also addressed.

Cleary[48] also reviewed these data as well as 24 reports written by other authors. The generic process described in Table 5.2 was judged to be generally applicable for use on Westinghouse steam generators.

Table 5.2 EPRI/SGOG Processes

Process Step	Chemistry	Temperature
Iron Sludge Removal	10% EDTA, 1% N_2H_4 0.5% CCI-801, pH 7.0/NH_3	195°F ± 5°F
Copper Sludge Removal	5% EDTA, 2% H_2O_2, pH 7.0/NH_3, pH 10/EDA	<100°F
Crevice Cleaning Solvent	20% EDTA, 1% CCI-801, pH 6.0/NH_3	250°F ± 5°F
Passivation Solution	NH_3 to pH 10.2, 300 ppm N_2H_4	200°F

In a very concise review, Welty[45] described the entire research and development process and results that led to the EPRI/SGOG procedures. Because these have been discussed at length here, the discussion will not be repeated. Two areas of concern are the effect of residual sulfur from CCI-801 and galvanic corrosion to various weldments and remain the subject of continued work.

A review of the various corrosion issues was presented by Leedy.[49] The discussion included the free corrosion rates on various metals (controlled successfully) as well as the galvanic corrosion of weldments. The authors concluded that the attack on the weldments would not be excessive in the geometries expected in steam generators. Processes based on the EPRI/SGOG development effort have been used to clean steam generators at Three Mile Island 1,[50] Byron 1, Sequoya 1 & 2, Surrey 1 and 2, Callaway and Kori (Korea), Duke Oconee, and Ontario Hydro Pickering.

Several alternative processes have been developed (or proposed) and used to clean steam generators. Oliver[51] presented a two-stage process that was used to remove 6300 lbs. of iron and 960 lbs. of copper from a

PWR in Spain. The iron solvent was described as an acetic acid base at high pH. It was applied at 360–430°F as the unit came off the line, with an exposure time of about 2 hours. The copper solvent, described as an ammonia base at high pH with air, was applied at <140°F (60°C). The authors claim that the corrosion rates were within accepted parameters.

Semmler[52] has provided details of the solvents tested to dissolve the mixed deposits found in a CANDU system. The solids contained iron oxides, copper, and copper oxides and some flakes were contaminated with silicates. Lab tests were conducted to determine the ability of EPRI-SGOG–type solvents to remove either powder or flake solids. Table 5.3 shows the lab results. These data demonstrated that the flake deposits (which contained zinc silicate) did not dissolve to the same degree as the powders that contained only iron and copper salts.

Table 5.3 Solvation of CANDU Steam Generator Solids

Cleaning Step	Solvent	Temperature (°C)	Duration, hrs.	% Dissolved (Powder)	% Dissolved (Flake)
Copper Removal	5 wt.% EDTA, pH 9-10/NH_4OH +ethylene diamine 3% H_2O_2	Ambient	3	35.5	8.6
Iron Removal	15 wt.% EDTA, pH 7/NH_4OH 1% N_2H_4/N_2 purge 1 vol % CCI-801	93	50	62.3	47.6
Crevice Cleaning	15 wt.% EDTA, pH 6/NH_4OH 1 vol % CCI-801	120	100	Not determined	Not determined

Veysset[53] described the development and use of a citric acid/gluconic acid process developed by Électricité de France. A gluconic acid (7.5%) and citric acid (4%) mixture at a pH of 3.3 (ammonia) and inhibited with "P6" (a sulfur-containing inhibitor) was applied at 85°C (185°F) to remove the iron oxide deposits. The same solution (after the pH was raised to 10 with ammonia) was used to oxidize copper. Air provided the oxidizing power. A final passivating and inhibitor removal stage consisted of citric acid, MEA, and butanediol (pH 4). Operational experiences at three French and Belgian nuclear plants were detailed. The solvents removed 280–1500 kg of iron oxide and 2–38 kg of copper from the generators. IAEA[36] claims that 170 hours of the iron solvent was sufficient to clean the whole tube bundle, but some crevices could not be cleaned.

Roofthooft[54] described several techniques that have been used in Belgium to clean the secondary sides of steam generators. These include the EDF and EPRI/SGOG processes described above, as well as the Siemens process (also called KWU[55] chemical cleaning). This process is based on modified EDTA chemistry. The iron solvent is used at concentrations between 1% and 6% EDTA, and the copper process uses EDTA, EDA, and air at an alkaline pH value. General procedures were used on different reactors. None of them were completely successful for cleaning the crevices. The Siemens process was judged to be easier to apply because it was performed during the shutdown time of the plant and took a shorter time to apply than the other procedures.

Additional process improvements were suggested by McSweeney.[43] The solvents contained hydrogen peroxide, and a parametric study of the peroxide degradation reaction was conducted:

$$2H_2O_2 \rightarrow 2H_2O + \bullet O_2 \qquad \text{(Equation [5.8])}$$

The reaction results in the production of gas (oxygen). The peroxide degradation and foaming was related to the amount of ammonia and iron and the temperature of the process. The authors recommended that EDA be added to diammonium EDTA to raise the pH to 9.5, that the iron concentration be limited to <100 ppm, and that the temperature should be <35°C. Note that although hydrogen peroxide has been used extensively in cleaning utility power boilers (see Section 4.2), this peroxide is difficult to handle, and the catalytic degradation of the chemical is always a safety/operational issue.

High-temperature chemical cleaning solvent (HTTC-Areva[56]) has been claimed to be useable without a corrosion inhibitor. This solvent is an EDTA-based cleaning solvent and is slightly alkaline (morpholine is added). The temperatures during the iron removal step are achieved by using the primary side heat in plant shutdown or startup conditions.

This creates a heat flux on the tube surfaces, promoting iron oxide dissolution on tube surfaces and particularly in the crevices (tube/tube support plate area).[56] The corrosion associated with the process is claimed to be limited by stepwise chemical injection, under strongly reducing conditions achieved by use of hydrazine and a slightly alkaline pH range. In this way, the reduction of Fe^{3+} of magnetite to Fe^{2+} is promoted, and the oxidation of Fe° (base metal) to Fe^{2+} is counteracted. Thus, the authors claim that no corrosion inhibitors are required. Owing to step-wise injection, the steam generator internals are exposed to limited amounts of free chelate of the cleaning solvent only for short time periods.

Odar[57] claims that four chemical cleanings were performed in 2007 using this process. On each unit, significant masses of deposits were removed (9500 kg–11800 kg), mainly from the free span of the tubes.

IAEA[36] claims that more than 62 applications of the AREVA HT process have been performed to clean more than 250 steam generators worldwide. This publication also recognizes two permutations of the HTTC-Areva[56] process: deposit accumulation reduction technology-high temperature (DART HT) and DART-low temperature (DART LT). The authors state that they used sub-stoichiometric amounts of complexing agents to clean tube support plates at different temperatures.

A study of the chemistry of the application of EDTA to dissolve steam generator sludge from nuclear systems in India was conducted by Rufus et al.[58] The study was done using samples of sludge taken from various steam generator units as well as synthetic sludge and metal that simulated various fouling conditions. The steam generator sludge was analyzed to understand the effects of various additives, such as pH adjusting and reducing and oxidizing agents, on the kinetics of dissolution of various phases present in the sludge. The sludge deposits (Table 5.4) contained large amounts of iron oxides, copper, and copper oxides as well as nickel deposits (nickel ferrite)

This chapter provides a detailed review of this paper because the tests parallel many of the tests that were conducted in the United States in the 1970s and 1980s and thus help to verify the various chemistries.

Table 5.4 Composition of Sludge in Steam Generator Deposits

	Tube Sheet Deposits			Tube Deposits	
Elemental Composition (%)	Reactor 1	Reactor 2	Reactor 3	Reactor 2	Reactor 3
Fe	11.6	30.0	11	37.5	17
Cu	22.8	26.8	21	22.5	25
Ni	10.5	5.6	2	7.3	10
Ca	5	7.3	3	4.8	0.6
Mg	2.5	0.3	2	0.5	2
Zn	--	1.3	4	1.7	4.5
Si as SiO_2	16	0.5	20	0.5	2
Insoluble	15	4	5	5	8

Dissolution curves for a sludge deposit are shown in Figure 5.6(a). Rufus et al.[58] claim that the increased rate of dissolution is due to the reductive dissolution of the iron oxide by hydrazine. A mechanism proposed is that under acidic conditions, hydrazine exists as $H_2H_5^+$ ions that compete with the H^+ ions of the solution for the adsorption site on the oxide (ferrous ion in the oxide lattice) and form $[>O\text{-}Fe(II)N_2H_4]^+$ complex, which decomposes to give Fe^{2+}, N_2 and $e^-_{(aq)}$. The electrons thus released reduce Fe^{3+} present in the lattice. Once Fe^{3+} is reduced to Fe^{2+}, the ionic radius increases, which destabilizes the lattice and thereby effects the easy dissolution of the oxide.

Tests with different bases to reach pH 6 (NH_3, NaOH, and EDA) produced the same dissolution rate, but Na_2H_2EDTA did not dissolve as much scale because EDA and ammonium ions have available protons to continue the reaction with the oxide. See Equation (5.7).

Rufus et al.[58] also conducted tests to dissolve metallic copper (Cu°). The reagents (Figure 5.6 [b]) included ammonium EDTA solutions with and without oxidants. The data show that H_2O_2 was a much more efficient oxidant for copper compared with air, and the control dissolved little copper. These data are in line with the work described in Sections 4.2.1 and 4.2.3 and are similar to work in the non-nuclear industry.

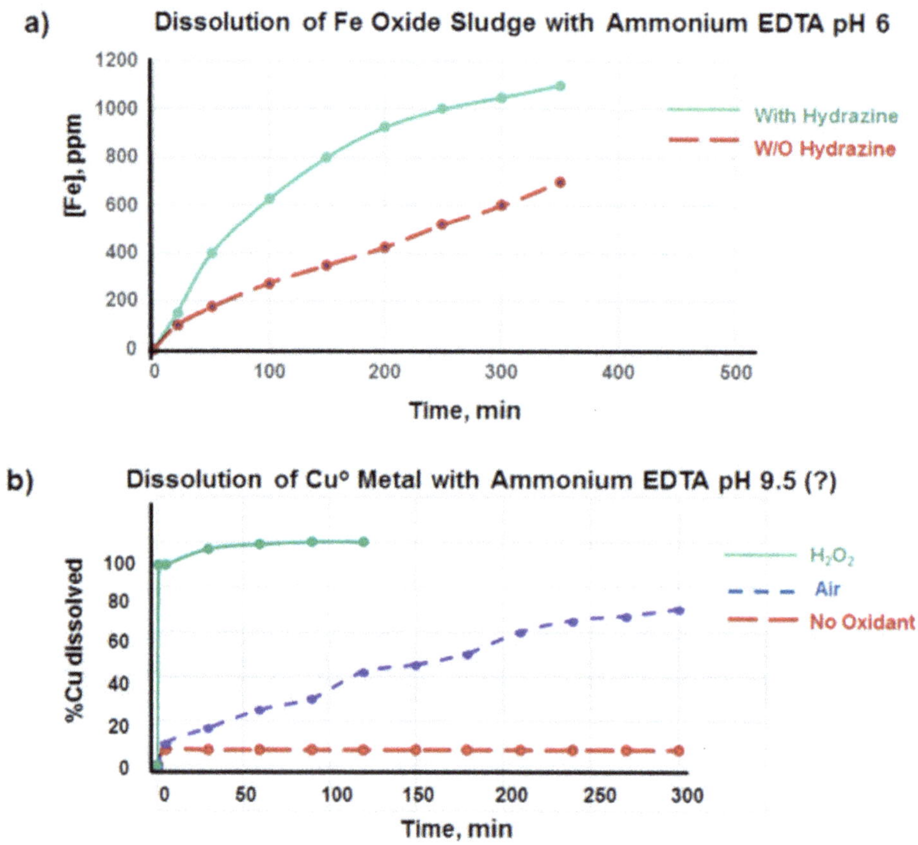

Figure 5.6 Iron and Copper Dissolution Curves in Ammonium EDTA[58]

Gmurczyk et al.[59] describe experiments carried out on actual corrosion products from steam generators. The authors suggest that 15% acetohydroxamic acid solutions may have similar effectiveness to that of the EPRI/SGOG iron solvent (which contains 15% EDTA) with respect to certain magnetite-rich deposits. The addition of malonic acid and hydrazine enhanced the dissolution of magnetite in acetohydroxamic acid solution. The structures of the complexers are shown in Figure 5.7.

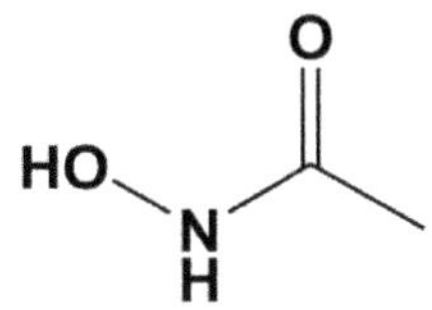

Acetohydroxamic acid

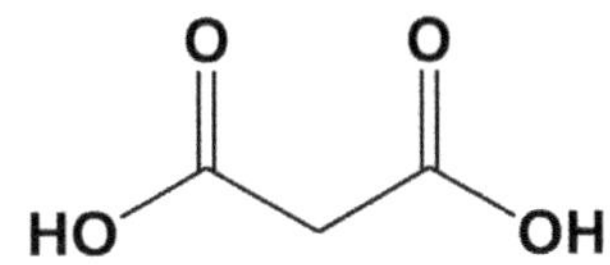

Malonic acid

Figure 5.7 Structures of Biodegradable Steam Generator Cleaning Chemicals

These authors also found that the EDTA and acetohydroxamic acid were similarly effective in the decontamination of stainless steel surfaces superficially contaminated with ^{60}Co. Results showed that the acetohydroxamic acid in the spent solvents was completely decomposed through acid-catalyzed hydrolysis using nitric or hydrochloric acid. Complete decomposition can also be achieved through oxidation of acetohydroxamic acid with active manganese dioxide or potassium permanganate.

In addition, Gmurczyk, et al.[59] found that spent, contaminated acetohydroxamic acid solutions were regenerated using a cation exchange resin to remove ferric or ferrous ions from the solution. Cobalt ions were removed from such solutions using charcoal. On the basis of these experimental findings, acetohydroxamic acid solutions appear to be promising candidates for use in decontamination or chemical cleaning, followed by decomposition or regeneration of the spent solutions. However, it has not been confirmed if this system was used to clean nuclear steam generators.

5.3.2. "Soft" Steam Generator Cleaning Processes

A recent trend in steam generator cleaning is the use of diluted or different solvents to remove *some* of the corrosion products and leave some of the product on the tubes. These are frequently called "soft" processes (compared with the high-strength or "hard" processes; see Section 5.3.1. These processes[60] are claimed to offer several advantages:

- Less time, equipment, and chemicals are used, so waste and other storage issues are minimized.
- These processes can be performed more often during short outages of the units.
- The residual scale on the tubing is claimed to enhance "nucleate boiling" and may actually increase heat transfer.[60]

Several different chemistries are in use.

A deposit minimization treatment (DMT) process[61,62] uses oxalic acid to remove smaller amounts of iron oxides than the "hard" treatments. The overall process chemistry is

$$Fe_3O_4 + 4H_2C_2O_4 \rightarrow Fe[Fe(C_2O_4)] + FeC_2O_4 \downarrow + 4H_2O \quad \text{(Equation [5.9])}$$

The authors claim that after draining the cleaning solvent, iron(II)-oxalate (FeC_2O_4) remains in the steam generator. This oxalate (a green solid) is mobilized and removed in a subsequent step, which also leads to protection of the carbon steel surfaces by iron oxides. Evans[62] recommends a copper removal step before the final rinses on the steam generators when the copper composition of the sludge is $\geq$2%, or the expected iron deposit removal will result in an "as left" deposit copper concentration of $\geq$2%. Like the iron and conversion steps, a copper step with a blend of oxalic acid, hydrogen peroxide, and ammonium hydroxide is used to remove available copper and copper oxides. IAEA[36] claim that as of the date of their report, DMT has been applied successfully in the United States.

Evans[62] also notes that in the case of the soft cleaning processes, which are designed to be applied periodically, the iron content and amount of generated waste still affect the total cleaning cost. An oxalic acid–based process solution is claimed to be readily decomposed to significantly smaller waste streams, which meet governmental regulations and consequently reduce the total costs of waste disposal. A diagram of these authors' process is shown in Figure 5.8.

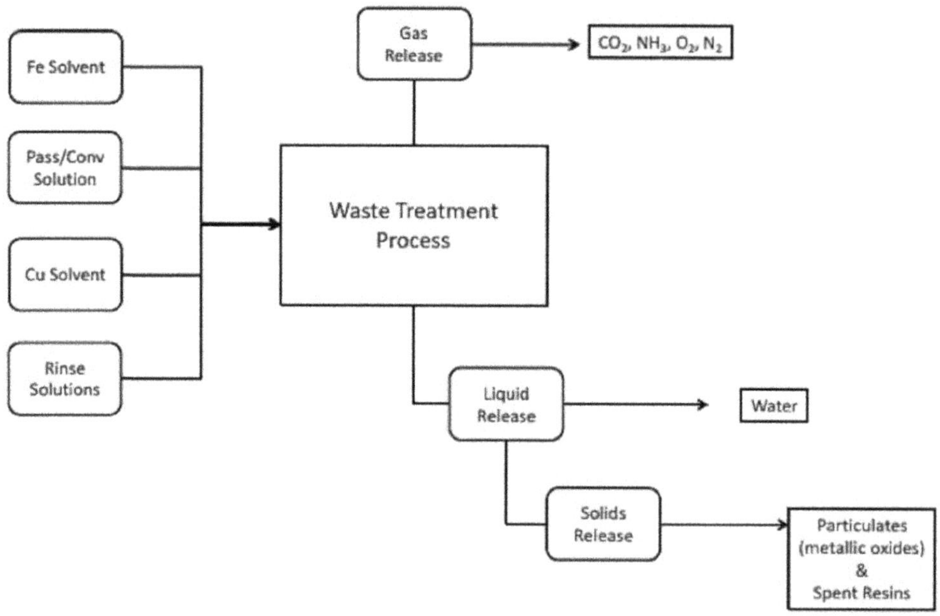

Figure 5.8 Soft Chemical Cleaning Flow Chart

Weiss[61] also claims that during an EDTA treatment, a soluble product is formed, whereas in the case of oxalic acid, a highly insoluble product is generated that forms an adherent closely protective iron(II)-oxalate layer on the base metal surface and consequently prevents further corrosion.

A low-strength version of the HTTC-Areva process[56] is described by Weiss.[61] The DART LT process is also claimed to be a soft process that uses diluted HTTC chemicals at a temperature range of 80°C to 120°C. An AREVA proprietary sulfur-free inhibitor can be applied to control carbon steel corrosion. The authors claim that the sulfur-free inhibitor prevents the ingress of detrimental impurities into the plant system.

Little[37] and Varrin[63] describe a technology called advanced scale conditioning applications (ASCA) processes. The authors claim that the ASCA solvent chemistry is largely based on the solvent chemistry used in the EPRI/SGOG steam generator chemical cleaning process. Many of the same chemical constituents used in EPRI/SGOG solvents are also used in ASCA solutions, albeit at lower concentrations. Thus, it is claimed that ASCA applications are related to a significant body of favorable steam generator cleaning experience within the industry.

One patent[63] notes that the process generally comprises these steps:

- Introducing a dilute (>2%) aqueous solution of additives into the interior of a heat exchanger vessel (for example, the secondary side of a nuclear steam generator)
- The solution introduced into the heat exchanger reacts with the deposits to partially dissolve the densified magnetite in a controlled manner and modify the structure of the remaining deposits such that they are more easily removed using known hydro-mechanical cleaning techniques.

The process is claimed to achieve partial dissolution of the magnetite and modification of the deposit structure without causing excessive corrosion of carbon and low-alloy steel structural components within the steam generator. Additional patents by Roothman[64] and Varrin[65] provide more information on the use of dilute chelant–based ASCA processes. The Roothman[64] patent describes a solvent based on

- A chelant
- A reducing agent
- A pH control agent
- A non-ionic surfactant

An example of the dissolution and re-precipitation of copper during lab-scale ASCA tests was reported by Marks et al.[3] However, neither the exact conditions of sludge/solvent ratios nor the test temperatures of these particular tests were revealed. Part of this paper describes the modeling of the dissolution process for copper and iron as well as the kinetics of deposition of copper.

Their analyses of the solvent produced a profile of copper dissolution during the iron dissolution step, and the curves are shown in Figure 5.9(a). Also, shown on the graph are empirical fits to the data using the following general equation:

$$[Cu](t) = [Cu]_{total}\left(1 - e^{-t/\tau_{diss}}\right)\left(e^{-t/\tau_{plat}}\right) \qquad \text{(Equation [5.10])}$$

The time constants τ_{diss} and τ_{plat} are the characteristic times for the initial dissolution of copper from the deposits and the subsequent plating, the two steps that transform copper in the deposits as a dispersed phase into a plated copper film on the metal surfaces in the system.

On the basis of Equation (5.10), the amount of time required to form a breachable copper film (See Figure 5.1) on the corrosion coupons was calculated. According to the authors, these calculations demonstrate that, depending on the factors influencing the rates of copper dissolution and plating, different test conditions have a different minimum exposure time required before free span non-uniform corrosion can begin. This is a significant factor in the modeling discussed below, which focuses on predicting the maximum penetration depth.

The bulk chemistry model also incorporates a model for the concentration of iron during the process using the following equation.

$$[Fe](t) = [Fe]_{max}\left(1 - e^{-t/\tau}\right) \qquad \text{(Equation [5.11])}$$

As illustrated in Figure 5.9(b), the concentration of iron in the bulk solution increases significantly over the course of the cleaning process. In both plots, the data and plots of the equations attest to the strength of the models. Note that the equations as well as the kinetic plots are similar in form to those developed for iron oxide dissolution and copper deposition in EDTA solutions. See Sections 4.2.1, 4.2.3, and 4.3.2.

a) Cu dissolution and plating

b) Fe dissolution

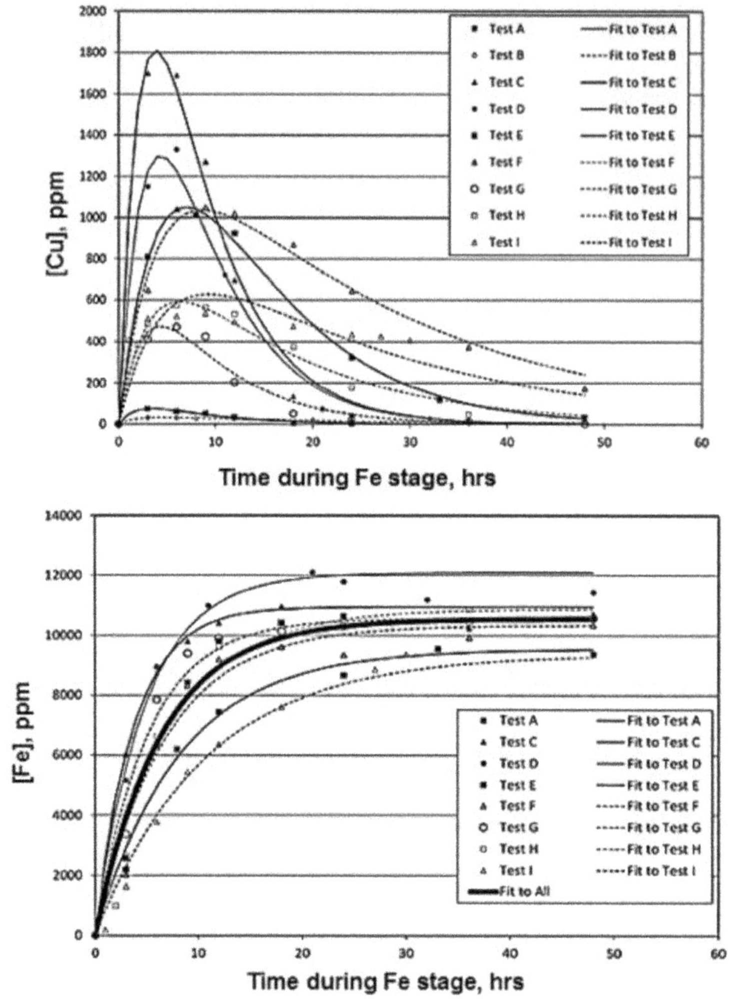

Figure 5.9 Dissolution of Copper and Replating (a) and Iron Dissolution (b) during ASCA Cleaning Tests

These treatments use both iron dissolution and copper dissolution stages. Figure 5.10 shows the surfaces of steam generator tube segments before and after a Cu-ASCA cleaning treatment. The removal of >200 lb. of copper was accomplished in one cleaning of a steam generator. On a subsequent outage, both iron and copper stages were accomplished. The treatment removed more than 600 lb. of magnetite and decreased the steam generator fouling factor by 23 µh-ft^2-ºF/Btu. The authors claim that more than 25 units have been treated in several countries without problems associated with the technology.

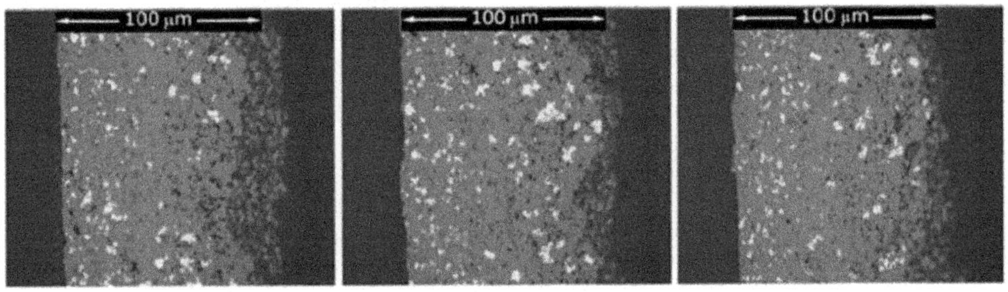

Plant SG Tube Scale Collected prior to ASCA Treatment

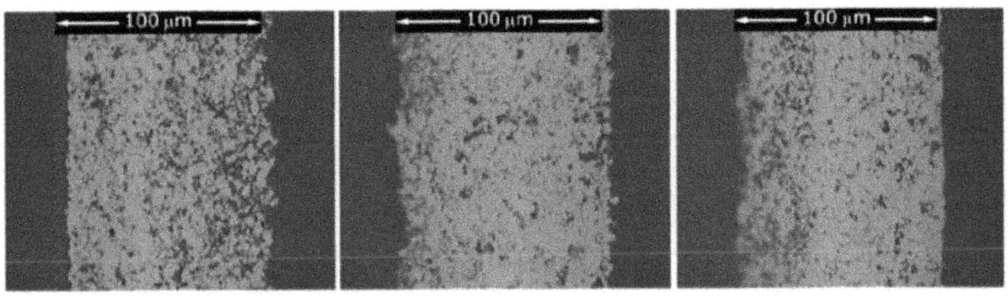

Plant SG Tube Scale Collected Following to Cu ASCA Treatment

Figure 5.10 Steam Generator Tubes before and after Cu-ASCA Treatments

Marks et al.[2] claim that an increase in porosity of a scale on a steam generator may occur after ASCA treatments, and this change may increase the heat transfer efficiency of the steam generators. The investigators claim that laboratory testing by numerous independent organizations conducted during the development and qualification of ASCA technology has demonstrated that this cleaning approach alters the tube scale morphology by increasing the porosity through much or all of the tube scale thickness via controlled dissolution. Also see the discussions in Battaglia and Rogosky[66] and Little et al.[67]

Two examples of the increase in tube scale porosity observed following ASCA solution exposure in the laboratory are shown in Figure 5.11 (a–d); each figure panel reflects two scale samples collected from a single PWR unit. The image on the left side in each figure panel reflects a tube scale flake as collected from the steam generator, and the image on the right side reflects a different flake collected from the same unit at the same time, but following exposure to an ASCA solution.

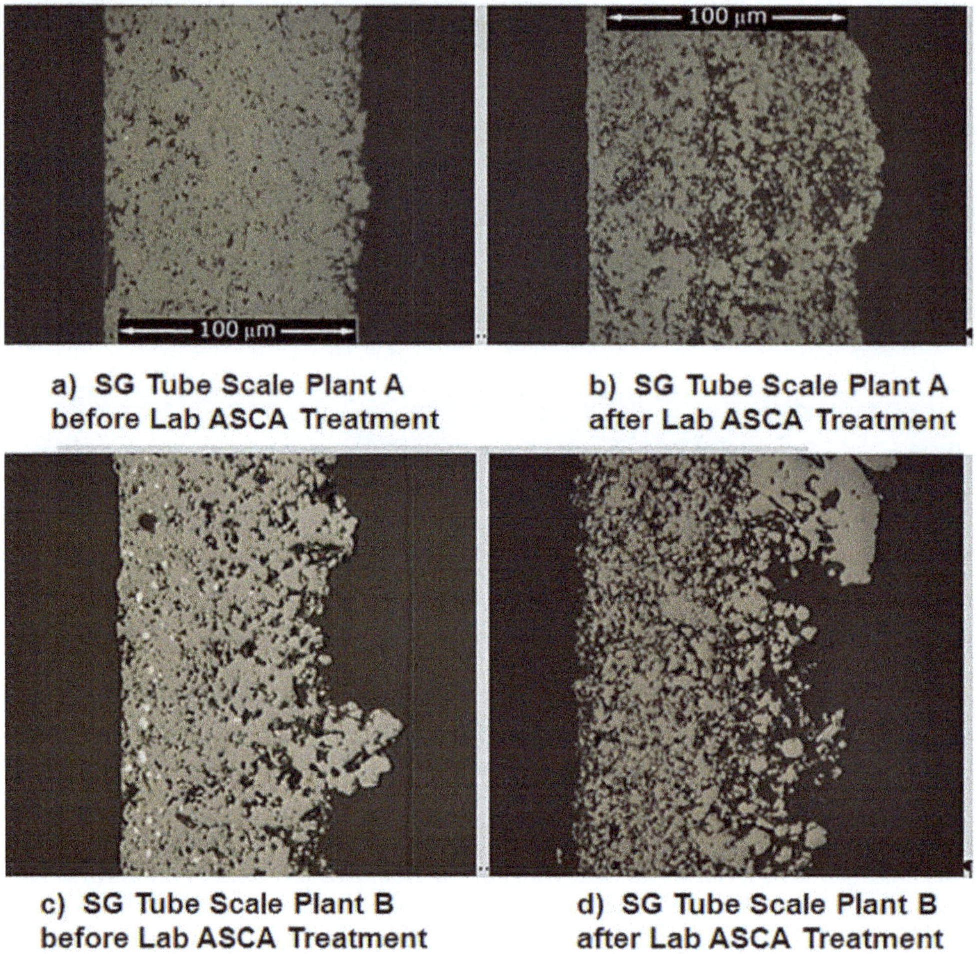

Figure 5.11 Condition of Steam Generator (SG) Tube Scale Samples before and after ASCA Treatments

The images shown in Figure 5.11 (a–d) illustrate the increases in scale porosity caused by exposure to the ASCA solutions. It should be noted that tube scale samples exhibit a wide variety of structures owing to spatial variation within the tube bundles. Accordingly, scale samples with substantially larger and smaller levels of porosity than shown in the photographs were observed both before and after exposure to ASCA solutions. However, systematic image analysis of 20 to 80 different scale samples from each unit both before and after ASCA testing revealed an average increase in porosity following the ASCA exposure in all cases. For the two units represented in Figure 5.11 (a–d), the changes were found to be

- Plant A: 17% to 30% → increase of 13%
- Plant B: 26% to 49% → increase of 23%

According to Marks et al.,[2] the net benefit for one plant following three ASCA applications as shown in Figure 5.12 is more than 40 µh-ft²-°F/Btu reduction in the fouling factor.

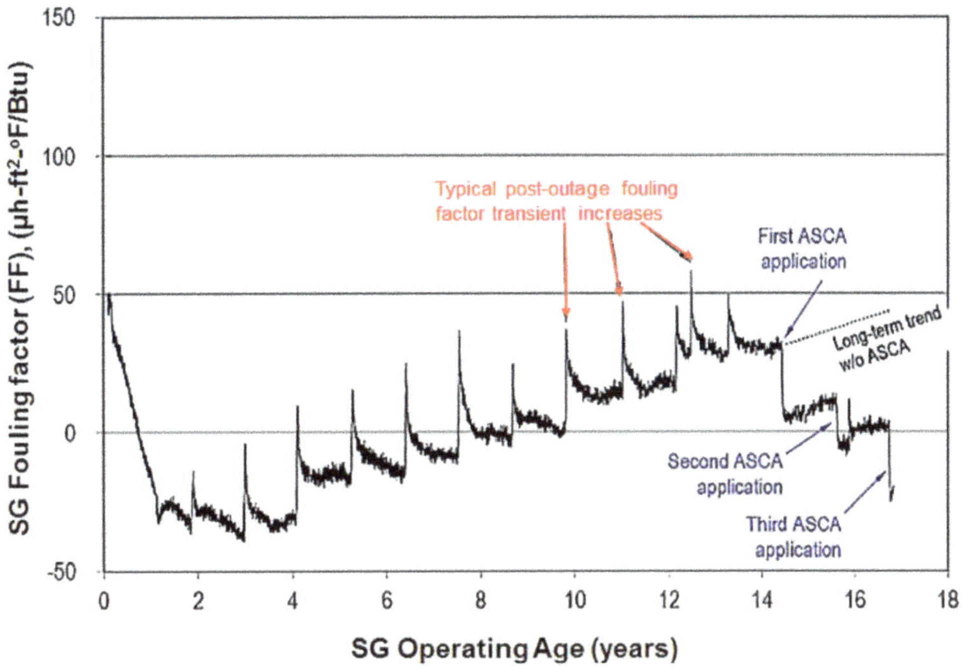

Figure 5.12 ASCA-Induced Changes in Plant B Steam Generator (SG) Fouling Factor

Marks et al.[2] also note that the hard processes usually cannot completely clean the tube and support structure region owing to the presence of hard, tenacious collars that form around the tubes at this location. See the discussion in Section 3.8 and Figure 3.16.

As an additional method to address this issue, the authors developed a cleaning process known as consolidated deposit extraction (CODE) removal that is described in Little et al.[67]

The CODE process is designed to selectively remove binding species such as oxides of aluminum and silicon that promote the consolidation of hardened scale in the collars while leaving carbon steel, stainless steel, and nickel-alloy steam generator components unaffected. At some plants, these binding species comprise substantial fractions of the hard deposit collar mass (up to 30%–50%), and this material often remains in place after hard cleaning processes, serving as initiation sites for rapid collar scale reformation.

Marks et al.[2] give an example of the changes induced in hard collars; see Figure 5.13, which depicts an actual deposit collar collected from a PWR steam generator tube sheet before and after laboratory exposure to a cleaning process that includes an iron removal step and a CODE step. Mass measurements indicated that approximately 20% of the total collar mass was dissolved by the process, and approximately 30% of the available aluminum was removed. This removal of binding species considerably weakens the residual collar structure, facilitating its subsequent breakup and removal through routine water lancing.

A patent by Varrin Jr. et al.[68] describes a process that uses tetramethyl ammonium hydroxide (TMAH) solutions that also may contain chelating agents to dissolve aluminum oxides and some silica. A formulation of TMAH and EDTA could dissolve significant amounts of AlOOH and smaller amounts of silicon dioxide.

Interest in solvents for silica and silicates is increasing. Note the possibly similar processes described in Section 4.4.2 of this book.

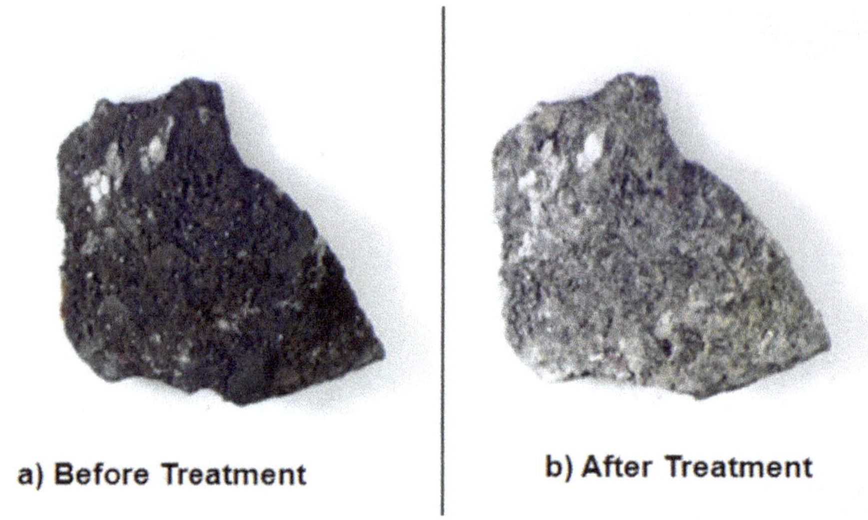

Figure 5.13 Hard Steam Generator Collar Deposit before (a) and after (b) Exposure to CODE Solution

5.3.3. Mechanical and Alternative Cleaning Processes for Nuclear Components

Vepsalainen[69] claims that the three most common mechanical methods for removal of sludge from steam generators are

- Blowdown
- Gas/liquid pressure (pressure pulse methods)
- Lancing

The blowdown method usually removes only sludge that exists as suspended particles and is ineffective at removing hardened sludge. It is used only for recirculating steam generators and uses draining system built into the steam generator.

Two gas pressure methods use quick-acting valves that release nitrogen below the water surface at approximately 60-bar pressure. Violent displacement of water (also known as a water hammer) with gas flushes and dislodges deposits. Deposits removed from the steam generator are extracted using filters or special draining procedures.

A patent by Rootham[70] describes an improvement of the pressure pulse cleaning method that includes the addition of alkaline amine chemicals into the aqueous fluid. The preferred method comprises introducing an aqueous solution of water soluble, strongly basic amines into the heat exchanger vessel. Then, a series of pressure pulses are generated within this aqueous solution inside the secondary side of the nuclear steam generator to create shock waves that will displace the solution and dislodge, dissolve, and fluidize the sludge. The preferred chemical cleaning agent is formed from a combination of either a non-corrosive lower alkyl C1-C6 amine, a non-corrosive lower hydroxyalkyl C1-C6 amine, or a non-corrosive lower C1-C6 alkoxy C1-C6 alkyl (or carrier agent).

Sludge lancing uses high-pressure water jets to remove sludge from steam generators. In this method, a device with water nozzles is inserted into the steam generator through inspection ports and high-pressure jets of water are used to dislodge the sludge. The sludge-water mixture is pumped out from the steam generator and filtered to remove suspended sludge particles.

Sludge lancing can remove soft and some sticky sludge from steam generators. Consolidated sludge may be difficult to remove with lancing techniques. The tube sheet lancing technique has also been extended to reach locations higher in the tube bundle. Water lancing is a well-known technique that is routinely applied in refineries; see Section 8.3.1 for details and illustrations.

Reinhardt et al.[71] describe a water jetting device for cleaning the upper bundles of a PWR. They claim that sludge lancing removes deposits from the tube sheet area of the secondary side. It is estimated (by the authors of this paper) that only about one-tenth of the secondary side deposits are in the tube sheet region. The balance is in the upper region of the steam generator, which is inaccessible to existing sludge lancing systems. Chemical cleaning is effective in this upper bundle region but has several disadvantages, including high cost and the generation of a large quantity of potentially hazardous waste.

They describe a system for cleaning the upper bundle region of PWR steam generators. This system employs high-pressure water jets to remove sludge and scale from tube and support plate surfaces and from the broach holes throughout the upper bundle region of steam generators. The cleaning equipment is installed through existing hand holes in the tube sheet region and accesses the upper bundle through flow slots in the tube support plates. The system includes a bulk cleaning head that directs high-pressure water jets into the tube bundle, a vertical deployment system that delivers and positions the head at the desired locations in the upper bundle region, and a high-pressure water pumping and sludge-processing system. A drawing of an upper bundle hydraulic cleaning system is shown in Figure 5.14.

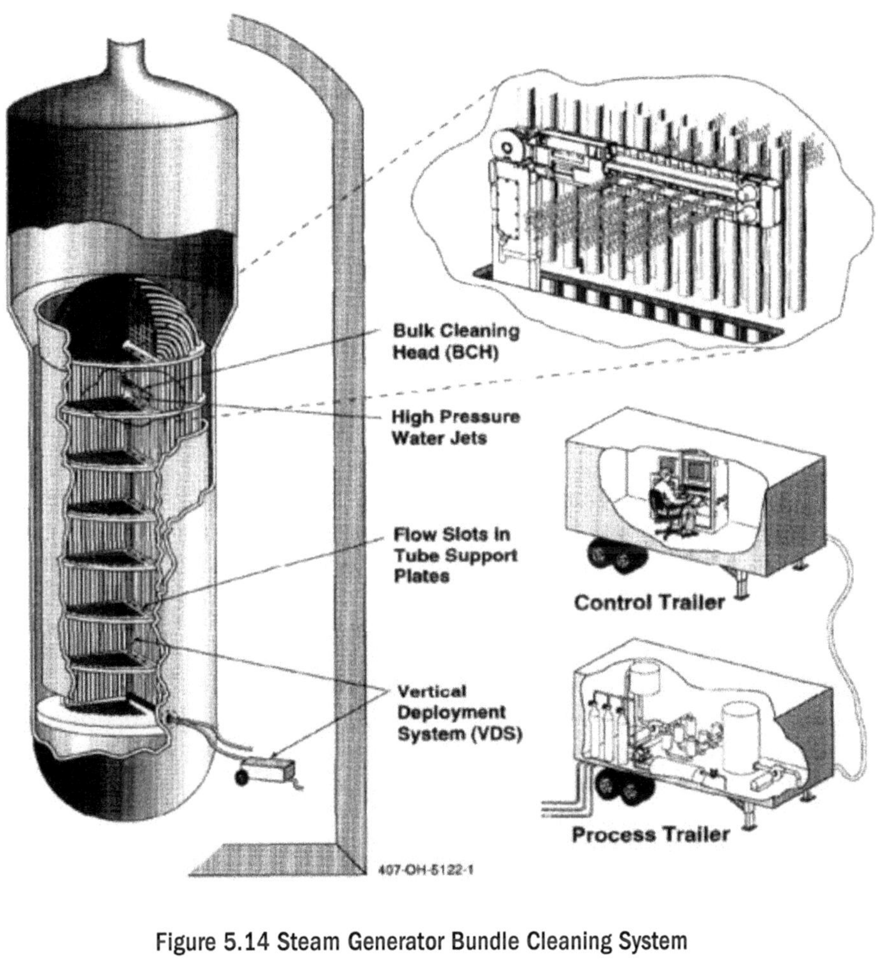

Figure 5.14 Steam Generator Bundle Cleaning System

A patent by Auld et al.[72] describes a process that uses both mechanical and chemical methods (described in Section 5.3) to remove sludge and scale from steam generators. Three steps are claimed to produce the best results:

- Sludge lancing
- Initiation of pressure pulses within a liquid chelate-containing chemical cleaning agent after the agent has been introduced into the interior of the vessel to create shock waves in the liquid for dislodging, dissolving, and fluidizing sludge and corrosion products. The pressure pulses are created by injecting pulses of gas into the liquid chemical agent at a pressure of between about 300 and 900 psi (the pressure increases as the chemical cleaning agent fills the vessel).
- Simultaneously recirculating the chemical cleaning agent through a filter assembly during the pressure pulsing operation to remove fluidized sludge and corrosion products dislodged by the pressure pulsing

Several additional mechanical processes have been described. Varrin[73] detailed the development and use of a water jet nozzle array for removing adherent scale from the secondary side tube surfaces. The array can deliver water at pressures up to 4500 psi and was tested at Indian Point 2 nuclear plant. A mixing process was described by Buford.[74] The system includes pumps, educators, and gas injection to improve mixing and agitation of the steam generator cleaning solvents.

Varrin and Little[75, 76] suggest the use of steam to inject cleaning chemicals into a nuclear steam generator:

- Introduce a volume of water into the secondary side of the steam generator
- Introduce a predetermined quantity of a chemical cleaning reagent into the water to form a liquid cleaning solution on the secondary side of the steam generator. The chemical cleaning reagent is selected from a group of chelates, complexing agents, oxidizing agents, reducing agents, and mixtures thereof. The liquid cleaning solution forms a large bath that substantially fills open areas of the secondary side below a given elevation.
- Inject steam through the temporary adapter and directly into the liquid cleaning solution in the secondary side of the steam generator to heat the steam generator and liquid cleaning solution to a target cleaning temperature range. Maintain the steam generator and the liquid cleaning solution within the cleaning target temperature range during a cleaning period by controlling a flow rate of the injected steam.

This process requires less equipment, maintains the cleaning temperatures needed for the process, and generates less waste than the solvent processes described in Sections 5.3.1 and 5.3.2. A very useful drawing from the patent is shown in Figure 5.15. Note that steam or vapor cleaning of refinery equipment has been practiced since about 1960[77] and is described in Section 7.6.

Kaneda et al.[78] describe a process that uses ultrasonic transducers suspended in a steam generator to help remove some of the scale. These methods are also used to clean refinery heat exchangers, and the technology is described in Section 8.3.2.

Remark[79] describes a nuclear steam supply system mode 5 (using primary side heat) steam generator chemical cleaning system, which applies chemical cleaning solvent (formulated to remove steam generator deposits) to the secondary side of the steam generator using primary side–supplied heat during NSS Mode 5 shutdown to maintain the desired application temperature. Mixing is achieved via nitrogen sparging. The solvents are a modified EPRI/SGOG EDTA-based fluid.

An experimental mechanical system is described by Jeong[80] that consists of an ultrasonic cleaning technology for a nuclear steam generators. The developers used lab-produced disks of simulated steam generator sludge (mostly magnetite) to test the application of ultrasonic energy to help remove compacted sludge. They report a 20% improvement in sludge removal compared with control tests. The solution used was not described in detail. Note that this type of ultrasonic process is employed commercially to clean refinery-type heat exchangers. See Section 8.3.2.

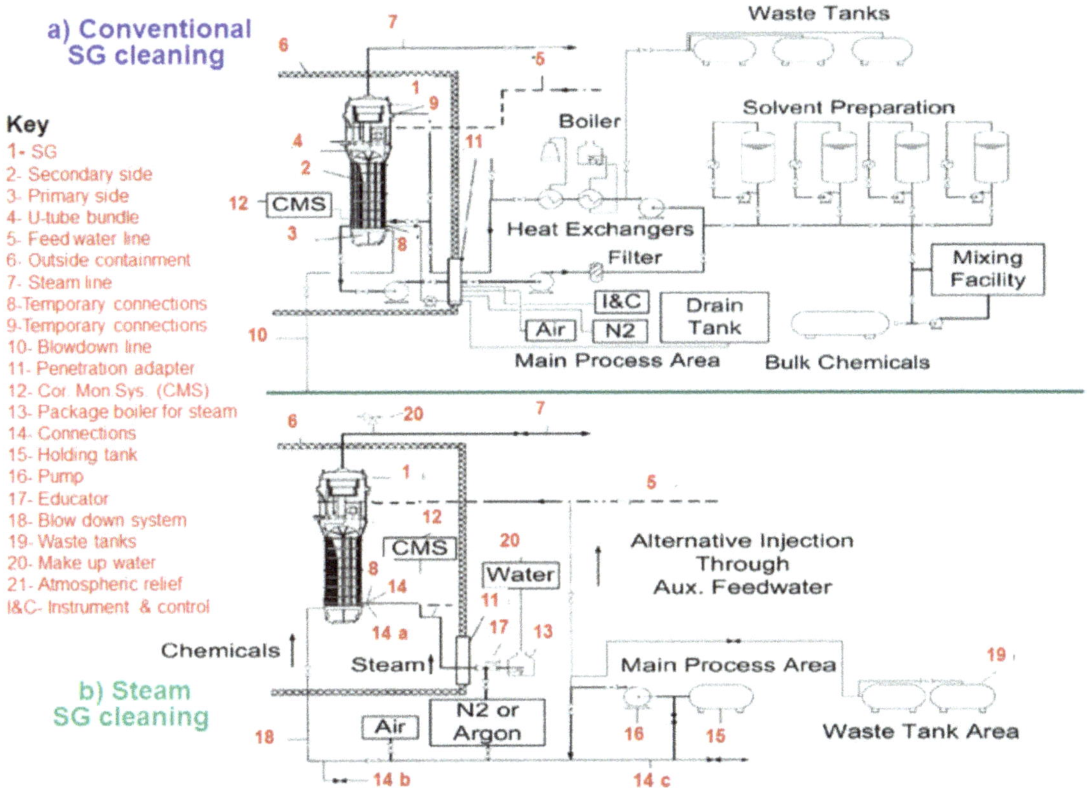

Figure 5.15 Comparison of Liquid (Conventional) and Steam Cleaning Setup for Steam Generators

5.4. Cleaning Auxiliary Equipment in a Nuclear Plant

A nuclear power plant has a wide variety of other units that may be inside or outside of the containment structure that may become fouled and in need of cleaning services. These include very large cooling water systems and condensers. This short section notes articles that are associated with the auxiliary units. While conducted inside the secured plant limits and thus possibly subject to greater control, cleaning auxiliary equipment may use the same technologies used in other industrial plants. See Chapter 4.

The largest known (at the time of the article) service water system chemical cleaning[81] was performed at the Gulf States Utilities Company, River Bend Station. Brice[81] described the process, which consisted of an iron oxide dissolution stage, alkaline stage (AS), copper removal stage, and a lay-up stage. The first iron stage used sodium HEDTA (Table 4.1) at an acidic pH with a commercial corrosion inhibitor. This stage lasted about 100 hours. The AS consisted of TSP and a surfactant. The copper removal stage used an alkaline chelating agent and a liquid oxidizing agent. The final lay-up stage used high pH water containing an oxygen scavenger. The cleaning removed approximately 17,000 lb. of iron and 700 lb. of copper.

Ruiz[82] describes the work in several French nuclear plants. Inspections on moisture separator reheaters in the French CPO reactor series have shown evidence of significant clogging owing to deposits of magnetite inside tube bundle tubes whose presence could damage equipment and harm productivity. They have developed a procedure that consists of a mechanical pre-cleaning of all tubes of the moisture separator reheaters to unblock them, followed by chemical cleaning in which magnetite is dissolved and flushed out of the tube bundle. The chemicals used have included citric acid, EDTA, and/or formic acid at low concentrations, and an inhibitor. This is followed by a passivation step that increases the pH to 9.5/ammonia and oxidation of the fluid with H_2O_2. Note that this process also was described for cleaning utility boilers (Sections 4.1.2 and 4.2.3).

CONCO[83] described the cleaning of a large condenser in a South Texas Nuclear power plant. There are 96,234 titanium condenser tubes (size 3/4" O.D.) in the multistage condenser. The tubes in the condenser were fouled with calcium carbonate. The crystalline scale had thickness values that ranged from 7.6 mils to 31.2 mils, with an average of 19.06 mils. The scale was removed using tube drilling and hydro assisted tube cleaners that are like those described in Section 8.3.1. According to the website, 7.5 cubic yards of calcium carbonate, weighing over 5,000 pounds, were removed from the condenser tubes in unit 1. Because of the scale removal, the plant reports a significant increase of approximately 3 megawatts of power.

Because the systems in nuclear plants are water based, they may be subject to water-born deposits similar to those in other industries, such as corrosion, scale, and biofouling (including microbiologically induced corrosion).

Terry[84] documents a biofilm episode in the main condenser of a nuclear power plant. This report discusses all the factors leading up to the problem and the measures taken to solve the problem. The most difficult aspect of the problem was identifying the onset of biofilm buildup.

By sampling the deposits they found that

- All the deposits had microbiological activity
- The LOI was between 20% and 50% (see Section 2.6)
- There were very small amounts of $CaCO_3$
- The primary inorganic deposits were silt and a calcium phosphate

The deposits were not found in the hottest parts of the condenser, indicating that precipitation was not the cause.

The problem was reversed by using a NaBr/(NaOCl-bleach) mixture, which is an oxidizing biocide (see Frenier[85]), and a wetting/penetrating agent, along with pH adjustments and other measures.

Alternative monitoring techniques are discussed by Terry,[84] and the treatment and reversal of the problem are documented. The rationale for differentiating between biofouling and deposition caused by inorganic precipitation is also discussed.

5.5. Summary and Best Practices

5.5.1. Summary

A wide variety of chemical and mechanical methods are in use (or have been proposed) for cleaning various parts of nuclear power plants. This chapter reviews chemical and mechanical methods for removing various fouling or hazardous (radioactive) solids.

The industry has used various acronyms and code words to describe the various cleaning fluids or processes. Table 5.5 summarizes the various chemistries described in this chapter. All the chemical data are from the literature cited, and no proprietary information has been disclosed. Note that with the exceptions of the LOMI solvents, the chemistry is very similar to that used in other parts of the chemical cleaning industries. The difference is the rigor that is required for testing and for quality control of the field solvents to satisfy requirements of the US NRC and other laws.

Table 5.5 Summary of Nuclear Chemical Services

Acronym or Code	Problem Serviced	Chemistry (from Literature)	Comments
AP	Oxidize radioactive sludge Cr(III)-Cr(VI)	Alkaline potassium permanganate	
AC	Remove iron oxide	Ammonium citrate	
CITROX	Remove iron oxide	Ammonium citrate and OA	
CANDECON	Decontaminate CANDU reactors	Weak (0.1%) solutions of organic acids (including OA) that are regenerated using ion exchange	
CANDEREM	Decontaminate CANDU reactors	CANDECON with OA replaced with EDTA used with AP stages	
AP/CITROX	Decontaminate steam generators and BWR surfaces	Two-stage process of AP then CITROX	
APO	Decontamination	AP followed with OA	
NP	Oxidize radioactive sludge Cr(III)-Cr(VI)	Nitric acid/potassium permanganate	
Low oxidation (state) metal Ions (LOMI) solvent	Remove iron oxides or nickel ferrite, especially in BWRs	V(II) bipyridyl in formic acid	
OPG	Remove UO_2 solids	OA, sodium gluconate, gluconic acid, 8-hydroxyquinoline, and hydrogen peroxide	Used at TMI-2
OA	Decontaminate radioactive waste tanks and as part of other solvents	OA solutions and may be used with mineral acids	Waste tank cleaning at Savannah River plant
NS-1	Decontaminate BWR surfaces	A proprietary EDTA-based solvent	Used at Dresden-1
EPRI-SGOG solvent system	Steam generator hard cleaning solvent process	See Table 5.2	A very well-tested and frequently used system
KWU- Siemens	Steam generator hard cleaning solvent process	Ammonium-based EDTA is used as a chelate, and a high concentration of EDTA solution is injected several times. The iron step is applied for several hours in the range of 433°K to 448°K without an inhibitor.	Has been tested for cleaning Japanese steam generators
CCI-801	Inhibitor for EPRI-SGOG iron stages and for other related solvents	Proprietary materials	Well-tested inhibitor
Citric/Gluconic -EDF	Cleaning EDF steam generators	Mixture of citric and gluconic acids and ammonia solution with oxidant for copper	
HTTC-Areva	Cleaning steam generators in Europe and elsewhere	EDTA-based cleaning solvent, slightly alkaline (morpholine added)	
DMT	Soft iron oxide solvent for steam generators	OA	Leaves a green precipitate that is washed out
Deposit Accumulation Reduction Technology (DART LT)	Soft steam generator cleaning system	Dilute HTTC	Soft systems are claimed to be more economical and easy to use and may enhance power due to nucleate boiling
Advanced Scale Conditioning Applications (ASCA)	Soft steam generator cleaning system	A dilute version of EPRI-SGOG solvent system	
Copper removal systems	Remove copper from steam generators	Various systems derived for the ammonia/oxidant or iron-EDTA technologies	See Sections 4.2.1 and 4.2.3
Consolidated Deposit Extraction (CODE) removal process	Solvent to remove part of hard scale that contains Al and Si components	Possibly a formulation of trimethyl ammonium hydroxide and EDTA	Solvent used as part of a Soft treatment. See Varrin Jr. et al.[68]

Abbreviations: OA, oxalic acid; BWR, boiling water reactors.

5.5.2. Best Practices

Hudson[1] proposes a method for designing chemical cleaning systems for nuclear plants. Over time in nuclear power plant operation, corrosion products, such as copper and iron from piping and feedwater systems, deposit on the tube sheet, tube bundle, and in crevices in the secondary side of steam generators. These deposits can harm the steam generator and steam generator tubes by inducing tube degradation such as stress corrosion cracking and tube denting. To remove corrosion deposits that have grown over several cycles and to mitigate further tube damage, developers must design and apply a chemical cleaning system to remove bulk deposit loading with targeted regions of cleaning. The design and performance of a chemical cleaning system and application is tailored to the individual utility, since deposit loadings and characteristics, steam generator designs, and utility objectives differ. A successful chemical cleaning requires much planning and design with consideration of safety, environmental, and performance factors. These three factors drive system modifications and design improvements as utilities seek to decrease costs with improved efficiency while performing a safe chemical cleaning that has a low impact on schedule.

Hudson[1] claims that a significant amount of planning and design is required when performing a chemical cleaning with the intent to remove as much deposit as possible in a safe and environmentally friendly way. Safety, environmental, and performance issues are the core factors that are considered when designing a chemical cleaning system and process sequence that is cost and time effective, safe, and efficient. Each one of these factors is thoroughly processed and discussed individually to provide a successful chemical cleaning.

Evans[86] reviewed several steam generator chemical treatments. Since 1998, several steam generator chemical cleanings have been performed in North America and worldwide with the objective of reducing the corrosive impact of steam generator secondary side deposits on the steam generator tubes and improving the heat transfer performance of the steam generator. During the last 4 years, cleanings have been performed on several different types of steam generators. These cleanings used various cleaning processes, including high-temperature plant heat processes, external heat crevice cleaning processes, and bulk deposit removal processes. Each process was tailored to cleaning objectives, deposits characteristics, system design, and application strategy. This paper describes the experiences and results from four of the most recent steam generator chemical cleaning projects in the United States since 1998. A table from this report is reproduced here (Table 5.6) to show the amounts of deposits removed with a hard cleaning process.

Table 5.6 Salem Unit 2 Chemical Cleaning Deposit Removed, Value in Pounds (Kilograms)

Process Step	Magnetite, Fe_3O_4 Lb.(Kg)	Nickel Oxide, NiO_2 Lb.(Kg)	Zinc Oxide, ZnO Lb.(Kg)	Copper, Cu Lb.(Kg)
Iron	11,019 (4,998)	182 (83)	6 (2.7)	13 (5.9)
Crevice	1,711 (776)	38 (17)	1 (0.5)	0 (0)
Cu/Passivation	14 (6.4)	1 (0.5)	0 (0)	279 (127)
Totals	12,744 (5,781)	221 (100)	7 (3.2)	292 (132)
Total from chem cleaning		13,264 (6,017)		
Total from sludge lancing		1,233 (6,576)		
Grand total		14,497 (6,576)		

Little[37] described the application of ASCA cleaning processes discussed in Section 5.3.2. They report that a total of 25 ASCA applications had been performed before 2010. ASCA applications are generally categorized into two types: full-bundle and top of tube sheet (TTS). More than 80% of ASCA applications performed to date have involved a full-bundle process, with the remainder consisting of TTS processes.

The implementation details and cleaning objectives for full-bundle and TTS applications may differ slightly depending on plant-specific cleaning objectives and considerations. Full-bundle ASCA applications can be applied as online processes with plant heat, which significantly reduces equipment and operational complexity, or can be applied when the plant is offline. When plant heat is used, ASCA processes are typically applied during plant shutdown at the start of a refueling outage and are designed to be easily integrated with the plant's normal time-temperature profile during a shutdown sequence to minimize support required from plant operations and avoid affecting the critical path outage schedule (see additional discussion below). External heating is also possible during full-bundle ASCA cleaning applications. This method of heating increases the complexity of the equipment setup (relative to a plant heat configuration) but allows the process to be applied at any time during the outage. To date, all full-bundle ASCA applications have been performed with plant heat, but a full-bundle ASCA application with external heating is planned.

Fujiwara et al.[87] investigated the application of chemical cleaning to dissolve and remove scales and sludge by chemicals at the Japanese PWR plant to maintain a high heat transfer condition and to prevent steam generator tube degradation. This paper investigated fundamental characteristics and cleaning effectiveness of the EPRI process and the KWU (Siemens) process, which are typical cleaning processes. Both processes showed a satisfactory cleaning effectiveness for sludges and scales, and the cleaning effectiveness of the standard KWU process was improved by tailoring it to crevice cleaning conditions. This report gives a lot of details about the preliminary testing and setup of jobs.

References

[1]W.H. Hudson, J.B. Delrue, S.W. Evans, J.F. Remark, "Designing a Chemical Cleaning System: A Conceptional Design," CORROSION/2006, paper no. NACE 06534 (Houston, TX: NACE, 2006).

[2]C. Marks, M.A. Kreider, M.J. Little, R.D. Varrin Jr, "Benefits of Partial Removal of Corrosion Deposits from Nuclear Steam Generators: ASCA and CODE Applications," CORROSION/2017, paper no. C2017-9481 (Houston, TX: NACE, 2017).

[3]C. Marks, R.D. Varrin Jr, A. Pellman, M.J. Little, "Stochastic Modeling of Non-Uniform Corrosion of Carbon and Low Alloy Steel During Chemical Cleaning," CORROSION/2017, paper no. NACE C2017-9489 (Houston, TX: NACE, 2017).

[4]C.J. Wood, "A Review of the Application of Chemical Decontamination Technology in the United States," *Progress in Nuclear Energy* 23 (1990): 35-80.

[5]L.E. Boing, "Decontamination Technologies," International Atomic Energy Agency, Decommissioning of Nuclear Facilities, Manila, Philippines, Oct. 16-20, 2006.

[6]J.F. Remark, "Decontamination Experience and Technology: A Review," Decontamination of Nuclear Facilities, International Joint Topical Meeting ANS-CNA, Niagara Falls, Canada, Sept. 19-22, 1982.

[7]W.D. King, M.S. Hay, "Alternative and Enhanced Chemical Cleaning: Basic Study Results Fy 2010" (Aiken, SC: Savannah River National Laboratory, 2011).

[8]B.J. Wiersma, "Hydrogen Generation During the Corrosion of Carbon Steel in Oxalic Acid," (Aiken, SC: Westinghouse Savannah River Company, 2004).

[9]R.W. Staehle, "Protecting Dresden 1 Reactor Coolant Materials from Corrosion During Decontamination and after Return to Service," American Power Conference on Decontamination of Nuclear Plants, Columbus, OH, 1975.

[10]C.J. Wood, "Recent Developments in Chemical Decontamination Technology," In Proceedings of the Third International Workshop on the Implementation of ARKL (Washington, DC: U.S. Nuclear Regulatory Commission, 1995): pp. 717-725.

[11]WNOR, Decommissioning Nuclear Facilities, 2016, World Nuclear Association, London, http://www.world-nuclear.org/information-library/nuclear-fuel-cycle/nuclear-wastes/decommissioning-nuclear-facilities.aspx (Sept. 2016).

[12]J.A. Ayers, *Decontamination of Nuclear Reactors and Equipment* (New York, NY: Ronald, 1970).

[13]K. Ishida, K. Anazawa, Y. Takamori, N. Uetake, M. Nagase, H. Yoshikawa, T. Tamagawa, "Method of Chemical Decontamination," U.S. Patent 6549603, 2003.

[14]M.G. Segal, W.J. Williams, "Kinetics of Metal Oxide Dissolution. Oxidative Dissolution of Chromium(III) Oxide by Potassium Permanganate," *Journal of the Chemical Society, Faraday Transactions 1: Physical Chemistry in Condensed Phases* 82, 10 (1986): 3245-3254.

[15]K. Ishida, H. Hosokawa, M. Aizawa, "Method for Chemical Decontamination for Carbon Steel Members of Nuclear Power Plant," U.S. Patent 9230699, 2016.

[16]T.G. Bengel, J.F. Remark, "Alkaline Permanganate Process," U.S. Patent 5278743, 1994.

[17]D. Bradbury, M.G. Segal, R.M. Sellers, T. Swan, C.J. Wood, "Development of LOMI Chemical Decontamination Technology. Final Report. [PWR; BWR]" (Palo Alto, CA: Electric Power Research Institute, 1983).

[18]D. Panias, M. Taxiarchou, I. Paspaliaris, A. Kontopoulos, "Mechanisms of Dissolution of Iron Oxides in Aqueous Oxalic Acid Solutions," *Hydrometallurgy* 42 (1996): 257-265.

[19]F.A. Cotton, G. Wilkinson, *Advanced Inorganic Chemistry* (New York, NY: John Wiley & Sons, 1980).

[20]C.J. Wood, D. Bradbury, G.R. Elder, "Regenerative LOMI Decontamination Process," U.S. Patent 5805654, 1998.

[21]R.D. Weed, "Decontamination Solution Method," U.S. Patent 3258429, 1966.

[22]A.P. Murray, "Method for Decontaminating a Pressurized Water Nuclear Reactor System," U.S. Patent 5024805, 1991.

[23]P.J. Pettit, "Decontamination of CANDU Primary Coolant System," Chalk River Nuclear Labs Canada, 1988.

[24]R.A. Speranzini, R.L. Tapping, D.J. Disney, "Corrosiveness of Decontamination Solutions to Sensitized Aisi 304 Stainless," *Corrosion* 43, 10 (1987): 632-641.

[25]R.A. Speranzini, *Water Chemistry of Nuclear Reactor Systems* (London, U.K.: British Nuclear Energy Society, 1989).

[26]J. Semmler, L. Stolberg, P.V. Balakrishnan, R.A. Speranzini, "Chemical Cleaning Processes: Present and Future," COG/IAEA 4th PHWR Conference, Kyongju, Korea, April 22-27, 1996.

[27]R.A. Speranzini, R. Voit, M. Helms, "Decontamination Effectiveness of Mixtures of Citric Acid, Oxalic Acid and EDTA," *Nuclear Engr. Int.* 35, 434 (1990): 52.

[28]N.A. Al-Mobarak, "Effect of Oxalic Acid on the Dissolution of Magnetite Coupled with Iron of Various Surface Area," *Int. J. Electrochem. Sci.* 3 (2008): 666-675.

[29]A.P. Murray, D.A. Eckhardt, S.L. Weisberg, "Dilute Chemical Decontamination Process for Pressurized and Boiling Water Reactor Applications," *Nuclear Technology* 71, 2 (1985): 482-496.

[30]G.R. Choppin, R.L. Dillon, B. Griggs, A.B.J. Johnson, J.F. Remark, A.E. Martell, "Literature Review of Dilute Chemical Decontamination Processes for Water-Cooled Nuclear Reactors" (Palo Alto, CA: Electric Power Research Institute, 1975).

[31]M.A. Blesa, H.A. Marinovich, E.C. Baumgartner, A.J.G. Maroto, "Mechanism of Dissolution of Magnetite by Oxalic Acid-Ferrous Ion Solutions," *Inorganic Chemistry* 26, 22 (1987): 3713-3717.

[32]D.E. Harmer, F.P.J. Frauson, M.A. Snyder, O.U. Anders, J.J. Holloway, "Developing a Solvent, Process, Equipment, and Procedures for Decontaminating the Dresden-1 Reactor," American Power Conference, Chicago, IL, April 21, 1975, vol. 37: pp. 282-288.

[33]M.R. Poirier, S.D. Fink, "Analysis of Samples from Tank 6f Chemical Cleaning" (Aiken, SC: Savannah River National Laboratory, 2010).

[34]B.J. Wiersma, J.I. Mickalonis, K.H. Subramanian, E.T. Ketusky, "Corrosion Testing of Carbon Steel in Oxalic Acid Chemical Cleaning Solutions," CORROSION/2012, paper no. NACE-2012-1538 (Houston, TX: NACE, 2012).

[35]Iron(II) Oxalate, Wikipedia, Dec. 19, 2015, https://en.wikipedia.org/wiki/Iron(II)_oxalate (Jan. 11, 2016).

[36]IAEA, "Assessment and Management of Ageing of Major Nuclear Power Plant Components Important to Safety: Steam Generators" (Vienna, Austria: International Atomic Energy Agency, 2011).

[37]M.J. Little, R.D. Varrin Jr., A.T. Pellman, M.A. Kreider, "Advanced Scale Conditioning Agent (ASCA) Applications: 2010 Experience Update," EPRI SGMP 2010 SG Secondary Side Management Conference, San Antonio, TX, Mar. 2–4, 2010.

[38]E.C. Pitzer, "Neutral Metal Cleaning Compositions Containing Hydrazine and a Polyaminocarboxylic Acid," U.S. Patent 3297580, 1967.
[39]F.J. Pocock, W.S. Leedy, "Chemical Cleaning Research for Steam Generators," 32nd International Water Conference, Pittsburgh, PA, Nov. 2-4, 1971.
[40]P.J. Pettit, P.F. McTigue, "Chemical Cleaning of Pressurized Water Reactor Steam Generator to Prevent Denting," CORROSION/78, paper no. 78208 (Houston, TX: NACE, 1978).
[41]C.R. Frost, P. Walsmsley, P.V. Balakrishnan, P. McSweeney, "Chemical Cleaning of the Nuclear Power Demonstration Unit Boiler," *Nuclear Technology* 55, 1 (1981): 207-217.
[42]P.V. Balakrishnan, P. McSweeney, C.R. Frost, P. Walmsley, "A Chemical Cleaning Process for Nuclear Steam Generators," *Nuclear Technology* 55, 2 (1981): 349-361.
[43]P. McSweeney, "Removal of Copper from Nuclear Steam Generators Using Hydrogen Peroxide," CORROSION/93, paper no. NACE 93387 (Houston, TX: NACE, 1993).
[44]R.H. Hausler, "Role of Hydrazine in EDTA Containing Chemical Solvents," CORROSION/83, paper no. NACE 83030 (Houston, TX: NACE, 1982).
[45]C.S. Welty, J.M. Jevec, W.S. Leedy, "Laboratory Studies for Removing Sludge from Secondary Side of Nuclear Steam Generators," CORROSION/1982, paper no. NACE 82034 (Houston, TX: NACE, 1982).
[46]R.H. Hausler, NP-3030, "Developed the Only Qualified Corrosion Inhibitor for Nuclear Steam Generator Cleaning" (Palo Alto, CA: Electric Power Research Institute, 1983).
[47]J.M. Jevec, W.S. Leedy, "Chemical-Cleaning Solvent and Process Testing. Final Report [Pwr]" (Palo Alto, CA: Electric Power Research Institute, 1983).
[48]W.F. Cleary, G.B. Gockley, Report NP 2976 (Palo Alto, CA: Electric Power Research Institute, 1983).
[49]W.S. Leedy, J.M. Jevec, M.E. Scott, *Environmental Degradation of Materials in Nuclear Power Systems-Water Reactors*, eds. G.J. Theus, J.R. Weeks (Toronto, Ontario: Metallurgical Society, 1988).
[50]R.P. Dow, P.M. Knollmeyer, M.C. Thaller, "Three Mile Island Station General Chemical Cleaning and In situ Corrosion Monitoring," CORROSION/93, paper no. NACE 93370 (Houston, TX: NACE, 1993).
[51]T.W. Oliver, K. Froehlich, ASME/IEEE Power Generation Conference, New York, NY, 1987.
[52]J. Semmler, P.J. Leinonen, "Characterization and Dissolution Studies of Bruce Unit 3 Steam Generator Secondary Side Deposits," CORROSION/98, paper no. NACE 98324 (Houston, TX: NACE, 1998).
[53]J.P. Veysset, P. Saurin, "Experience with EDF Chemical Cleaning Process in Nuclear Power Plants," CORROSION/93, paper no. NACE 93367 (Houston, TX: NACE, 1993).
[54]R. Roofthooft, L. Laborelec, "Secondary Side Chemical Cleaning of Steam Generators," International Conference on Chemistry in Water Reactors: Operating Experience and New Developments, Nice, France, 1994.
[55]S. Odar, K. Kuhnke, "The KWU Chemical Cleaning Process as a Remedial Measure to Improve Steam Generator Performance," Proc. Int. Conf. Water Chemistry of Nuclear Systems, Water Chemistry 6, British Nuclear Energy Society, 113, 1992.
[56]M. Dijoux, O.d. Bouvier, S. Mercier, D. Pages, J.L. Bretelle, P. Leclercq, F. Tempier, F. Roumiguiere, "Application of Areva Inhibitor-Free High Temperature Chemical Cleaning Process against Blockages on Sg Tube Support Plates," NPC International Conference on Water Chemistry of Nuclear Reactor Systems, Berlin, Germany, Sept. 15-18, 2008.
[57]S. Odar, "Field Experience with KWU Chemical Cleaning Process," Water Chemistry of Nuclear Reactor Systems, 5th BNES, London, U.K., 1989.
[58]A.L. Rufus, V.S. Sathyaseelan, M.P. Srinivasan, P.S. Kumar, S.N. Veena, S. Velmrugan, S.V. Narasimhan, "Chemistry Aspects Pertaining to the Application of Steam Generator Chemical Cleaning Formulation Based on Ethylene Diamine Tetra Acetic Acid," *Progress in Nuclear Energy* 39, 3-4 (2001): 285-303.
[59]M.U. Gmurczyk, A. Barkatt, L. May, S. Olszowka, "Degradable Chelating Agents for Decontamination and Chemical Cleaning," CORROSION/98, paper no. NACE-98327 (Houston, TX: NACE, 1998).
[60]P.J. Battaglia, D.L. Rogosky, "The Use of Advanced Scale Conditioning Agents for Maintenance of the Secondary Side of Nuclear Plant Steam Generators," International Conference Nuclear Energy for New Europe, Portorož, Slovenia, 2006.
[61]S. Weiss, A. Drexler, J. Fandrich, "Steam Generator Maintenance Measures as Part of an Integrated Management in PWRs," 3rd International Conference on NPP Life Management (PLIM) for Long Term Operations (LTO), Salt Lake City, UT, 2012.

[62]S.E. Evans, S.D. Jones, "Deposit Minimization Treatment in Secondary Side Steam Generators for Nuclear Plants," CORROSION/2015, paper no. NACE-2015-5970 (Houston, TX: NACE, 2015).
[63]R.D. Varrin, M.W. Rootham, "Scale Conditioning Agents and Treatment Agents," U.S. Patent 7344602, 2008.
[64]M.W. Rootham, R.D. Varrin Jr., "Scale Conditioning Agents," U.S. Patent 6740168, 2004.
[65]R.D. Varrin Jr., S. Kaneda, N. Savada, "Scale Conditioning Agents and Treatment Methods," U.S. Patent 7857911, 2010.
[66]P. Battaglia, D. Rogosky, "The Use of Advanced Scale Conditioning Agents for Maintenance of the Secondary Side of Nuclear Plant Steam Generators," Proceedings of the 5th CNS International Steam Generator Conference, ISBN 0-919784-87-9 (Toronto, Canada: Canadian Nuclear Society/Société Nucléaire Canadienne, 2006).
[67]M. Little, A. McCree, R. Varrin Jr, A. Pellman, M. Kreider, "Advanced Scale Conditioning Agent (ASCA) Applications: 2014 Experience Update," paper no. 10104, NPC 2014: Nuclear Plant Chemistry Conference, Sapporo, Japan, The Nuclear Institute, Oct. 26–31, 2014.
[68]R.D. Varrin Jr., M.J. Little, C.E. Anderson, "Method and Compositions for Removing Deposits," U.S. Patent 8728249, 2014.
[69]M. Vepsalainen, "Deposit Formation in PWR Steam Generators" Finland, VTT-R-00135-10, 2010.
[70]M.W. Rootham, "Chemical Claiming Method for the Removal of Scale Sludge and Other Deposits from Nuclear Steam Generators," U.S. Patent 5764717, 1998.
[71]M. Reinhardt, G. Lewis, "Full Bundle Mechanical Cleaning of Pressurized Water Reactor Steam Generators," CORROSION/98, paper no. 98331 (Houston, TX: NACE, 1998).
[72]G.D. Auld, Allen J. Baum, Judith B. Esposito, W.J. Stenger, "Chemical Cleaning Method for Steam Generators Utilizing Pressure Pulsing," U.S. Patent 5154197, 1992.
[73]R.D. Varrin, C.A. Babbit, "Removal of Scale from Steam Generator Tubing with an In-Bundle Lancing System," CORROSION/93, paper no. NACE 93380 (Houston, TX: NACE, 1993).
[74]A.C. Buford, D.W. Moore, S.J. Weems, J.B. Mason, "Steam Generator Chemical Solvent Mixing System and Method," U.S. Patent 5257296, 1993.
[75]R.D. Varrin, M.J. Little, "Chemical Cleaning Method and System with Steam Injection," U.S. Patent 8459277, 2013.
[76]R.D. Varrin Jr., M.J. Little, "Chemical Cleaning Method and System with Steam Injection," U.S. Patent 8999072, 2015.
[77]W.B. Brown, C.M. Loucks, "Chemical Cleaning of Metal Surfaces Employing Steam," U.S. Patent 3084076, 1960.
[78]S. Kaneda, N. Sasada, R.D. Varrin Jr., M.W. Rootham, "Ultrasonic Cleaning Method," U.S. Patent 6572709, 2003.
[79]J.F. Remark, S.E. Evans, "Chemical Cleaning of Steam Generator During Mode 5 Generator Shut Down," U.S. Patent 7302917, 2007.
[80]W.-T. Jeong, S.-T. Kim, J.-K. Oh, S.-J. Yoon, "Development of Ultrasonic Cleaning Technology for a Nuclear Steam Generator," Transactions of the Korean Nuclear Society Spring Meeting, Taebaek, Korea, May 26-27, 2011.
[81]O.T. Brice, W.A. Glover, "Service Water Chemical Cleaning at River Bends Gets Results," CORROSION/94, paper no. NACE 94251 (Houston, TX: NACE, 1994).
[82]J.T. Ruiz, P. Guerra, C. Carreres, "Heat Exchanger Cleaning at French Nuclear Plants," *Nuclear Engineering International* May, 27 (2013).
[83]CONCO, "South Texas Project Nuclear Power Plant Unit 1 Condenser" (Verona, PA: CONCO Services Corporation, 2015).
[84]J.P. Terry, G. Schultz, S. Ewens, "Control and Reversal of Biofilm Episode in a Nuclear Power Plant Main Condenser System," CORROSION/96, paper no. NACE-96175 (Houston, TX: NACE, 1996).
[85]W.W. Frenier, *Chemical Methods for Controlling Corrosion During Oil and Gas Activities* (Houston, TX: NACE International, 2017).
[86]S.E. Evans, J.F. Remark, S.L. Watson, F. Ammann, "Steam Generator Chemical Cleaning Experiences 1998-2002," CORROSION/2003, paper no. NACE-03579 (Houston, TX: NACE, 2003).
[87]K. Fujiwara, H. Kawamura, H. Kanbe, H. Hirano, H. Takiguchi, K. Yoshino, S. Yamamoto, T. Shibata, K. Ishigure, "Applicability of Chemical Cleaning Process to Steam Generator Secondary Side, (I) Outline of the Investigation and Cleaning EĐectiveness," *J. Nuclear Science and Technology* 41, 1 (2004): 44-54.

6 Corrosion Inhibitors for Aqueous Cleaning Solvents

Essentially all the solvents employed for dissolving inorganic deposits (Chapters 4 and 5) and some of the solvents for organic/mixed deposits are corrosive to some metals of construction. In addition, it may be necessary to passivate equipment after most chemical treatments to reduce additional corrosion from the atmosphere or to facilitate special processes such as some copper removal techniques. While the definitions can become blurred, passivation is thought to involve the formation of a very thin oxide coating, whereas most of corrosion inhibitors used in cleaning formulations contain complex organic molecules that adsorb/react at the metal surface and form barrier films.

Passivation is particularly important during chelant cleaning to remove iron and copper deposits (Section 4.2.3). Passivation treatments are described in Section 4.2.4. Corrosive solvents are used in both non-nuclear and nuclear cleaning operations. Chapter 6 describes the chemistries needed to inhibit mineral acids (Section 6.2) and the organic acid/chelating agent solvents (Section 6.3) used to remove acid-soluble mineral scales. Special problems (Section 6.4) and toxicity issues (Section 6.5) are also discussed. Section 6.6 details the test methods used to select the correct inhibitor for a solvent at the needed temperature for specific metals of construction.

6.1. Introduction to Industrial Cleaning Corrosion Inhibitors

Most industrial equipment requiring cleaning is constructed of iron-based alloys, such as carbon and alloy steels, or copper-based alloys. These metals are attacked by aggressive solvents (mineral acids, organic acids, and chelating agents) unless the cleaners contain corrosion inhibitors. An uninhibited solvent would damage the equipment being cleaned, and the corrosion would consume the expensive solvent. This is true even for neutral pH solvents.

Thus, without the use of corrosion inhibitors, most chemical de-scaling operations should not be performed. This section briefly reviews the technology required to protect industrial equipment from devastating attack during application of descaling fluids. Chemicals currently available to protect the metals in units to be cleaned may work in ambient temperature to about 350°F, but the time of contact, the type of descaling fluid, and the metallurgy must be carefully considered before selecting the treatment. Note that the cleaning temperatures for mineral acids are usually much lower than for organic acids and chelating agents. See Table 4.1 for a list and properties of most of the cleaning chemicals. In most cleaning operations, the corrosion reaction that the inhibitor must minimize is acid attack of some type. The electrochemical equation is

$$2H^{+} + Fe^{0} = H_{2} + Fe^{2+} \qquad \text{(Equation [6.1])}$$

This type of acid attack takes place in hydrochloric acid, organic acids, and chelating agents, including formulations that have pH values as high as 9, especially if they contain ammonium (NH_4^+) ions. The rate of corrosion can vary by several orders of magnitude depending on the acid strength and temperature. In very general terms, the amount of damage potential increases with acidity and temperature, but most chemical scale removers are corrosive and require use of corrosion inhibitors. The references cited below give much more information about corrosion inhibitors that may be used in aqueous descaling formulations. Other types of corrosion, such as pitting and stress corrosion cracking, also may occur but usually are initiated by acid attack.

The following subsections review details of the development of corrosion inhibitors for mineral acids, organic acids, and chelating agent scale solvents. Different types of corrosion inhibitors are required for these types of fluids. In addition, special inhibitors are required to protect some types of boiler steels. Each subsection also has information on the mechanisms of action of each general inhibitor type. Table 6.1 at the end of Section 6.3 summarizes the inhibitor choices based on solvent type, metallurgy, and temperature.

An inhibitor formulation that is used in a chemical cleaning treatment has several components including

1) The active ingredient that provides most of the inhibitor protection
2) A surfactant to help disperse the inhibitor
3) Solvents to make a one-component mixture
4) Inhibitor aids to improve performance in special conditions (such as high-alloy steel)

These components affect the use range (such as chemical type, concentration, and temperature) and the overall effect of the inhibitor on the environment. An important consideration is that cleaning fluid inhibitors should NOT leave an oily film on the cleaned surfaces, whereas in the oil and gas industries, an oily film may be desirable for prolonging the inhibitors' life on the metal.

6.2. Inhibitor Technologies for HCl-Containing Fluids

This section very generally reviews advances in corrosion inhibition technology that are applicable to chemical treating fluids that appeared in the literature since the early 1970s. For clarity, a few earlier references also are included. Much of the work was done for oilfield situations, but industrial cleaning (IC) corrosion inhibitors come from some of the same labs and have similar chemistries.

Acid corrosion inhibition technology was reviewed by Riggs,[1] and applications to chemical cleaning processes were reviewed by Oakes[2] in 1972. Also refer to the excellent reviews in *Corrosion Inhibitors*.[3]

This chapter will discuss chemicals and processes that are used with mineral acids as well as with organic acids and chelating agents, with an emphasis on inhibitors designed to protect iron and steels. This is not intended to be an exhaustive review but rather is a sampling of the most important recent corrosion inhibition technology. For other general information on chemical cleaning science and technology, see NACE TPC Publication 6 (1985),[3] the EPRI manual,[4] and the NACE Chemical Cleaning Manual.[5] Additional information on inhibitors is included in the compilations of Robinson,[6] Nathan,[7] Raman,[8,9] and Rozenfeld.[10] See also Frenier and Growcock (1989)[11] and a NACE book by Frenier[12] for more details of the various corrosion inhibitors developed for oil and gas treatments.

Acid cleaning of inorganic deposits in all types of equipment originated from technology used in the stimulation of oil wells. Olbertz[13] explored various facets of steel corrosion inhibition and hydrogen absorption by steel in hydrochloric acid (HCl). He describes some extensive mechanistic research on mixtures of propargyl alcohol and quaternary nitrogen salts as well as many individual quats, aldehydes, and ketones. He also provides an extensive compendium of inhibitors for hydrogen absorption, especially developments in the Soviet Union through 1979. Sanyal[14] gave a historical account of corrosion inhibitors through 1979 that classifies organic inhibitors for iron dissolution as amines and imines, sulfur compounds, onium compounds, and unsaturated compounds containing double and triple bonds. Schmitt[15] described organic inhibitors as primarily nitrogen compounds and acetylenic and, to a lesser extent, sulfur compounds for HCl, HF, and H_3PO_4; sulfur compounds and, to a lesser extent, acetylenic for H_2SO_4; and sulfur compounds for HNO_3.

HCl is the solvent most often employed for scale removal applications because it is economical and attacks a wide range of scales (see Section 4.1.1). More than 100 papers on organic corrosion inhibitors for steel or copper in HCl have appeared in print since 1970. More details are in Frenier.[16,12] The main goal of this chapter is to discuss the major classes of compounds, organized by heteroatom: nitrogen, oxygen, and mixed heteroatoms. The chemical structures of some of these classes are given in Figure 6.1.

If a deposit contains silica or silicates and iron oxides, HF or more usually ammonium difluoride (NH_4HF_2) may be added to the HCl solvent. This additive may affect the corrosion inhibition provided by the corrosion inhibitor formulation described in this chapter. The effects can be positive or negative. Bradley and Smith[17] reported that corrosion rates of exposed steel were much lower when mixtures of HF and HCl were used to dissolve magnetite, compared with HCl alone. In both cases, an organic corrosion inhibitor was used to control the attack of the acid on the steel. Lower corrosion rates also were reported for HF solutions compared with similar concentrations of HCl. Similar results were noted by Alfandry.[18] Poff[19] also described the benefits of adding ammonium difluoride (ABF) to acidic solvents. This compound decreased the corrosion rates of the acidic solvents.

However, corrosion rates in the initial solvents that are pumped in a unit may be higher than those without HF. Thus, the specific solvent for a job must be evaluated using the methods described in Section 6.6.

6.2.1. Nitrogen Compounds

This heteroatom is present in the largest classes of organic inhibitors for HCl. These can be very effective at moderate temperatures (<149°F [<65°C]) and acid concentrations (<15 wt.% HCl). Subramanyan et al.[20] studied various amines (aromatic and aliphatic) and triazoles in various acids, including HCl, and found them to be highly effective. In a study of polymethylene imines and diamines, McCafferty et al.[21] discovered that the inhibition was due to the formation of polymers. Alfandary et al.[22] claimed that hexamine (i.e., hexamethylenetetramine) is effective even in the presence of Fe^{3+}, and Dhirendra and Sanyal[23] showed that hexamine works well even on AISI 321 SS.

Aromatic nitrogen salts have been extensively investigated. Halides of n-alkyl pyridine,[24] n-alkyl benzylpyridine (Katapin in the Soviet Union),[25] n-alkylisoquinoline,[26] and alkylbenzylquinolines[27] are highly effective steel corrosion inhibitors. Most of these compounds are excellent inhibitors not only for iron dissolution but also for hydrogen absorption and appear to work well even in the presence of H_2S.

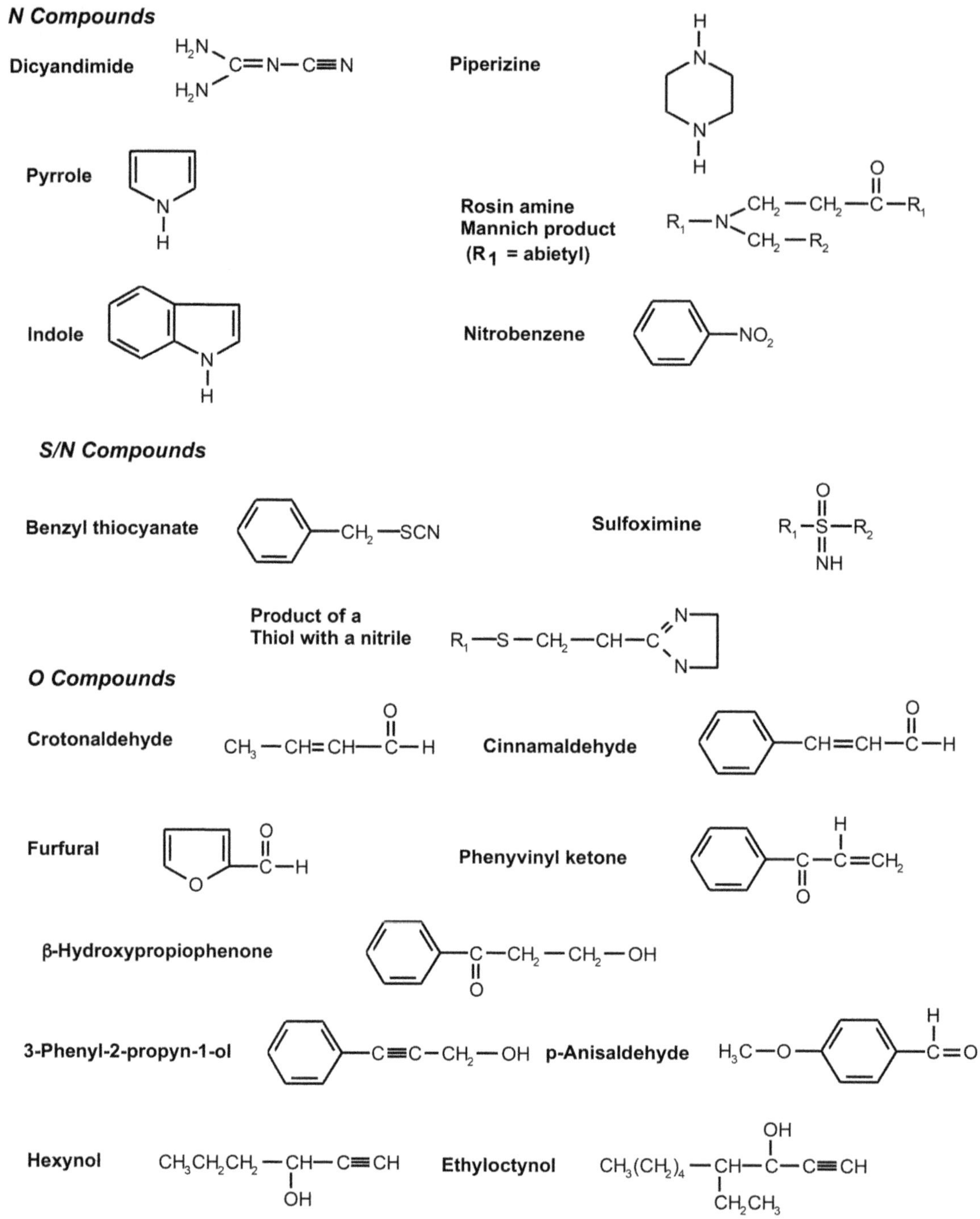

Figure 6.1 Common Corrosion Inhibitor Structures

Schmitt and Bedbur[28] systematically tested several pyridinium and quinolinium derivatives on iron and steel in deaerated 10% HCl; the best corrosion inhibitor was found to be l-naphthylmethylquinolinium chloride. Interestingly, none of the inhibitors performed as well on carbon steel as on iron. See Figure 6.2 for some aromatic nitrogen structures. Since these types of compounds are used in several commercial formulations, some details of the syntheses are described.

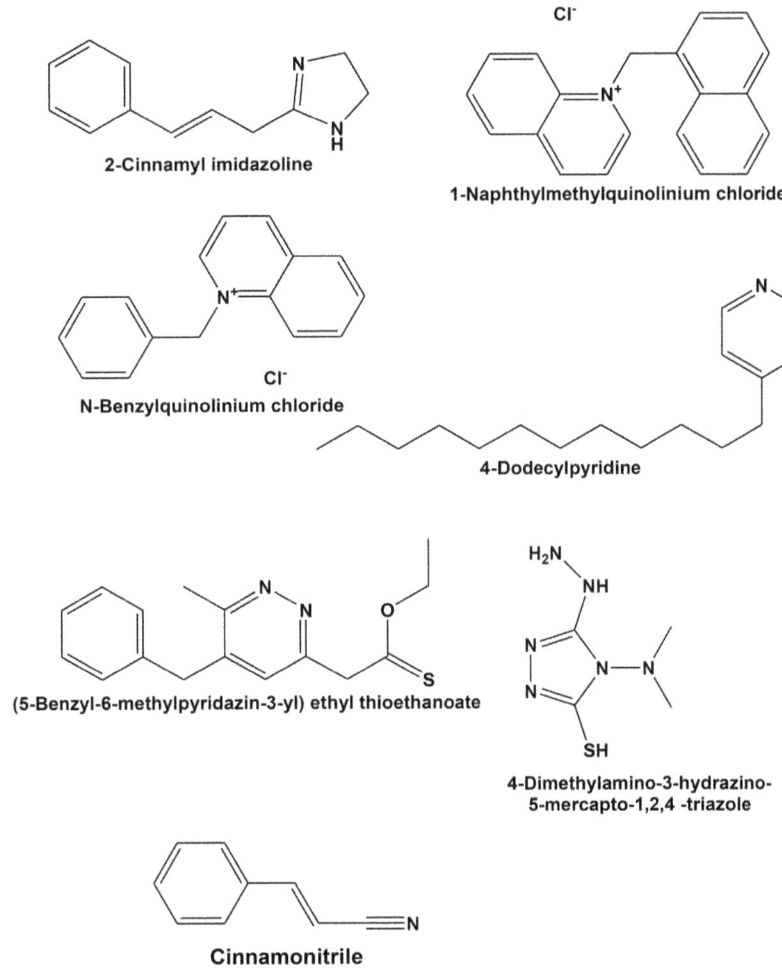

Figure 6.2 Examples of Aromatic Nitrogen-Based HCl Inhibitors

The reaction of pyridine with an alkyl bromide (bromododecane [BDD], in this case) is known as the Menshutkin reaction. It was used as the model reaction to determine the effect of the reaction solvent on the production of a pyridinium bromide salt that could then be formulated to produce some of the test inhibitors. The dodecyl pyridinium bromide (DDPB) reaction was run using equimolar amounts of pyridine and BDD in 43% (wt.) solvent at 200°F. Marek[29] has described the synthesis of a family of n-alkyl pyridinium bromide salts with chain length of C_8-C_{20}.

The reaction is

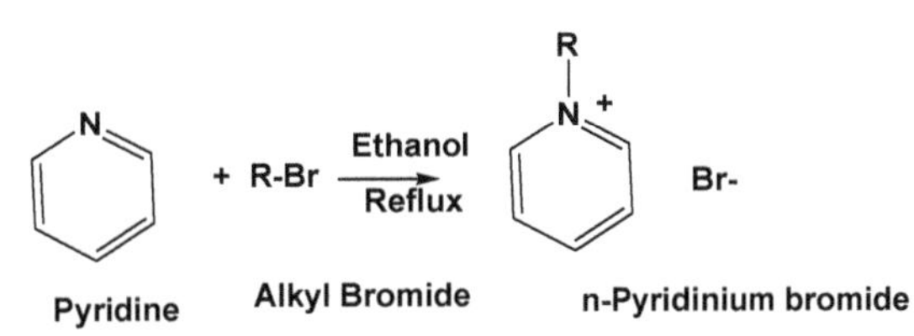

(Equation [6.2])

These authors found that the medium chain length quat preps (C_{12}-C_{18}) produced the highest yields in the range of 82% to 95%. Note that all of these preps were performed in ethanol as the solvent, which has a low flash point.

This process also has been used to produce less hazardous commercial inhibitor products. Frenier[30] describes the production of both pyridinium and quinolinium quats. See Figure 6.3 for several of the basic reactions.

For example, DDPB was prepared in a stirred three-neck flask (250 mL) fitted with a reflux controlled with a digital temperature controller. In each test, 64.2 g of alcohol, glycol, or glycol ether; 20.7 g of pyridine (0.26 mole); and 65.1 g (0.26 mole) of 1-bromododecane were placed in the 5 flask. The flask was heated with stirring to 85°C. An exotherm was encountered, which increased the temperature to about 100°C. The temperature controller was adjusted to 95°C, and this temperature was maintained throughout the preparation. At set times, samples were drawn, and the bromide concentration was determined on an autotitrator using the chloride specific ion silver nitrate method.

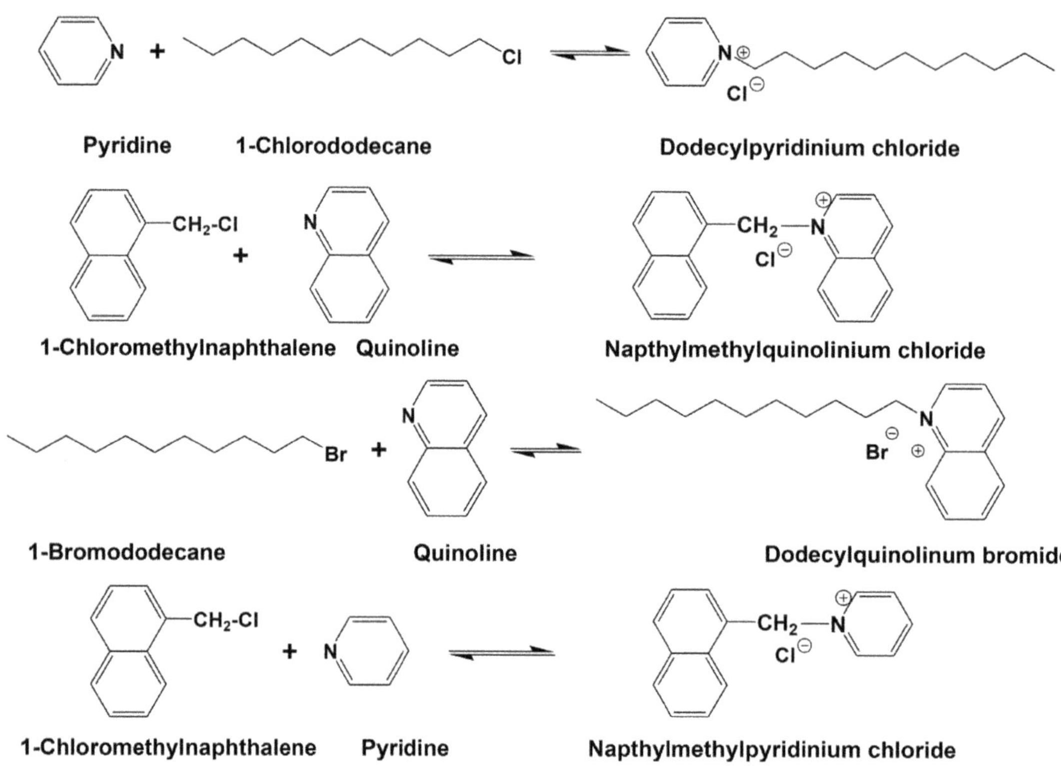

Figure 6.3 Nitrogen Quats Corrosion Inhibitor Components and Synthesis

The reaction was terminated when the bromide concentration reached 96% of the theoretical value. Part of the research was to test both the leaving group as well as the solvents. These were isopropyl alcohol (IPA), propylene glycol (PG), dipropylene glycol methyl ether (DPM), tripropylene glycol methyl ether (TPM), and monopropylene glycol methyl ether (PM).

Where DPM and TPM were used as solvents, it was necessary to add 10 mL of water at the 6-hour time mark to drive the reaction to completion. The concentration of DDPB in IPA, PG, and DPM was 57 wt %, whereas the concentration in DPM and TPM was 54 wt %. Figure 6.4 illustrates the rates and percent completion of these preparations. All the preparations reached at least 95% completion in about 8 hours at 95°C. The comparative preparation in IPA took the longest to reach completion. The three preparations in glycol ethers were essentially equivalent, whereas the preparation in PG was the fastest of the five reactions. It must be noted that the preparations in DPM and TPM required the addition of water to force the reaction to proceed to greater than 95% completion. Because of the exothermic nature of the reaction, it may be necessary to add the second compound, 1-bromododecane, slowly or incrementally during the reaction process.

Similar preps of NMQCl from 1-(chloromethyl) naphthalene and quinoline were made using the same molar ratio. If DPM is used, the reflux is at about 220°F (as well as with other combinations), and the reaction also may be exothermic. A graph of comparative reaction rates is shown in Figure 6.5.[16] Note that the chemical structures of both the cholor/bromo compound and the nitrogen compound are important. The halide should be a good "leaving group" in the reaction, and this is a function of the organic halide as well as the aromatic nitrogen compound.

The use of DPM as a solvent also produces a quat inhibitor product that has a high (>180°F) flash point. Note that corrosion inhibitors based on these reactions have been used commercially in different chemical cleaning formulations.

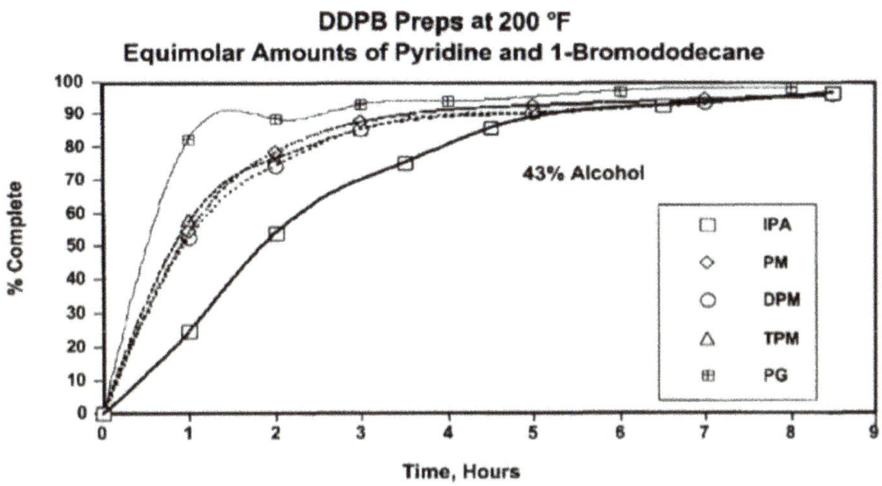

Figure 6.4 DDPB Preps

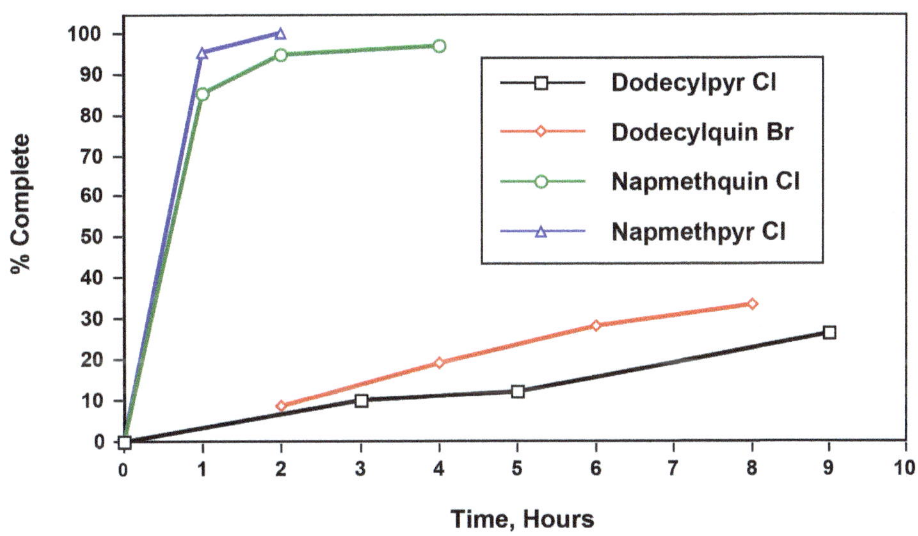

Figure 6.5 Kinetics of Alternate Quat Preps at 230°F in DPM

6.2.2. Oxygen-Based Industrial Cleaning Inhibitors

Most of the important oxygen compounds used for HCl cleaning fluids have unsaturated groups. Many of these are acetylenic alcohols (α-alkynols), which are widely known for their superior performance at elevated temperatures. Several acetylenic alcohol structures are shown in Figure 6.6. Propargyl alcohol has been a staple of many commercial inhibitor formulations but is toxic, especially with skin contact.[11, 31]

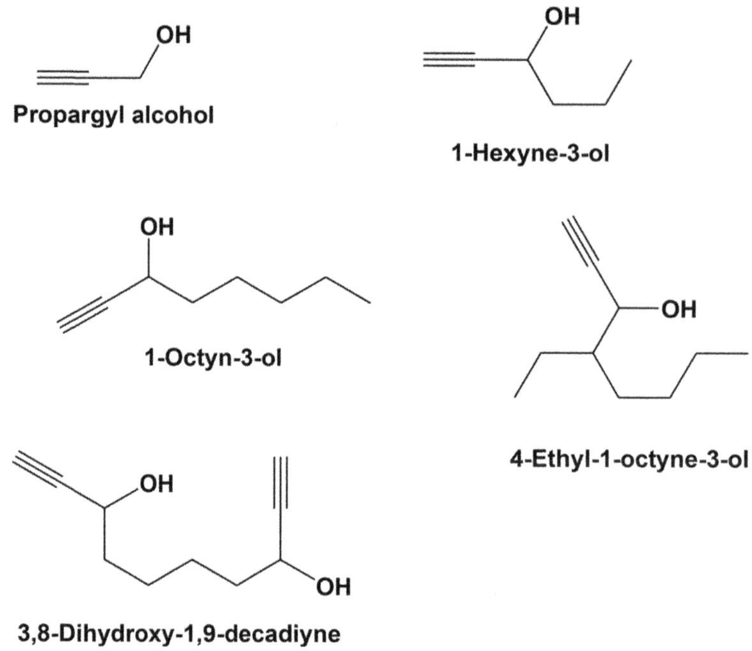

Figure 6.6 Acetylenic Alcohols Used in Inhibitor Formulations

Tedeschi et al.[32] developed several highly effective aliphatic acetylenic alcohols. Growcock et al.[33] and Growcock and Lopp[34] studied phenyl propynols. Generally, the aromatic acetylenic inhibitors appear to be just as active as aliphatic inhibitors and perform especially well in the presence of surfactants. Allabergenov et al.[35] synthesized various propargyl ethers of substituted phenols, which also proved to be quite effective (18% HCl at 160°F [71°C]). One such product, 1-phenoxy-2-butyn-4-ol, is so exceptional that it has been the subject of subsequent studies.[36]

Other oxygen compounds that have been investigated include pyrrole, indole, and their dimethyl derivatives.[37] These compounds not only inhibit the corrosion of steel relatively well but also accelerate the rate of the dissolution of magnetite; however, in that study, the magnetite and steel were not coupled, so it is difficult to surmise what the effect might be on magnetite-covered steel. The α,β–unsaturated aldehydes and α-alkenylphenones were also investigated.

Of the aldehydes, aromatic species such as cinnamaldehyde are the most effective. Devarajan and Balakrishnan[38] investigated several aromatic aldehydes in four mineral acids, including HCl. Growcock and Frenier[39] observed high levels of protection when the aldehydes were used with a surfactant, such as an ethoxylated phenol or alkanol. The best results were obtained with a quaternary nitrogen salt as the surfactant. Frenier et al.[40] discovered that benzoylallyl alcohol, one of the simplest α-alkenylphenones, is an effective inhibitor in both 15% and 28% HCl. Also, on exposure to the acid, several seemingly unrelated compounds, such as 5-benzoyl-1,3-dioxane and p-hydroxypropiophenone, convert to the alcohol. Surfactants enhance the inhibition of these compounds, particularly in strong HCl.

However, some of the acetylenic alcohols are quite toxic to mammals, and the lower alcohols (propargyl alcohol and hexynol) are readily absorbed through the skin.[11, 31] These chemicals are not usually commodity products and thus also may be too expensive to be used as the only active ingredient in a corrosion inhibitor formulation.

For these reasons, there is a growing necessity to produce corrosion inhibitors that are less toxic than current formulations and that are available commercially. Less toxic, conjugated unsaturated oxygen inhibitors (especially trans-cinnamaldehyde [TCA]; see Figure 6.7[a]) can be effective when mixed with surfactants, especially nitrogen compounds.[16, 40–44]

Growcock[45] demonstrated that surfactants greatly improve the performance of TCA as a corrosion inhibitor for J55 steel in HCl by enhancing adsorption at the surface; increased solubility or dispensability of the inhibitor is an incidental effect. The cationic surfactant DDPB is effective in this regard far below its critical micelle concentration, probably as a result of electrostatic adsorption of the monomeric form of DDPM; this leads to formation of a hydrophobic monolayer, which attracts the inhibitor. On the other hand, Growcock notes that an ethoxylated nonylphenol (nonionic surfactant) acts by incorporating the inhibitor into micelles, which themselves adsorb on the steel surface and facilitate the adsorption of TCA. Surfactants are routinely present in all inhibitor formulations, but the actual role of an additional surface active agent is rarely considered.

A patent application by Walker[46] proposes a mixture of aldehydes and ammonium iodide (a potential aid described later in this section). A preferred formulation includes

Cinnamaldehyde
Formaldehyde
Benzaldehyde
C_{14}-C_{15} alcohol with 10 moles of ethoxylation

Work (proprietary in Schlumberger labs) on the details of the Mannich reaction (see next subsection) indicated that at least part of the mechanism was the formation of unsaturated aromatic ketones (from acetophenone).

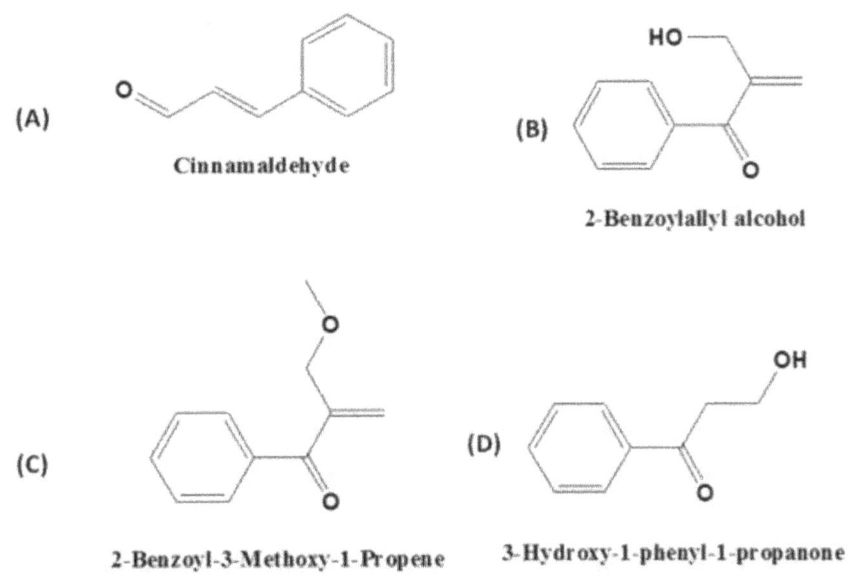

Figure 6.7 Unsaturated Oxygen Compound Inhibitors

6.2.3. Mixed Inhibitor Molecules (N and O) and Formulations

Many commercial acid inhibitors used in well acidizing services and IC have been based on the Mannich condensation reactions introduced by Saukaitis.[47] See also Mansfield[48] and Monroe.[49] This process uses formaldehyde, an amine, and a ketone to produce a Mannich bas (Figure 6.8). The Mannich base has high electron-density on the aromatic rings and nitrogen bases that should adsorb onto an iron surface. However, note that the final reaction products are complex;[50] for more information, see the "mechanism" sections. Also, the structure noted in Figure 6.8 may not be the primary inhibitor.

Because this reaction rarely goes to completion, some formaldehyde will remain in the reaction product that is formulated as the commercial inhibitor. Formaldehyde is an animal carcinogen and appears on the Occupational Safety and Health Administration Specifically Regulated Substance List and the National Toxicity Program Source List 9C (compilation of carcinogens). Therefore, it is desirable to remove this substance from inhibitor formulations. The following patents claim to address this issue.

A patent by McCormick[51] has proposed a series of inhibitors based on the reaction of a phenolics polymer, an amine and an aldehyde, or a ketone, or a mixture of them. In addition, Lindert[52] has tested inhibitors based on the Mannich reactions,[47–49] but then the excess formaldehyde was controlled using sodium sulfite. These are claimed to be less toxic than the standard Mannich process. They then claim that the base formulations were compounded for HCl or organic acid fluids. In many of the patents and applications, the final product will be mixed with additional oxygen-containing molecules or additional nitrogen-based chemicals.

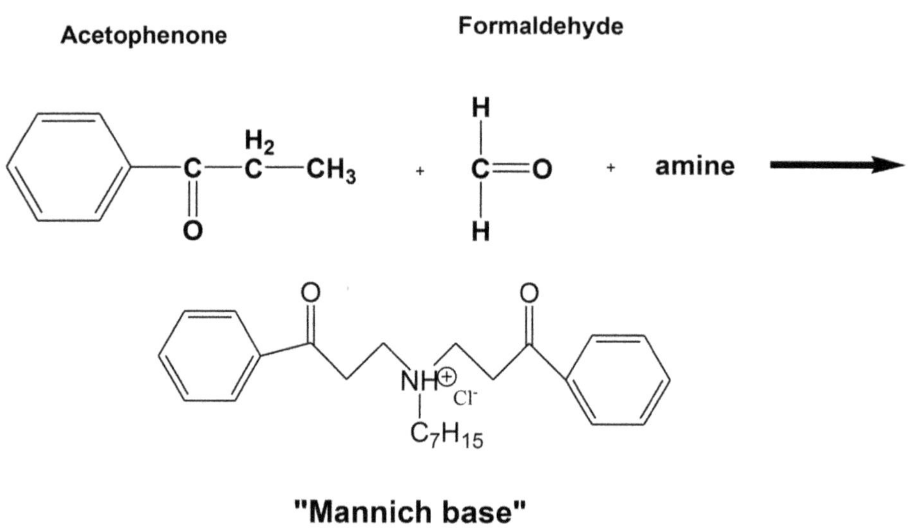

Figure 6.8 Mannich Inhibitor Reaction

Natural products also have been studied. Kumar et al.[53] tested gelatin (a polypeptide composed primarily of glycine, proline, and hydroxyproline) as a steel corrosion inhibitor in HCl and found it to be quite effective at moderate acid concentrations and temperatures. Surprisingly, gelatin lowers the apparent activation energy of the corrosion reaction and the rate; with an inhibitor loading of 1%, levels of protection remain high (80%–94%), even at acid concentrations as high as 32%.

Nitrogen compounds and acetylenic alcohols are often used together to give high levels of protection at elevated temperatures and high acid concentrations. One very effective mixture, which supposedly is effective on zinc, steel, and copper, consists of N,N'-dicyclohexylthiourea, ethynylcyclohexanol, and polyaminomethylene phosphonate.[54]

Nichols[55] patented a similar mixture consisting of a heterocyclic nitrogen compound, an acetylenic alcohol, and a dialkyl thiourea. Mixtures of quaternary nitrogen salts, imines, thiourea derivatives, and acetylenic alcohols are often used with good results, and Mannich bases mixed with acetylenic alcohols are also quite effective.

The performance of amine-base inhibitors can be substantially improved by adding metal halides, especially CuI.[56] Hexamethylenetetramine (HMTA or urotropine) was used for years in mixtures with Sb^{3+} and/or Cu^{2+} salts or KI (the Soviet inhibitors SU, SUM, U-1, and UM series).[13] Nichols[56] discovered that a mixture of an acetylenic alcohol, an aromatic nitrogen salt (quaternary ammonium compound), an aromatic hydrocarbon, and an antimony salt are very effective on ferrous metals in 15% HCl (500 psig) at 325°F (163°C); he claims this mixture will work well up to 500°F (260°C).

Silica and silicate deposits cannot be removed with HCl.[19] When such scales are a problem, hydrofluoric acid (HF) is used either alone or in conjunction with HCl (or other acids). In the latter case, HF is usually generated in situ via the addition of NH_4HF_2. The HCl/HF mixture has the added features of being more aggressive toward iron scales and reducing base metal corrosion by Fe^{3+} in magnetite and hematite deposits, a result attributable to the formation of a strong complex between HF and Fe^{3+}.[17] Generally, inhibitors effective for HCl are also effective for HF.

6.2.4. HCl Corrosion Inhibitor Mechanisms

The mechanisms by which many corrosion inhibitors function were elucidated over the past 50 years. See Damaskin[57] for an excellent introduction to this subject. The illustration in Figure 6.9 shows a generic mechanism in which two different types of inhibitors protect steel in a solution containing HCl, an acetylenic alcohol, and a pyridinium quat.

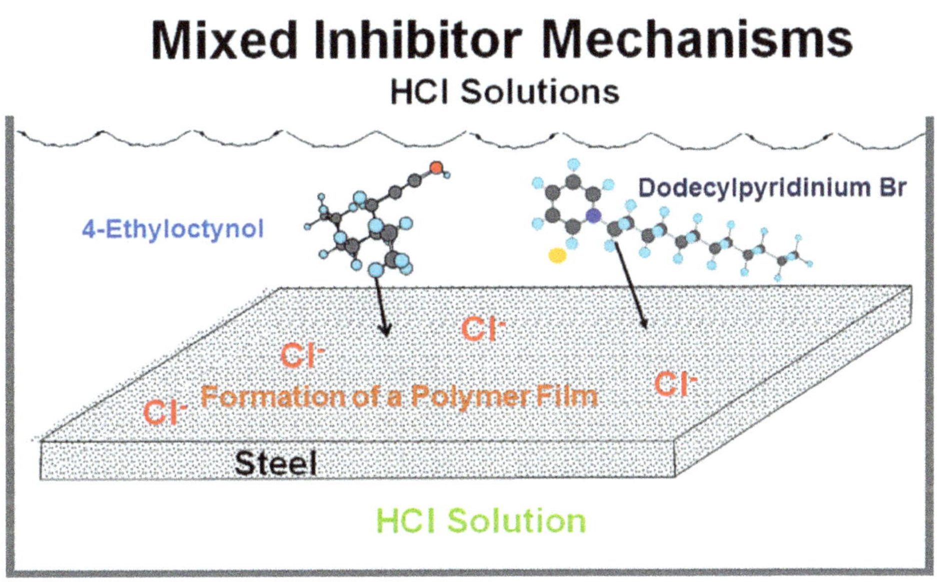

Figure 6.9 Mixed Inhibitor Mechanisms in HCl

Most neutral nitrogen compounds, such as benzotriazole or quinoline, are thought to function by an acid/base reaction that involves sharing of the lone pair of electrons on the nitrogen with a metal ion or atom; the effect most often observed in potentiodynamic tests is suppression of the anodic reaction.[58] On the other hand, charged species, such as pyridinium or quinolinium salts, or readily protonated amines preferentially suppress the cathodic reaction.[59] The latter are thought to function by the formation of an ion pair with Cl^- on the surface, but acid/base interactions cannot be completely discounted. As discussed below, in strong acid, even the neutral nitrogen compounds tend to become protonated; thus, the surface-inhibitor bonds of these species may take on more of an ionic character with increasing acid concentration.

Some inhibitors are reacted on the surface to yield secondary products, which are, in themselves, the active inhibitors. "Filming" amines, such as polymethylene imines and diamines, almost certainly undergo reduction on the surface; protonation may very well be involved as an initiating step. Quaternary nitrogen salts may also undergo reduction reactions, but currently there is no evidence to that effect.

Oxygen compounds can undergo protonation in mineral acids, a reaction critical to film formation. Aldehydes apparently protonate and appear to co-adsorb with SH^- ions, thus functioning in a synergistic manner. Foroulis[60] and Olbertz[13] studied many aliphatic aldehydes and benzaldehyde derivatives with and without H_2S and found a substantial synergistic effect in 16% HCl.

To check for the formation of thioaldehydes on the surface, Olbertz[13] tested several ketones and found no synergistic effect whatsoever, indicating that thioaldehydes are not formed in the former case but rather that SH- and the protonated aldehyde undergo ion pair formation on the surface. Cinnamaldehyde also appears to undergo protonation, either in solution or on the surface, and shows great improvement in inhibitor effectiveness when halide ions are present:[28] in 19% H_2SO_4 (4.4N), a surfactant/cinnamaldehyde mixture provides only 50% protection to mild steel (149°F [65°C]), whereas in 15% HCl (4.4N) it provides 98% protection.

Acetylenic alcohols also appear to be protonated in strong mineral acids, either in solution or on the steel surface. It is now well established that acetylenic alcohols function primarily by chemisorption.[3] Growcock[61] investigated 1-octyn-3-ol and found that chemisorption involves donor/acceptor bonding of the protonated form of the inhibitor and that halide ions are necessary for good inhibitor performance. Furthermore, octynol undergoes reduction reactions on the surface to yield conjugated dienes and ultimately a film comprised of low-molecular-weight hydrocarbon (2-5 monomer units long) with some double-bond unsaturation.

Pathania and Poling's electrochemical and infrared reflectance experiments[62] showed that the chemisorbed layer of other acetylenic alcohols, which is only a few Angstroms thick, reduces the Fe/Fe^{2+} exchange current density by 100 times and that growth of the polymer film results in significant additional reduction of the current. Using in situ ellipsometry, Jovancicevic[63] confirmed that the chemisorption/film formation mechanism is operative even in relatively weak HCl (0.01 N); a chemisorption/film formation mechanism is also operative in H_2SO_4, but the details differ from the HCl mechanism. Using AC impedance spectroscopy, Epelboin[64] showed that 2-butyn-1,4-diol on iron in H_2SO_4 increases the charge transfer resistance with increasing exposure time. Also, Hausler[65] showed that 2-butyn-1,4-diol can be just as effective in highly concentrated HCl (>20%) as it is at lower concentrations, e.g., 5%. The system is sensitive to O_2, but the effect is complex: over certain ranges of temperature and acid/inhibitor concentration, O_2 hinders the inhibition, whereas elsewhere it can substantially aid the inhibition.

The aliphatic acetylenic alcohols are relatively stable in acid (in protonated form) and undergo reaction only in the presence of an electroactive metal surface. Aromatic acetylenic alcohols tend to react further with acid and with the metal surface, forming products that themselves are active inhibitors.[32] These secondary products may function as intermediates and continue to react, either on the surface or in solution, to yield an adherent low molecular-weight polymer film. For example, 3-phenyl-2-propyn-1-ol undergoes acid-catalyzed hydrolysis (protonation and hydration) in the solution to form p-hydroxypropiophenone (possibly protonated), which can dehydrate.

Mechanistic studies have started to shed some light on the synergistic effects observed between quaternary ammonium compounds and acetylenic/aldehydes, between nitrogen-containing compounds and sulfur-containing compounds, and between sulfur compounds and acetylenic/aldehydes. In all these cases, the mechanisms appear to involve co-adsorption. Cases involving sulfur compounds may form a structure on the surface akin to SH^-, thus providing a negatively charged surface on which cationic quaternary nitrogen compounds (nitrogen salts) or protonated acetylenic/aldehydes can form ion pairs. In a manner analogous to the halide/nitrogen salt synergism described above, the sulfur compound and protonated oxygen or nitrogen compound competitively co-adsorb. In the cases involving quaternary nitrogen compounds and protonated acetylenic/aldehydes, both adsorbing species are cationic, and the adsorption may be dominated by attractive interactions among the chains of the adsorbed species.

Structure activity studies have not revealed many surprises. Chakraborty[66] studied four cyanoguanidine (guanidine: (NH_2-C=NH)) derivatives to determine the effect of electron density distribution on inhibitor performance for mild steel in 6% HCl. Using Hickel molecular orbital calculations, Chakraborty determined that the inhibition directly correlated with the ionization potential and inversely correlated with the electron affinity. Thus, biguanide monohydrochloride, at a concentration of 0.02 M, gave the highest protection (95.5% in a 1-hour test) and had the highest ionization potential and the lowest electron affinity. Szklarska-Simalowska[67] determined that the performance of an inhibitor that acts solely by covering surface sites is a function of its solubility, the presence of functional groups, and interaction among the adsorbed molecules.

Linear free energy relationships for correlating molecular structure with inhibiting ability are in use. One such relationship is

$$Log(I_x / I_j) = a\Sigma\sigma + bE_s + C + d \qquad \text{(Equation [6.3])}$$

This equation[68] describes the relative effectiveness of an inhibitor $\mathbf{I_i}$ to that of its parent molecule, $\mathbf{I_j}$, as a function of dipolar contributions, **a**, steric effects, **Es**, and electron delocalization, **C.** Here, **a**, **b,** and **d** are constants. The general trend found in the past for the effect of an alkyl chain length appears to be almost universal: optimum inhibitor efficiency for relatively polar parent molecules results with a carbon chain length usually in the range of C_8-C_{12}. This effect is attributable to the opposing effects of reduced solubility and increased adsorbate interaction with increasing chain length. Some substituents provide a second site with high electron density (a Lewis base site), such as an atom with a non-bonded pair of electrons. Other substituents, such as alkyl groups, can produce inductive effects, whereas others may lead to stabilization or destabilization of reaction intermediates or products, e.g., resonance stabilization of aromatic compounds or their reaction products.

In summary, many types of chemicals have been tested for use in HCl. However, most commercial formulations contain a nitrogen quat or amine and an unsaturated oxygen compound. The oxygen compound polymerizes at the HCl-iron surface, and the nitrogen compound helps to hold the polymer film on the surface. Depending on the fluid, other compounds such as thiourea derivatives also may be present.

For use in IC formulations, the inhibitor *must not* leave a heavy organic film on the cleaned surfaces. Thus, careful formulation and testing of the final commercial package must be a part of the development process. A book by Frenier[12] has specific examples of the inhibitor development process.

6.3. Inhibitors for Organic Acid and Chelating Agent Fluids

A different type of corrosion inhibitor package is required for fluids containing organic acids or chelating agents. These are becoming very important in cleaning formulations that are used in boiler and nuclear systems. These chemicals also can be used with corrosion-resistant alloys. In addition, they generally are less toxic and can be used at higher temperatures than the mineral acids.

Petit[69] tested several organic acid mixtures with carbon and alloy steels as part of a program to develop solvents, which would remove magnetite and sludge from nuclear pressurized water reactors (PWR). The solvents tested were ammonium citrate, citrate/ ethylenediaminetetraacetic acid (EDTA), ammonium hydroxyethylethylenediaminetriacetic acid (HEDTA), citric/HEDTA, and citric/ascorbic acid. Diethylthiourea (DETU) or a proprietary inhibitor was used. The DETU was adequate at 185°F (85°C) for most of the solvents, but the proprietary inhibitor was required at 257°F (125°C). In some cases, the A516 carbon steel was pitted by solutions inhibited by DETU.

Some generalized structures are shown in Figure 6.10.

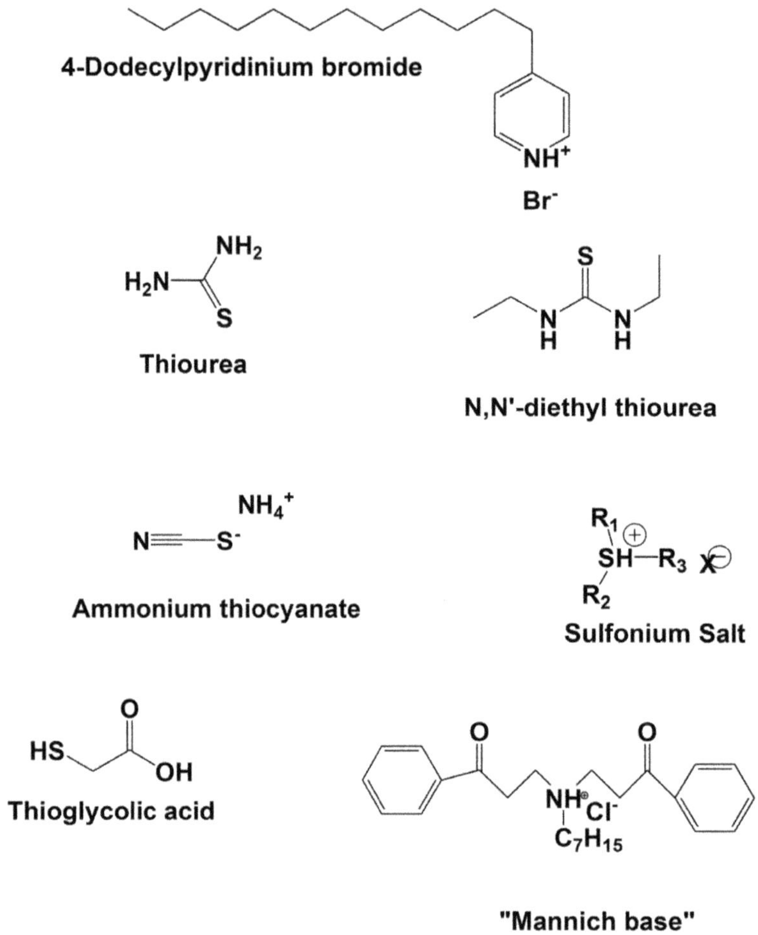

Figure 6.10 Inhibitor Structures for Organic Acid and Chelating Agent Solvents

Kennedy[70] described a versatile inhibitor for organic acids and chelating agents. This formulation consists of an alkypyridinium or quinolinium salt such as DDPB and sulfur compounds such as thiourea or ammonium thiocyanate. The author claims that this mixture will protect various carbon and alloy steels in solvents that include HEDTA, ammonium EDTA (pH of 5 and 9), EDTA/citric acid, EDTA/formic acid, and hydroxyacetic/formic acid mixtures. The temperatures range from 203°F to 302°F (95–150°C). Proprietary inhibitors containing nitrogen and sulfur compounds also are reported to protect carbon and alloy steels from attack by HEDTA acid (pH of < 3) solutions.[71] The solvents were tested at 149°F and 203°F (65°C and 95°C), and corrosion rates were below 0.005 g/cm^2/day for 1018 CS, SA209T1A, and SA213T22 alloy steels. Frenier[30] has proposed a low-toxicity inhibitor containing quats and sulfur compounds.

Brunet[72] made an extensive evaluation of corrosion and inhibition in EDTA solutions containing citric acid and/or hydrazine. Inhibitors tested include benzotriazole, benzimidazole, benzimidazole with gelatin, and a commercial inhibitor. All these inhibitors reduce the corrosion of several types of steels. When the steel is coupled to magnetite, the potential is shifted in the anodic direction. With a large enough ratio between magnetite and bare steel, most of the inhibitors desorb. Only the commercial inhibitor was effective at a pH of 9 in the presence of magnetite.

Hausler[73] tested 150 compounds in a solution of 10% EDTA and 1% N_2H_4, at a pH of 7.0 with NH_3. These tests were conducted to produce an acceptable corrosion inhibitor for cleaning nuclear steam generators (see Section 5.3). The solution also contained 17.5 g/L Fe_3O_4 to simulate dissolving scale. Only sulfur-containing compounds, such as dibutylthiourea, TIA, and alkylpyridines gave more than 90% protection in the initial screen. The sulfur compounds and the alkylpyridines were rejected owing to solubility problems. A new class of compounds (imino-amido condensation products) containing thioether linkages was synthesized and found to give >90% protection at 154°F and 257°F (68°C and 125°C). A final experimental formulation was made. The TIA was rejected because it would not protect SA-533-A steel at 257°F (125°C). The experimental formulation was then used in a pilot size "pot boiler" and model boiler with EDTA/hydrazine to clean sludge and remove magnetite from simulated crevices.[181] Some concern was expressed about the sulfur content (~1.25%) in Hausler's experimental formulation because the residual sulfur may cause cracking of the Alloy 600 tubes found in nuclear steam generators. Note (Table 5.4) that one commercial inhibitor formulation is still in use in the nuclear steam generator cleaning industry.

The most successful inhibitor formulations for organic acids and chelating agents contain amines, reduced sulfur compounds, or combinations of a nitrogen compound (amine, quats, or polyfunctional compound) and a sulfur compound. The only major exceptions are the sulfonium compounds. Jofa[74] investigated the mechanism of mixtures of quaternary ammonium or alkyl ammonium compounds used with sulfur compounds, such as thiourea. He concluded that the sulfur compounds form HS^- ions in solution. These adsorb onto the iron surface, thus attracting the cationic amine inhibitor.

$$\begin{aligned}&\text{A) } R - SH = HS^- \\ &\text{B) } Fe + HS^- = Fe(HS^-) \\ &\text{C) } \text{Inhibitor}^+ + Fe(HS^-) \rightarrow \text{Inhibitor - } Fe(HS)\end{aligned} \qquad \text{(Equation [6.4])}$$

Frenier proposes the theory that the nitrogen/sulfur inhibitors function in a similar manner in the organic acid type of solvents. A diagram (Figure 6.11) shows an illustration of this type of mechanism.

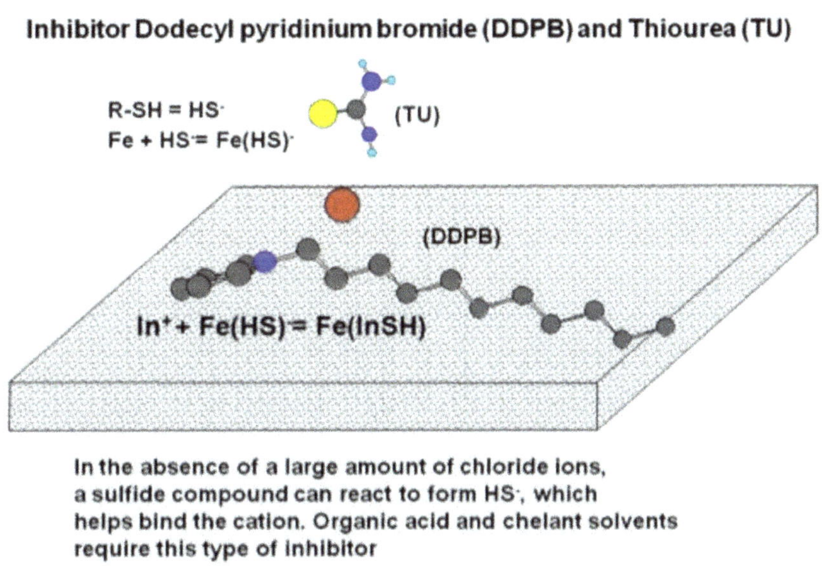

Figure 6.11 Mechanism for Organic Acids and Chelating Solvents

Several exceptions to the sulfide/nitrogen inhibitors have been proposed. Sulfonium compounds have been employed to inhibit corrosion in EDTA solutions at low (<150°F) temperatures. Frenier[75,76] demonstrated that benzyl sulfonium salts (Figure 6.12[a]) were very effective in pH 5 ammonium EDTA solutions, even in the presence of ferric (Fe^{3+}) ions. This work also showed a secondary hydrocarbon layer (b and c) after a corrosion test. This was attributed to the electrochemical reaction of the sulfonium molecule with Fe on the steel's surface. These compounds in the EDTA solvents were employed commercially to clean cooling water systems in refineries or oil and gas facilities.

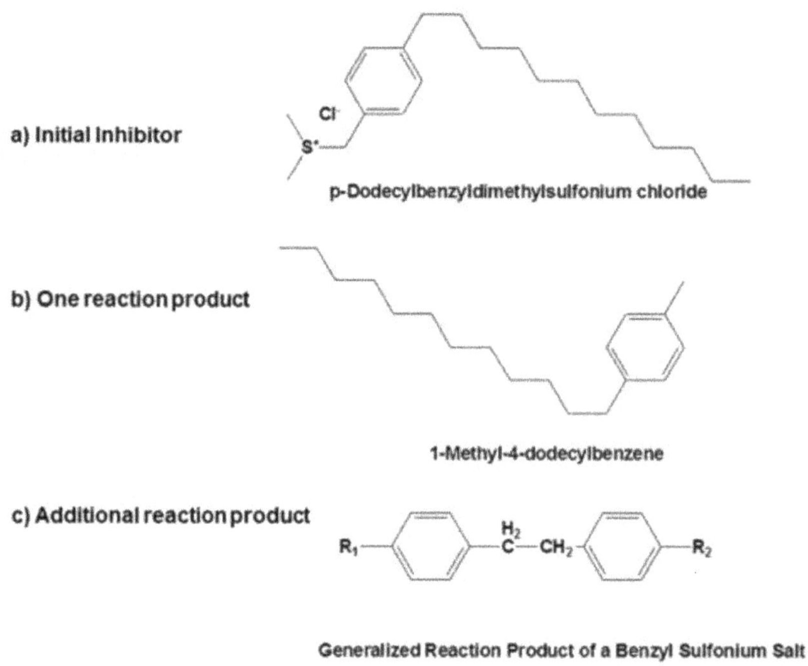

Figure 6.12 Sulfonium Salts as Corrosion Inhibitors in Ammonium EDTA (pH 5)

According to the literature, hydrochloric acid solvents require quaternary nitrogen or amine-based formulations plus an unsaturated oxygen compound. All of the other scale removal formulations require a nitrogen compound plus sulfur-containing molecules. Virtually all commercial scale-cleaning inhibitor formulations contain these active ingredients.

For ecotox reasons, quaternary ammonium salts usually are synthesized in a polar solvent such as IPA. Although this chemical is an excellent solvent for the reactants and the final formulation, the inhibitor has a flash point < 140°F and thus is labeled as flammable. Thus, one of the goals of the project was to eliminate this alcohol from the formulation. Duffin[77] has shown that the solvent can drastically affect the kinetics of a quaternization reaction, so it was not clear that a different solvent would allow the reaction to proceed to completion. See Figure 6.3.

Table 6.1 gives a short review of corrosion inhibition requirements and the use of different types of inhibitors for different metals, solvents, and temperatures.

Table 6.1 Inhibitor Selection for Industrial Scale Removal Fluids

Fluid	Temp. Max., °F	Metallurgy	Generic Inhibitor Type
HCl	200	All	Ammonium quat and unsaturated oxygen compound
HCl/HF	200	All	Different ammonium quat and unsaturated oxygen compound
Organic acids	200	All	Ammonium compound and reduced sulfur compound
Chelating agents	350	All	Ammonium compound and reduced sulfur compound

Corrosion inhibitors used in descaling fluids must be rigorously tested using high-pressure equipment and the actual metals that may be found in the treatment path. Unfortunately, there is no standard method for evaluating inhibitors and fluids. Each vendor uses slightly different procedures. Hill [78] described some of the procedures required for development of inhibitors.

6.4. Special Problems Associated with Industrial Cleaning Solvents

Several problems unique to the cleaning industries are described in this section.

6.4.1. Corrosion-Resistant Alloys

Boilers and nuclear steam generators frequently are tubed with alloys that have small amounts of chromium and molybdenum. Austenitic and ferritic stainless steels (Series 300 and 400 steels with chromium and nickel) also are encountered in addition to carbon steels. The inhibitors for cleaning fluids usually are developed using the less expensive carbon steels, and these inhibitors may not protect the chromium steels, or higher concentrations of inhibitors may be required.

Frenier and Loescher[79] studied the corrosion of six different power boiler metals in ammonium EDTA (pH of 9.2) at 347°F (175°C), HCl at 154°F (68°C), and HCl with thiourea (a copper complexer) at 154°F (68°C) (see Table 6.2, Table 6.3, and Table 6.4). All three solvents were inhibited with commercial inhibitors that contained nitrogen and sulfur compounds. For the EDTA and HCl/thiourea solvents, the chromium/molybdenum steels required more inhibitor or underwent more corrosion than the low-carbon steels (SA210 and SA178C). In HCl without thiourea, there was no significant difference between the steels. Frenier and Loescher also determined corrosion rates and performed metallurgical examinations of welded carbon steels. Only a poor quality "stick" (arc) weld underwent any perceptible preferential attack. The high-quality tungsten inert gas and metal inert gas arc welds did not undergo significant preferential attack. Note that s/v is the surface/volume ratio in cm^2/cm^3 units, and, as noted, higher concentrations of the corrosion inhibitor may be needed at higher s/v values.

If a utility boiler has copper and if the EDTA/oxidant process described in Section 4.2.3 is used, the inhibitor must be specially formulated and tested to ensure that the sulfur part of the corrosion inhibitor does not interfere with the copper removal process. See Frenier.[80]

Table 6.2 Static Corrosion Rates (lb./ft²/day) in 4.5% Ammonium EDTA (pH of 9.2) at 325°F (163°C). Inhibitor "A" Used

A. s/v = 0.6 cm^{-1}

Metal	0.1% A	0.15% A	0.2% A	0.3% A	0.4% A
SA213T11	0.094	0.082	0.033	0.027	
SA209T1a	0.084	0.064	0.037	0.015	
SA178C	0.070	0.040	0.023	0.012	
SA210C	0.068	0.063	0.036	0.016	
B. s/v = 0.8 cm^{-1}					
SA213T11			0.060	0.030	0.019
SA209T1a			0.062	0.017	
SA178C			0.026	0.017	
SA210C				0.013	
C. s/v = 1.5 cm^{-1}					
SA213T11				0.028	0.017
SA209T1a				0.012	0.011

Table 6.3 Static Corrosion Rates (lb./ft²/day) in 5% HCl With 1% Thiourea at 160°F (71°C) 0.1% Inhibitor "C" Used

Metal	s/v = 0.3 cm^{-1}	s/v = 0.7 cm^{-1}
SA213T22	0.016	0.039
SA213T11	0.020	0.047
SA210C	0.008	0.024
1018CS	-	0.015

Table 6.4 Static Corrosion Rates (lb./ft²/day) in 10% HCl at 160°F (71°C) 0.2% Inhibitor "C" Used

Metal	s/v = 0.3 cm^{-1}	s/v = 0.7 cm^{-1}
SA213T22	0.0063	0.0091
SA213T11	0.0090	0.016
SA210C	0.0023	0.016
1018CS	-	0.0064

Nuclear steam generators and nuclear boiling water reactor units have many parts that are composed with various corrosion-resistant alloys. See Figure 1.9 and Table 1.1 for a list of some of the alloys found in steam generators.

Preferential attack on special metals (i.e., alloy 600) and stressed and welded carbon steels (SA210 GR-C and SA508-CL-II) was of special concern to the developers of the steam generator–cleaning solvent containing EDTA, hydrazine, and a sulfur/nitrogen inhibitor.[81] There were significant differences in the response of various carbon steel alloys in the EDTA/hydrazine solvent. The A-533-B ring material underwent the most corrosion of the metals tested and was very sensitive to the amount of hydrazine present. The corrosion always increased with the increasing concentration of hydrazine. These solvents always contained magnetite to simulate the dissolution of scale. Also note that Hausler[73] had to modify the cleaning inhibitor used in the EDTA/hydrazine solvent because of the high corrosion rates of the SA-533-A steel.

Poulson and Robinson[82] assessed the possibility of stress corrosion cracking for mild steel in inhibited HCl and mild steel and 9Cr-1Mo steel in inhibited ammonium citrate (commercial proprietary inhibitors were used). A sensitized 316 SS also was tested in inhibited ammonium citrate. The stress corrosion cracking tests were performed using "C" rings deformed above the yield stress, and slow-strain-rate tests also were performed. Mild steel did not crack in either HCl or citric acid. Cracking could be induced in the 9Cr-1Mo using very harsh conditions but not under actual use conditions (only if improperly heat treated). The 316 SS cracked only in boiling citric acid.

6.4.2. Ferric Ion Corrosion

Any time iron oxides are dissolved, iron ions are released into the cleaning solution. In most industrial equipment, the iron is partially or totally in the 3+ oxidation state. When dissolved by the solvent, the ferric ions will reduce when in contact with the bare metal surfaces as they are exposed by the cleaning solutions.

$$2Fe^{3+} + Fe \rightarrow 3Fe^{2+} \quad \text{(Equation [6.5])}$$

If the scale is quite thin or is very electrically conducting, such as magnetite, a reductive dissolution process may occur, and the iron ions may come into solution directly as ferrous ions. See Equations (4.5 and 4.12).

If the scale is thick or is predominantly ferric oxide (red rust or one of the hydrated iron oxides), a portion of free ferric ions is released into the cleaning solvent and may cause corrosion damage at diverse places in the unit being cleaned.

Three methods have been devised to reduce the additional corrosion caused by the ferric ions in the scale being dissolved:

1) Special ferric ion inhibitors
2) Reducing agents
3) Complexers

Most organic corrosion inhibitors used in cleaning solutions will not prevent ferric ion corrosion. Special chemicals have been developed to reduce this problem. Streicher[83] reported that ferric ion corrosion of steel in sulfuric acid can be reduced by a mixture of diorthotolylthiourea and an anionic surfactant, such as an alkyl-aryl sodium sulfonate. Sulfonium salts also have been used as ferric ion inhibitors.[76,75] These inhibitors are effective in various mineral and organic acids but are most effective in acidic (pH 5) EDTA solutions. The sulfonium salts (Figure 6.12) were shown to electro reduce at the corrosion potential to form a hydrophobic product (hydrocarbon) that may enhance the protection of the steel from oxidation by the ferric ion.

$$2[R_1 - S^+ - R_2R_3Cl^-] + Fe \rightarrow R_1 - R_1 + 2R_2R_3 - S + FeCl_2 \quad \text{(Equation [6.6])}$$

Ferric ion inhibitors are important in many cleaning processes and are essential to clean open cooling water systems. If the cooling tower in such systems cannot be bypassed, ferric ions would be continually regenerated during every pass of the cleaning solvent over the tower (which is a very efficient aerator).

$$\begin{aligned} &\text{A)}\ 8Fe^{3+} + 4Fe \rightarrow 12Fe^{2+} \\ &\text{B)}\ 12Fe^{2+} + 6O_2 + 12H^+ \rightarrow 12Fe^{3+} + 6H_2O \end{aligned} \quad \text{(Equation [6.7])}$$

Reducing agents (described below) would not be effective in such a highly oxygenated system. However, these agents are probably the inhibitors of choice during cleaning of closed systems. Hydrazine was added to EDTA solutions to speed the dissolution of magnetite that was not in close electrical contact with steel. The nuclear steam generator cleaning solutions described above are examples of the use of a reducing agent. In these cases, the hydrazine speeds the dissolution of the scale through a reductive dissolution mechanism and also reduces ferric ion corrosion. Shigehiro and Ueki[84] describe the use of erythorbic acid (an isomer of ascorbic acid) with another organic inhibitor (not described) as a system for protecting steel from ferric attack in HCl and citric acid. This reducing agent also can speed the dissolution of iron oxide. Figure 6.13 shows the rate of dissolution of iron oxide with and without the reducing agent. The erythorbic acid was found to be more effective than other reducing agents, such as stannous chloride or hydrazine.

A proprietary reducing agent was described by Lawson et al.[85] This reducing agent was tested and found to be effective in HCl and hydroxyacetic/formic acid as well as in sulfamic acid solutions. Commercial organic inhibitors also were used in the cleaning solutions. Hydrogen sulfide (H_2S) also reduces ferric ions in acidic solutions. An aldehyde-containing organic corrosion inhibitor was also present to control the corrosion caused by the H_2S. Walker[86] described the use of hydroxylamine and copper as reducing agents in oil well acid. Hill[87] proposed several reducing agents that are associated with hydrazine. Note that more information on the reductive dissolution mechanism of the removal of magnetite from steel is in Section 4.3.

The effectiveness of the reducing agents depends on the number of electrons available per molecule. Stannous chloride provides 2, sulfide 2, and hydrazine 4, but erythorbic acid may provide as many as 8 (depending on the pH).[88] Some structures are shown in Figure 6.14.

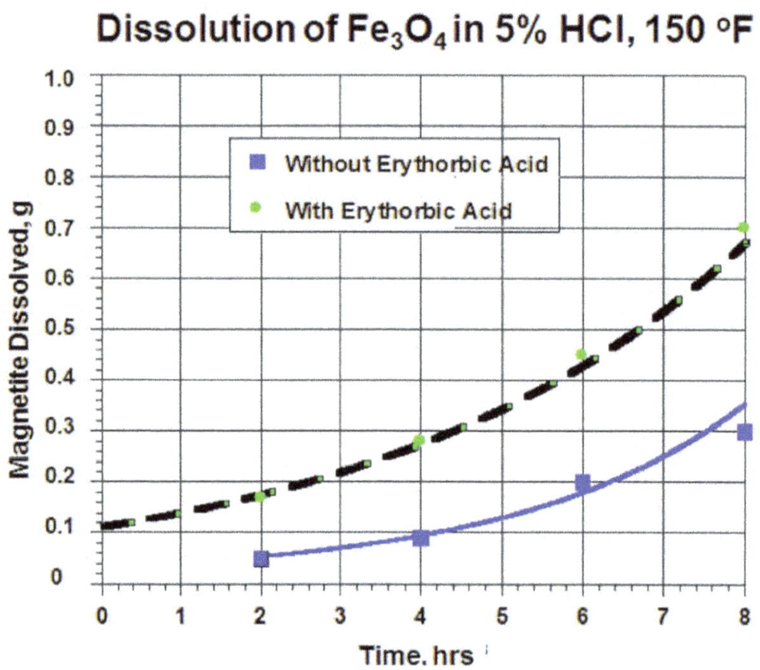

Figure 6.13 Effect of Reducing Agent on Dissolution of Iron Oxide in HCl

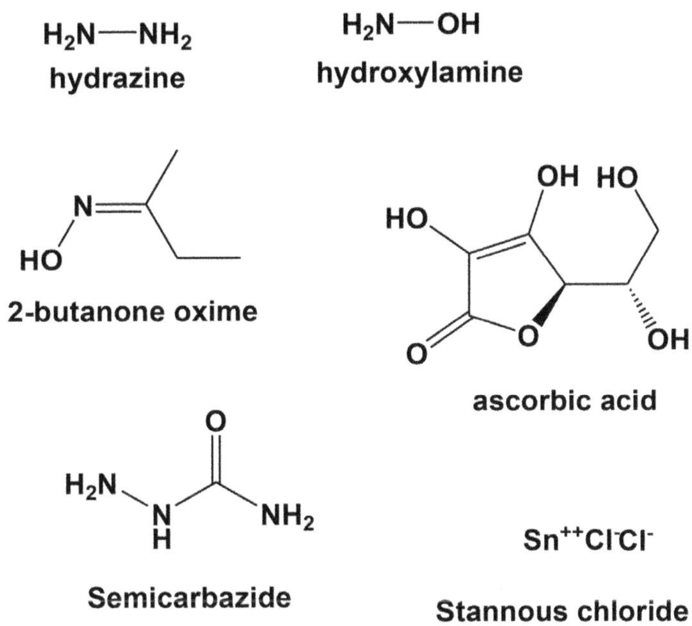

Figure 6.14 Reducing Agents Proposed to Control Ferric Iron

Fluoride ions have been reported to complex ferric in acidic solutions to such an extent that ferric corrosion is reduced or eliminated.

$$3F + Fe^{3+} + FeF_2 \qquad \text{(Equation [6.8])}$$
$$\log k = 11.5$$

Many of the complexers and reducing agents used in chemical cleaning also were used in oilfield acidizing fluids to control the precipitation of ferric hydroxide precipitation as the acid dissolves and spends on carbonate-containing formations (see Frenier[89]).

6.5. Toxicity Issues and Green Inhibitors

Along with improvements in understanding corrosion inhibitor mechanisms, the importance of reducing the toxicity and environmental impact of these chemicals has dominated the development of new materials for the last 25 years.

Interest in "green chemistry" or other health, safety, and environment issues is a recent phenomenon. At SEIC (European Symposium on Corrosion Inhibitors) 6 in 1985, there were no papers with "toxicity" or "green" in the title (even though some described materials with lower toxicity values). Likewise, the first NACE International symposium that considered green inhibitors was held in 1995. Since then, additional symposia addressed this topic until it became clear that most inhibitor papers would have some ECO/Tox content. Likewise, many of the papers at SEIC 7 and 8 were concerned with toxicity or environmental concerns. The patent literature also has started to describe formulations that disclose toxicity values or environmental improvement. At SEIC 9, Frenier[16] presented a plenary lecture on the industries' progress as of 2000.

In 2004, Gupta[90] reviewed the progress in green inhibitors (as part of the green chemistry movement). The following general principles for green corrosion control chemistries are from Anastas,[91] ECI (Sandstein Conference),[92] Gupta,[90] and Frenier.

- **Prevention**: It is better to prevent waste than to treat or clean up waste after it has been created.
- **Multifunctional**: Synthetic methods should be designed to maximize the incorporation of all materials used in the process into the final product. This is a common practice in the corrosion inhibitor industry. See Frenier[30] for one example.
- **Less hazardous chemistry**: Wherever practicable, synthetic methods should be designed to use and generate substances that possess little or no toxicity to human health and the environment.
- **Design low-toxicity products:** Chemical products should be designed to preserve the efficiency of function while minimizing toxicity.
- **Benign solvents:** The use of auxiliary substances (e.g., solvents or separation agents) should be made unnecessary wherever possible and innocuous when used.
- **Design for energy efficiency:** Energy requirements of chemical processes should be recognized for their environmental and economic effects and should be minimized. Synthetic methods should be conducted at ambient temperature and pressure.
- **Use of renewable feed stocks**: Wherever technically or economically practicable, a raw material or feed stock should be renewable rather than depleting resources.
- **Reduce derivatives**: Unnecessary derivatization (use of blocking groups, protection/deprotection, and temporary modification of physical/chemical processes) should be minimized or avoided if possible because such steps require additional reagents and can generate waste.
- **Catalysis:** Catalytic reagents (as selective as possible) are superior to stoichiometric reagents.
- **Design for degradation**: Chemical products should be designed so that at the end of their function they break down into innocuous degradation products and do not persist in the environment.
- **Real-time analysis for pollution prevention**: Analytical methodologies need to be further developed to allow for real-time, in-process monitoring and control before the formation of hazardous substances.
- **Inherently safe chemistry**: Substances and the form of a substance used in a chemical process should be chosen to minimize the potential for chemical accidents, including releases, explosions, and fires.
- **Use life-cycle:** Engineering ideas should look at the product from conception to final disposal of all materials.
- **Community engagement:** Local communities should be actively engaged in the use of chemical products and production plants.

The Gupta[93] paper also reviews areas of progress such as the removal of chromate, the use of natural products, and many other advances that are described in subsequent sections.

Annually, the USEPA[94] uses these principles as some of the criteria for the Presidential Green Chemistry Challenge. The goals are to promote green chemistry in the design of chemical products and processes that reduce or eliminate the generation of hazardous substances. The EPA's efforts to speed the adoption of this revolutionary and diverse discipline have led to significant environmental benefits, innovation, and a strengthened economy

Today, many articles or presentations involving corrosion inhibition mention an assessment of health, safety, and the environment and include many of the principles noted above.

6.5.1. Drivers for Eco/Tox Changes

The changes in attitude on toxicity of various corrosion inhibitor chemicals coincide with the societal changes in many countries in the 1970s and 1980s. The First World Earth Day was held in 1970 and has been held at varying intervals since that time.

In the United States, the Clean Water Act[95] is the primary federal law governing water pollution. Its objective is to restore and maintain the chemical, physical, and biological integrity of the nation's waters by preventing point and nonpoint pollution sources, providing assistance to publicly owned treatment works for the improvement of wastewater treatment, and maintaining the integrity of wetlands. It is one of the United States' first and most influential modern environmental laws. As with many other major U.S. federal environmental statutes, it is administered by the EPA in coordination with state governments. Its implementing regulations are codified at 40 C.F.R. Subchapters D, N, and O (Parts 100-140, 401-471, and 501-503).

Technically, the name of the law is the Federal Water Pollution Control Act (FWPCA). The first FWPCA was enacted in 1948 but took its modern form when completely rewritten as the FWPCA Amendments of 1972.[96] Major changes have subsequently been introduced via amendments including the Clean Water Act of 1977 and the Water Quality Act of 1987. Other amendments were introduced in 2002.

The Occupational Safety and Health Act of 1970 created the Occupational Safety and Health Administration (OSHA)[97] to ensure safe and healthful working conditions by setting and enforcing standards and by providing training, outreach, education, and assistance. Workplace toxicity issues are monitored by this agency. One of the outcomes of the EPA and OSHA are lists of chemical substances that are toxic to human life and to the environment. Some of these are noted in later sections.

Foreign governments, especially those associated with the European Union, also have detailed protocols for developing green chemicals, especially chemicals used in the North Sea.[98]

6.5.2. General Chemical Approaches for Green Corrosion Inhibitors

The work on mechanisms of actions of corrosion inhibitors is the basis for developing green materials. The major mechanism is the adsorption theory described in Section 6.1.

A major advancement in the production of green petroleum production inhibitors has come from the work of Singh[99] and Singh et al.[100,101] This work was the result of a consortium formed at Texas A&M University to develop less toxic petroleum production inhibitors in a systematic manner by applying the types of chemical computational methods that have been successful in the pharmaceutical industry.

Toxicity prediction parameters and corrosion inhibition prediction parameters have been developed. The first model of toxicity assumed that the inhibitor must pass through the lipid bilayer of an organism's cell. Thus, a standard measure of the partition of a molecule between water and octanol (logP) was an early toxicity predictor. Many potential inhibitor compounds fit the following model:

$$\log\left(\frac{1}{LC_{50}(\text{fathead minnow})}\right) = a\log P + b \qquad \text{(Equation [6.9])}$$

See Frenier[31] for the definition of the toxicity values (LC_{50}). The authors[100, 101] claim "r" values > 0.90. The major restriction is that different equations must be developed for sets of similar chemicals. That is, non-polar chemicals, polar aliphatic, and polar aromatics must be treated separately. The logP value also can be used to estimate the bioaccumulation of the chemical in aquatic species. There are also methods for estimating toxicity using non-empirical values of the molecules (molecular connectivity or molecular orbital values); however, they work well only for materials that have a narcotic effect. For a large range of corrosion inhibitor molecules, toxicity increases as the length of a side chain increases. There are good fits for amines, acetylenic alcohols, fatty acids, and nitriles.

Quantitative structure activity relationships (QSAR) also were developed for predicting corrosion inhibitor data. A large range of molecular descriptors as well as adsorbility trends was used to screen (mathematically) many inhibitor molecules that fall into four classes. These data were used to identify 30 new molecules. Lukovits[102] described the use of QSAR equations and techniques for predicting the corrosion behavior of thiourea derivatives. However, no toxicity data were listed, even though the title suggested that these were less toxic inhibitors.

Gece[103] used quantum methods of the electron density distributions in a hypothetical molecule to bind to a surface (iron or an iron salt). This author claims that quantum chemically derived parameters are fundamentally different from experimentally measured quantities, although there is some natural overlap. They claim that unlike experimental measurements there is no statistical error in quantum chemical calculations. However, there is inherent error associated with the assumptions required to facilitate the calculations. Even with modern programs and computers, these assumptions can be significant.

Gece[103] includes numerous references and should be read to understand the various methods and the molecular orbital parameters that can be measured. Note that the basic methods have been used since at least 1980 by Frenier[104] and Growcock[105] to study inhibitor mechanisms. Also see the articles by Ramachandran[106] and Rodriguez-Valdez[107] for developing new materials.

Several authors have suggested investigating natural products or chemicals based on natural products. Cinnamaldehyde is present in oil of cinnamon and has been noted in earlier parts of this book (see Section 6.2.2 and Frenier[108]). Choudhary[109] has proposed the use of extracts from a perennial bush plant available in many parts of the world. The root of the chicory plant can be roasted and ground for use as a coffee substitute or additive. The main active ingredient is inulin (two different types of inulin are shown in Figure 6.15).

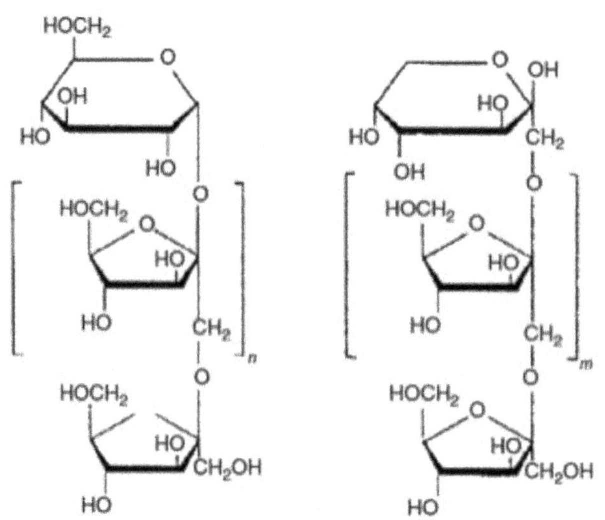

Figure 6.15 Two Depictions of the Inulin Molecular Structure

The authors claim that chicory is environmentally acceptable and is biodegradable. This study claims that chicory can provide corrosion protection for alloys such as N-80, 13Cr-L80, and 1010 steel in the presence of either inorganic or organic acids at temperatures up to 250°F (121°C). This publication has numerous references to other natural products that have been tested as corrosion inhibitors in HCl solutions.

Toxicity issues will continue to be some of the most important aspects of corrosion inhibitor technology, and meeting the demands for less toxic inhibitors will consume a large portion of the funds available to develop new formulations. Other issues such as flash point and presence of heavy metals will limit the palette available to the inhibitor formulator. A comprehensive review of green inhibitor technology was presented at the 2000 Ferrara inhibitor conference.[16]

6.5.3. Examples of More Eco/Tox Friendly Cleaning Inhibitors

Many commercial acid inhibitors used in chemical cleaning services were based on the Mannich condensation reactions (Figure 6.8) introduced by Saukaitis and Gardner,[47] Gardner and Saukaitis,[110] Mansfield,[48] and Monroe.[49] This process uses formaldehyde, an amine, and a ketone to produce a Mannich base. Because this reaction rarely goes to completion, some formaldehyde will remain in the reaction product that is formulated as the commercial inhibitor. Formaldehyde is an animal carcinogen and appears on the OSHA Specifically Regulated Substance List and the National Toxicity Program Source List 9C (compilation of carcinogens); therefore, it is desirable to remove this substance from inhibitor formulations.

In addition to a condensation reaction product, commercial acid inhibitor formulations frequently contain acetylenic alcohols such as propargyl alcohol, hexynol, or ethyloctynol (see Figure 6.1 and Figure 6.8 for chemical structures). While these materials can produce excellent corrosion inhibitor packages, many acetylenic alcohols are quite toxic to mammals (Table 6.5 and Sax[111]), and the lower alcohols (propargyl alcohol and hexynol) are readily absorbed through the skin. For these reasons, there is a growing necessity to produce corrosion inhibitors that are less toxic than current formulations.

Table 6.5 Mammal Toxicity Studies: Oral LD_{50} Values for Rats

Compound	LD_{50}, mg/kg	Source
Propargyl Alcohol	55	Sax Handbook[206]
Hexynol	34	Sax Handbook
Cinnamaldehyde	2200	Sax Handbook
Formaldehyde	800	Sax Handbook
Dodecylpyridinium Br	320	Dow Chemical Co.
Naphthylmethylquinolinium Cl	644	Dowell Schlumberger, Inc.
Nonylphenol-ethylene oxide Surfactants	1310	Sax Handbook

Frenier and Growcock[39] described formulations of α,β-unsaturated aldehydes with surfactants that were shown to be effective corrosion inhibitors for steel in HCl. On the basis of this technology, the aromatic compound t-cinnamaldehyde was proposed as an ingredient in low-toxicity inhibitor formulations.[68] According to Growcock and Frenier[39] and Growcock,[45] this aldehyde adsorbs onto the surface of the steel and then forms a low-molecular-weight polymer film that enhances the corrosion protection. A low-toxicity commercial inhibitor[16] formulation based on cinnamaldehyde was developed for use in HCl-based cleaning formulations. It is claimed to be suitable for use in IC solvents containing HCl or HCl and ABF. The inhibitor is claimed to protect various steel alloys as well as or better than currently used commercial acid inhibitors. The compositions are listed in Table 6.6, and the results of tests with 1018 CS are listed in Table 6.7. The production of quats using low-toxicity solvents are shown in Figure 6.3, Figure 6.4, and Figure 6.5.

Table 6.6 Hydrochloric Acid Inhibitor Formulations

Formulation Code	Quat	Oxygen comp.	Surfactant	Pour Point, °F	Flash Point, °F
C	DDPB	Cinnamaldehyde	LAE	<20	> 210
D	NMQCl	Cinnamaldehyde	LAE	<20	198
E	DDPB	Octynol	LAE	<20	>210
F	NMQCl	Octynol	LAE	20	200

Abbreviations: LAE, linear alcohol exthoxalate; DDPB, dodecylpyridinium bromide; NMQCl, napthmethylquinolinium chloride.

Table 6.7 Corrosion Rates in Hydrochloric Acid 150°F, 1018 CS, 0.1% Inhibitor

Inhibitor	Acid Strength, %	Rate, lb./ft²/day
Comm. B	5	0.002
C	5	0.001
D	5	0.001
E	5	0.002
F	5	0.001
Comm. B	7.5	0.002
C	7.5	0.001
D	7.5	0.001
E	7.5	0.003
F	7.5	0.001

Table 6.8 compares the static corrosion rates for the low-toxicity (C; based on chemistry in Table 6.6) with a classic Mannich-base inhibitor (A) that contains formaldehyde and hexynol. In most of the tests, the low-toxicity inhibitor provided similar or superior inhibition compared with the classic acid inhibitor. Inhibitor C also was used to clean a 600-ft length of carbon steel piping that had a light coating of iron oxide in a chemical plant.

A solution of 7.5% HCl inhibited with 0.1% of Inhibitor B was heated to 140°F (60°C) and then was pumped back and forth between two truck tanks. The iron and acid-strength curves during the cleaning are shown in Figure 6.16. The acid and iron concentrations reached stable values after 3 hours of contact. The leveling trend observed for acid concentration and iron concentration confirms that the inhibitor provided adequate protection to the piping being cleaned. After the acid had been flushed from the line, a soda ash/sodium nitrite passivation stage was used to prepare the piping for use by the client.

Table 6.8 Corrosion Rates (lb./ft.²/day) without Additives at 151°F (66°C) and with 0.1% Inhibitor

Metal	[HCl], %	Inh. A	Inh. C
C1010	5.0		0.0014
C1010	7.5	0.0010	0.0011
C1018	5.0	0.0013	0.0007
A-515-Gr70	5.0		0.0014
A-515-Gr70	7.5	0.0045	0.0015
A-213-T22	5.0		0.0011
A-213-T22	7.5	0.0018	0.0021
A-213-T11	5.0		0.0019
A-213-T11	7.5	0.0013	0.0012
A-106	5.0		0.0015
A-106	7.5		0.0019

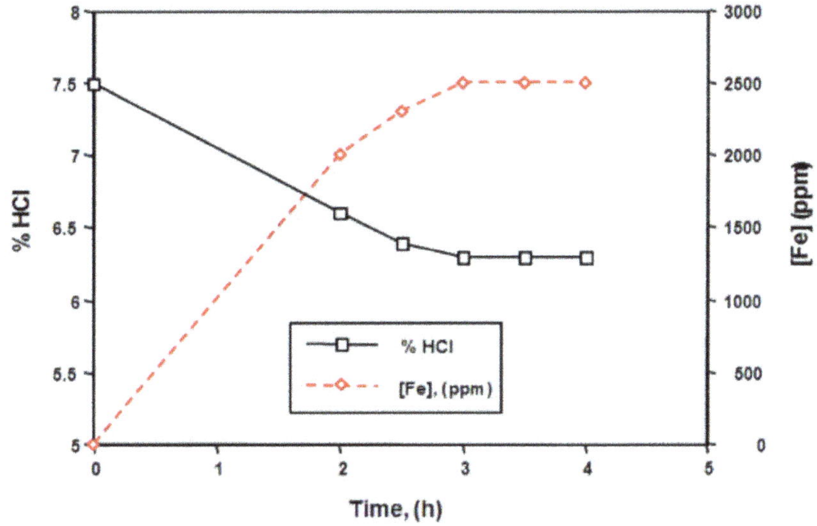

Figure 6.16 Dissolution of Iron Oxide from Piping in a Chemical Plant at 140°F (60°C). HCl Inhibited with 0.1% a Low-Toxicity Corrosion Inhibitor

A low-toxicity acid inhibitor was proposed by Lindert and Johnston.[112] This composition contains low-toxicity amines or quaternary nitrogen salts (abietyl amine or pyridine-containing rings), a divalent sulfur compound (such as ammonium thiocyanate), or a dialkyl thiourea derivative. Enhancer salts such as potassium iodide and solvents (propylene glycol) also may be present. High levels of inhibition in HCl solutions are claimed. Low-toxicity amides of fatty acids were described by Jiricek et al.[113] When tested in 5% HCl at 113°F (45°C), very few of the amides provided >80% protection for carbon steel. Inhibitors for use in cleaning water wells were proposed by Anderson and Seigle.[114] The HCl (up to 15%) solution's attack on steel was inhibited using a partially hydrolyzed protein such as gelatin or casein. The useful temperature range is from about 32°F (0°C) to about 90°F (32°C). A low-toxicity composition for use in HCl was patented by Carandang.[115] This complex mixture contains gelatin, a sugar-derived polymer such as sorbitol, and an iodide compound such as KI. Phosphoric acid, sodium sulfate, and propylene glycol also may be present.

Sitz[44] discussed the preparation of an acid inhibitor that contained only oxygen-based inhibitors (see Figure 6.1) that could be used to inhibit HCl to >200°F. Shank[116] conducted a series of tests using an acid inhibitor that did not contain propargyl alcohol compared with commercial inhibitors that do contain this toxic chemical. The tests showed that HCl used for chemical cleaning could be inhibited by low-toxicity formulations. Low-toxicity corrosion inhibitors for organic acid/chelating agent formulations, including those by Frenier,[117] contain quat/sulfur formulations with improved solvents.

In addition, several exceptions to the sulfide/nitrogen inhibitors have been proposed. Sulfonium compounds have been employed to inhibit corrosion in EDTA solutions at low temperatures (<150°F). See the work by Frenier[75, 76] that demonstrated that benzyl sulfonium salts (Figure 6.12) were very effective in pH 5 ammonium EDTA solutions, even in the presence of ferric (Fe^{3+}) ions. Their work also found a secondary hydrocarbon layer (b and c) after a corrosion test. This was attributed to the electrochemical reaction of the sulfonium molecule with Fe on the steel's surface. These compounds in the EDTA solvents were employed commercially to clean cooling water systems in refineries and oil and gas facilities.

An additional corrosion inhibitor formulation has been proposed. Jenkins[118] filed a patent application for a combination of a propoxylated acetylenic alcohol and a mercaptan. See Figure 6.17 for a preferred combination. The author of this patent claims it will successfully inhibit 22% FA up to about 180°C.

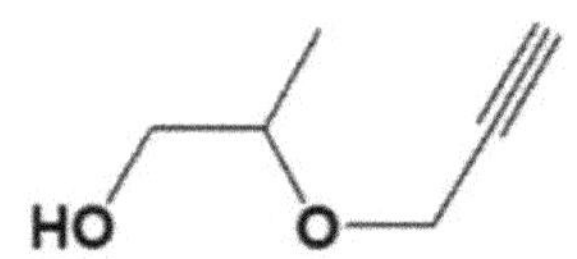

Propoxylated prop-2-yn-1-ol
2-(Prop-2-yn-1-yloxy)propan-1-ol

2-Mercapto ethanol

Figure 6.17 Organic Acid Inhibitor Combination[118]

6.6. Test Methods for Evaluating Corrosion and Inhibitors in Industrial Cleaning Solvents

This section describes currently used test methods for determining the corrosion rates and types of corrosion observed in the fluids that may transit the industrial units during chemical cleaning. These methods are used to determine rates and manifestations of corrosion with and without inhibitors or any other process that may reduce the severity of attacks. Note that in many evaluations, tests without inhibitors (called blanks or controls) are run to determine the effectiveness of the interventions and to determine the basic mechanisms of the corrosive attacks. Also, a company standard formulation or a well-known molecule may be used as an additional standard for comparison when evaluating a new corrosion inhibitor. Note that there are not any recognized standard methods for evaluating inhibitors in the chemical cleaning industry. However, NACE does have some reports[119,120] on general techniques. The NACE[120] report has methodologies especially for cleaning solvents, but it cannot be considered to be a standard in NACE terminology.

This section reviews general methods that may be used in laboratory screening as well as field monitoring and compares specific lab and field methods. Two NACE International committee reports[121,119] have reviewed a wide range of methods for determining corrosion rates with and without inhibitors in field operations. These documents are referenced where appropriate in the following subsections of this book. However, note that each lab may use a local modification of a test and datasets will vary from site to site.

The different methods for determining corrosion rates and corrosion damage include using corrosion coupons and various electrical and electrochemical methods for measuring corrosion rates and possibly providing evidence of pitting. Hydrogen emissions also have been measured in pipeline operations to indicate corrosive situations. These types of measurements are commonly used in laboratories as well as in field operations for general corrosion assessments and for evaluation of inhibitors and cleaning solvents.

Many of the methods that are described in subsequent sections are used to determine a corrosion mass loss rate to estimate a projected lifetime of a pipe or other piece of equipment. If a corrosion inhibitor is present, a corrosion mass loss rate may be determined and compared with the inhibited rate to provide a value of the percent inhibition efficiency (also called the percent protection):

$$IE = \frac{[CR]_{no\ Inh} - [CR]_{Inh}}{[CR]_{no\ Inh}} \bullet 100 \qquad \text{(Equation [6.10])}$$

6.6.1. Metal Specimens (Coupons) Weight Loss Testing

Coupons (pieces of metal, often from field-use alloys) can be cut and machined into various sizes and shapes. These provide the basis for lab and field corrosion tests. In some cases for screening, research, or field uses, commercial grades of metals can be purchased that simulate field stocks. Note that because some materials are not produced to a chemical specification but to a material specification (e.g., strength or hardness), corrosion susceptibilities may significantly vary. Reducing the materials variability by using a large stock of similar metals may be desirable. RP0497-2004[122] provides general guidelines for preparing corrosion specimens for both lab and field uses. A small example of commercial samples from a metal coupon supplier[123] is provided in Figure 6.18, which shows types of coupons and examples of coupon probe holders and insertion equipment.

These techniques are known as weight loss (or mass loss) methods because the specimens are weighed before and after exposure to a corrosive fluid, and the change in mass is expressed as a corrosion rate.

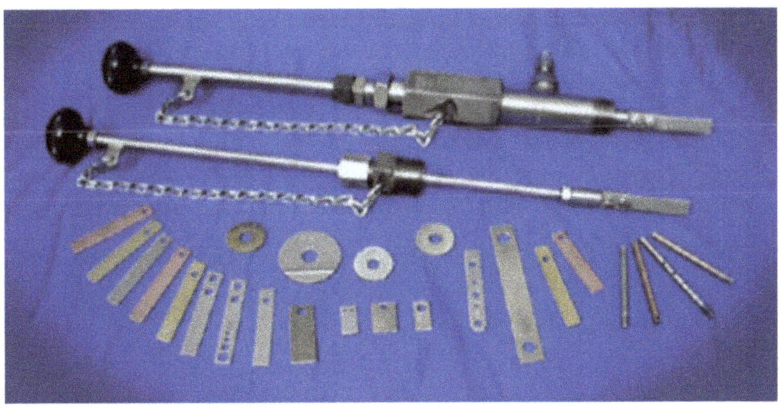

Figure 6.18 Corrosion Coupons from Commercial Metals[123]

Important ideas concerning the use of corrosion coupons are

- One-point of data from each test is provided
- It is an integral method (average of mass loss over time of test)
- Integral over surface of the coupon unless a pit depth measurement also is performed
- Usually employs simple lab equipment and can be placed in many types of equipment
- Easily adaptable to many types of specimens
- Non-uniform corrosion and scale can be seen and analyzed

Note also that numerous materials and variations of a single material can be simultaneously evaluated. In addition, alloy chemistry variations and metallurgical variations (i.e., the effect of heat treatment, microstructure, welding, and stress) can be considered. Coupon tests can be easily adapted to evaluate specific types of corrosion, such as crevice corrosion and galvanic corrosion. Special coupons have been produced with specific weld and weld types for evaluation of weld bead and heat affected zone corrosion. See Frenier.[79,124]

The output datum from a coupon weight loss corrosion test usually is a corrosion weight loss rate expressed as a mass of metal lost per a specific area over a specific time period. One common rate value is mils (in/10^3) penetration into the metal/year (mpy).

The following equation is used to calculate the corrosion rate from metal loss penetration (initially in mm/y):

$$mm/y = \frac{8.67W}{DAT}$$ (Equation [6.11])

Here, W is the weight loss in mg, D is the metal density in g/cm^3, A is the area of sample in cm^2, and T is the time of exposure of the metal sample in hours.

Then,

$$1\,mpy = 0.254\,mm/y = 25.4\,\mu m/y$$ (Equation [6.12])

However, because many expressions for the rates are reported in the literature,[123] it is vital for the reader to know the system and make comparisons to known examples. Different output values can be calculated from the same dataset as long as measurements of mass, density of the metal, surface area, and time of exposure values are recorded.

The most common corrosion rate used in cleaning solutions is lb./ft^2/day. A NACE report[125] gives general details for conducting lab tests for cleaning chemicals. More information is in following sections of this book.

For lab tests, the use of standards and controls is mandatory. As noted in the previous paragraph, a physical examination of the coupon after the test may provide critical data on the mode of attack, scale depth of pits, etc. A coupon manufacturer[123] provides information on various types of coupons as well as preparation and interpretation of coupon test results. Also, since no two labs have exactly the same operation conditions, only general comparisons are possible.

6.6.2. Electrical and Electrochemical Methods

The next sections describe electrical as well as electrochemical methods for determining corrosion rates.

Electrical Resistance (ER) Method. When a current is caused to flow through a sensing element (a piece of test metal) of an ER probe (see Figure 6.19 for one configuration) a resistance value (R) is measured, based on Ohm's law:

$$R = \frac{E}{I}$$ (Equation [6.13])

Here, E is electrical potential difference in volts, I is current in amps, and R is DC resistance in ohms.

As the probe is corroded and the mass is reduced, R increases, and when R is measured, the change is noted by the instrument package. Because this may happen slowly, this is a long-term indication, not an instantaneous response, and a corrosion rate is calculated from the slope of the change in R vs. time. The probe also contains a comparison resistor to calibrate the current change at a fixed voltage.

ER probes are the most flexible of the electronic methods in terms of their possible applications. Because the technique is not dependent on an electrochemical reaction, a current-carrying electrolyte is not necessary. ER monitoring can be used to measure virtually all environments, such as solids, non-aqueous liquids, gases, and vapors, all of which can be of high or low conductivity. By complementing probe readings with corrosion coupons exposed over the same time interval, the two measurement techniques will verify the qualitative results. ER probes can indicate changes in corrosion rates and will show when the changes took place, which is helpful in evaluating the process. See Heidersbach[126] for more examples of ER probes.

Figure 6.19 An Electrical Resistance Probe

In summary, this type of method

- Is useful for monitoring and recording long-term corrosion and changes in an environment
- Can be used in almost any environment
- Is sensitive to both corrosion and erosion
- May be slow to respond to a *slightly* more corrosive environment but is useful for large changes in the corrosivity of the fluid
- May fail at very high corrosion rates and thus cease to provide data
- May provide some qualitative information about the corrosion processes via examination of the active portion of the probe

Direct Current Electrochemical Corrosion Monitoring. These methods comprise a very well-respected set of techniques for determining instantaneous corrosion rates. Both direct current (DC) and alternating current (AC) test methods are used in labs as well as field operations. The current section describes DC methods, and the next section discusses some AC methods.

The corrosion process is depicted in Figure 4.15, which shows a theoretical polarization curve for corrosion of iron in an acid that illustrates the electrochemical process. The key values at equilibrium (I_{cor} and E_{cor}) are noted. Changes in concentrations and reactants can alter the equilibrium and shift the curves. In electrochemical terminology, the concentration and type of oxidizing agents reacting on the metal surface cause the metal to be polarized in either the anodic or cathodic direction, and this change will cause a change in the corrosion rate of the metal (I_{cor}).

Using an instrument called a potentiostat, researchers can generate this type of curve in a laboratory setting and obtain information about the system. The mass of metal lost vs. time (that is the *corrosion rate*) can be calculated from I_{cor} using Faraday's law:

$$m = \left(\frac{Q}{F}\right)\left(\frac{M}{z}\right) \quad \text{(Equation [6.14])}$$

Here,

- m is the mass of the substance liberated at an electrode in grams
- Q is the total electric charge passed through the substance (from I_{cor})
- $F = 96485\ C\ mol^{-1}$ is the Faraday constant
- C = coulomb
- M is the molar mass of the substance
- z is the valance number of ions of the substance (electrons transferred per ion)

The anodic and cathodic slopes define the Tafel (named after Julius Tafel, 1862–1918) constants (β_{an} and β_c) seen in Figure 4.15.

Fontana[127] has provided a very readable explanation of electrochemical corrosion and measurements of corrosion rates using DC methods. Information shown in Figure 2.6 indicates that both anodic (metal loss) and cathodic (oxidation) reactions occur simultaneously on the surface of a metal during the corrosion process. When the corrosion process is at equilibrium, the oxidation and the reduction reactions proceed at the same rate and at the same potential (voltage) on the metal surface. Using a polarization cell, such as Figure 6.20, that is connected to a potential/current controller (usually called a potentiostat or galvanostat), researchers can isolate the two reactions by changing the potential of the surface in either the anodic or cathodic direction.

If both the current and the potential (vs. a reference such as calomel or Ag/AgCl) are determined, then a set of curves such as those in Figure 4.15 can be generated. This plot is an example of activation polarization (ap), in which the rate-determining step is the surface reaction (in the cleaning solutions, this usually is proton transfer and reaction). The second type of rate control is concentration polarization (cp), in which the diffusion of the oxidant to the surface is rate controlling.

Using this plot, researchers can determine the corrosion rate (I_{cor}) and the corrosion potential (E_c), by extrapolation from the point where the lines (called the anodic and cathodic Tafel lines [β_a and β_c]) intersect.

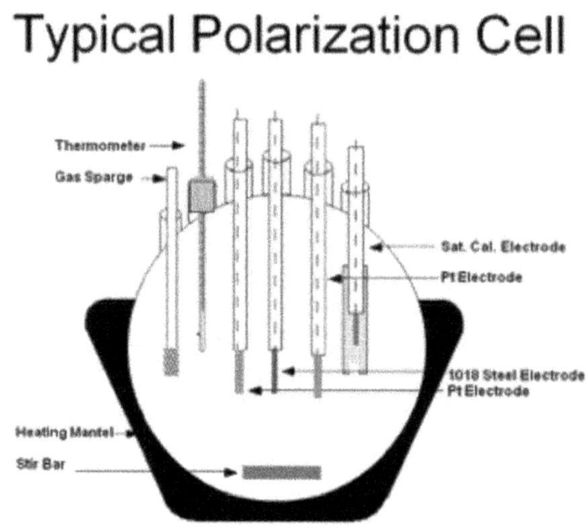

Figure 6.20 Polarization Cell

The Butler-Volmer equation can also be used to calculate these values:

$$I = I_{corr}\left(e^{(2.3(E-E_{corr})/\beta_a)} - e^{-(2.3(E-E_{corr})/\beta_c)}\right) \quad \text{(Equation [6.15])}$$

Thus, the corrosion rate at any instant as well as the corrosion potential can be calculated. Note that this is a semi log plot and takes significant time to generate, even though knowledge of the entire set of curves may provide additional information about the corrosion reaction. Using automated (computer) equipment, investigators can conduct these types of studies fairly rapidly. Figure 6.21 shows an example of an automated scan of a steel electrode in dilute HCl; from the data fit, the corrosion rate, corrosion potential, and the Tafel slopes are readily calculated.

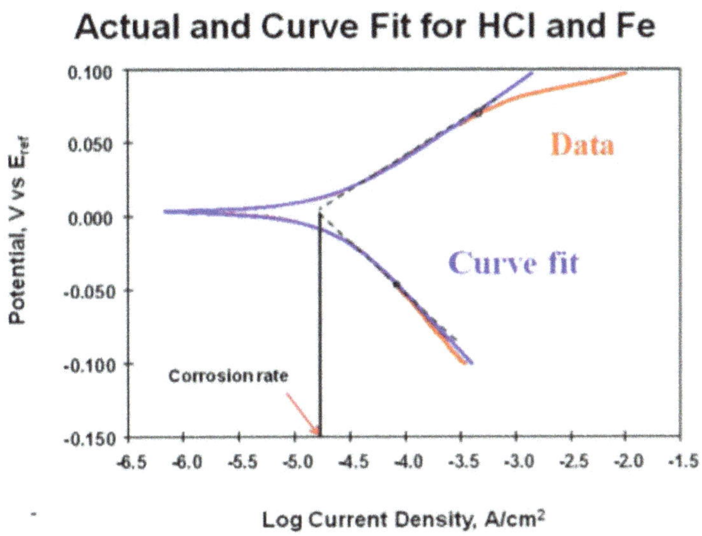

Figure 6.21 Automated Corrosion Scan of Steel in HCl, 25°C

Stern[128,129] has also observed that within about 10 mv of the corrosion (also known as the open circuit) potential, the corrosion current and the potential change behave in a linear relationship. These observations led to the Stern-Geary equation:

$$I_{corr} = \frac{(1/R_p)(\beta_a \beta_c)}{2.3(\beta_a + \beta_c)} \qquad \text{(Equation [6.16])}$$

Here, R_p is called the polarization resistance, and from R = E/I (within 10 mV of the corrosion potential) the instantaneous corrosion rate is determined. The linear polarization resistance (LPR) method is now the basis of many automated lab and field instruments used in pipelines and many other industrial installations. Note that the Tafel slopes for a particular corrosion reaction are needed for the calculation, but these usually also can be determined or estimated using lab tests.

Rearranging Equation (6.16) yields

$$I_{corr} = \frac{B}{R_p}, \text{ where } B = \frac{(\beta_a \bullet \beta_c)}{2.3(\beta_a + \beta_c)} \qquad \text{(Equation [6.17])}$$

Some observations of EC tests include the following:

- EC tests provide instantaneous rates and can be averaged over time to obtain rates comparable to weight-loss methods.
- All LP rates depend on selection/determination of Tafel β values. In many lab and field uses of LPR, "B" is assumed to be 26 (assuming both slopes to be 120 mV).
- Large over-voltage changes can corrode/damage the electrode.
- Current distribution in the cell can be critical.
- Equipment and electrodes may be expensive.
- One can see changes with time.
- They can provide a fast test.
- Some mechanistic data may be extracted.
- EC tests usually use small specimens.
- EC test specimens may be fouled by scale or oil.

- EC tests depend on the fluid having a conductivity above about 10 μmohs/cm and thus are not useful in oil.[130]
- EC tests cannot detect localized corrosion without an examination of the electrodes.
- EC tests allow easy storage of data on computers and provide automated process control.

Jasinski[131] performed a large series of tests over a wide range of temperatures, corrosive fluids, and metals to compare mass loss (blue squares), LPR (red triangles), and Tafel slope (green circles) methods of corrosion rate measurements. There was good qualitative correlation between the methods but with a lot of scatter. The data are provided in Figure 6.22. The black line would indicate a 1:1 corelation between the mass loss and electrochemical methods.

A NACE document[119] also has reviewed various electrochemical data and claims that general corrosion rates obtained using the LPR method agree with those of the mass loss coupon method more than 75% of the time. One large error is from using an estimate of the Tafel slope in Equations (6.15 and 6.17).

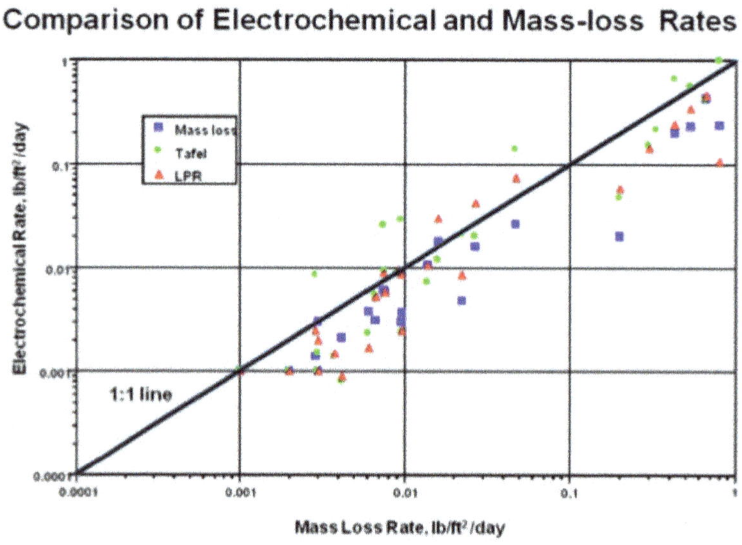

Figure 6.22 Comparison of Electrochemical and Mass Loss Rates

This NACE report[119] provides other cautions for using EC corrosion tests:

- All electrochemical techniques measure the corrosion rate of the probe, and this corrosion rate is an indication only of the corrosivity of the environment at the time and location of the measurement. This corrosion rate does not represent the corrosion rate of the entire infrastructure.
- To obtain meaningful data from electrochemical techniques, the probes are placed in the conductive phase (the aqueous phase, not the oil phase).
- If the probe surface is covered with crude oil, abnormally low corrosion rates typically result. Also, if the probe surface is in the gas phase (or a gas bubble from the aqueous phase covers its surface), the data may show a significant amount of scatter.
- Corrosion rates calculated from some electrochemical measurements (e.g., LPR, EN, or EIS) are theoretically valid only if the reaction is under charge-transfer control (i.e., ap); the rate of change of electrons at the metal/electrolyte interface is the slowest and rate-determining step, compared with a diffusion-controlled process (i.e., cp)

This report is very comprehensive and should be referenced by those needing information about oil/gas field corrosion measurements.

6.6.3. Laboratory Tests for Studying Various Corrosion Manifestations

The electronic and the coupon tests described in Section 6.6.2 can be used in the laboratory and in some field monitoring applications. Wet chemical, instrumental, and surface analytical methods are also widely used to help elucidate various corrosion manifestations. Some of these are described briefly in this section. This section notes several types of laboratory corrosion test methods and devices. Many of these methods may be conducted with and without inhibitors to determine the effects of the added chemicals to the corrosion protection rate.

Useful and effective lab tests should model field conditions, including changes in the aqueous phase, hydrocarbon phase, temperatures, and different inhibitors.[132] Lab experiments also are much less expensive and easier to perform than field trials. However, closely modeling field conditions is almost *never* completely possible.[133] Note that large numbers of testing devices have been proposed and used for flow assurance studies and corrosion studies. However, none of these can represent all possible solid deposition or corrosion scenarios. Instead, each is built to meet a specific need or to mimic a specific process. Some of the challenges in replicating the field environment in a test facility include

- Large dimensions of the field operation
- Large volumes of liquid and gases
- Extreme pressures and temperatures
- Identification of corrosion mechanisms
- Limited analogs of non-aqueous liquids
- Multiphase flow
- Corrosion of the test equipment
- Assessment of pitting/localized corrosion rate
- Pass/fail criteria for secondary tests (e.g., emulsion, foaming, partition testing, or compatibility with other additives)
- Safety and disposal issues owing to use of flammable/hazardous materials

Conventional scaling laws[134] for designing a reduced scale laboratory model are difficult to apply for flow assurance/corrosion studies because only a limited set of field parameters can be matched in a reduced scale laboratory model. For example, for deposition or corrosion in a flow line, the flow regime (laminar or turbulent) *and* the shear rate at the pipe wall are both critical. However, in most reduced scale laboratory models, it is only possible to match one of them. For flow of a single phase Newtonian fluid in a pipe, the flow regime is determined by the Reynolds number:

$$Re = \frac{\rho u d}{\mu} \qquad \text{(Equation [6.18])}$$

Whereas the shear rate at the wall is given by

$$-\left(\frac{dv}{dr}\right)_w = \frac{8u}{d} \qquad \text{(Equation [6.19])}$$

In the above equations, d is the pipe diameter, ρ is the fluid density, u is the average fluid velocity, μ is the fluid viscosity, and v is the fluid velocity at radius r. Note that the Reynolds number is proportional to ud, whereas shear rate at the wall is proportional to u/d. Therefore, to match both these parameters in a laboratory model with a smaller pipe diameter, the fluid viscosity and/or density must be adjusted. However, it is not acceptable to do so, as this would inevitably mean a change in fluid composition or temperature, which may alter the precipitation chemistry of the fluid. Most of the flow assurance issues encountered in the field involve multiple phases. Therefore, it is even a greater challenge to create a representative laboratory model as a larger set of parameters must be matched than those for the single-phase problem described above.

Loop tests. Several large flow loops exist in the industry today to circumvent issues concerning a reduced scale model. Faclone et al.[135] have reviewed multiphase flow loops available in research facilities worldwide. Only a few of these loops are dedicated for flow assurance or corrosion research. However, a large flow loop is not a panacea and has its own set of problems. First, most of the large flow loops are built to operate at relatively low pressures, as the cost of building a large, high-pressure loop is significantly higher. These low-pressure loops are not suitable for studying precipitation of asphaltenes from light oils and hydrates from gas streams because these phenomena usually occur at high pressure. Secondly, large fluid volumes are required for operation of large flow loops, and practically only stock tank oil or recombined oil can be used. However, to mimic this process in a large flow loop would require use of enormous volumes of live downhole fluid and would be prohibitively expensive.

In many laboratory models, a compromise is usually made in matching field parameters, and one needs to have a reasonable understanding of the deposition process to ensure that the critical parameters are not missed. Chemical parameters that are required in these studies include

- Chemistry of the fluids
- Oil chemistry (if required)
- Metallurgy of corrosion specimens
- Appropriate introduced gases (CO_2, H_2S, removal of O_2)

A drawing of a generic low-pressure loop test[136] is shown in Figure 6.23. The test section (G) in the paper cited consisted of an EIS system; however, a loop with a similar layout could be used with a pipe section of a container with coupons or other EC probes.

Several additional commonly used tests are reviewed in this section. Some of these test setups can be employed with coupons, ER probes, or EC methods. Note that loops for studying cleaning progress also are used (see Section 2.6.4) and are like this design.

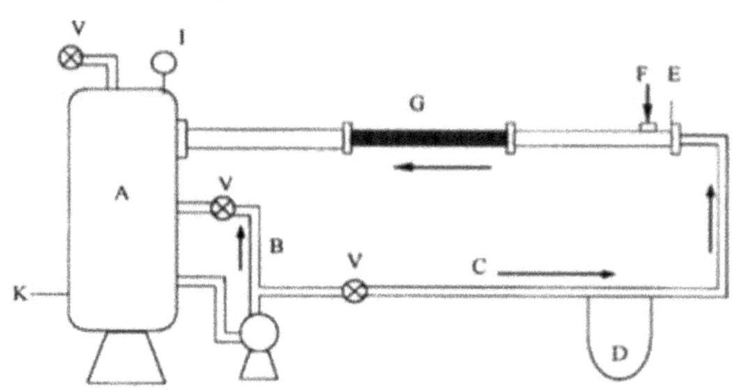

Figure 6.23 Low-Pressure Loop Test Equipment

Figure 6.24 shows a drawing of the configuration of 3-electrode LPR probes for lab or field use (see explanations in Section 6.6.2). Many variations are possible. The most important difference between this device and the polarization cell in Figure 6.20 is that the LPR probe has three electrodes made from the same metal or two probes of the test metal and a probe (possibly the case) that can act as the auxiliary (collector) electrode. The computer in the controller selects one as the working electrode, one as the auxiliary electrode, and the third as the reference electrode. Thus, only potential differences can be determined compared with a full 3-electrode system as seen in Figure 6.20.

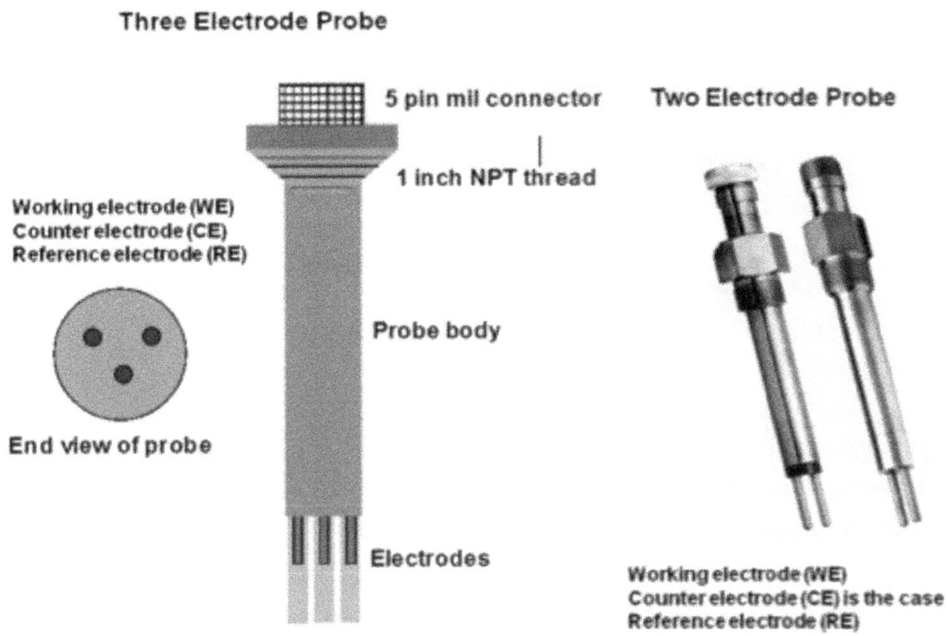

Figure 6.24 LPR Probes

Glassware tests. These test setups can use an electrochemical method for determining the corrosion rate of an aqueous cleaning fluid; however, coupons also can be used instead of or in addition to the EC probe. Similar glassware tests also are run with coupons to test chemical cleaning fluids. They are limited to about 200°F, even if they have a condenser.

Figure 6.20 shows one example of a simple heated glass cell that has a working electrode (the metal), a reference electrode, and two Pt auxiliary electrodes to complete the electrical circuit. Gases can be continuously purged (bubbled) through the flask, and the reactor also can be stirred. Unless very heavy glass is used, these systems are useful only for low-pressure/temperature applications.

Glassware tests also can be fitted with an LPR probe or possibly a hydrogen probe. Figure 6.25 from Chen[137] shows a series of test rigs set up for multiple LPR or coupon tests. Multiple tests with LPR probes can be computer automated.

Figure 6.25 Glassware Test Setups

Corrosion Autoclave Tests. To perform corrosion tests at elevated temperatures, various autoclave methods are used. One of the types uses the NACE/API cell depicted in Figure 6.26. These types of single test cells can be used for testing acids and other production and pipeline fluids using coupons. A back-pressure regulator maintains a constant test pressure. Similar cells also could be fitted with LPR probes for automated testing.

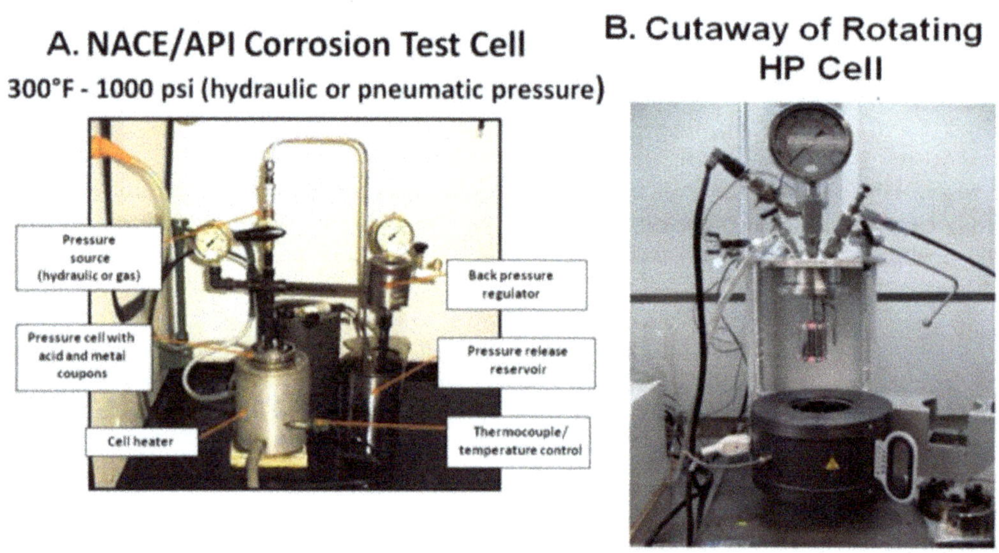

Figure 6.26 Autoclave Cells. A.NACE Cell. B. Chen

A setup for the use of a rotational pressurized cell is shown in Figure 6.27, and various rotational cells are shown in Figure 6.28.

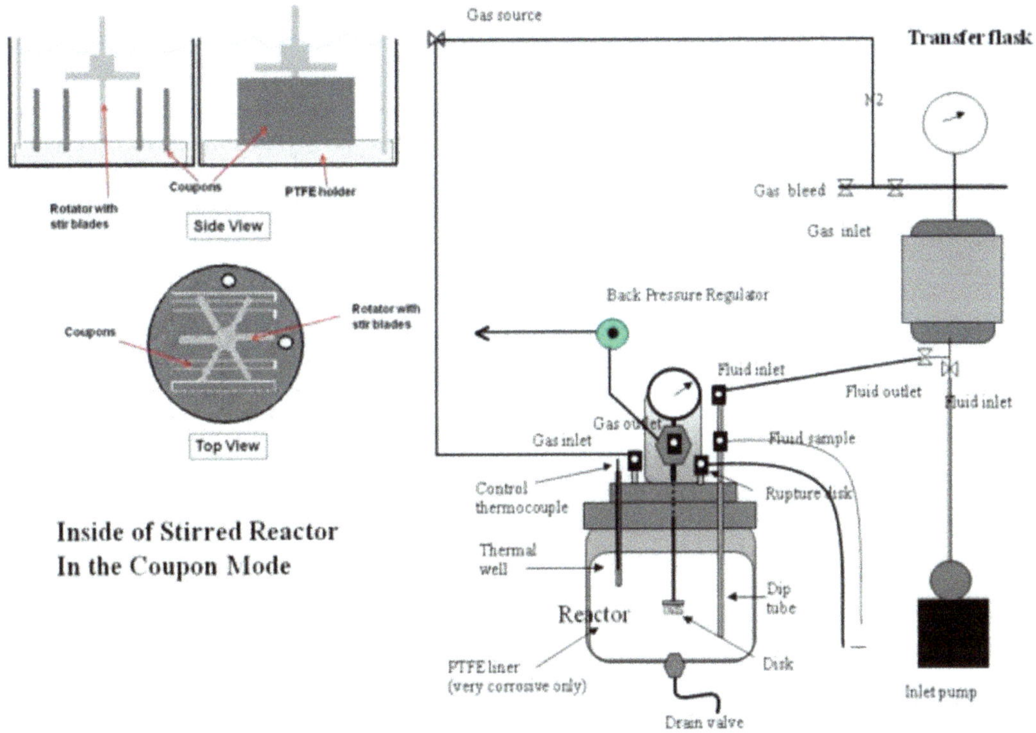

Figure 6.27 Setup of Parr Type Rotational Cell

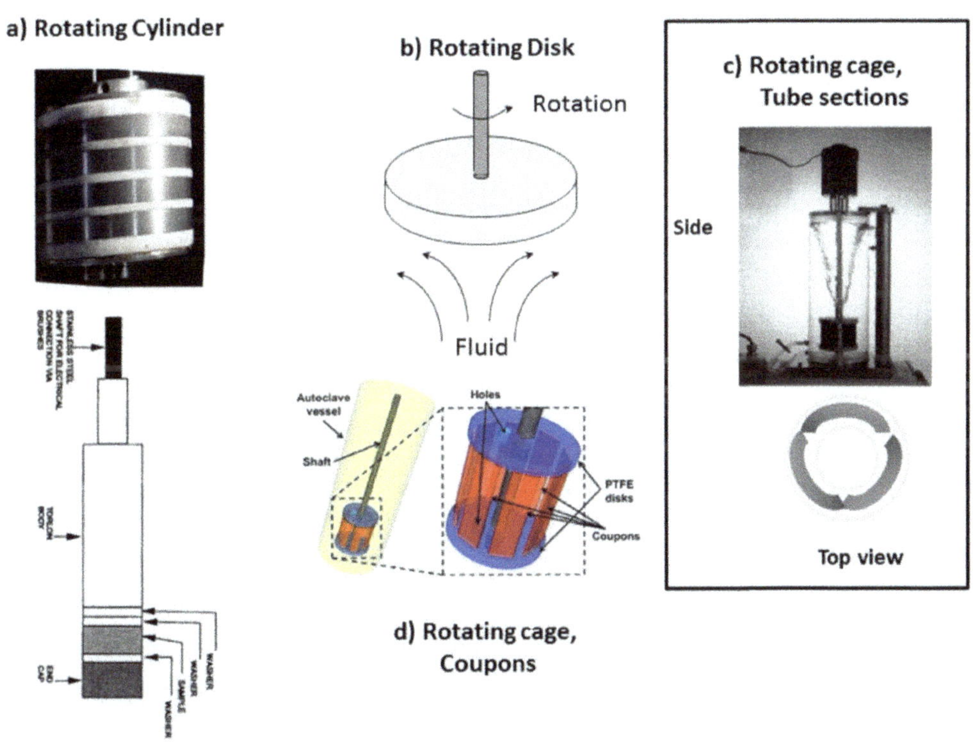

Figure 6.28 Rotating Elements for Rotational Tests

Note that the rotating elements can be applied in many different test scenarios.

6.6.4. Pit Depth and Morphology Measurements

A wide range of methods and devices are used to measure the pit depth and area and to determine the number of pits/crevices on a pipe or another surface. They range from hand-held or visual to lab instruments and inline inspection (ILI) devices for pipelines. The general types with references are noted.

Direct measurements. Some labs use standard coupons that were exposed to different corrosive conditions, and the pitting index based on a visual comparison and ranking (0–10) is recorded. Informal specifications (such as in NACE[120]) can be used to rate inhibitors.

A wide range of pit gauges and measurement devices are used to determine the depth and area of a pit. This is a very diverse area; see WI[138] for images and details.

Pit/Crevice Imaging. A wide range of direct optical and microscopic examination of pits can be automated to give a profile of number, depth, and possibly morphology of pits. Ribeiro[139] has described a process for imaging pits using an inverted metallurgical microscope coupled to a color digital camera. Proprietary software produces plots of depth, width, morphology, and number of pits on SS surfaces. The investigators claim that quantitative parameters such as area at pit mouths, pit density, and corroded area have been systematically determined. Additionally, they claim that scanning electron microscopy and energy-dispersive x-ray (EDX) combined techniques have been used to identify elemental composition of grains and grain boundaries. This surface analysis also has been carried out using a scanning electron microscope with a coupled energy dispersive spectrometer. ASTM[140] has a standard for using microscopic examination of pits.

Pessu[141] reports that pit depth measurements were conducted in alignment with ASTM G46-94. Also, an NPFLEX 3D interferometer was used in this study for defining the discrete geometry of pits on almost the entire steel sample surface area. Pits were identified based on carefully chosen thresholds with distinct pit depths, diameters, and areas being quantified.

Several investigators have tested the use of laser devices to scan pits. He[142] describes the use of three- and two-dimensional laser profilometry[143] scans that were performed on two sets of Alloy 22 crevice corrosion specimens to accurately and quantitatively characterize corrosion damage profiles. One set of specimens had multiple crevice contacts, and the other set had single crevice contact. Both sets of specimens showed that laser scans are more sensitive to variations in penetration sites than those previously obtained by optical microscope, allowing for measurement of larger maximum crevice corrosion penetration depths. According to the authors, the laser scan measurements also captured the depth variation around grain boundaries.

6.6.5. Physical and Chemical Examination of Surfaces

Several microscopic and spectroscopic methods are used to understand corrosion processes. Surfaces of corroded metals can be probed using SEM to observe pits and the EDS function to determine elemental compositions. Short reviews of these methods are in Frenier,[144] and more detailed information, including many references, is provided on the U.S. Department of Energy website.[145] The following methods are often used in corrosion studies:

- Metallurgical microscopic analyses using light or electron techniques
- Various x-ray diffraction (XRD) and scattering methods to determine scale compositions or elemental analyses
- Organic compound surface analyses are conducted using Fourier transform infrared (FT-IR) or FT-Raman spectroscopic methods

Several examples of surface analysis of corrosion-related processes in pipeline operations are presented.

Pyshmintsev[146] has investigated the use of light microscopic as well as XRD to probe the microstructure of pipeline steels intended for use in arctic regions. On the basis of these data, metallurgical factors, such as possible ductile fracture, may be assessed and compared with the burst strength of a pipe. Figure 6.29 shows micrographs of the microstructures of different oilfield steels (from Akeer[147]).

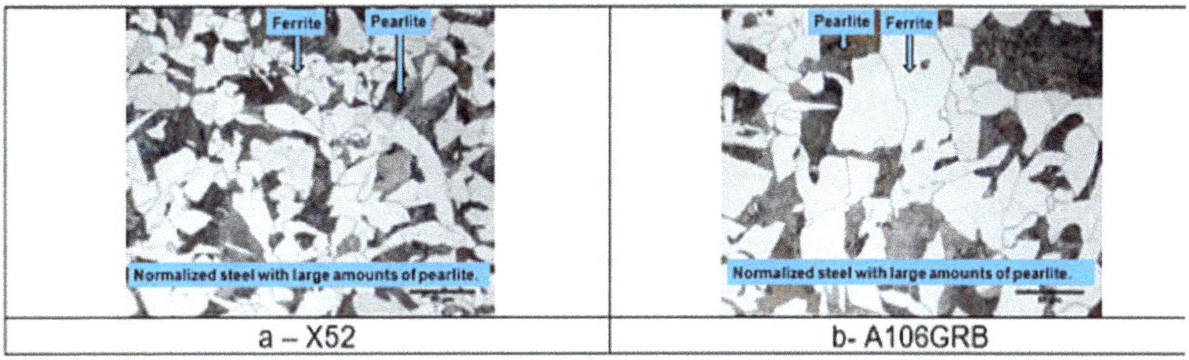

Figure 6.29 Normalized Steels (a) X52 and (b) A106GRB

Umoru[148] studied the metallurgy of several candidate metals to combat various corrosion-related problems in a production system. They used EDX spectroscopy to analyze surface deposits and SEM to study the surface of the corroded stainless steels to determine the form of corrosion. Pitting was found, and an austenitic SS was found to be more suitable than a ferritic steel alloy.

Sun[149] reported that understanding and controlling hydrogen sulfide (H_2S) corrosion becomes increasingly important in the refining industry. Iron sulfide is formed on the internal pipeline surface as a corrosion product of carbon steel in an environment containing H_2S. Some of the iron sulfide particles are suspended in the liquid phase or deposited at the bottom of a pipeline. Iron sulfide decreases the corrosion inhibition efficiency owing to the adsorption of corrosion inhibitor on the surface of iron sulfide particles. The objectives of this study were to investigate the effect of iron sulfide on the efficiency of a corrosion inhibitor in a highly sour environment and to explore effective technical methodologies to study H_2S corrosion and inhibition. Experiments were performed in a series of autoclaves. Iron sulfide effects were studied separately in two different cases: directly deposited on the steel surface or suspended in the test solution. In the experiments, weight loss was applied to study the corrosion inhibition efficiency. The corrosion product films/scales formed on the surface of the steel were analyzed using SEM and EDX.

Frenier[79] used light microscopic studies of dissected tube sections to observe corrosion attack on welds. Metallurgical microscopic analyses were conducted by Ajmal et al.[150] They also observed pitting on several parts of the exposed portions of a ruptured pipeline. Mechanical testing was carried out to check the quality of the pipeline material. Optical and scanning electron microscopic studies were also carried out. The metallographic microscopic analyses revealed inter-granular fracture. The soil analysis indicated a high pH (i.e., about 9–10). Additional stresses at the rupture area of the pipeline due to lowering/submerging of the pipe line appear to have contributed to this rupture. Figure 6.30 shows micrographs of two different corrosion manifestations.

Growcock[61] used FT-IR analyses of surface films to determine the compositions of corrosion inhibitor films formed from inhibited HCl solutions. Griffin[151] also used FT-IR and other surface analytical methods to examine the structure of scale inhibitors and measured the chemical effect of temperature on scale inhibitors via molecular weight determination, thermogravimetric analysis, pH change, and FT-IR analysis. The authors measured the performance of these inhibitors under static and dynamic conditions for inhibiting barium sulfate scale. These results help to further the knowledge of inhibitor degradation due to thermal effects and indicate the direction for further product development of thermally stable scale inhibitors.

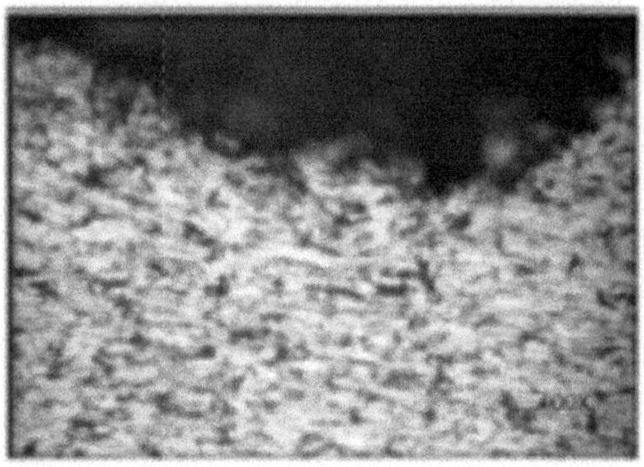

Micrograph of a Pit at 400x Magnification

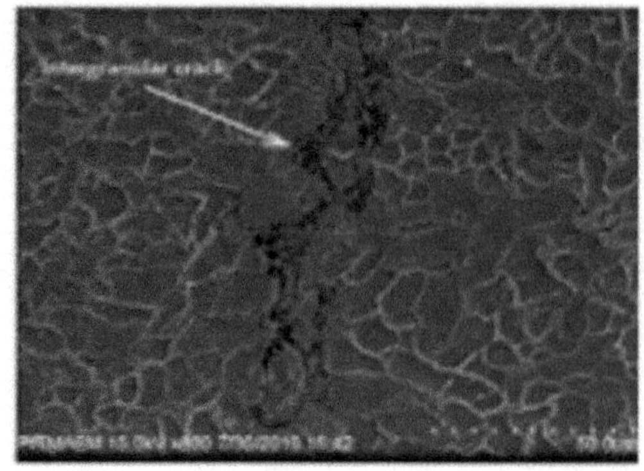

SEM Image of Intergranular Crack

Figure 6.30 Micrographs of Corrosion Damage[150]

6.6.6. Field Monitoring of Corrosion Processes

Field monitoring/testing can include corrosion monitoring as well as chemical testing of gases and liquids. Some of the analyses can be accomplished on site, but many tests require that a sample (or coupon) be returned to a lab for processing. Thus, the sampling process (especially of fluids) is critical because a representative sample must be obtained and preserved.

This is an inherently dangerous process because fluids are under pressure and are often very hot. A generalized set of instructions are in Petrowiki,[152] and local safety rules must be obeyed. The details also are dependent on the information needed:

- Is the sampling location pertinent to the information needed?
- Is a suitable container available to preserve dissolved gases (if needed) or other components?
- Will the sample require onsite treatment to preserve the property/chemical to be analyzed?
- Is there a shelf-life for the analyte?
- Is the lab capable of doing the testes required?

Solid samples for analysis can be obtained from scraping opened vessels and often from the debris after a pigging run (i.e., "pig trash"; see Frenier[124]). Kilbane[153] claims that microbiologically induced corrosion samples can be obtained from these solids, but the microbiological samples may be heterogeneously dispersed in the "trash," and it should be sampled in several places. Taking samples from multiple pipeline locations also improves the overall analysis of the conditions.

Field Corrosion Monitoring

Various corrosion test methods must be adapted for field use in flowing pipelines and surface facilities for the basic tasks of integrity and flow assurance and are mandated by government and company directives.

NACE[121,119] and Wold[154] have reviewed field monitoring methods including intrusive and non-intrusive procedures.

Intrusive methods include those that must penetrate the walls of the pipe or well tubing:

- Coupons (Section 6.6.3, Figure 6.18, and Figure 6.31)
- ER probes (Figure 6.19)
- Electrochemical probes (Figure 6.24)

Non-intrusive methods include

- Methods based on exterior ultrasonic techniques[124]
- Interior methods based on ILI tools[124]

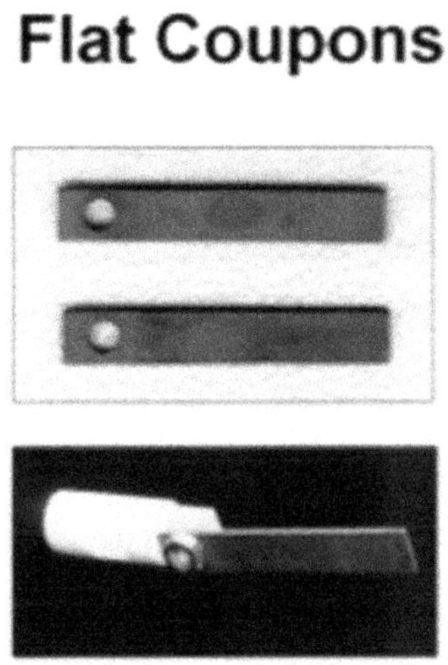

Figure 6.31 Corrosion Coupons for Field Use

The intrusive methods must have an access port drilled into the pipe/vessel wall at a specific point that is suspected to be susceptible to corrosion. The various probes or coupon sets (Figure 6.31) must be inserted through the port, connected to monitoring electronics (for electronic methods), and removed at specific times for inspection (coupons and the electronic probes). Thus, physical access to the port is required. Figure 6.32 shows two types of ports (hydraulic and mechanical) with probes. Note that probes monitor the corrosivity of the fluid passing over it, *not* the actual condition of the pipe surfaces. These conditions may be monitored by inline inspection tools.

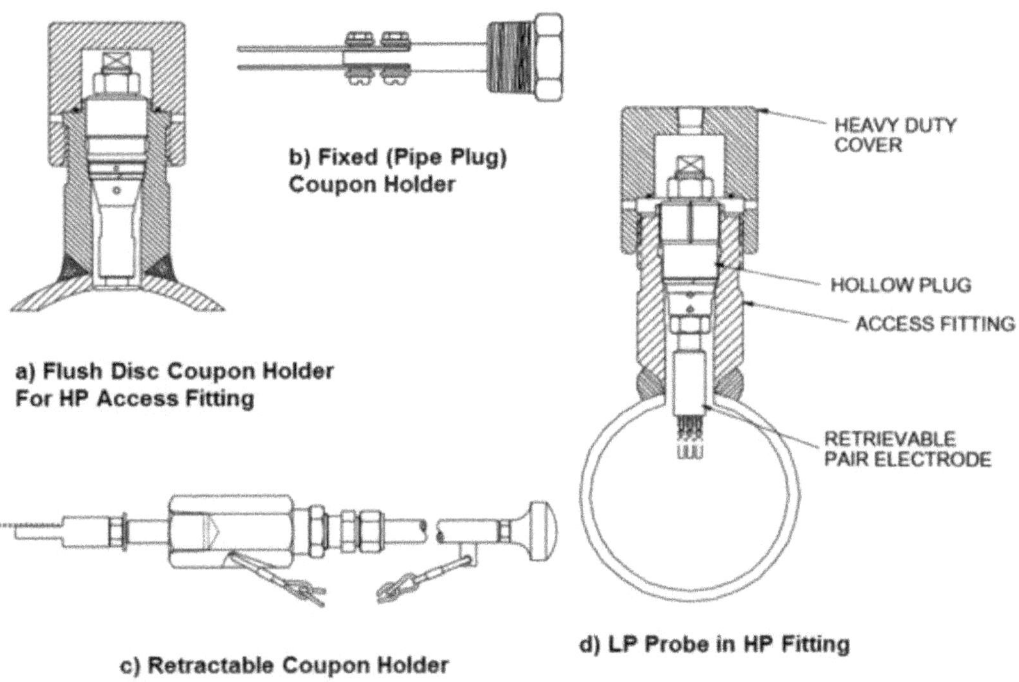

Figure 6.32 Corrosion Probe Mounts

The diagrams in Figure 6.32 show flush-mounted sensing elements that provide less interference with fluid flow than probes mounted deeper into the flow path.

NACE[119] notes that the coupons or LPR probes must be placed in the water phase to determine the corrosivity of the liquid mixture. For complete monitoring of the piping or vessel diameter, a tree of coupons can be used. See Figure 6.33.

Note that data can be collected only when coupons are removed. See NACE[122] for details of coupon preparation and post treatment. Retrieved coupons also can be examined to determine the composition of scale or for microscopic tests of pits and cracks.

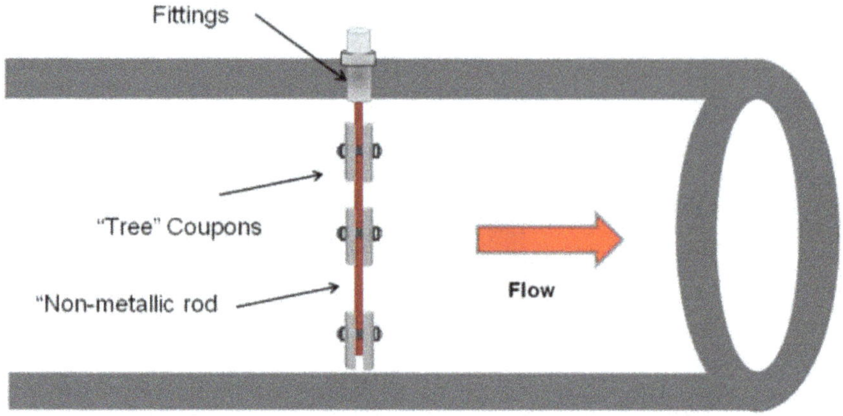

Figure 6.33 Coupons in a Line on a "Tree-like" Holder

Instruments are selected to perform the necessary measurement(s) such as LPR, ECN, EIS, and galvanic current. However, additional consideration should be given to how the data are collected and managed. In some cases, it is sufficient for an operator to take periodic readings manually, whereas automated data generation, control, and storage devices are also used.

6.7. Selection of Corrosion Inhibitors for Industrial Cleaning Fluids

Selection of the most applicable corrosion inhibitor is an important process. On the basis of the references in this section as well as in NACE,[120,155] several selection criteria are listed below:

- The chemistry of the formulations must be able to provide an industry-accepted level of protection to the metals in the system based on an appropriate test (see Section 6.6). Although there are no universally accepted standards, a rate of < 0.01lb/ft^2/day is cited in some documents. See Table 6.2 and Table 6.4.
- A test formulation must contain all of the additives required for a particular job. The additives must include the corrosion inhibitors, surfactants, other acids (such as HF), solvents, and any other component needed for a particular job.
- The formulation (the mixture of active ingredients and additives) must be compatible with the specific cleaning fluid.
- The formulation must leave the metal surfaces clean without an obvious organic film.
- The formulation must be tolerant to solids and penetrate (at least to some degree) a scale layer.
- The formulation must be effective at the temperatures of the unit for the time of the treatment.

- Environmental characteristics must be compatible with the regulations in the use locations (Sections 6.5.3 and 10.3).
- The formulation must be able to be stored for long times in the application areas (very hot and/or very cold) and must be pourable/pumpable. Formulations may be customized for very cold storage conditions and must flow from the tanks under arctic conditions.[156]
- The inhibitor material must be available in the use area and be cost effective compared with other corrosion control methods.
- The formulation must be non-corrosive to the storage tanks, seals, and elastomers.
- The formulation should not stabilize emulsions or foams.
- A schedule (tables or a computer program) must be assembled to select the proper corrosion inhibitor type and concentration for all of the usage solvents in a company's inventory. For example, to find the most effective combination, developers can run a series of tests, and a computer calculation[157] can produce a response surface that may enable optimization of the treating mixture. An example from Jasinski[158] for only two components is shown in this paper, but more complex models[157] have also been used to optimize mixtures.

6.8. Summary and Best Practices

6.8.1. Summary

This chapter describes corrosion inhibitors for use in HCl, organic acids, chelating agents, and nuclear cleaning formulations. The types of chemicals discussed include

- Amines
- Unsaturated and aromatic nitrogen compounds (especially quaternary salts)
- Unsaturated oxygen compounds
- Sulfur compounds
- Mixtures of these chemicals

The formulation must be appropriate for the base acid (HCl and organic acid corrosion inhibitors have different compositions).

Testing methods include

- Coupons
- Electrical-electrochemical
- Pitting
- Surface analyses

6.8.2. Best Practices

Frenier and Loescher[79] discuss the tests needed to evaluate various boiler metals and different inhibitors in cleaning fluids.

Hudson[159] gives recommendations for the use of coupons, LPR probes, and ZRA devices to determine galvanic corrosion. They also note that acceptable chemical cleaning allowances are based on the corrosion rates measured, and this will define the allowed duration of the treatment.

Perdomo et al.[160] describe work to determine optimal cleaning practices for Kraft digesters, evaporators, and other equipment in paper mills during scheduled shutdowns to minimize cleaning costs, maximize effectiveness during cleaning, and ultimately to minimize corrosion (metal loss) during scale removal. This article presents results from a laboratory study of the variables to develop cleaning solvents and describes tests to find the best corrosion inhibitor for each solvent. They evaluated the inhibition of sulfamic acid as well as HCl solutions.

References

[1]O.L.J. Riggs, "Theoretical Aspects of Corrosion Inhibitors and Inhibition" in *Corrosion Inhibitors*, ed. C.C. Nathan (Houston, TX: NACE, 1973).

[2]B.D. Oakes, "Corrosion Inhibitors" (Washington, DC: The Edison Electric Institute, 1972).

[3]"Corrosion Inhibitors" (London, U.K.: The Institute of Metals, 1994).

[4]NACE, *Bibliography on Chemical Cleaning of Metals, vol. 2: 1970-1982* (Houston, TX: NACE International, 1985).

[5]S.T. Powell, Report CS-3289, "Manual on Chemical Cleaning of Fossil-Fueled Steam Generation Equipment. Final Report" (Palo Alto, CA: Electric Power Research Institute, 1984).

[6]J.S. Robinson, *Corrosion Inhibitors: Recent Developments* (Park Ridge, NJ: Noyes Data Corp, 1979).

[7]C.C. Nathan, *Corrosion Inhibitors* (Houston, TX: NACE International, 1973).

[8]A. Raman, P. Labine, *Reviews of Corrosion Inhibitor Science and Technology* (Houston, TX: NACE International, 1993).

[9]A. Raman, P. Labine, *Reviews of Corrosion Inhibitor Science and Technology, vol. 2* (Houston, TX: NACE International, 1996).

[10]I.L. Rozenfeld, *Corrosion Inhibitors* (New York, NY: McGraw-Hill, 1981).

[11]W.W. Frenier, F.B. Growcock, "Corrosion Inhibitors for Chemical Cleaning Solvents: A Review of Recent Literature," CORROSION/89, paper no. NACE 440 (Houston, TX: NACE, 1989).

[12]W.W. Frenier, *Chemical Methods for Controlling Corrosion in Oil and Gas Activities* (Houston, TX: NACE International, 2017).

[13]B. Olbertz. "Acid Inhibitors. II. Influence of Quaternary Ammonium Salts on the Hydrogen Absorption of Mild Steel in H_2S-Free and H_2S-Saturated Hydrochloric Acid" (Rheinisch-Westfalischen Technischen Hochschule, 1980).

[14]B. Sanyal, "Organic Compounds as Corrosion Inhibitors in Different Environments a Review," *Prog. in Org. Coat.* 9, 165 (1981).

[15]G. Schmitt, "Application of Inhibitors for Acid Media," *Br Corros J.* 19 (1984): 165.

[16]W.W. Frenier, "Review of Green Chemistry Corrosion Inhibitors for Aqueous Systems," 9th European Symposium on Corrosion Inhibitors (Ferrara, Italy: University of Ferrara, 2000).

[17]G.W. Bradley, J.A. Smith, "Removal of Iron Oxide Deposits without Ferric Ion Corrosion," *MP* 12 (1973): 48-52.

[18]M. Alfandry et al., Ferrara Univ. Annali, Sez. 5 (1971): 137-146.

[19]G. Poff, "Effect of Ammonium Bifluoride on the Corrosion Rates of Some Common Chemical Cleaning Solvents," *MP* 17, 10 (1978): 24-26.

[20]N. Subramanyan, S.K. Rangarajan, K. Balakrishnan, S.V. Iyer, S. Venkatesan, B. Sathianandhan, "Thiophenol as an Inhibitor of Corrosion," 3rd European Symposium Corrosion Inhibitors, Ferrara, Italy, 1970.

[21]E. McCafferty, N. Hackerman, "Double Layer Capacitance of Iron and Corrosion Inhibition with Polymethylene Diamines," *Electrochem. Soc.* 119 (1972): 146.

[22]M. Alfandry et al., *Ferrara Univ. Annali* 5 (1971): 137-146.

[23]G.N. Dhirendra, B. Sanyal, "Hexamine as an Inhibitor in HCl," *Kanpr. Sympos.* (1979):215.

[24]F.B. Growcock, W.W. Frenier. "Corrosion of Mild Steel and J55 Oil-Field Steel in Uninhibited and Inhibited HCl," CORROSION/84, paper no. 121 (Houston, TX: NACE, 1984).

[25]L.A. Nanova, N.G. Pulieva, S. Dimitrova, "Corrosion Inhibiting Composition," *Khim. Ind., Sofia* 47, 108 (1975).

[26]R.J. Meakins, "Inhibition of the Corrosion of Steel by Long-Chain N-Alkylisoquinolinium Compounds," *Br. Corros. J.* 8 (1973): 230.

[27]G. Schmitt, B. Olbertz, "Acid Inhibitors. II. Influence of Quaternary Ammonium Salts on the Hydrogen Absorption of Mild Steel in H_2S-Free and H_2S-Saturated Hydrochloric Acid," 5th European Symposium Corrosion Inhibitors, Ferrara, Italy, Sept. 15-19, 1980.

[28]G. Schmitt, K. Bedbur, "Inhibition of Corrosion and Hydrogen Embrittlement" *Werkst. Koros.* 36 (1985): 273-280.

[29]J. Marek, P. Stodulka, J. Cabal, O. Soukup, M. Pohanka, J. Korabecny, K. Musilek, K. Kuca, "Preparation of the Pyridinium Salts Differing in the Length of the N-Alkyl Substituent," *Molecules* 15 (2010): 1967-1972.
[30]W.W. Frenier, "Low Hazard Corrosion Inhibitors and Cleaning Solutions Using Quaternary Ammonium Salts," U.S. Patent 6521028, 2003.
[31]W.W. Frenier, "Introduction to Toxicity Testing," CORROSION/96, paper no. NACE 96154 (Houston, TX: NACE, 1996).
[32]R.J. Tedeschi, "Acetylenic Corrosion Inhibitors," *Corrosion* 31, 4 (1975): 130-135.
[33]F.B. Growcock, V.R. Lopp, R.J. Jasinski, "Corrosion Protection of Oilfield Steel with 1-Phenyl-2-Propyn-1-ol," *J. Electrochem. Soc.* 135 (1988): 823.
[34]F.B. Growcock, V.R. Lopp, "The Inhibition of Steel Corrosion in Hydrochloric Acid with 3-Phenyl-2-Propyn-1-ol," *Corros. Sci.* 28, 4 (1988): 397-410.
[35]K.D. Allabergenov, F.K. Kurbanov, A.B. Kuchkarov, S.A. Sarankina, *Dokl. Akad. Nauk. Uzb. SSR* 11 (1976): 41.
[36]S.F. Karaev, Z.M. Tsalikova, N.I. Podobaev, I.A. Shikhiev, *Korroz. Zashch. Neftegazov* (1974): 3.
[37]C.W.N. Cumper, R. Grzeskowiak, P. Newton, "Effect of Pyrrole, Indole and Their Dimethyl Derivatives on the Dissolution of Magnetite," *Corros. Sci.* 22 (1982): 551.
[38]G. Devarajan, K. Balakrishnan, "Aldimines: Effective Corrosion Inhibitors for Mild Steel in Hydrochloric Acid Solution," *Trans. SAEST* 19 (1984): 71.
[39]F.B. Growcock, W.W. Frenier, "Kinetics of Steel Corrosion in Hydrochloric Acid Inhibited with Trans-Cinnamaldehyde," *J. Electrochem. Soc.* 135 (1988): 817.
[40]W.W. Frenier, F.B. Growcock, V.R. Lopp, "Alkenylphenones: A New Class of Acid Corrosion Inhibitors," *Corrosion* 44, 9 (1988): 590.
[41]W.W. Frenier, F.B. Growcock, "Cinnamaldehyde: A Low Toxicity Acid Corrosion Inhibitor," 7th European Symposium on Corrosion Inhibitors, Ferrara, Italy, Sept. 17-21, 1990.
[42]W.W. Frenier, F.B. Growcock, V.R. Lopp, R.E. Dixon, "Process and Composition for Inhibiting Iron and Steel Corrosion," U.S. Patent 5013483, 1991.
[43]G.A. Scherubel, R. Reid, K.S. Fauke, K. Schwartz, "Environmentally Improved Acid Corrosion Inhibitor," U.S. Patent 5854180, 1998.
[44]C. Sitz, W. Frenier, C. Vallejo, "Acid Corrosion Inhibitors with Improved Environmental Profiles," Paper SPE 155966, SPE International Conference & Exhibition on Oilfield Corrosion, Aberdeen, U.K., May 28-29, 2012.
[45]F.B. Growcock, "Surfactants Can Affect Corrosion Inhibition of Oil-Field Steel," Paper SPE-16265-MS, presented at the SPE International Symposium on Oilfield Chemistry, San Antonio, TX, Feb. 4-6, 1987.
[46]M.L. Walker, "Corrosion Inhibitor for Acid Stimulation Systems," U.S. Application 2011/0065614 A1, 2011.
[47]A.J. Saukaitis, G.S. Gardner, "Derivatives of Rosin Amines," U.S. Patent 2758970, 1956.
[48]R.C. Mansfield, J.G. Morrison, C.J. Schmidle, "Corrosion Inhibiting Compositions," U.S. Patent 2874119, 1959.
[49]R.F. Monroe, C.H. Kucera, B.D. Oakes, N.G. Johnston, "Composition for Inhibiting Corrosion," U.S. Patent 3077454, 1963.
[50]W.W. Frenier, D.G. Hill, R.J. Jasinski, "Corrosion Inhibitors for Acid Jobs," *Oilfield Review* 1, 2 (1989): 15-21.
[51]D.R. McCormick, A. Lindert, J.R. Pierce, "Aqueous Compositions Containing Polyphenol Copolymers and Processes for Their Preparation," U.S. Patent 5891952, 1999.
[52]A. Lindert, W.G. Johnston, "New Low Toxicity Corrosion Inhibitors for Industrial Cleaning Operations," CORROSION/99, paper no. NACE-99107 (Houston, TX: NACE, 1999).
[53]Y. Kumar, B. Sanyal, G.N. Pandey, "Corrosion Rates Are Obtained in the Presence of Thiourea," *Metal Finish.* 81, 4 (1983): 45-48.
[54]H.D. Clark, "Corrosive Inhibitor Compositions," U.S. Patent 3969260, 1976.
[55]J.D. Nichols, R. Derby, G.T. von dem Bussche, D.A. Hannum, "Inhibiting Acid Corrosion of Metals," U.S. Patent 4557838, 1985.
[56]M. Walker, "Method and Composition for Acidizing Subterranean Formations," U.S. Patent 4498997, 1985.
[57]B.B. Damaskin, O.A. Petrii, V.V. Batrakov, *Adsorption of Organic Compounds on Electrodes* (New York, NY: Plenum Press, 1971).
[58]P.N.G. Shankar, K.I. Vasu, *J. Electrochem. Soc., India* 32 (1981): 47.

[59]A. Frignani, F. Zucchi, C. Monticelli, "Inhibition of Iron Corrosion in Different Acid Media by N-Decyl-Pyridinium Derivatives," *Br. Corros. J.* 18 (1983): 19.
[60]Z.A. Foroulis, "Inhibiting and Accelerating Effects of Aminophenols on the Corrosion and Permeation of Hydrogen through Mild Steel in Acidic Solutions," 4th European Symposium Corrosion Inhibitors. Ferrara, Italy, 1975.
[61]F.B. Growcock, W.W. Frenier, V.R. Lopp, "Corrosion Inhibition of Steel in HCl by 1-Octyn-3-Ol. I Kinetics," 6th European Symposium Corrosion Inhibitors, Ferrara, Italy, Sept. 16-20, 1985.
[62]R.S. Pathania, G.W. Poling. "Inhibition of Iron Corrosion in 1 M HCl. Part II. Study of the Protective Properties of 2-Aminothiophenol and 2-Aminophenyl Disulfide by Impedance Spectroscopy," 5th International Congress on Metallic Corrosion, Houston, TX, May 1972.
[63]V. Jovancicevic, B. Yang, J.O.M. Bockris, "Adsorption, Orientation, and Polymerization of 1-Octyne-3-Ol on Iron," *J. Electrochem. Soc.* 135 (1988): 94.
[64]I. Epelboin, P.H. Morel, H. Takenouti, "Use of Impedance Measurements for the Determination of the Instant Rate of Metal Corrosion," 3rd European Symposium Corrosion Inhibitors, Ferrara, Italy, 1970.
[65]R.H. Hausler, *Corrosion Chemistry* (Washington, DC: American Chemical Society, 1979).
[66]A. Chakrabarti, "Quantum-Chemical Study of the Corrosion Inhibition of Mild Steel in 6 Percent HCl," *Br. Corros. J.* 19 (1984): 124.
[67]Z. Szklarska-Smialowska, G. Wieczorek, "Adsorption Isotherms on Mild Steel in H2so4 Solutions for Primary Aliphatic Compounds Differing in Length of the Chain," *Corros. Sci.* 11 (1971): 843.
[68]K. Aramaki, *Boshoku Gijutsu* 46 (1978): 86.
[69]P.J. Petit, P.F. McTigue, "Chemical Cleaning of Pressurized-Water Reactor Steam Generators to Prevent Denting," CORROSION/78, paper no. NACE 78208 (Houston, TX: NACE, 1978).
[70]W.C. Kennedy, "Corrosion Inhibitors for Cleaning Solutions," U.S. Patent 4637899, 1987.
[71]W.W. Frenier, E.F. Hoy, "HEDTA: A Versatile Chemical for Cleaning Steel Surfaces," *MP* 25, 4 (1986): 18-24.
[72]S. Brunet, G. Turluer, "Ferrous Metals," 5th European Symposium Corrosion Inhibitors, Ferrara, Italy, 513-538, 1980.
[73]R.H. Hausler, NP-3030, "Developed the Only Qualified Corrosion Inhibitor for Nuclear Steam Generator Cleaning" (Palo Alto, CA: Electric Power Research Institute, 1983).
[74]Z.A. Jofa, "Adsorption of Inhibitors of the Acid Corrosion of Iron & the Mechanism of Their Action," 2nd European Symposium on Corrosion Inhibitors, Ferrara, Italy, Sept. 13-17, 1965.
[75]W.W. Frenier, W.J. Settineri, "Sulfonium Compound as a Corrosion Inhibitor in Aqueous Acid," U.S. Patent 3764543, 1973.
[76]W.W. Frenier, W.J. Settineri, "A Mechanistic Study of Sulfonium Salts Acting as Ferric Ion Corrosion Inhibitors," *Corrosion* 36, 7 (1980): 323-327.
[77]G. Duffin, "The Quaternization of Heterocyclic Compounds," in *Advances in Heterocyclic Chemistry, vol. 3*, ed. A.R. Katritzky (New York, NY: Academic Press, 1964).
[78]D.G. Hill, K. Dismuke, W. Shepherd, I. Witt, H. Romijn, W. Frenier, M. Parris. "Development Practices and Achievements for Reducing the Risk of Oilfield Chemicals," Paper SPE 80593, SPE/EPA/DOE Exploration and Production Environmental Conference, San Antonio, TX, Mar. 10-12, 2003.
[79]W.W. Frenier, K.J. Loescher, "A Study of Corrosion of Boiler Metals in Inhibited Chemical Cleaning Solvents," Paper IWC 80-30, 41st International Water Conference, Pittsburgh, PA, Oct. 20-21, 1980.
[80]W.W. Frenier, "Development and Testing of a Low Hazard Corrosion Inhibitor for Use in Organic Acid and Chelant-Based Cleaning Solvents," Paper IWC-97-43, Engineers' Soc. of West. PA, Pittsburgh, PA, Nov. 10-11, 1997.
[81]J.M. Jevec, W.S. Leedy, "Chemical Cleaning Solvent and Process Testing" (Palo Alto, CA: Electric Power Research Institute, 1983).
[82]B. Poulson, R. Robinson, "Assessing the Possibility of Stress Cracking During Acid Cleaning," *Corrosion Science* 20, 6 (1980): 687-706.
[83]M.A. Streicher, "Synergistic Inhibition of Ferric Ion Corrosion During Chemical Cleaning," CORROSION/72, NACE 71100(Houston, TX: NACE, 1972).
[84]M. Shigehiro, K. Ueki, Japanese Patent Application 39073 (1965).

[85]M.B. Lawson, W.P. Banks, L.D. Martin, "Control of Ferric Ion Corrosion During Chemical Cleaning," CORROSION/83, paper no. NACE 83220 (Houston, TX: NACE, 1983).
[86]M.L. Walker, W.G.F. Ford, W.R. Dill, and R.D. Gdanski, "Composition and Method of Stimulating Subterranean Formations," U.S. Patent 4683954, 1987.
[87]D.G. Hill, "Gelled Acid," U.S. Application 2005/0065041, 2005.
[88]C.W. Crowe, "Evaluation of Agents for Preventing Precipitation of Ferric Hydroxide from Spent Treating Acid," *Journal of Petroleum Technology* 37, 4 (1985): 691.
[89]W.W. Frenier, D. Wilson, D. Crump, L. Jones. "Use of Highly Acid-Soluble Chelating Agents in Well Stimulation Services," Paper SPE 63242, SPE Annual Technical Conference and Exhibition, Dallas, TX, Oct. 1-4, 2000.
[90]D.V.S. Gupta, "Green Inhibitors: Where Are We?" CORROSION/2004, paper no. NACE-04406 (Houston, TX: NACE, 2004).
[91]P.J. Anastas, J.C. Warner, *Green Chemistry: Theory and Practice* (New York: Oxford University Press, 1998).
[92]ECI, "Defining Green Engineering: Multidisciplinary Group Set of Principles for Practice," Engineering Conferences International, New York, 2003.
[93]D.V.S. Gupta, "What Is Green?" CORROSION/98, paper no. NACE 98206 (Houston, TX: NACE, 1998).
[94]U.S. Environmental Protection Agency, Green Chemistry Award, 2015, Washington DC, http://www2.epa.gov/greenchemistry
[95]Clean Water Act, Wikipedia, June 30, 2015, https://en.wikipedia.org/wiki/Clean_Water_Act (Aug. 11, 2015).
[96]33 U.S.C 1251 et seq., "Federal Water Pollution Control Act" (1972).
[97]OSHA 3132, Occupational Safety and Health Administration Information Booklet (Washington, DC: Occupational Safety and Health Administration, U.S. Department of Labor, 2000).
[98]European Council, Regulation (EC) No 1907/2006, "Registration, Evaluation, Authorization, and Restriction of Chemicals (REACH)" (Brussels, Belgium: European Parliament and Council, 2006).
[99]J.O. Bockris, J. Ahmed, W.P. Singh, G.H. Lin, Y. Kang, "About a Chemical Computational Approach to the Design of Green Inhibitors," CORROSION/95, paper no. NACE 95033 (Houston, TX: NACE, 1995).
[100]W.P. Singh, J.O. Bockris, "Toxicity Issues of Organic Corrosion Inhibitor: Application of QSAR Model," CORROSION/96, paper no. NACE 96225 (Houston, TX: NACE, 1996).
[101]W.P. Singh, J.O. Bockris, G.H. Lin, Y. Kang, "Designing Green Corrosion Inhibitors Using Chemical Computational Methods," CORROSION/98, paper no. NACE 98208 (Houston, TX: NACE, 1998).
[102]I. Lukovits, et al, "The Way to Less Toxic Inhibitors: A Model of the Inhibition Mechanism of Thiourea Compounds," CORROSION/96, paper no. 148 (Houston, TX: NACE, 1996).
[103]G. Gece, "The Use of Quantum Chemical Methods in Corrosion Inhibitor Studies," *Corrosion Science* 50, 11 (2008): 2981-2992
[104] W.W. Frenier, P.A. Anderozzi, "A Study of Inhibitor Aids Used in High Temperature HCl," in *Surfaces, Inhibition, and Passivation: Electrochemical Society, Proceedings 86-7*, eds. E. McCafferty, R.J. Brodd (Princeton, NJ: Electrochemical Society, 1987).
[105]F.B. Growcock, W.W. Frenier, P.A. Andreozzi, "Inhibition of Steel Corrosion in HCl by Derivatives of Cinnamaldehyde: Part II. Structure–Activity Correlations," *Corrosion* 45, 12 (1989): 1007-1015.
[106]S. Ramachandran, V. Jovanciecvic, "Molecular Modeling of the Inhibition of Mild Steel Co2 Corrosion by Imidazolines," CORROSION/98, paper no. NACE 98017 (Houston, TX: NACE, 1998).
[107]L.M. Rodrıguez-Valdez, W. Villamisar, M. Casales, J.G. Gonzalez-Rodriguez, A. Martınez-Villafane, L. Martinez, D. Glossman-Mitnik, "Computational Simulations of the Molecular Structure and Corrosion Properties of Amidoethyl, Aminoethyl, and Hydroxyethyl Imidazolines Inhibitors," *Corrosion Science* 48, 2006 (2006): 4053-4064.
[108]W.W. Frenier, F.B. Growcock, "Mixtures of Unsaturated Aldehydes and Surface Active Agents Used as Corrosion Inhibitors in Aqueous Fluids," U.S. Patent 4734259, 1988.
[109]Y.K. Choudhary, A. Sabhapondit, A. Kumar, "Application of Chicory as Corrosion Inhibitor for Acidic Environments," *SPE Production & Operations* 28, 3 (2013): 268-276.
[110]G.S. Gardner, A.J. Saukaitis, "Pickling Inhibitors," U.S. Patent 2807585, 1957.
[111]N.I. Sax, *Dangerous Properties of Industrial Materials,* 6th ed. (New York, NY: Van Nostrand -Reinhold, 1984).
[112]A. Lindert, W.G. Johnston, "A New Low Toxicity Corrosion Inhibitor for Industrial Cleaning Operations," CORROSION/99, paper no. 107 (Houston, TX: NACE, 1999).

[113]I. Jiricek, J. Vosta, J. Macak, J. Fikar, "Ecologically Acceptable N-Based Compounds in Corrosion Inhibition in Acidic Media," (Prague, Czech Republic: Inst. of Chemical Technology, 1995): 235-244.
[114]J.D. Anderson, S.K. Seigle, "Corrosion Inhibited Acid Cleaners," U.S. Patent 4780150, 1988.
[115]C.M. Carandang, "Non-Toxic Acid Cleaner Corrosion Inhibitors," U.S. Patent 4851149, 1989.
[116]R.A. Shank, T.R. McCartney, "Comparative Study of Commercially Available Propargyl Alcohol-Free Corrosion Inhibitors for Hydrochloric Acid Systems," CORROSION/2013, paper no. NACE 2013-2760 (Houston, TX: NACE, 2013).
[117]W.W. Frenier, "Strategies for Formulating Low Hazard Corrosion Inhibitors for Use in Chemical Cleaning Solvents," CORROSION/98, paper no. NACE 98214 (Houston, TX: NACE, 1998).
[118]A. Jenkins, "Performance of High-Temperature, Biodegradable Corrosion Inhibitors," CORROSION/2011, NACE 11272 (Houston, TX: NACE, 2011).
[119]NACE 2014-03-03, TG 070, "Field Evaluation of Corrosion Inhibitors Using Electrochemical Techniques" (Houston, TX: NACE International, 2014).
[120]NACE Standard TM0193-2000, "Laboratory Corrosion Testing of Metals in Static Chemical Cleaning Solutions at Temperatures Below 93°C (200°F)" (Houston, TX: NACE International, 2000).
[121]NACE Publication 3T199, "Techniques for Monitoring Corrosion and Related Parameters in Field Applications" (Houston, TX: NACE International, 1999).
[122]NACE Standard RP0497-2004, "Field Corrosion Evaluation Using Metallic Test Specimens" (Houston, TX: NACE International, 2004).
[123]MS, Metal Samples Corrosion Monitoring Systems Alabama Specialty Products, Inc., Munford, AL, http://www.alspi.com/corrosion.htm (Oct. 30, 2012).
[124]W.W. Frenier, *Chemical and Mechanical Methods for Pipeline Integrity* (Richardson, TX: Society of Petroleum Engineers, 2015).
[125]NACE TM0193-2000, "Laboratory Corrosion Testing of Metals in Static Chemical Cleaning Solutions at Temperatures Below 93°C (200°F)" (Houston, TX: NACE International, 2000)..
[126]R. Heidersbach, *Metallurgy and Corrosion Control in Oil and Gas Production* (Hoboken, NJ: John Wiley & Sons, Inc., 2011).
[127]M.G. Fontana, *Corrosion Engineering,* 3rd ed. (Boston, MA: McGraw-Hill, 1986).
[128]M. Stern, A.L. Geary, "Electrochemical Polarization: I. A Theoretical Analysis of the Shape of Polarization Curves," *J. of the Electrochemical Society* 104 (1957): 1.
[129]M. Stern, "A Method for Determining Corrosion Rates from Linear Polarization Data," *Corrosion* 14, 9 (1958): 440-444.
[130]NACE, "Underdeposit Corrosion (UDC) Testing and Mitigation Methods in the Oil and Gas Industry (Report)" (Houston, TX: NACE International, 2013).
[131]R. Jasinski, DL-10237, "Chemistry of Steel Surfaces" (Sugar Land, TX: Schlumberger Technology Corporation, 1987).
[132]A. Kirkpatrick, M. Reid, D. Machray, A. Adelusi, "Understanding the Impact of Field Development and Changing Fluid Conditions on Corrosion Inhibitor Selection and Performance," Paper SPE-169613-MS, SPE International Oilfield Corrosion Conference and Exhibition, Aberdeen, Scotland, May 12-13, 2014.
[133]W.W. Frenier, M. Ziauddin, *Formation, Removal, and Inhibition of Inorganic Scale in the Oilfield Environment* (Richardson, TX: Society of Petroleum Engineers, 2008).
[134]R.L. Nielsen, M.R. Tek, "Evaluation of Scale-up Laws for Two-Phase Flow through Porous Media," *SPEJ* 3, 2 (1963): 164-176.
[135]G. Faclone, C. Teodoriu, K.M. Reinicke, O.O. Bello, "Multiphase-Flow Modeling Based on Experimental Testing: An Overview of Research Facilities Worldwide and the Need for Future Developments," *SPE Project, Facilities and Construction* 3, 3 (2008): 1-10.
[136]H.B. Wang, H. Shi, C. Kang, W.P. Jepson, "Characterization of Inhibitor and Corrosion Product Film Using Electrochemical Impedance Spectroscopy (EIS)," CORROSION/2001, paper no. NACE 01023 (Houston, TX: NACE, 2001).
[137]L. Chen, N. Obeyesekere, J. Wylde, "The Effect of Surfactant Additions on the Mechanism of Sulfur Corrosion Inhibition in Sour Systems," CORROSION/2014, paper no. NACE-2014-3828 (Houston, TX: NACE, 2014).
[138]WI, Pit Gauges (Brochure) (Sturgeon, Alberta CA: Western Instruments, Inc., 2007).

[139]R.B. Ribeiro, G. Silva, A.H.J. Araujo, N.T. Matias, J.W.D.J. Silva, B.S.M. Domingos, "Morphology Characterization of Pitting Corrosion on Sensitized Austenitic Stainless Steel by Digital Image Analysis," in *Developments in Corrosion Protection* (Rijeka, Croatia: Intech, 2014).
[140]ASTM G46–94, "Metals Test Methods and Analytical Procedures. Standard Guide for Examination and Evaluation of Pitting Corrosion," Volume 3.02: Wear and Erosion–Metal Corrosion (West Conshohocken, PA: ASTM International, 2004).
[141]F. Pessu, R. Barker, A. Neville, "Early Stages of Pitting Corrosion of UNS K03014 Carbon Steel in Sour Corrosion Environments: The Influence of Co2, H2s and Temperature," CORROSION/2015, paper NACE 5583 (Houston, TX: NACE, 2015).
[142]X. He, T. Mintz, "Quantitative Characterization of Alloy 22 Crevice Corrosion Damage Profiles," CORROSION/2010, (Houston, TX: NACE, 2010).
[143]R.D. Roberts, "Laser Profilometry as a Inspection Method for Reformer Catalyst Tubes," *NDT.net* 4, 2 http://www.ndt.net/article/v04n02/roberts4/roberts4.htm (1999).
[144]W.W. Frenier, M. Ziauddin, R. Venkatesan, *Organic Deposits in Oil and Gas Production* (Richardson, TX: Society of Petroleum Engineers, 2010).
[145]"Tools for Surface Analysis," U.S. Department of Energy, Washington, DC, http://www.emsl.pnl.gov/capabilities/spectroscopy/surface_analysis/ (June 29, 2013).
[146]I. Pyshmintsev, A. Gervasyev, A. Struin, T. Yesiev, A. Arabey, "X80 Pipelines in Arctic Environment: Prediction of the Long-Distance Ductile Fracture Propagation/Arrest," *Pipeline Technology Journal,* 2013, 29-37, http://www.pipeline-journal.com/pdf/ptj-1-2013.pdf, (2013).
[147]E. Akeer, B. Brown, S. Nesic, "The Influence of Mild Steel Metallurgy on the Initiation of Localized Co2 Corrosion in Flowing Conditions," CORROSION/2013, paper no. NACE 2013-2383 (Houston, TX: NACE, 2013).
[148]L.E. Umoru, A.A. Afonja, B. Ademodi, "Corrosion Study of AISI 304, AISI 321, and AISI430 Stainless Steels in a Tar Sand Digester," *Journal of Minerals & Materials Characterization & Engineering* 7, 4 (2008): 291-299.
[149]H. Sun, H. Fang, J. Davis, R. Hudgins, "Investigation of Effects of Iron Sulfide on Corrosion and Inhibition of Carbon Steel in H_2s Containing Conditions," CORROSION/2012, paper no. NACE 2012-1651 (Houston, TX: NACE, 2012).
[150]M. Ajmal, B. Saleem, A.R. Saleemi, "Stress Corrosion Cracking of 18-Inch (45.72cm) Diameter Pipeline (X52)," CORROSION/2012, paper no. NACE 2012-1130 (Houston, TX: NACE, 2012).
[151]J.R. Griffin, J.R. Johnstone, T.E. Cotter, A.E. O'Brien, "Factors Affecting the Stability of Scale Inhibitors Used for Capillary Injection in High Temperature Wells," Paper SPE 164128, SPE International Symposium on Oilfield Chemistry, The Woodlands, TX, April 8–10, 2013.
[152]Petrowiki, "Fluid Sampling," Society of Petroleum Engineers, Richardson, TX, Aug. 16, 2016, http://petrowiki.org/Fluid_sampling (Sept. 4, 2013).
[153]J.J. Kilbane, "Monitoring Pipelines for Microbiologically Influenced Corrosion," NACE International 2014 Central Area Conference, Tulsa, OK, Aug. 24-27, 2014.
[154]K. Wold, R. Stoen, "Monitoring Internal Corrosion in Pipelines," ASME India Oil and Gas Pipeline Conference, Goa, India, Feb. 4-5, 2011.
[155]NACE TG076, "Prediction of Environmental Aggressiveness in Oilfield Systems from System Conditions" (Houston, TX: NACE International, 2013).
[156]A. Jenkins. "The Challenges Associated with the Development and Application of Oil and Gas Corrosion Inhibitors," Paper SPE-169615-MS, SPE International Oilfield Corrosion Conference and Exhibition, Aberdeen, Scotland, May 12-13, 2014.
[157]A.I. Khuri, S. Mukhopadhyay, "Response Surface Methodology," *WIREs Comp Stat.* 2 (2010): 128-149.
[158]R. Jasinski, W. Frenier, S. Grannan, "Inhibiting HCl Corrosion of High Chrome Tubular Steels," CORROSION/88, Paper NACE 88188 (Houston, TX: NACE, 1988).
[159]W.H. Hudson, J.B. Delrue, S.W. Evans, J.F. Remark, "Designing a Chemical Cleaning System: A Conceptional Design," CORROSION/2006, paper no. NACE 06534 (Houston, TX: NACE, 2006).
[160]J. Perdomo, P.M. Singh, P. Conde, J. Mahmood, "Corrosion Prevention During Acid Cleaning in Digesters and Evaporators," CORROSION/2004, paper no. NACE-04246 (Houston, TX: NACE, 2004).

7 Formation and Removal of Organic and Mixed Deposits from Chemical Plants and Refineries

7.1. Introduction

The cleaning of refinery, petrochemical, and chemical plant process equipment can be much different than cleaning operations in power generating plants owing to the complex organic deposits and mixed organic/inorganic deposits in some units. However, the water-contacted sides of units may have only the inorganic deposits described in Chapter 3 and can be dissolved using the solvents described in Chapter 4.

Chapter 7 reviews chemical processes concerned with the formation and removal of organic and mixed foulants from industrial equipment that have been used over the past 30 years. This includes new innovations for removing organic solids and the frequently mixed deposits, as well as the dissolution of iron sulfide scales. Also see the report on formation and removal of oil/gas production organic solids by Frenier.[1] Chapter 8 describes the mechanical methods (possibly used with a solvent) for removing less-soluble solids or those that cannot be easily circulated/flushed with a chemical solvent.

7.2. Introduction to Chemical Plant Operations

Refineries are a specialized type of chemical plant in which the primary (but very complex) job is converting crude oil (in all its almost infinite varieties) into useful product groups. A large portion of the product stream are fuels for automobiles, trucks, aircraft, and many other modes of transportation. An additional part of a refinery's product stream provides feed stocks for other operations that produce additional products to industries and to public consumers.

Some chemical plants are directly (or indirectly) associated with the processing of petroleum feed stocks and are frequently called "petrochemical" operations. Major products produced include various resins and plastics.

There are an almost endless variety of other chemical plants that are not directly associated with petroleum feed stocks (though they frequently use natural gas as a fuel or feed stock).

These plants include paper mills, metal-producing entities, and pharmaceuticals and have many similar operations that may necessitate chemical/mechanical treatments to remove fouling deposits. The general types of unit operations that take place in all of these plants include the following:

- Various feed stocks come in by pipeline, rail, ship, or trucks.
- The feed stocks are frequently stored in large tanks.
- They are combined in chemical reactors of all sorts to produce the desired products.
- Heating, cooling, pumping, and transferring products and reactants using piping almost always are a part of the operations.
- Boilers of every variety provide direct heat or steam for processing or electrical power generation.
- Massive amounts of water, air, or other fluids are required to control the temperatures of the chemical reactions.
- Heat exchange and exchangers are usually needed to conserve and transmit energy.
- Piping and the various reactors and heat exchangers are frequently constructed of steel and other metals that may become corroded.

See Figures 1.13-1.16 for examples of some types of process equipment. Table 7.1 shows a list of various types of heat exchange units and their function in the plant's operations.

Figure 7.1 shows some of the units listed in this Table. Usually the water sides will be fouled with mostly inorganic scales, whereas the process sides may have organic deposits containing many different chemicals. Leakages can cause mixed deposits to form on both sides of a heat exchanger unit.

Figure 7.2 shows drawings of two types of towers in chemical plants or refineries. These examples are given to illustrate the vast variety of equipment that may be fouled by organic deposits and/or inorganic scales. The types of scales found will depend on the fluids that contact hot or cold surfaces and the desired and unwanted reaction products of the processes.

Table 7.1. Heat Exchanger Types

Name	Function
Cooler	Cools a process fluid, usually using water
Chiller	Uses a refrigerant to cool a process fluid to a temperature below that obtainable with water.
Condenser	Condenses a vapor or a mixture of vapors either alone or with a noncondensable gas
Exchanger	Performs a double function: heats a cold process fluid by using a hot fluid, which it cools.
Heater	Uses steam to heat either water or a process fluid
Vaporizer	Vaporizes part of a fluid
Evaporator	Evaporates water using steam as a heating medium
Steam generator	Produces steam from water using a process fluid as the heating medium
Reboiler	Provides the heat necessary for distillation when attached to the bottom of a fractionating tower
Waste Heat Boiler	Provides the heat necessary for distillation when attached to the bottom of a fractionating tower

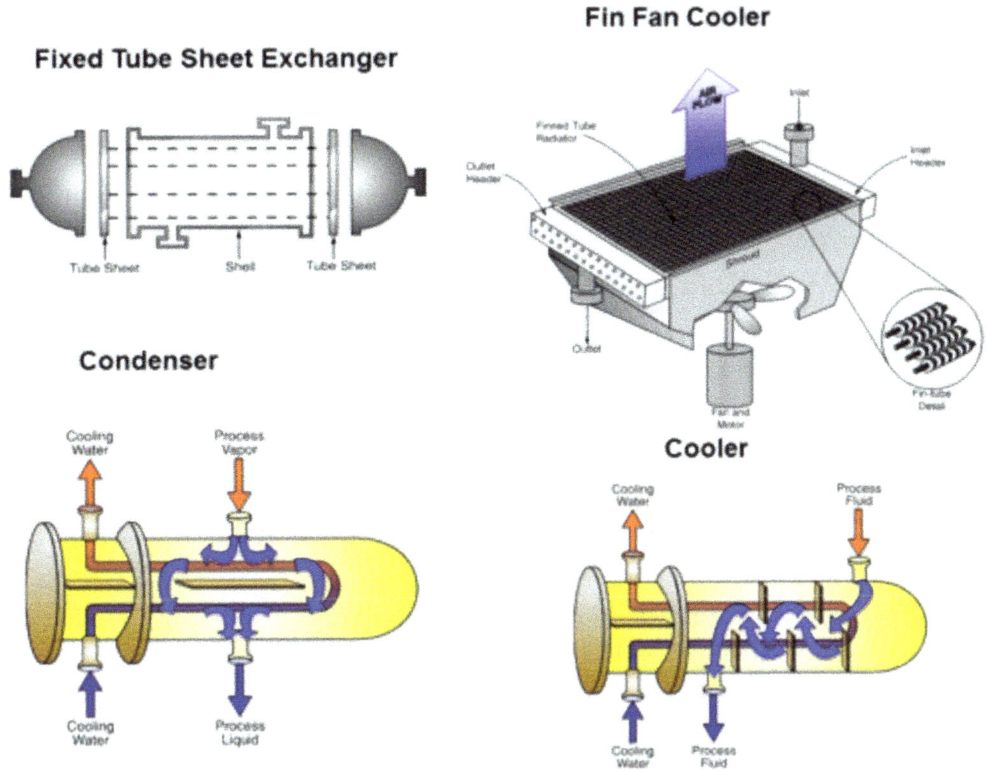

Figure 7.1 Views of Heat Exchange Types of Process Equipment

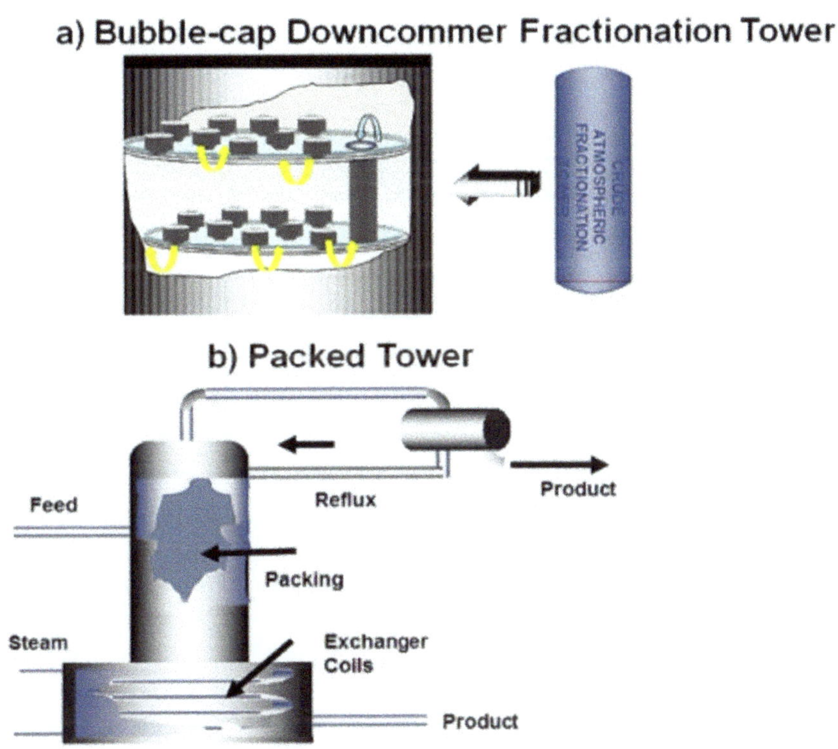

Figure 7.2 Two Types of Towers in Refineries or Chemical Plants

7.3. Introduction to Refinery Operations

This section describes the layout and processes that occur in oil refineries. Note that a refinery also is a model in many aspects for all chemical plants (Section 7.2). Thus, all refineries are chemical plants, but not all chemical plants are refineries; many unit operations and fouling deposits are similar. A detailed description of the typical deposits is in Section 7.4. Note that all of the figures and other information provided can only approximate onsite refinery operations and equipment, which can be much more complex than the illustrations in this book.

Refineries are primarily designed to convert crude oil into its components. The chemical makeup of the crude oil that enters the refinery is a major determinant for the units and chemistries needed and thus the possible deposit that may form on surfaces. During the production and transport processes, a large array of chemicals that were *not* originally in the reservoirs may be added to the fluids. These include corrosion inhibitors, scale inhibitors, dispersants, and solvents. See Frenier[2, 3] and Kelland[4] for more comprehensive lists. Any of these chemicals may affect the quality of the oil that enters a refinery.

7.3.1. Characteristics of Crude Oil

Understanding the crude oil streams that may enter a refinery is the first step in understanding refinery processes, deposits, and cleaning methods. Petroleum crude is a highly complex mixture often containing hundreds of hydrocarbon compounds in any number of combinations, as well as inorganic solids and trapped gases.[5]

Hydrocarbons found in crude oil can range anywhere from 1 to approximately 60 carbon atoms per molecule, with each compound having different properties, such as viscosity, volatility, and boiling point. Figure 7.3 shows examples of specific molecules that fall into these classes of organic compounds. This figure also shows the progression from the least polar compounds (paraffins) through the more polar monoaromatic and polyaromatic asphaltene molecules. Table 7.2 from Shank et al.[6] shows some of the characteristics of the major crude fractions.

Table 7.2 Fractional Distillation Characteristics of Crude Oil Products

Hydrocarbon	Number of Carbon Atoms	Boiling Range
Volatile Gases	<5	<20°C (<70°F)
Petroleum Naphthas	6-12	140-200°C (280-400°F)
Kerosene	11-18	150-300°C (300-575°F)
Diesel Fuel Oil	11-18	200-300°C (390-575°F)
Lubricants, Greases, and Waxes	18-30	300-500°C (575-930°F)
Residual Fuel Oils	20-27	350-450°C (660-940°F)
Bitumen, Asphalt, Tar	>35	>500°C (>930°F)

Figure 7.3 also demonstrates that not all the compounds in crude oil are completely hydrocarbon. Different heteroatoms and especially nitrogen sulfur and oxygen-containing compounds may exist in a particular crude oil. These compounds, such as aromatic nitrogen and sulfides, can greatly affect the corrosivity and solids formed during processing.

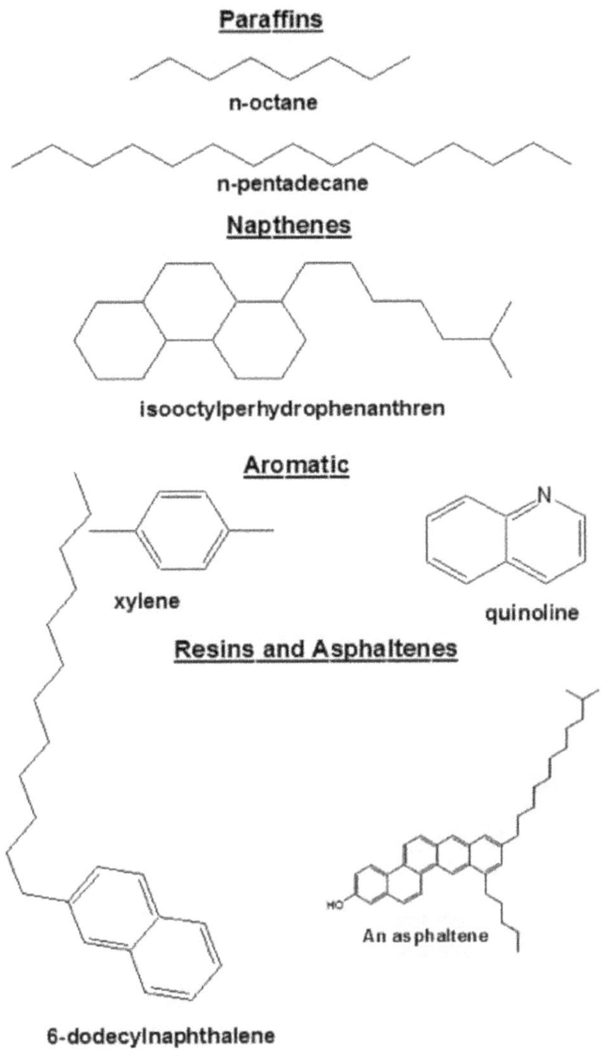

Figure 7.3 Model Chemical Species in Crude Oil

The petroleum industry classifies crude oil using several different systems. These include classification by the location of its origin (e.g., West Texas Intermediate, WTI [a.k.a. Texas Light Sweet] or Brent) and often by its relative weight (API gravity, defined below) and viscosity. Note that there are also pricing benchmarks for crude oils on the basis of other characteristics but are typical from the region of origin. See the definitions in Wikipedia.[7] Refiners may also refer to the fluid as sweet, which means it contains relatively little sulfur, or as sour, which means it contains substantial amounts of hydrogen sulfide or mercaptans and requires more refining in order to meet current product specifications. Low-sulfur crude oil contains less than 0.5% sulfur, and high-sulfur oil may contain more than 2% sulfur.[8]

Each crude oil has unique molecular characteristics that can be determined by various analyses in petroleum laboratories (described in detail in Frenier[1]). The petroleum production industry developed in stages, and different terms have been, and continue to be, used to describe and characterize petroleum crude oil.

The API gravity of petroleum liquids is frequently used as a descriptor and is defined as

$$\text{API gravity} = \left(\frac{141.5}{\text{Sp. Gr. (at 60 }^{\circ}F)}\right) - 131.5 \qquad \text{(Equation [7.1])}$$

A temperature of 60°F (or 15.6°C) is used as the reference temperature for the specific gravity for calculations. Water with a specific gravity of 1.0 has an API gravity of (141.5/1.0) - 131.5 = 10.0 °API. It must be noted that the term "API gravity" is applied only for stock tank oils (STO), that is, oil that is on the surface and not in the reservoir.

Usually, oils with higher API gravity values have a greater commercial value, and lower API gravity value crude oils have lower commercial value. This general rule only applies to up to 45° API gravity; beyond this value, the fraction of useable motor fuel that can be distilled from the liquid diminishes or is too volatile. Crude oil is also classified as light, medium, or heavy, per its measured API gravity.

Light crude oil is defined as having API gravity higher than about 30°API. Medium oil is defined as having API gravity between 20°API and 30°API. Heavy oil is defined as having API gravity below 20°API. Extra heavy oils are defined as having API gravity below 10°API. In 1982, UNITAR[9] (an international working group) defined heavy oil to be gas-free oil between 100 cP and 10,000 cP at original reservoir temperature, with a density between 10°API and 20°API gravity.

At a given pressure, the flow rate of a fluid is inversely proportional to the viscosity, so viscosity of the crude oil is an important value. The viscosity of the live oil is typically correlated with the gas content, API gravity, and temperature. Oil with high API gravity usually has low viscosity values. Oil that will not flow at normal temperatures or without dilution, usually called bitumen, may have viscosities of >10,000 cP, and the API gravity is generally less than 10°API. Bitumen derived from the oil sands deposits in the Alberta, Canada area has an API gravity of around 8°API. It is upgraded (in a limited function refinery) to an API gravity of 31°API to 33°API, and the upgraded oil is known as synthetic oil. These upgraded fluids have viscosity values less than 100 cP.[10, 11]

Simanzhenkov and Idem[8] describe the crude oil fractions as paraffinic, naphthenic (cycloparaffins), aromatic, and asphaltic (Figure 7.3). These authors[8] also state that the very wide range of naphthenes (polycyclic non-aromatic compounds) make up the largest class of compounds in most crude oil streams. Resins are another class of aromatic molecule that was first described by Nellensteyn[12,13] and usually contribute to the asphaltene component.

These molecules have been associated with various theories of asphaltene stability.[14,15] Hammami and Ratulowski[16] also state that resins are thought to be molecular precursors of asphaltenes and are the fraction that is liquid and soluble in the oil phase. They also have been described[17] as organic surfactants. A chemical structure of 6-dodecylnaphthalene that is a model resin molecule[5] is shown in Figure 7.3. Porte et al.[18] contend that asphaltenes and resins are part of a continuum and that there must be a significant overlap between the properties.

The complexity of a refineries' crude stream affects the specific deposits that may form in any of the individual processing units and connecting piping. The various chemical properties described above determine whether a refinery can process a specific crude stream; however, many unit operations are similar.

Zellou[19] noted that the source of the crude oil (that is characteristic of its composition) requires different refineries that are able to accept a particular hydrocarbon blend. The report showed that the very light (high in low carbon number molecules, i.e., high API) must be sent from North Dakota to the east coast, whereas heavier (low API) mixtures are frequently sent to the U.S. gulf coast for refining.

All of this information is provided to help explain this very complex industry and how different crude streams may yield different types of organic fouling deposits and, thus, different treating methods.

7.3.2. Description of Refinery Operations

The major operations of a refinery separate the various fractions, which together make up crude oil, by using pressures from partial vacuums to many hundreds of pounds. In these separations, heat is applied to distill the various fractions. The chemical compositions and structures also may be changed. These reactions include alkylation and cracking processes that maximize the production of gasoline and other products. Advances in refinery operations may change some of the details in the subsections of this treatment.

Notice the layout of typical refinery[6] operations shown in Figure 1.13 and Figure 7.4. See Bland[20] and Shank et al.[6] for more details. This is a simplified view, but it shows the complexity of a refinery. This drawing shows the major unit operations as well as typical products that may be sent to additional industries for final processing and sale. In this book, the initial distillation processes are referred to as the crude unit because mostly unprocessed crude oil is fractionated, and the rest of the refinery is referred to as the chemistry units, in which the molecular structures are modified.

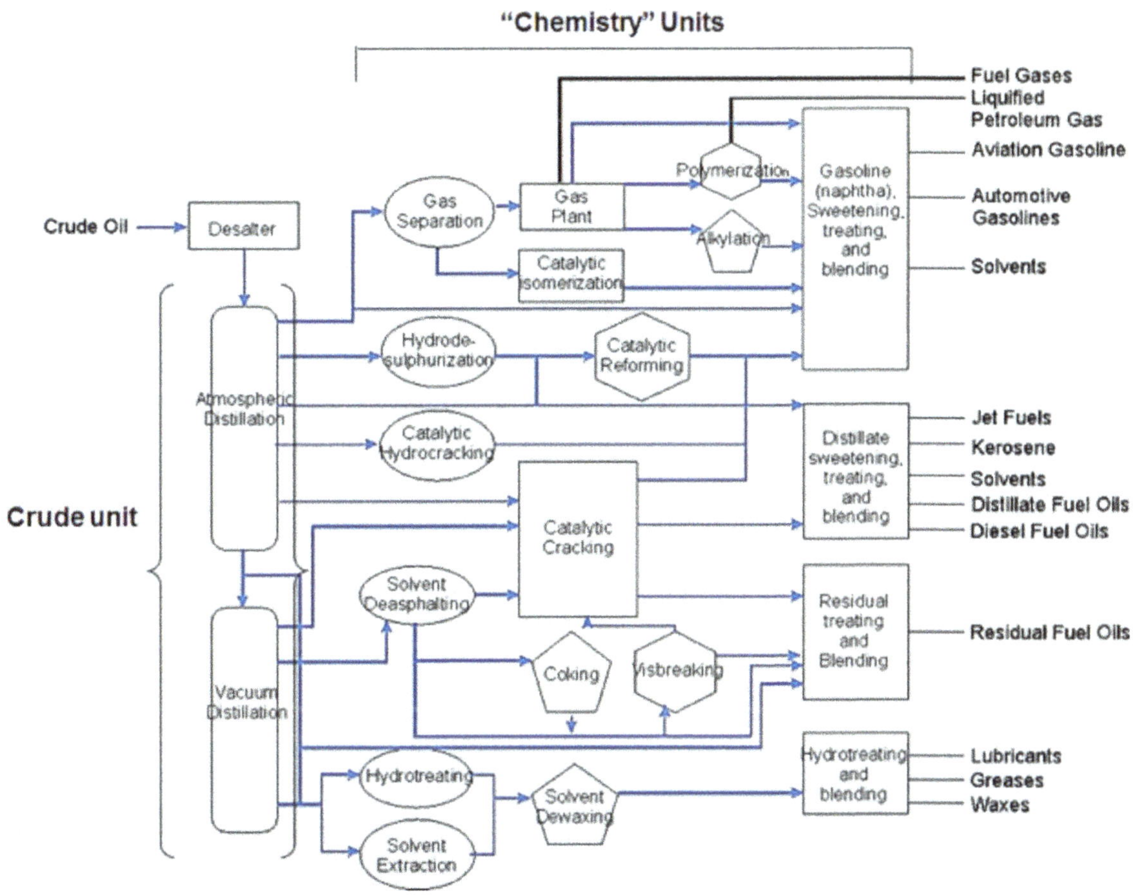

Figure 7.4 Generalized Refinery Layout with Products[6]

This section describes the major units, the system, and problem areas that may require chemical cleaning. However, refineries are not all alike and often vary quite substantially. Also, note that some of the same unit operations occur during the treatment of oil and gas from production wells, before the products are shipped to comprehensive refineries. Additional details of deposits are in Section. 7.4.

When the crude oil arrives at the refinery by pipeline, truck, or barge, it is placed in a storage tank, and some of the solids settle out. While in the tank, the crude oil is tested for API gravity, water content, and metal and mineral content and to determine the compositions of its oil fractions. This testing helps to determine the parameters required to process the crude oil.

Once the testing is completed, the crude oil is transferred to the desalter. In this unit, the crude oil is washed with water to remove soluble salts and particulates. Chemicals and/or electrical charges also are used to remove contaminants. These units can be cleaned by circulating aromatic-containing solvents,[21] emulsions, detergents, and acids to remove the tars and corrosion products that foul the drums. The desalter drums are very large, horizontal vessels that do not lend themselves to typical circulation methods.

Then the oil is transferred to crude processing units[22] in which the first actual separations occur. The feed charge passes through the preheat bundles, generally on the shell side, where it picks up heat from an already processed stream. These bundles are where one type of (undesirable) refinery coke is formed. The temperatures are not high enough to form true carbon, but the deposit is still extremely hard to chemically remove. Deposit removal would be easier if unit operators would clean early enough, but this very seldom happens. These deposits have a loss on ignition (LOI) of 40% to 100% and contain iron (0%–50%). These bundles are normally carbon steel, but alloys also are used. These exchangers are among the most susceptible to fouling in a refinery, so up to two in-place spares may be found per bundle. A comprehensive discussion of the cleaning of refinery heat exchangers has been published by Frenier and Barber.[22] Additional information for using mechanical cleaning processes is in Section 8.1.

The crude oil is then heated in a fired furnace to 650–700°F (343–371°C) and is transferred to the atmospheric distillation (ADU) tower. Here, it is broken into four major fractions: gas and naphtha, gas oil, kerosene, and bottoms (the heavier fraction that is removed from the bottom of the tower).

Each of the fractions is further conditioned to remove water in the gas stream or to further purify the individual streams. Quite often, additional towers will be used for this extra treating, with the bottoms being recycled back to the atmospheric tower. Should heavy end processing be required, the atmospheric tower bottoms will be charged to a vacuum distillation (VDU) tower, where low pressures are used to separate any lighter materials that may be present.

The bottoms sent to the VDU pass through a heat exchanger into the column. The bottoms are heated by exchanging heat with some of the hot, distilled fractions and other streams before entering the column itself. The vacuum on the column creates a low pressure within the system (approximately 10–40 Torr [mmHg]), which serves to limit the operating temperature of the column to below 370–380°C (700–720°F). See Figure 7.5[6] for more detailed diagrams of an ADU and a VDU.

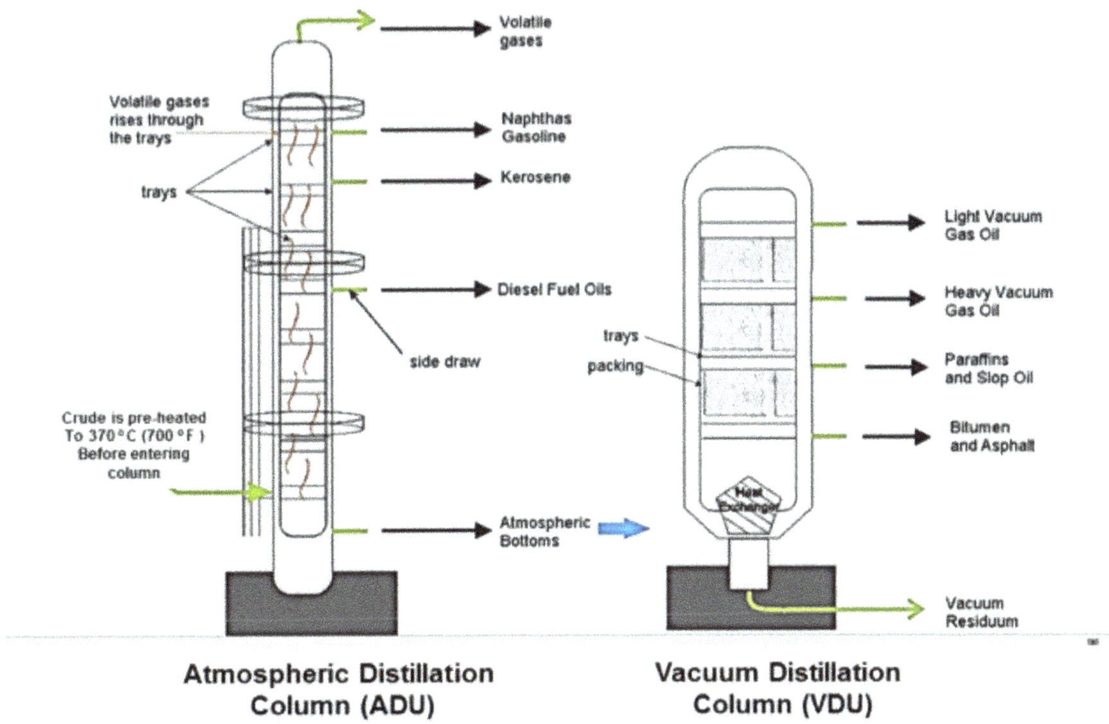

Figure 7.5 Petroleum Refinery Atmospheric and Vacuum Distillation Units (ADU) and (VDU)[6]

The products from the vacuum tower are different grades of lube oil and bottoms that will go to a coking unit, a cracking unit, an asphalt plant, or a heavy oil upgrading process (designated as chemistry units in this book). The boiling points of heavy residuals are lowered by the reduced pressure to avoid cracking the molecules while separating the lighter hydrocarbons. Some of the different types of operations after the primary distillation towers are shown in Figure 7.4 and Figure 7.5. Cracking processes use heat and catalysis to reduce the molecular weight of some of the large molecules shown in Figure 7.3.

Alkylation processes use various catalysts (usually acids described in Section 7.4.1) to increase the carbon chain length of smaller molecules. The goal is to produce the needed range of product streams with the correct mixtures of carbon molecules required by the consumers. Transportation fuels of many types are the primary product line for many refineries, but the number and variety of product streams are vast and defy easy illustration.

The amount of sulfur in the crude stream also affects different refining operations and thus the deposits formed. Another important consideration is the crude oil stream that is available to a refinery; this crude stream composition affects the specific unit operations that are needed. For example, the very light hydrocarbon streams produced in the Bakken may be shipped to the refineries in the northeast United States, whereas heavier crude streams are processed on the U.S. gulf coast.[19]

7.4. Deposits in Refinery Units and Chemical Plants

The deposits in the various parts of a model refinery are described in this section. It is most important that these are recognized to be approximations, and the actual deposits must be identified by sampling and chemical analysis methods described in Section 2.6. Depending on the products processed, similar types of deposits form in chemical plants that have similar unit operations.

Metallurgy in units varies from carbon steel to stainless steels to Monel. Depending on which types of crudes are being processed, the potential for naphthenic acid corrosion, hydrochloric acid corrosion, hydrogen sulfide corrosion, sulfide penetration, polythionic acid corrosion, and erosion exists. Therefore, it is necessary to know the conditions of these units before treating with any aggressive solvents. In general, the lighter the product, the more likely is the chance for corrosion.

7.4.1. Details of Formation of Refinery Deposits

Note that more information is provided in Frenier[23] and Shank et al.[6]

Deposits in the crude oil unit. These vary from the coke on the preheat bundles to iron sulfide with light oils in the atmospheric tower overhead to heavy lube oils and asphalt in the vacuum towers. Most refineries are already set up to circulate hydrocarbon solvents throughout the system and to steam/air decoke the furnaces. This decoking involves passing a steam and air mixture through the product coils in the furnace to burn off the deposits. The air/steam ratio is constantly adjusted throughout this decoking to control the rate of combustion. There has been success with both the application of alkaline solutions or alkaline solutions in conjunction with hydrocarbon solvents to clean the low heat areas (parts of some of the towers). The light product condensers contain iron sulfides with light organic coatings, and the heavier ends contain asphaltenes and tars.

Some ideas on refinery fouling deposits, especially in the vital VDU columns, are abstracted from Shank et al.[6] They note that distillation column fouling encompasses many fouling mechanisms, including vaporization of volatile components, sedimentation, crystallization and precipitation, chemical reactions, and corrosion. Fouling in distillation columns can occur in either the liquid or the vapor phase, but the heaviest fouling tends to be near the bottom of the column. Complex fouling problems arise when two or more of the fouling mechanisms occur at the same time.

In most cases, the authors claim that fouling of the VDU results if the volatile components can vaporize. In the case of a vacuum tower, the vaporization of the volatile components is desired for the proper functioning of the unit. However, if the column is placed on recirculation without the addition of fresh feed, the volatile components are removed, and the viscosity of the residual bottoms increases beyond the pour point.

The authors[6] claim that the remaining high-viscosity heavy residuum greatly reduces the throughput and flow rate of the system, which can introduce hot spots in the system. These hot spots can be much hotter than the average temperature of the system, which can lead to coking of the oil in the VDU. Coke particulates can then precipitate onto the exchanger, reducing the heat transfer efficiency of the system, or into tubes creating blockages that can stall the entire refining process.

Shank et al.[6] tested deposits found in a VDU. A sample deposit from a VDU (Figure 7.6) was analyzed to determine the composition of the foulant. The sample was subjected to both organic extraction and inorganic extraction (see details in this paper as well as the methods in Section 2.6). The results are shown in Table 7.3.

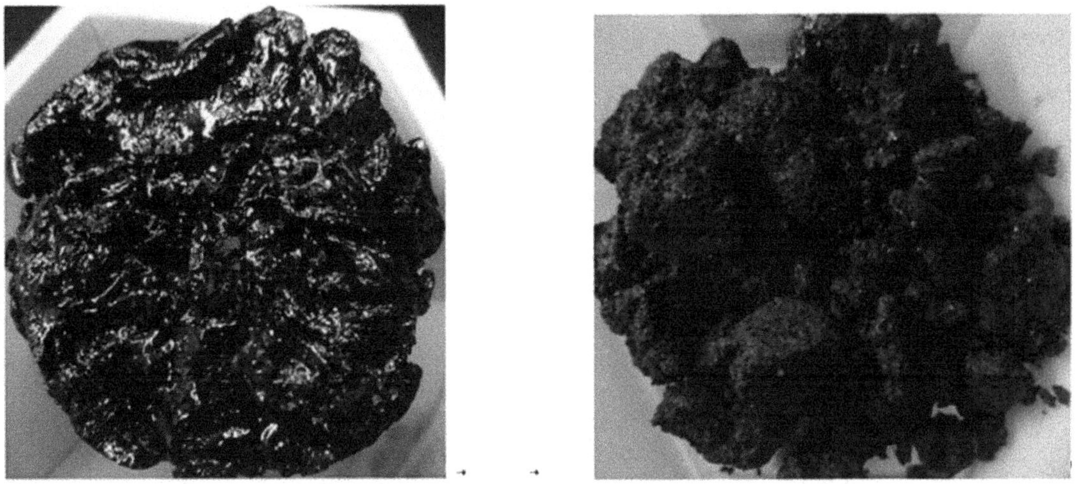

(a) Shell side deposit (b) A crumbly deposit

Figure 7.6 Scale from the Vacuum Distillation Unit Heat Exchanger
(a) Shell Side Deposit is a Hard, Shiny Organic Deposit. (b) Crumbly Deposit that is Primarily Inorganic

Table 7.3 Analysis of Composition of VDU Solids (Figure 7.6)

Constituent	Tube Side Deposit	Shell Side Deposit	Interpretation
IPA soluble	7.08%	38.08%	Light hydrocarbon, glycol, and water
Toluene soluble	60.67%	10.17%	Heavy hydrocarbon and asphaltenes
Hydrochloric acid soluble	18.53%	19.33%	Iron oxides and iron sulfides
Insoluble residue	13.72%	32.47%	Polysulfides and coke

Shank et al.[6] also point out that many processes produce suspended solids (such as silica sand, metal oxides, inorganic salts, and coke fines) that can settle out on the trays and within the packing of the column via sedimentation. Furthermore, precipitation and crystallization of soluble components can occur when the process fluid becomes supersaturated or through the introduction of chemical reactions of inorganic salts with the carbon steel surface.

Generally, once the process streams leave the crude area, they are sweetened (if needed) in a hydrotreater.[20, 24] In a hydrotreater, the feed is combined with hydrogen and passed over a catalyst at elevated temperatures and pressures; the hydrogen links with sulfur to produce H_2S and with nitrogen compounds to produce ammonia. These gases are separated from the oils and are sent to other units for processing (See Sections 1.3.4 and 7.6). The hydrogen may also combine with the hydrocarbon to produce a more valuable product. Iron sulfide deposits and organic sludge may form. Note that additional acid gas removal processes may be required on the various gas streams within a refinery. See Section 7.4.2.

The oil feed is preheated in cross exchangers (See Sections 7.4.4 and 8.3) to a temperature usually between 400°F and 700°F (204–371°C). It then passes to a fixed bed reactor vessel together with hydrogen-rich gas. In the reactor, the feed is treated with hydrogen and a regenerable metal oxide catalyst at pressures usually between 200 and 1000 psig. After leaving the reactor, the product stream is cooled before entering a low-pressure separator. This procedure allows the removal of dissolved gases before the next stage, a stripper column, in which the product is separated from hydrogen sulfide and any light ends. Part of the reactor effluent stream may be used to supply heat to the product stripper reboiler, or steam may be alternatively used.

Deposits in Chemical Reaction Units. The initial processing units separate fractions on the basis of the molecular weight of the hydrocarbons. The next units alter the chemicals to provided needed products. The next units the oils will encounter, and possibly the most important units in some refineries, are the cracking units. These units use catalytic cracking, thermal cracking, or hydrocracking processes to reduce the molecular sizes of the various chemicals in the crude stream. Many refineries have all three types, with each one feeding the other or operating in parallel. The purpose of these units is to convert high-boiling non-gasoline hydrocarbons to lower boiling gasoline components.

Hydrocracking. This is essentially a hydroformer, which operates at high temperatures (650–800°F [343–427°C]) and pressures (1200–3000 psi). Almost all the materials in these units are alloys ranging from Series 304 to 430 stainless steels. The deposits are similar to those found in hydrotreaters, such as iron sulfide and heavy organic deposits. The feedstock is similar to heating oil. These units will generally only run in summer months.

Thermal Cracking. The thermal process was the earliest method used by refineries to produce gasoline components. High temperatures (over 600°F [316°C]) and residence time were used to determine the yields of lighter products. The advent of catalytic cracking processes essentially replaced thermal cracking, but it is still used in some places. Today, thermal cracking is used mostly in coking units. There are two basic types of coking units: delayed coking and fluid coking. In delayed coking, the charge is fed directly into a fractionator, where it is combined with the recycle and is pumped to the coke heater, where it is heated (900–950°F [482–510°C]) to produce partial vaporization and mild coking. This vapor/liquid mixture then enters the coke drum, where the vapor experiences successive cracking and polymerization until it is converted to vapor and coke.

The coke drum overhead vapor enters the fractionator, where it is separated into gas, light and heavy gas oils, and recycles oil that combines with fresh feed. These units will have at least two drums: one is in coking service, and the other is decoked with high-pressure water jets.

In fluid coking, the feed is pumped through a furnace where it is heated (1110–1200°F [599–649°C]), and then it proceeds to a reactor where it is thermally cracked (900–1050°F [482–566°C]). Products other than coke are quenched overhead in a scrubber, where entrained coke is removed. The lighter fractions go overhead to be treated elsewhere, and a heavy fraction with entrained coke fines is recycled to the reactor. Coke from the reactor circulates back to the heater and then passes again through the reactor until it finally leaves the process unit.

Other than high-pressure jetting, there is little to offer for the process side of these units. The auxiliary systems associated with these units, especially fluid coking, offer applications for cleaning services. Emulsions of caustic soda and diesel fuel have been used to clean a fractionator tower, and caustic foam has been used to clear the air-cooled exchangers. See the information in Section 7.6.

There are many lube and seal oil systems in these units. The fluid cokers may have up to three cyclones or venturi to separate the coke fines from the overhead products. Because the products leaving these units must be cooled, waterside problems will occur at these temperatures (formation of iron oxides and hardness deposits will require chemical treatment). The furnace itself will be steam/air decoked, but the fuel lines will need to be periodically cleaned to reduce burner fouling.

So far, this section has followed the path of the bottom cuts from the crude processing unit. We are now left with the naphtha and gas cuts from the atmospheric distillation tower. As both these cuts leave the area, they are sweetened to remove sulfides and any organic sulfur in a hydrotreater or amine unit before further processing. More details of the amine units are in Section 7.4.2.

The naphtha goes to a catalytic reforming unit.[20, 24] These units are designed to increase the antiknock quality of motor fuel blending stocks. The principal reaction in this process is dehydrogenation of naphthenes to form aromatics. Basically, a straight-chain naphtha is heated and passed through a catalyst bed to form a ring structure. During this reaction, hydrogen is produced—some of which is recycled to sustain reactor pressure and to reduce coke formation. The rest of the hydrogen is sent offsite to be used in other processes.

During desulfurization, the process streams are not very corrosive, so carbon steel or lowgrade alloys are used in the construction of reformers. Because the feed is very clean to protect the catalyst, there is very little fouling of these units. Deposits include cyclic naphtha, which is the second type of refinery coke. These deposits require an oxidizing solution (such as chromic acid or permanganate). There may also be iron sulfides, if the upstream processes are not properly operating.

One item, which should be looked at during an outage, is the piping going through the furnaces to the reactors and the hydrogen supply and recycle lines. During an outage, these lines will oxidize and form rust. Should these lines be directly placed into service, there will be an immediate drop in pressure across the catalyst beds that will result in higher temperatures and reduced flow rates. Because these lines will generally not stand a complete liquid fill, foam or vapor phase cleaning should be considered. The catalyst regeneration piping also falls in this category. Other fouling problems in a reformer are the lube oil system on the hydrogen recycle compressor and the cooling water systems. The fuel lines on the furnaces will also eventually foul.

Alkylation Units

The product from the reformer is blended with the products from the alkylation units to make gasoline (a complex blend with mostly C_4-C_{12} hydrocarbons). These alkylation units[20] process the gas fraction of the crude oil. These units use a catalyst to combine olefins, paraffin, or aromatics to form a high-octane product to blend into gasoline. There are two types of alkylation units that derive their name from the catalyst used: sulfuric acid and hydrofluoric acid. McCoy[25] claims that most units in the United States use HF as the catalyst.

In the sulfuric acid alkylation units, the feed is mixed with a recycle stream of acid and is fed to the contactor. These units and the HF alkylation processes increase the length (and complexity) of the hydrocarbons to produce higher grade gasoline streams.

The liquid contents of the reactor are circulated at high velocities, and an extremely large amount of interfacial area is exposed between the reacting hydrocarbon and the acid catalyst. The entire volume of the liquid in the contactor is maintained at a uniform temperature. In fact, the maximum deviation between any two points within the reaction mass is less than 1°F (-17°C). The hydrocarbon/acid emulsion proceeds to a settler, where gravity separation of the hydrocarbon from the acid occurs. The settled acid is recycled to the contactor.

The light ends are flashed off, washed, and sent offsite. The bottoms also are washed (separately) and go to the deisobutanizer tower. The overhead from the deisobutanizer and effluent refrigerant cycle normally constitute the total isobutane recycle to the reaction zone. The isobutanes and all other hydrocarbons are maintained in the liquid phase as diluents throughout the reactor, thereby serving to promote the alkylation reaction.

Several units in a sulfuric alkylation system may require cleaning. The depropanizer tower will probably contain iron sulfides that can be removed by an acid/aldehyde[26] solvent (see Section 7.7). There may be sludge in the acid storage tank and the caustic wash tank that can be cleaned by high-volume flushing or by a rotojet. The best opportunities will be in the cooling water system and the refrigeration system. The refrigeration system is cleaned and dried before initial startup, but a seal oil leak into the system may necessitate operational servicing. In general, all these operational deposits can be removed by a degreasing solution followed by acid or chelant. An emulsion solvent of a hydrocarbon with an acid should also be necessary. Most scale in these units will generate a dilute acid and heat on contact with water, so care should be taken when filling, pressure testing, etc.

The process flow of an HF alkylation unit[20] is somewhat similar. The inlet charge is intimately mixed with the HF catalyst and then goes to a settler. The acid phase is recycled to the reactor; occasionally, a small portion is regenerated. The settler overhead product is freed of isobutane and lighter components in the isostripper, the bottom product of which is a motor blending alkylate. A portion of the isostripper overhead vapors is depropanized, and the propane is separated from acid in the HF stripper. Note that because of the toxicity of HF, these types of units may be replaced with other technologies.

The deposits are usually iron fluorides and, very infrequently, iron oxides (the fluid usually is too acidic for oxides to form). Dilute HCl is generally used to clean these units. Addition of boric acid sometimes aids the treatment. The cooling water system and lube and seal oil systems also need to be cleaned periodically. Because of the toxicity of HF, special production, handling, and storage of this acid is required.

McCoy[25] claims that in addition to HF and H_2SO_4, ionic liquids (IL) are more effective and possibly less toxic alkylation catalysts. All salts, including NaCl, become an IL if the temperature is raised above the melting point of the salt. Some low-temperature ILs can be compared to ionic solutions, liquids that contain both ions and neutral molecules, and in particular to the so-called deep eutectic solvents, mixtures of ionic and non-ionic solid substances that have much lower melting points than the pure compounds. These low-temperature melting salts[27] may act as very strong acids or have other useful properties. An example of two ILs along with HF and H_2SO_4 is shown in Figure 7.7. McCoy[25] also claims that a major producer has scheduled a plant that will use ILs in an alkylation process by 2020. The compositions have not been described.

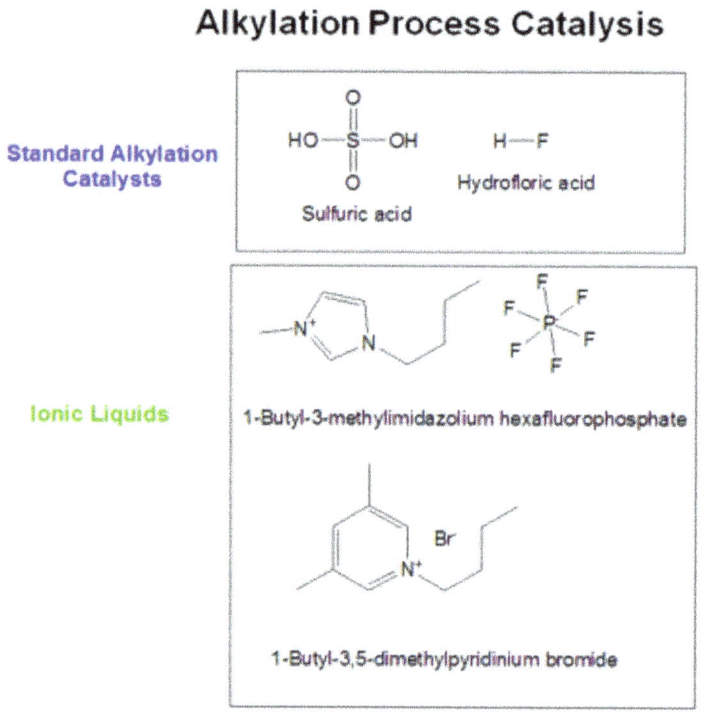

Figure 7.7 Alkylation Catalysts Structures

A summary of deposits and typical cleaning information is in the following Tables. Table 7.4 lists possible organic deposit types that may be found in refineries or other chemical plants. The progression form very light hydrocarbons to more complex materials is the result of temperature, contact time, and the specific chemical reaction taking place. Generalized cleaning processes are noted, and more details of chemistries are in Section 7.5.

Table 7.4 General Organic Deposits and Solvent Types Suggested

Foulant	Deposit Type	General Cleaning Solvents
F1	Light hydrocarbons, C1-C5	Degassing and light oil removal by detergents
F2	Gasoline, diesel and fuel oils, C4-C5	Degassing and light oil removal by detergents
F3	Motor oils, greases, bunker C, crude oil	Remove with alkaline detergents and light solvents
F4	Tar and asphaltic deposits	Solvents followed by alkaline detergents
F5	Black, hard solids with a polymer matrix	Alkaline detergents, organic solvents with accelerators
F6	Black, hard solids with a coke matrix	Mechanical methods
F7	Linear polyolefins	High-temperature oils or solvents followed by detergents
F8	Cross-linked polymers	Mechanical methods
F9	Paints, coatings	Alkaline detergents, solvents
F10	Cosmetic cleaning	Acidic or alkaline detergents

Table 7.5 shows deposits that may form in some specific units and classes of solvents that may be used to remove them. More details and discussion are in Section 7.5.

Table 7.5 Deposits/Solvents in Units Not Considered in Detail in Other Sections

Unit	Deposit	Solvents
Crude	Coke, mixed FeS, paraffin	NaOH, trisodium phosphate (TSP, sodium), metasilicate, acid/aldehyde, aromatic naphtha
Towers	Mixed organic	Emulsions, H_2SO_4
Overhead condensers	Mixed organic	Emulsions, solvents
Preheat exchangers	Coke, FeS, mixed organic	NaOH, TSP, sodium metasilicate, acid/aldehyde, aromatic naphtha
Coking units	Coke, mixed organic	NaOH, TSP, sodium metasilicate, emulsions
Cat reforming	Coke, FeS	Chromic acid, alkaline, acid/aldehyde, permanganate
H_2SO_4 alkylation	FeS, Fe sulfate	Acid/aldehyde, alkaline
Depropanizer	Iron sulfide	Chelants, emulsions, acid/aldehyde
Cooling water	Hardness, iron	Chelants, acids
HF alkylation	FeF_3, oils	HCl, reducing agent, boric acid

7.4.2. Deposits and Problems in Amine Systems

One additional important auxiliary unit in a refinery that requires cleaning is the amine unit. Although many processes can remove H_2S and/or CO_2 from gas streams (generated from various refinery operations), the most widely used are the amine systems that depend on the reactivity of H_2S with amino nitrogen for their absorption properties. The most used processes today are monoethanolamine (MEA), diethanolamine (DEA), diglycolamine (DGA), and diisopropylamine (DIPA). Also, note that these units exist in the oil/gas fields to perform the same processes and thus have similar deposits. The location of an amine unit within a specific refinery and its place in the unit operations varies with crude quality and refinery operations. The same types of units are used to remove acid gases from oilfield fluids in the production environment.[3]

These systems depend on the reactivity of H_2S with amino nitrogen. The units in refineries may be larger, but acid gas removal systems with similar chemistries and deposits are in the oil/gas production systems and require the same treatments.

The following reactions model the removal of acids and the regeneration of the amine. These are acid-based reactions that form salts that can be regenerated by heating. Note that the first step is the absorption of the acid gas by the liquid phase. The rate and mechanisms of this process must be understood for each gas and liquid pair.

With H_2S,

$$\begin{aligned} &2RNH_2 + H_2S \Leftrightarrow (RNH_3)_2S \\ &(RNH_3)_2S + H_2S \Leftrightarrow 2(RNH_3HS) \end{aligned} \qquad \text{(Equation [7.2])}$$

Most importantly, this reaction proceeds to the right at low temperature and high H_2S partial pressures. For this reason, the absorption is normally conducted at pressures of 300 psi or higher and at or near ambient temperature. Regeneration is normally performed at or near atmospheric pressure and at the solution boiling point.

With CO_2,

$$\begin{aligned} &2RNH_2 + CO_2 + H_2O \Leftrightarrow (RNH_3)_2CO_2 \\ &(RNH_3)_2CO_2 + H_2O + CO_2 \Leftrightarrow 2(RNH_3HCO_3) \end{aligned} \qquad \text{(Equation [7.3])}$$

MEA, a clear, colorless liquid that boils at 338°F (170°C) at atmospheric pressures,[23] is the most widely used of these compounds. Of all the amines used, it is the strongest base and so reacts most readily with acid gases. However, MDEA is used for H_2S removal because it is a tertiary amine and cannot react with CO_2 but will form an amine complex. The CO_2 can be removed with another amine or by carbonate/bicarbonate absorption.

The sour gas containing H_2S and/or CO_2 enters the plant through a scrubber, which removes any free liquids and/or entrained solids. The sour gas then enters the bottom of the absorber (Figure 7.2) and flows upward in countercurrent contact with the descending aqueous amine solution. Sweetened gas leaves the top of the absorber and flows to a dehydration unit, where saturation water from the aqueous amine solution is removed. The fluids are potentially very corrosive, and choosing the proper metallurgies is critical. Breakdown of the reactants and production of fouling solids can also occur.

The structures of the most common amines are shown in Figure 7.8.

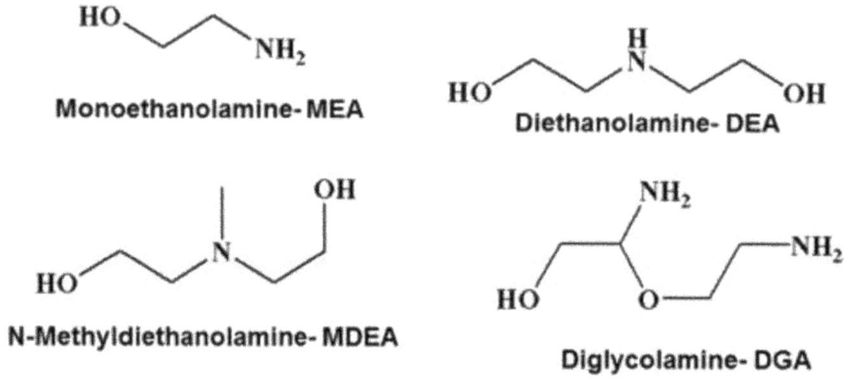

Figure 7.8 Acid Gas Absorption Chemicals

The processes are basically the same with some variation in flow scheme. The sour gas containing H_2S and/or CO_2 enters the plant through a scrubber, which removes any free liquids and/or entrained solids. The sour gas then enters the bottom of the absorber and flows upward in countercurrent contact with the descending aqueous amine solution. Sweetened gas leaves the top of the absorber and flows to a dehydration unit, where saturation water from the aqueous amine solution is removed.

The H_2S removed from the gas stream is then converted to elemental sulfur (S^0) using a chemical process. A major method is the Claus process,[28] which uses multiple steps that are conducted at high temperatures (>1800°F in a furnace, then 600°F in a catalytic reactor). The overall reaction is

$$2H_2S + O_2 \rightarrow 2S + 2H_2O \qquad \text{(Equation [7.4])}$$

Need for Cleaning and Deposits Present in Amine Systems. If these units are self-circulating, they can be cleaned using the system's pumps. These units operate on a pressure differential so some inert gas is needed. These units should be preoperationally cleaned. Oils, greases, and particulates will cause foaming of the amine by affecting surface tension. The resulting foaming will cause a loss of the amine off the top of the stripper. Although some iron sulfide is formed during operation, a smooth, passive surface may reduce the amount formed. Removal of iron oxides also generally means fewer filter changes during startup and less chance of particulates dropping out in low-flow areas. Iron oxide particles in the solution may also combine with sulfides during operation and possibly form ferrous sulfide in the stream that will turn the amine black.

The cleaning solutions should be filtered with system filters or with auxiliary filters. The amines themselves are good degreasers, so quite often the customer will circulate the systems themselves to clean them. The amount of amine lost owing to foaming and carryover may be in the price range of the chemical cleaning.

These units also can be cleaned to remove operational deposits. Because the amines are water soluble, these systems may not require a degreasing stage. However, some product will have entered the system and should be removed, but this should be handled on a case-by-case basis. Once the organic deposit has been removed, an acid/aldehyde mixture should be used as the solvent for FeS (See Section 7.6).

The entire unit should be cleaned to avoid contamination of the fresh amine, but the customer may not feel the need to do this. As a minimum, the stripper overhead, amine/amine exchangers, and water-cooled exchangers between the towers should be cleaned.

The stripper tower separates the acid gases from the amine systems, so its efficiency dictates the effectiveness of the system. Increased pressure drop caused by fouling means higher firing on the reboiler. This condition may lead to corrosion and more amine lost through carryover.

The cooler on the overhead system of the stripper tower controls the quantity of water and amine that is recovered from the overhead streams. This may be an air- or water-cooled exchanger or a combination. These must effectively cool if excessive amine is not to be lost. These exchangers also affect downstream corrosion because the corrosivity of these acid gases is increased by increasing temperature.

The amine/amine exchangers and the water-cooled exchangers have a significant effect on the operation of the system. As the absorber runs cool and the stripper runs hot, these exchangers need to be efficient to exchange the heat. The amine/amine exchangers play a large role in the corrosion found in the stripper tower and reboiler.

7.4.3. Effects of Oilfield Chemicals on Refinery Deposits

Kapusta[29] has described the impact of many oilfield chemicals (OFCs) on refinery deposits. See Frenier[3] and Kelland[4] for lists of OFCs and descriptions of their uses. Chemical additives play an important role in oil production. Chemicals are used to enhance oil production, control corrosion, prevent wax or paraffin deposition, enhance the flow characteristics of the crude oils, and aid in water separation. However, some of these chemicals may also cause processing problems in refineries. The most common problems associated with OFCs in the crude include

- Desalting upsets
- Corrosion in the overhead system of crude distillation units and in hydrotreaters
- Fouling of catalysts, foaming of amine treating units, and fouling of heat exchangers

In many cases, the actual relationship between OFCs and refinery upsets is hard to prove, as other operational or crude oil factors may account for the observed problems. An understanding of the impact of OFCs on refining operations is therefore of critical importance not only to refiners but also to producers since both are interested in minimizing any negative effect of these chemicals on the operation of the refinery.

Kapusta[29] claims to have updated information on the impact of OFCs on refinery operations. In particular, corrosion has been studied. One example of a new problem is rogue chlorides in certain crudes, which has led to severe corrosion problems in crude distillation units and naphtha hydrotreaters.

Table 7.6 from this report shows the possible effects on various refinery operations. Many of the chemicals could exacerbate corrosion and cause both organic and mixed scales to form.

Table 7.6 Possible Effects of Oil Field Chemicals on Refinery Operations

Problem	Type of Chemical Presumed to be Responsible	Alternative Causes
Poor tank farm dehydration	Surfactants, corrosion inhibitors	Presence of slops
Desalter upsets	Surfactants	Improper desalter operation
High salt content in crude	Higher Ca and Mg in crude due to acidizing wells	Improper desalter operation
Corrosion of pre-heat train	Presence of acidizing fluids	
Corrosion of furnace tubes	Presence of acidizing fluids	Formation of H_2S by high-temperature decomposition of crude
CDU overhead corrosion	Presence of acidizing fluids, other salts, decomposition of corrosion inhibitors	Improper desalter operation
CDU and NHT fouling and corrosion	Organic chlorides from well cleanup operations	Inorganic chlorides
Catalysts poisoning or bed fouling	Corrosion inhibitors	Iron sulfides in crude, silt from tank bottoms
Catalyst bed plugging	Silicon-containing antifoams	
High water content of hydro-treated naphtha	Glycol or glycol amines	Inadequate separation of water in the CDU overhead vessel
Poor bitumen quality	Silicon-containing antifoams	
Foaming in amine treaters	Corrosion inhibitors	Amine degradation, hydrocarbon carry over
Corrosion of amine regenerators	High levels of ammonia due to decomposition of amines or corrosion inhibitors	Ammonia in water used for desalting

7.4.4. Fouling Deposits in Heat Exchangers

Heat exchangers are some of the most ubiquitous and numerous types of units found in refineries and chemical plants. Table 7.1 lists some of the heat exchanger types in use. Muller-Steinhagen[30] describes some of the classes of deposits found and possible mechanism of formation. This book also notes that both inorganic and organic solids foul these units depending on whether it is the water side or the process fluid side where the deposits cause flow and heat exchange problems. Some heat exchanger devices may have water on both sides or organics on both sides. Muller-Steinhagen[30] describes categories of deposits that are formed due to crystallization, particulates, chemical reactions, or corrosion and biologically formed deposits and mixed fouling deposits. Several formation processes include

- Crystallization of inorganic salts (see Section 3.2 for details)
- Formation of particulates, i.e., small suspended particles of sand, clay, or corrosion products from upstream
- Chemical reactions that cause fouling causing polymers (described in Table 7.4)
- Corrosion product fouling (e.g., iron oxides and iron sulfides; see Section 3.3)
- Biological product fouling, including biological slime masses, mussels, algae, and microbiologically induced corrosion[31]
- Mixed fouling, with organic/inorganic solids

A general mechanism consists of initiation, mass transport, formation of a deposit on a surface, and then aging to produce a very stable mass that is difficult to penetrate, circulate, or remove by solvents or high-pressure water (see Section 8.3.1).

7.4.5. Pyrophoric Iron Sulfide

When FeS-containing solids are dried and then exposed to the air, they can spontaneously combust. This significant hazard is described in Section 3.3.2 (with references), and the reactions for forming these complex solids are shown in Equations (3.32–3.43). Various solvents for removing these solids safely are considered in Section 7.6.

7.5. Solvents for Cleaning Refinery and Chemical Plant Units

7.5.1. Introduction

This section describes chemical cleaning fluids used to dissolve or disperse organic and mixed deposits. The water and acid soluble inorganic salts found in refineries, chemical plants, and paper mills can be removed using the solvents described in Chapter 4 with the use of corrosion inhibitors discussed in Chapter 6.

Al-Darwish et al.[32] note that various mechanical and chemical methods can be used in refineries and crude oil treatment plants. The treatment selection is controlled by the types of foulants and by physical access to the units to be cleaned. Mechanical devices are discussed in Chapter 8. In addition, none of the chemical processes described in Chapters 4-7 can be accomplished without the various types and significant numbers of units described in Chapter 8.

Table 7.4 and Table 7.5 show the generalized types of organic deposits that may be found in industrial refinery/petrochemical equipment. The deposits noted as F1-F4 are mostly wax-type and asphaltic type of chemicals. These deposits may be mixed with residual oil and, in most cases, contain inorganic solids.

This section describes solvents and mixtures that were developed specifically to remove industrial organic solids. Products developed in the oil and gas production industries to remove waxes and asphaltenes also may be useful for removing industrial deposits and also are noted.[1] This section describes the large variety of chemical solvents that are in current use or have been proposed in the patent literature for removing organic fouling solids.

The dissolution of organic solids shares many of the aspects of inorganic scale dissolution.

Similarities include

- Necessity for identifying the deposit composition, amount, and location.
- Testing of the deposit to determine solubility in various fluids (assuming that adequate samples can be obtained)
- Determining the best application method
- Circulating/placing an adequate volume of solvent to achieve the goal of the treatment

However, a major difference between solvents for inorganic and organic scales is the degree of dissolution that is possible and needed. Using the appropriate amount and type of solvent, most inorganic scales can be removed completely. See the various discussions in books on inorganic scale.[2, 33]

Because of the heterogeneity of field organic foulants, complete solvency and thus removal may not be possible using a single solvent or even with several stages. In addition, there usually is a fraction of the deposit that is inorganic. Therefore, use of emulsions, dispersions, and co-solvents is often required for removing organic-containing deposits. Also, the thermodynamic requirement of total dissolution usually cannot be proven. Thus, the stability constant/solubility product nomenclature used in the inorganic scale literature (Frenier[2]) is usually not used to describe the dissolution process. However, some quasi-thermodynamic functions exist for some systems and are described later in this section. Ultimately, testing and developing custom formulations at the field level is a significant requirement for success. This section discusses the chemistry and types of commercial solvents and other fluids in current use to remove organic deposits.

Generalized methods for dissolving various types of organic scales found in these plants[23] are listed in Table 7.4. These deposits are found on refinery front-end equipment and in operational deposits. Solids found in the transport portion of the hydrocarbon supply chain usually are composed of the first four categories of organic deposits described in Table 7.4. The waxes frequently found in flow lines and production tubing might consist of higher carbon number alkanes compared with the refinery hydrocarbons but are soluble or dispersible using the same types of solvents as used in refineries. Thus, detergents or organic solvents should be able to remove most of the low-temperature formed organic deposits. In refineries and some chemical processing plants where heat is applied to effect a separation or reaction, fouling by higher molecular weight deposits (described as the F5-F8 categories) occurs. These types of foulants require high temperature solvents or mechanical methods for removal. See Section 8.3.

7.5.2. Chemistry of Organic Deposit Solvents

Several classes of organic deposit-removing solvent are described in this section.

Detergents. A vast variety of surfactants (i.e., detergents) are used to remove organic solids and disperse inorganic/mixed solids. The major categories are cationic, anionic, and nonionic. Simple molecules that are representative of these types are shown in Figure 7.9. Many amphoteric surfactants (see Frenier[34]) can be anionic or cationic depending on the pH of the aqueous phase.

A very wide range of nonionic surfactants is available and often is a first choice for at least a part of a cleaning formulation because they are compatible with other components. A range of materials with various hydrophilic/lipophilic balance (HLB) values are shown in Figure 7.10 (a and b). This HLB system allows for an estimation of the balance between the water-soluble parts (mostly ethylene oxide [EO]) and the hydrocarbon parts. This enables the formulator to try to match the systems to be cleaned. Part c) shows three alcohol ethoxylates that are in many detergent formulations.[35] See Frenier,[34] Griffin,[36] and Kelland[37] for more details of HLB calculations.

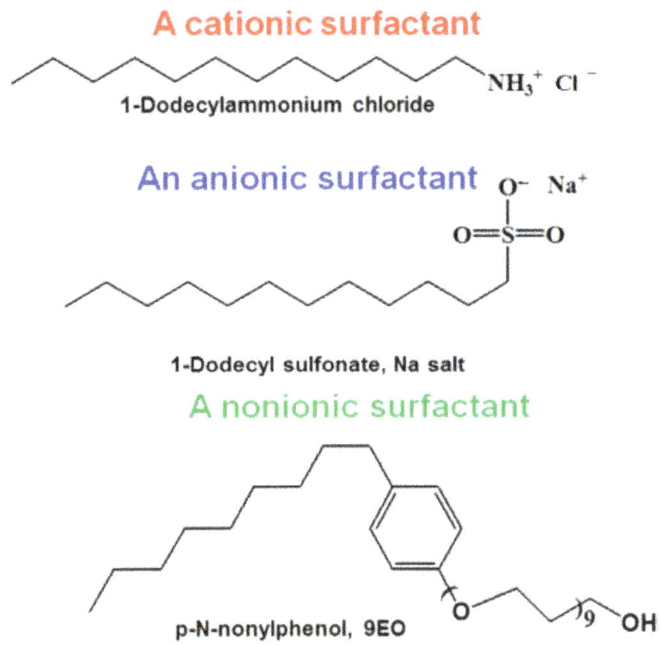

Figure 7.9 Surfactant Types

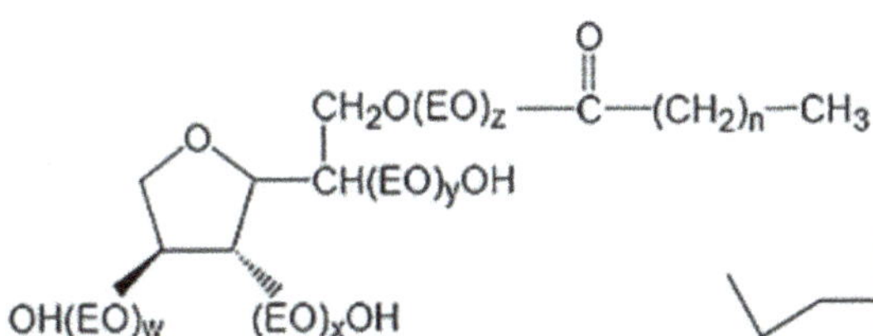

a) Polysorbates

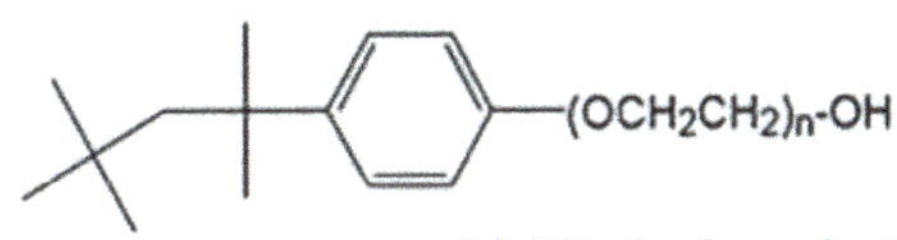

b) Alkyl phenyl ethoxylates

Product	HLB
TWEEN 20	16.7
TWEEN 21	13.3
TWEEN 40	15.6
TWEEN 80	15.0
TWEEN 81	10.0
TWEEN 85	11.0

Product	Average HLB
Triton-X 15	4.9
Triton-X 35	7.8
Triton-X 45	9.8
Triton-X 114	12.3
Triton-X 100	13.4
Triton-X 165	15.5

c) Alcohol ethoxylates

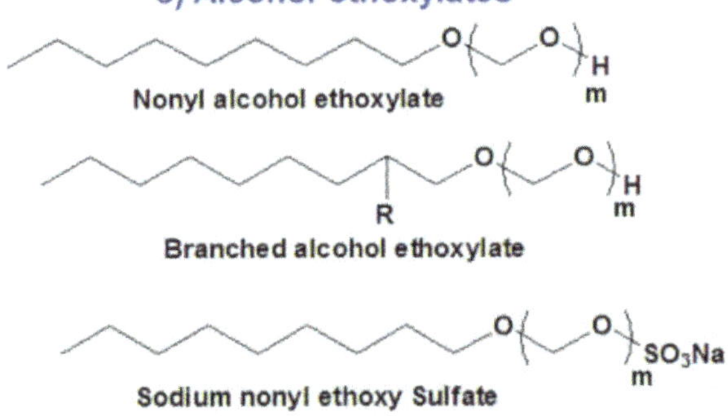

Figure 7.10 Common Ethoxylated Surfactants

Various types of polymeric surfactants that are based on both homopolymers and block and graft copolymers are used. See a review by Tadros[38] for more information. For many applications, environmentally benign (such as natural product–based) surfactants and solvents may be desirable. An example from Sassol[39] is shown in Figure 7.11. Panel (a) shows the base structure of a branched chain sulfonate (anionic detergent), and (b) shows several natural product alcohols that can be used to form various materials with varying solvency ranges for different types of deposits. TDW[40] also discusses the use of environmentally friendly surfactants for cleaning pipelines.

Many cleaning chemicals, including detergent cleaning mixtures, corrosion inhibitors, and defoamers, contain a mixture of different chemicals, including the active ingredients, solvents (water or alcohols), and possibly cosolvents (see a later section). Werff[41] reports on a proprietary solvent that contains concentrated liquid surfactants, polymers, and dispersants.

The formulation must be soluble/dispersible in the flowing stream and must be stable onsite and in unheated/cooled storage facilities. Many surfactant formulations have used methanol as a universal solvent (and possibly as an anti-foaming agent). However, this alcohol is toxic and very flammable and can be substituted by materials such as glycol ethers. See the discussions on less toxic formulation formulations by Frenier[42] and Sitz.[43]

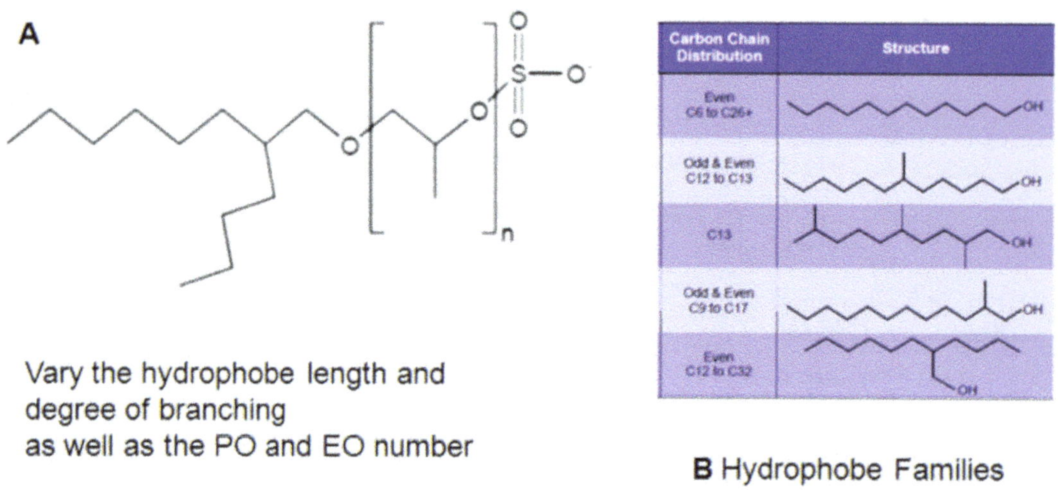

Figure 7.11 Surfactants from Natural Products[39]

Mechanism of Solvency. At a concentration of the surfactant in water that is known as the critical micelle concentration (CMC), the molecules associate to form a micelle. Figure 7.12 shows a diagram of a spherical micelle with the hydrophilic head at the water interface and the hydrophobic tails in the center of the micelle. In the presence of a hydrocarbon or fat molecule, these materials can be emulsified and thus removed from a surface.

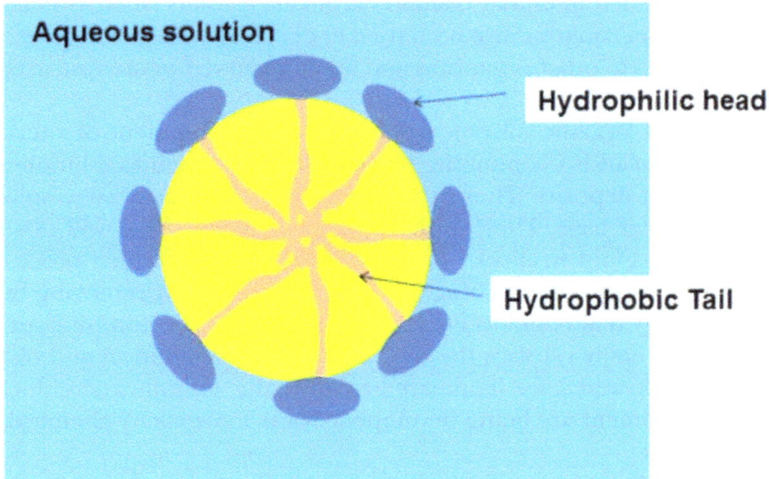

Figure 7.12 Surfactant Micelle

The cleaning action of detergents involves changing the interfacial surface tension (IST; see Figure 2.2) between the metal and water such that the surface becomes water-wet, not oil-wet. This action allows the organic compounds to be displaced. The remaining surfactant then forms an emulsion with the oil, preventing redeposition. The flow in the units keeps the organic soils suspended. See Figure 7.13 for an illustration of the process by which a surfactant removes and disperses (emulsifies) the organic deposit.

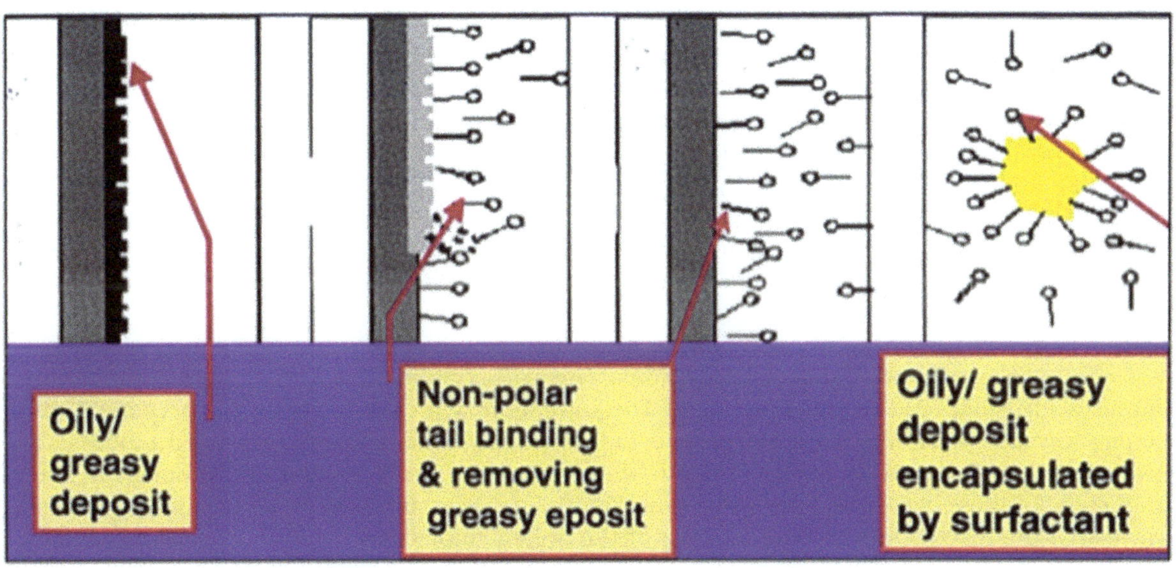

Figure 7.13 Illustration of Surfactant Removal of Organic Solid

Organic solvents or alkaline agents (sodium hydroxide, potassium hydroxide, sodium carbonate, or trisodium phosphate) often are added to help convert any fatty acids in the oil to a natural surfactant and improve detergency. From a commercial standpoint, these alkaline/surfactant mixtures are some of the most used (and least expensive) solvent compositions known.

General Organic Solvents. Table 7.4 notes that common aromatic solvents (such as toluene and xylene; Figure 7.14 [a, b]) and aromatic petroleum naphtha have been used to dissolve hydrocarbon-based solids. However, these materials are toxic to humans and to other organisms and are not allowed or are restricted in some markets.

Concerns about the use of toxic organic solvents prompted the development of alternative technologies to remove organic deposits. Formulations containing terpene solvents (such as d-limonene) were patented by Matta[44] to remove hydrocarbon deposits. These compositions include a terpene solvent, propylene glycol ethers, and surfactants in a water base. They are hard surface cleaners that will dissolve oils, asphaltenes, and heavy oils. Patents by Mehta and Krajicek[45] and Krajicek et al.[46] described processes that use terpenes and surfactants for cleaning and decontaminating refinery vessels and for removing benzene to allow maintenance. An alternate technology that contains enzymes and N-oxide type nonionic surfactants was reported by Mestetsky[47] as a solvent that easily releases the oil for enhanced separation and disposal of the wastewater. More ecologically acceptable and safer (compared to aromatic solvents and chlorinated hydrocarbons) methods to clean refinery equipment are being developed. Some low-toxicity chemicals are shown in Figure 7.14 (c–h).

Jansen[48] describes a chemical heat exchanger cleaning process. This is claimed to be a safe and effective process that can be accomplished in a very short period, without the need to disassemble the equipment and without the need to rinse contaminate from the equipment after cleaning. It uses a formulation of monocyclic saturated terpene mixed with a non-ionic surfactant package specifically suited to oil rinsing. The terpene-based chemical is injected into organically contaminated exchangers using a novel process involving high-pressure steam to form a very effective cleaning vapor. A long list of terpenes used with surfactants is noted in the patent text.

Lawson and Snyder[49] developed a solvent to remove heavy asphaltic materials. The composition consists of a heavy aromatic solvent and a compound with fused heterocyclic rings, which improves the cleaning provided by the organic solvent. Benzamidizole is an example of the fused ring compound.

Thus, formulations containing terpene solvents (such as d-limonene and turpentine-based solvents; see Figure 7.14) have been proposed. An example includes a patent by Matta[44] to remove hydrocarbon deposits.

Penny[50] patented a method for removing and preventing deposition of organic deposits by contacting deposits with a penetrating solvent and a cationic perfluoro compound. These materials are said to prevent or reduce the wetting of the surface by the hydrocarbons. Many solubilizing formulations have been proposed for removing waxes and asphaltenes. McCalflin[51] stated that a water-soluble ethoxylated alkyl phenol and a short-chain alcohol could be used to clean the well bore. Another common formulation is a surfactant (oil soluble) and a light (or aromatic) hydrocarbon. Kruka[52] also proposed such a formulation.

Jennings[53] claims that an exothermic reaction can occur between selected solvents and an acid, and the heat evolved may help to remove organic solid deposits. The acids may include organic acid compounds, such as sulfonic acids, sulfuric acid, and nitric acid. The solvents may include terpene and terpene derivative–containing solvents, including limonene, pinene, dipentene, myrcene, turpentines, and compounds having at least one double bond, such as methyl furan, dienes, styrene, and vinyl acetate. Although the reaction between some of these chemicals (some shown in Figure 7.14) was shown to generate substantial heat, it is not clear what the reaction does to the solvency of the organic molecules.

Emulsions and microemulsions (ME) of various organic solvents can be produced and are used to reduce the amount of organic solvent needed for deposit removal. ME are a special category of emulsions wherein the droplet size is reduced so that the emulsion appears to be thermodynamically stable. To form the ME, a specific ratio of surfactant, oil, and water as well as (usually) a cosolvent such as a butanol must be present. Figure 7.15 shows that only a very small area of the phase diagram will yield the ME, but these fluids can be powerful solvents with reduced environmental impacts in some situations.

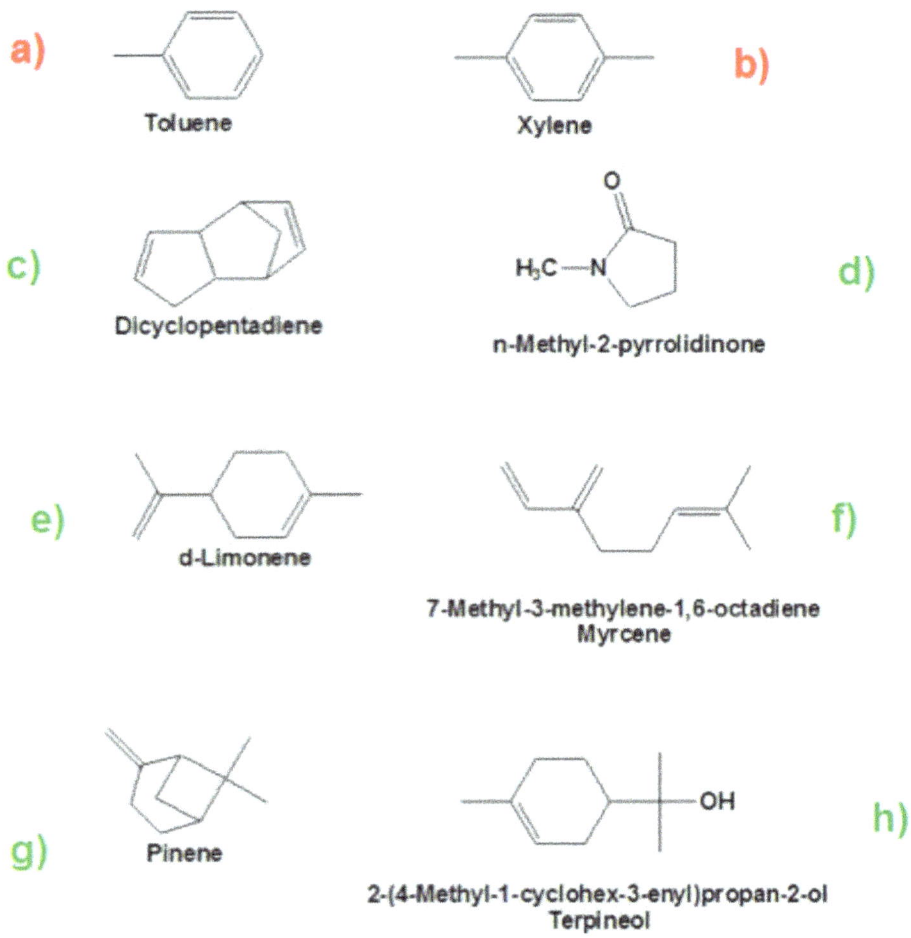

Figure 7.14 Standard (a, b) and Environmentally Improved (c–h) Solvent Molecules

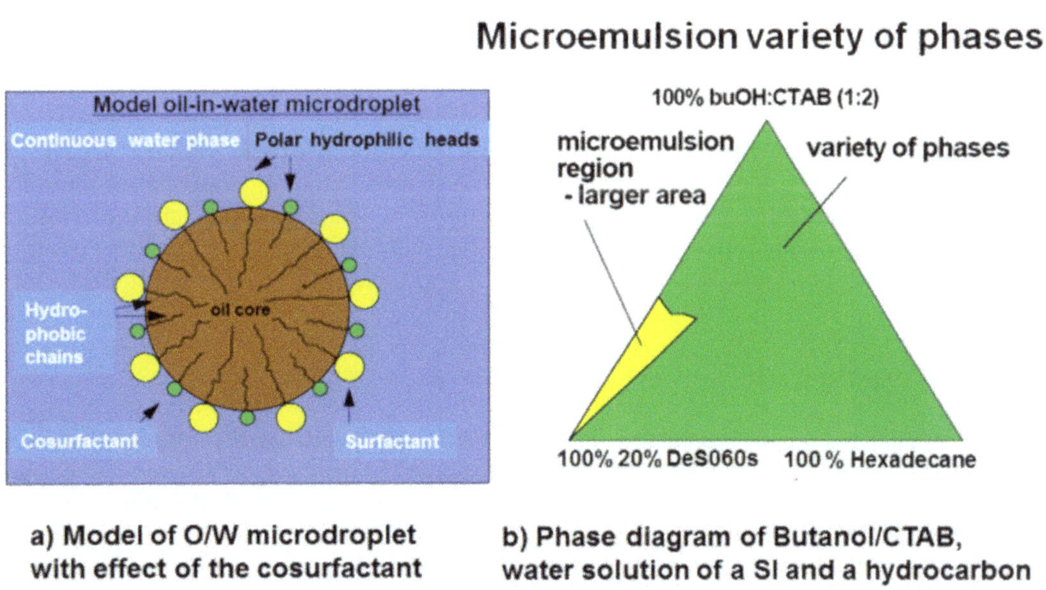

Figure 7.15 Microemulsion Phase Diagram

Collins[54] describes a ME comprising (i) an oil phase, (ii) an aqueous phase (an aqueous solution of a water-soluble oil field or gas field production chemical or an aqueous dispersion of a water-dispersible oil field or gas field production chemical), and (iii) at least one surfactant. The aqueous phase is distributed in the oil phase in the form of droplets having a diameter ranging from 1 mm to 1000 mm or in the form of microdomains having at least one dimension of length, breath, or thickness ranging from 1 nm to 1000 nm. This ME is claimed to be effective for removing various organic deposits in well environments. Emulsions using acids and chelating agents also can be used to remove mixed organic/inorganic deposits.

Solvents and techniques for specific types of deposits are reviewed in the next sections. Much of the text is derived from Frenier[1] and was developed for oil/gas production. However, these deposits may exist in lower temperature industrial units that handle petroleum products. Note that these are listed separately (mostly from the patent literature); however, the products also may be effective for a wider variety of organic refinery solids.

7.5.3. Solvents for Wax-Like Deposits

Bernadiner[55] extensively evaluated several surfactants and mixtures for removing wax and preventing re-deposition. The surfactant classes studied included alkyl sulfonates, alkylaryl sulfonates and polyethyleneglycol ether di-t-butylphenol materials. This detailed study looked at the effect of chemical type and flow by using rotating disks coated with wax. By plotting the square-root of the angular velocity (ω) of the disk vs. the dissolution rate, they found a significant effect of the detergent flow rate on the dissolution process in accordance with the Levich[56] relationships. Using mixtures of these surfactants at 0.4%–1.25%, the authors claim to have cleaned and prevented deposition of wax in wells located in Kazakhstan. These two papers (Fan and F.M. Llave[57] and Bernadiner[55]) are unusual in that they describe surface (mineral) dissolution kinetic approaches to look at organic dissolution mechanisms. They imply that the rates are diffusion limited at least under the test conditions.

Zhang et al.[58] claims a dewaxing fluid comprising

a. At least one paraffin-solubilizing organic solvent selected from the group consisting of aromatic hydrocarbons, terpenes, and isoparaffinic hydrocarbons (the paraffin-solubilizing organic solvent comprises about 25% to about 75% by volume of the composition)
b. At least one water-soluble polar organic solvent (about 25% to about 75% by volume of the composition)
c. At least one water-soluble surfactant (about 0.5% to about 20% by weight to volume of the composition)

The organic solvents in this patent include limonene, isoparaffinic hydrocarbon, and an alkylbenzene or dialkylbenzene. Examples of the polar organic include methanol, ethanol, isopropanol, butanol, acetone, ethylene glycol, and propylene glycol. A wide variety of water-soluble surfactants includes cationic, nonionic, or anionic materials, and the formulator has to develop the final formulation on the basis of all the ingredients chosen to provide a stable emulsion. This patent is a good example of typical oilfield formulations for wax removal.

Dyer[59] summarizes commonly used acid/detergent/emulsions for removing wax.

- Anionic surfactants used in detergents include fatty acid soaps, alpha olefin sulfonate, sulfonates, amine ethoxylates, amine salts or linear alkyl benzene sulfonic acid, aromatic sulfonates comprising cumene, xylene and toluene sulfonate, earth metal salts of olefin sulfonate and alcohol sulfates and sulfonates, as well as blends of such anionic surfactants.
- Nonionic surfactants suitable for use include ethoxylated nonionic surfactants selected from the group consisting of condensation products of ethylene oxide with aliphatic alcohols having from 8 to 22 carbon atoms in either straight or branched chain configuration as well as ethoxylated nonionic surfactants selected from the group consisting of condensation products of ethylene oxide with nonyl phenol,

phenol, butyl phenol, di-nonyl phenol, octyl phenol or other phenols, as well as blends of such nonionic surfactants.

- The acid constitutes from 0.1% by weight to about 15% by weight based upon the total weight of the emulsion (preferably from about 0.1% to about 5%). Suitable acids include hydrochloric, phosphoric, sulfuric, hydrofluoric, nitric, citric, oxalic, maleic, acetic, fumaric, malic, glutaric, or glutamic acids, as well as mixtures of these acids. The preferred acid is hydrochloric acid.

The hydrocarbon solvent constitutes from about 0.1% by weight to about 20.0% by weight based upon the total weight of the emulsion (preferably from about 0.1% to about 10%). The hydrocarbon solvents suitable for use in accordance with the present invention include kerosene, gasoline, diesel, jet fuel, xylene and mixtures thereof. The preferred solvent is kerosene.

A patent application by Dyer[60] describes a method for cleaning oil wells to increase the flow of oil by using a unique aqueous cleaning emulsion comprising water, hydrocarbon solvent, and detergent. This one-step method simultaneously cleans/removes asphaltene and/or paraffin and scale. This formulation contains water, a surfactant, and a hydrocarbon to form a stable emulsion and can be used to remove wax, asphaltenes, and scale. The specifications also mention adding a mineral acid to the emulsion.

Two authors have described solvents to remove wax (and possibly to inhibit wax formation) using biodegradable fluids. Tukenov[61] describes a solvent and a process that is claimed to be based on nanochemistry. This author specifically claims that the chemical works when mixed and activated with crude oil by a proprietary nanochemical mechanism and attacks wax differently. Tukenov claims that flushing wax with activated crude oil removes the wax by layers and creates a waxophobic (wax repelling) condition for an extended period. The chemistry is also environmentally friendly, is nontoxic, and is nonvolatile, and the activator is safe to handle. Spills can be washed down, and there are no special handling requirements. Neither the chemistry nor the mechanisms are revealed.

A patent by Andrecola[62] also describes the use of nanoemulsions of specific chemistries to efficiently remove wax deposits; however, the author does not claim residual inhibition of wax formation.

- The fluid system comprises a blend of micellar solution of fatty acids, a vegetable oil solvent, and one or more surfactants. The preferred fatty acids are from coconut oil, tall oil, or a blend of C8-C16 fatty acids such as isoleic or linoleic acids. Tall oil is the preferred fatty acid.
- The fluid system used in the invention is a nanoemulsion, defined as a multiphase system consisting of water, fatty acid, and vegetable oil blends, emulsifier(s), and alcohol and is a transparent and thermodynamically stable liquid solution.
- The nanoemulsion of droplet-type dispersions of oil-in-water has an average particulate size ranging from about 5 nm to about 50 nm in drop radius. Nanoemulsions of the current disclosure are prepared by providing an external energy input to the oil/water/surfactant system using high shear stress or inertial disruption to overcome the effect of interfacial tension to fragment large micro scale droplets into the nanoscale. Unlike ME, in which the average drop size grows continuously with time so that phase separation ultimately occurs, the nanoemulsions of the current disclosure are thermodynamically and kinetically stable. The drops of ME are generally large (>0.1 microns) and often exhibit a milky or cloudy, rather than a translucent, appearance as seen in nanoemulsions.

7.5.4. Heavy Asphalt-like Deposit Removal Methods

Because asphaltenes/asphalts contain more aromatic and unsaturated groups than paraffin and often contain a very wide mixture of different chemical species, the solvent tends to be more aromatic and more complex. Xylene is often used to remove asphaltenes, but other methods are described in the patent literature. Some of the references are from the oil and gas production literature, but the solvents could be used to remove heavy asphalt-like deposits in refinery equipment.

Preliminary dispersant and solvency tests were conducted by Thomas et al.[63] using an asphaltene dispersant test in hexane. Chemicals that provided promising results in dissolving and dispersing asphaltenes in the non-solvent medium of hexane were selected as candidates for field application. This consisted of making asphaltene pellets and noting how fast the brown color dispersed into hexanes in the presence of a dispersant or for additional testing in a core flow deposition removal test. The core flow test apparatus provided a method to introduce asphaltene fouling into a core and study its removal using chemical magnets.

Del Bianco and Stroppa[64] describe a composition that includes saturated species, alkylbenzenes, and polyaromatics that is useful for dissolving asphaltene residues in oil wells. The authors developed a graphical method for making the blends. The compositions are selected from those that were on a ternary chart and contained various polyaromatics, alkylbenzenes, and other hydrocarbons.

Solvent treatment of the oil is considered to be beneficial[65, 66] in some cases because it dilutes the crude oil and reduces the tendency of the heavy organics to precipitate. Solvent treatments may not be very successful largely because the solvents that can be used are limited to aromatic solvents. Xylene is generally the most common solvent selected for well stimulations, workovers, and heavy organics inhibition and cleaning. In some cases, xylene injection through the non-producing string may help to minimize the heavy organic deposition problem. They note that in oil fields with frequent need for aromatic wash, it may be necessary to design an aromatic solvent with stronger wash power and better economy for the particular deposit in mind.[67] Laboratory tests may be necessary to blend the most appropriate aromatic solvent and/or dispersant for a given oil field from the points of view of effectiveness, economy, and environmental friendliness.

Del Bianco and Stroppa[68] proposed a process for the dissolution of asphaltene sediments in oil wells that consists of introducing a solubilizing composition for the above asphaltenes into the oil wells. The solubilizing composition comprises

(a) A hydrocarbon fraction consisting of at least 70%, preferably at least 80% by weight, aromatic and alkyl aromatic hydrocarbons (alkyl group, C_1 to C_4); (b) a fraction consisting of quinoline and isoquinoline as such or alkyl substituted, preferably as such (alkyl group, C_1 to C_4). The weight ratio between fraction (a) and (b) ranges from 97.5/2.5 to 75/25, preferably from 97/3 to 90/10. For example, the fraction called "Wash Oil" (WO) deriving from the distillation of coal tar, which contains a certain content of quinoline (usually from 5% to 10% by weight), has proved to be particularly effective. WO has an initial distillation point (ASTM D2887) of between 198°C and 210°C, a final distillation point of between 294°C and 310°C, and 50% distills at a temperature lower than 230°C to 250°C (15TCAS15T number: 309-985-4; EINEX number: 101896-27-9). The WO can be used as such, or another quinoline or isoquinoline can be added as required.

Thorssen and Loree[69] proposed a wax and asphaltene solvation fluid for use in oil and gas wells that is derived as a residual fluid from a feed stock that includes a greater mass percentage of trimethylbenzene than decane and is preferably sour. Mass percentage of both aromatics and asphaltenes in the residual fluid is in the 30%–70% range, and a complex mixture of both is described. See also Loree.[70]

Several authors have proposed using various oilfield or refinery fluids for removing asphaltenes. Jamaluddin and Nazarko[71] developed a method that comprises the injection of deasphalted oil into the near-wellbore formation followed by a soaking period and a production period. This method has the great advantage of readily dissolving precipitated asphaltene in a well environment without other costly treatments, such as the use of solvents like xylene or toluene. Furthermore, this method does not require the incorporation of any additives in the deasphalted oil. Deasphalted oil is injected at a temperature ranging from 20°C–100°C to increase the solubilizing power of asphaltene particles. The deasphalted oil can be any oil from which the natural asphaltene or asphalt components have been removed. The deasphalted oil is obtained directly from the well to be treated, and the deasphalting of the crude oil is realized by precipitating asphaltene molecules with n-pentane. Deasphalted oil also can be obtained from oil-refining or heavy oil-upgrading processes.

Cimino et al.[72] note that monocyclic aromatic solvents such a toluene or xylene gave less than 50% dissolution of many asphaltenes and thus recommend inexpensive distilled refinery blends. A preferred mixture for testing could contain 1-methylnaphthalene, n-hexadecane, and toluene.

Buzelin and Lima[73] describe the use of a chemical product composed of diesel, isopropane, benzene, and naphthalene to clear subsea pipes clogged with mixtures of wax and asphaltenes. The pipes need to be treated with these chemicals for a minimum of 12 hours, and better results could be obtained by increasing the treatment period. A special filtration procedure is required to separate the liquids and solids from the material that is discarded from the pipes, and the liquids are recycled through the pipe again until the pipes are fully treated. In this process, via pumping from the surface, the cleaning mixture was introduced and then flushed through the pipeline.

Park et al.[74] proposed a process for cleaning fouled oil lines in refineries. They claim that cleaning agent can be introduced into the feed-cut process lines of oil refining plants or petrochemical plants and then circulated in the process lines with the aid of a pump without interrupting the process. Drawings in the patent (not reproduced here) show the lines in which the cleaning agent may be introduced.

The cleaning agents were mixtures of C_8, C_9, and C_{10} aromatic compounds, such as o-xylene, 1, 2, 4-trimethylbenzene, methyl-3-ethylbenzene, and 1-methyl-3-n-propylbenzene. The authors claim that about a 10%-20% volume of the mixture is then added to 80%-90% of a light cycle oil (LCO) or a light gas oil (LGO). These are hydrocarbon cuts that may be available in the refinery.

After introducing the cleaning agent into a feed line, it is closed, and the cleaning agent can be circulated through the feed line. Next, the cleaning agent is heated by use of a heat source present in the oil refining process to create thermal circulation while the feed line is being cleaned. The feed line is cleaned first with the aim of maximizing the cleaning effect by using a fresher cleaning agent to dislodge fouling formed in the main stream. The authors claim that mainstream fouling occurs more than in a side stream. During the thermal circulation of the cleaning agent (at about 250°F), fouling solids are dislodged.

The structures of these cleaning molecules are shown in Figure 7.16. These are not green molecules, but they will stay in the refinery processes and thus may not cause a significant environmental issue. The authors suggest that once used, the cleaning agent can be re-treated, together with crude oil, in a crude oil unit (CDU) or reused as a bunker-C oil blending material to avoid environmental pollution resulting from the treatment of waste oil. Note that the use of a refinery hydrocarbon stream with heavy aromatic chemicals is a recurring theme.

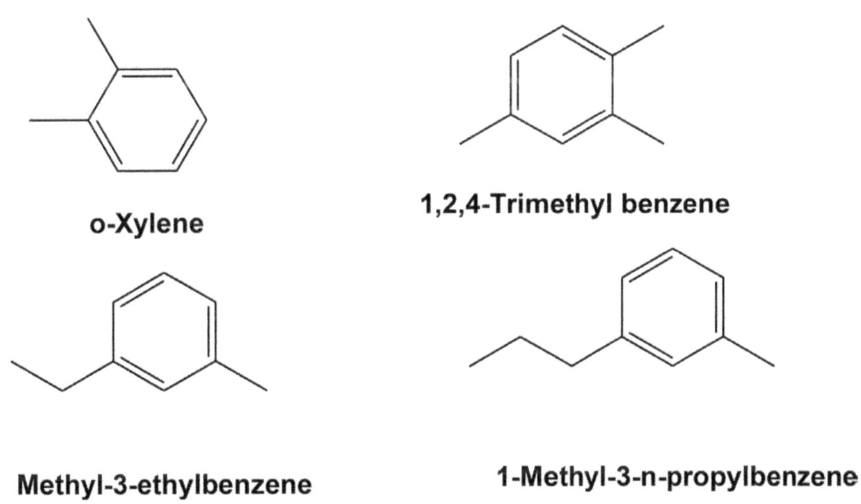

Figure 7.16 On-Line Solvents

7.5.5. Solvents for Mixed Deposits

Most if not all fouling deposits in pipelines, facilities, refineries, and petrochemical plants are mixtures of several (or many) different solids. If several organic solids and inorganic deposits are present, chemical methods may be needed to remove enough of the materials to allow corrosion assessments or other maintenance to be performed. It is most critical to conduct the deposit analyses described in Section 2.6.1 to determine the ratio of inorganic/organic solids in the deposit to start the solvent analysis process.

Wylde and Salyer[75] make some particularly useful observations about selecting solvents to remove deposits that are very mixed. Several chemical factors should be considered:

- Wetting of the solids by the solvents
- Solubilization (actual dissolution)
- Emulsification
- Dispersion
- Detergency
- Temperature

The point made in this article[75] is that several different chemicals (5-7) may be required, and these must be custom blended on the basis of the composition of the solids to be removed. More details of solvent testing and selection are in Section 2.6.

Removal of mixed organic/inorganic deposits and sludges poses unique problems because even more different chemistries must be addressed simultaneously. There are three broad strategies for removing mixed deposits:

Stages of Solvents. A very common strategy is to pump/spray an organic solvent to remove an organic solid and then follow it with an acid/chelant/detergent stage to dissolve or disperse the inorganic solids. An alkaline (NaOH + a phosphate/surfactant) wash also is often used as a first phase of a chelating process. Frequently, several cycles may be needed for complete removal of the deposit. Montgomery et al.[76] have described the use of various solvent stages for deposit removal. If the unit being cleaned is a pipeline segment, stages of solvents also can be used with pig trains described in Frenier[3] and in Section 8.2.

Emulsions or Dispersions of Organic Fluids with Aqueous Fluids. Oil outer phase and oil inner phase emulsions can be generated depending on the need. Many combinations of organic solvents and acids and/or chelating agents can be formed as an emulsion.[77,78] Many ME (that appear to be one phase) have also been produced.[54] The benefit of a single stage is that it is easier to control (from a placement perspective) than multiple stages. In all of these fluids, a major issue is the use of a properly tested corrosion inhibitor[2] to protect the base metal of the tubing and flow lines from attack from the aggressive fluids needed for removal of the mineral scale portion of the deposits. This is a difficult problem because the organic portion of the fluid may dissolve the inhibitor film if the formulation is not properly tested and formulated.

Aqueous Fluids Containing Surfactants and Cosolvents. Mixtures of aqueous fluids can be blended with relatively large amounts of alcohols (methanol, ethanol, or isopropyl alcohol) and ether-alcohols (ethyleneglycolmonobutyl ether or dipropyleneglycol methyl ether). See Figure 7.17. These types of fluids can dissolve some mixtures of organic/inorganic foulants but are not as aggressive as emulsions if there are very large concentrations of a heavy organic solid such as an asphaltene. Frenier and Brady[79] have proposed a mixture of several types of glycols with chelating agents and other solvents and surfactants that were effective for removing inorganic and organic deposits from the near well bore area of the formation. The fluids were a single phase and could be used with various corrosion inhibitors at high temperatures and in high salt content fluids. Organic acids such as acetic acid or stronger acids (HCl of H_2SO_4) also have been used with surfactants and mutual solvents to remove mixed foulants. McCoy[80] describes the development of alcohol-based solvents (and cosolvents; see right side of Figure 7.17) that are now used in many solid surface cleaners and detergents. The materials are based on natural products and are less toxic than ethylenegylcol monobutyl ether (left side of Figure 7.17). When used with surfactants, these materials may be used to clean many surfaces.

The choice of an application method as well as the choice of the individual solvent components depends on the total makeup of the fouling solids. The methods described in Section 2.6.4 can be used to choose and optimize the final formulation.

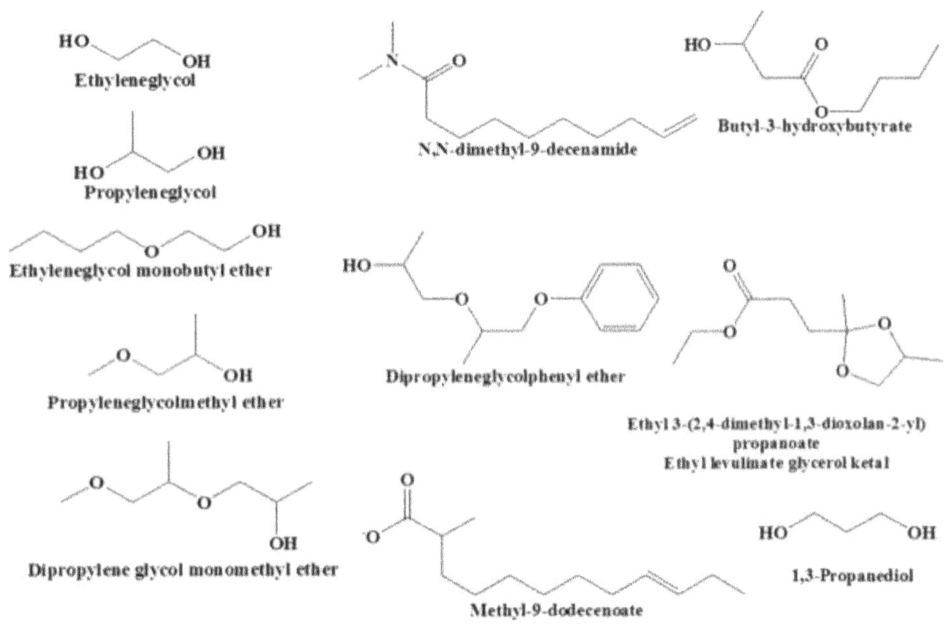

Figure 7.17 Alcohol -Type Solvents and Cosolvents

In conclusion, various mixtures of various organic solvents (Figure 7.14), surfactants (Figure 7.9), and cosolvents (Figure 7.17) can be formulated to wet, emulsify, and disperse mixtures of organic and inorganic solids from refineries and pipelines.

7.6. Decontamination and Cleaning of Refinery Vessels

When a refinery comes offline for maintenance, removal of toxic vapors and gases is required before any maintenance can be initiated on any of the process vessels. This is called decontamination (decon) or clearing for entry (CFE). This section describes the applications of some of the chemistries described in Section 7.5 for this important process of chemical cleaning. Since H_2S also may be one of the gases that may require removal, iron sulfide deposits may be present. The processes for removing these solids are reviewed in Section 7.7.

Eliminating dangerous volatile organic compounds (VOCs) and H_2S must be accomplished before personnel can safely enter any refinery vessel for maintenance, and this can be a slow process. Several short reports from various sources and patents will be reviewed to describe some of the approaches to this service.

Mason[81] describes parts of the decontamination process. He claims that following cool-down and evacuation of the residual oil, the refiner may begin steaming the equipment to purge gases out of the equipment and continue with liquid oil removal. In past years, the steaming process took days to meet the required personnel entry specifications (VOC reduction, H_2S removal, and pyrophoric iron control). More recently, Mason claims that contractors use the vapor-phase degassing process. This process is the introduction of chemicals (described here) into the steam or other fluid to reduce the time involved with equipment degassing.

This process uses steam that contains detergents or other solvents to help remove organic deposits from refinery or chemical process equipment. The heat and the surfactants in the fluid are effective in reducing the viscosity of organic deposits and then emulsifying them. This allows the surfaces to be cleaned. An important consideration is that the surfactants produce an unstable emulsion that readily breaks after the condensed steam and organic deposits flow out from the unit. This requirement means that shear forces must be continuous to maintain the emulsions until the deposits are removed from the unit. See Kokal.[82]

The currently used processes are based on a patent by Brown[83] and are called vapor phase cleaning. This process involves injecting the cleaning solvent into a flow of saturated steam. The mixture is then passed through the unit being cleaned. Generally, the patent claims that one part of cleaning solvent is injected for every four parts of steam on a weight basis. Several cleaning mechanisms have been claimed.

- Steam-volatile solvents (some organic acids) condense on the surface, and cleaning occurs by a mixture of solvent action and erosion by the velocity of the steam.
- Non-volatile solvents (surfactants) are atomized by the steam flow and contact the deposit in the form of small droplets and work by a combination of dissolving action and erosion.
- If viscous hydrocarbon deposits are being treated, the hot steam reduces the viscosity, and detergents/solvents in the steam can emulsify them.
- The steam also increases the reaction rates of dissolution of inorganic salts (iron oxides).

Note that the original process supported both organic and inorganic salt solvents. Any acid/chelate solvent would require an appropriate corrosion inhibitor (Chapter 6).

A more recent invention of Sears[84] uses terpenes (Figure 7.14). The process is a method of cleaning a contaminated vessel:

1) Providing a steam source
2) Providing a surfactant source
3) Providing an organic solvent source
4) Delivering steam from the steam source to the vessel
5) Introducing the organic solvent from the organic solvent source into the delivered steam
6) Introducing a surfactant from the surfactant source into the steam
7) Removing vaporous effluent from said vessel while the steam, organic solvent, and surfactant are being delivered to form a circuit
8) The fluids are drained

HydroChem[85] described a CFE process that uses non-hazardous chemicals and that has lower toxicity than most hand soaps. The decontamination chemicals are claimed to be detergents (Section 7.5.2) and do not contain substances deemed hazardous by the EPA. They are also claimed to be safe for all known metallurgies and have a minimum impact on wastewater treatment facilities.

A brochure from Halliburton[86] describes a decontamination fluid that can be used from about 75-100°C to remove volatile organics, crude oil, benzene, and hydrogen sulfide from vessels, heat exchangers, and process piping. The product is described as having a sweet, aromatic odor and a flash point > 95°C.

A patent by Mehta[45] describes a decontamination process in which an extractant and a surfactant are injected into steam to remove VOCs. The extractant is claimed to be a limonene and other similar materials, preferably the various terpenes (for example, dipentenes, cinenes, cajeputenes, diamylenes, the oils of bergamot, geranium, citronella, dill, and caraway) and related terpenes, such as hermiterpenes (isoprenes), sesquiterpenes (caryophyllenes), diterpenes, and polyterpenes. The extractant material should be non-toxic and non-hazardous and selected such that it has a high vapor pressure or boiling point within the range of the steam available at the plant for use. Several types of surfactants were noted, and Table 7.7 shows a preferred chemical additive mixture. Figure 7.14 shows the structure of several solvents.

Table 7.7 Composition of Decontamination Fluid

Component	Wt.%
D. Limonene	57
Surfactant (Block copolymer of ethylene oxide and propylene oxide having a molecular weight of 1950 and HLB of 12 to 18)	9
Surfactant (90% active ingredients of a 50/50 mixture of an alkanol-amide of a fatty acid and an alkanolarnine and a linear alkylbenzene sulfonic acid in 10% isopropanol)	17
Cosolvent (2-butoxy ethanol)	3
Surfactant (10 mole ethylene oxide) adduct of a nonylphenol	5
Water	9

Sears[84] describes a process for interior cleaning, removing noxious gas, and/or restoring the operating efficiency of organically contaminated hydrocarbon processing equipment in a safe and effective manner and in a very short period of time, without the need to manually enter an unsafe environment and mechanically remove organic contaminants. Sears used a formulation of non-aqueous, monocyclic saturated terpenes, mixed with a non-ionic surfactant package. The terpene-based chemical is injected into organically contaminated equipment using a novel process involving high-pressure steam for very effective cleaning.

A comprehensive review of various methods for decontaminating VDU towers is in the report by Shank et al.[6] The types of chemicals used in chemical cleaning blends may include various classes, including acids and bases, solvents, surfactants, oxidizers and reducing agents, chelants, and enzymes; however, chemical interactions and incompatibilities must be considered when formulating any chemical blend. For example, Shank et al.[6] claim that in the removal of asphaltic-based organic materials, aromatic solvents are particularly useful; however, paraffinic compounds, such as waxes and lubricants, require the use of more aliphatic solvents.

For the removal of many oxides and scales, they note that common acid cleaning is highly effective. However, it is claimed[6] that caustic cleaning agents are better suited to the removal of oils and grease from contaminated soils and are highly effective in the removal of sulfur deposits. These authors also propose that chelants may be used when the deposits are high in metal oxides or when electrochemical plating has occurred within the process equipment. Surfactants are highly effective in encapsulating particles and preventing redeposition of the deposit on the equipment surface.

Shank et al.[6] claim that chemical cleaning agents that are designed for scale and inorganic deposit removal can be delivered in an aqueous phase, which, in the case of the VDU column, requires that the column be de-oiled and degassed before cleaning.

However, they also note that organic solvents and solvent-based degreasers are traditionally not well suited for scale and deposit removal. Since both highly aromatic asphaltic sludge and pyrophoric iron sulfide scale are present in the system, a multi-stage approach must be taken to decontaminate the column. Some chemical cleaning programmers are designed to work as an emulsion, with both an aqueous and organic phase operating within the system simultaneously. However, for this type of chemical cleaning to work, the emulsion must be stable under the conditions within the column, and a storage tank must be provided where the emulsion can later break, either on its own or with the aid of an emulsion breaker. Details of these chemicals are in Section 7.5.

Shank et al.[6] also show that chemical cleaning of the VDU column can take several forms:

- Cascading of the chemical
- Recirculation of the chemical
- Vapor-phase steaming of the chemical

These methods and the direction of flow of the chemical within the system are depicted in Figure 7.18.

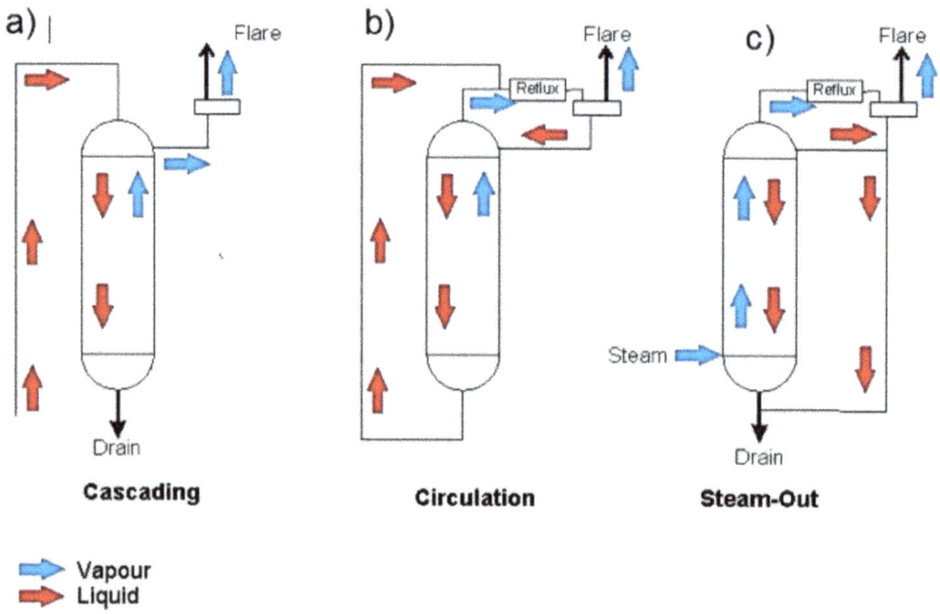

Figure 7.18 Chemical Cleaning Methods for Cleaning of a Vacuum Distillation Unit (VDU) Column with the Heat Exchanger Cleaned In-line. a) Cascading of the chemical through the system with a constant fresh chemical supply. b) Circulation of a chemical blend through the system, and c) Steam-out cleaning of the column using chemical in the vapor phase[6]

Cascading of the chemical cleaning blend, shown in Figure 7.18, occurs when it is not possible to pump the chemical back through the column, according to Shank et al.[6] A consequence is that a large amount of chemical and wash water is required to ensure that the column has been sufficiently cleaned. The chemical is applied into the wash water at the top of the distillation column or reflux line. However, steam can be added to supply more turbulence on the surface of the trays and within the packing. The chemical cascades down the column along with the wash water, contacting the packing, trays, and the walls of the column and finally reaching the exchanger at the bottom of the column. The chemical then passes through the drain and is pumped away for disposal. Volatiles rise to the top of the column, where they are vented to flare.

7.7. Iron Sulfide Dissolution Processes

7.7.1. Introduction

In many of the hydrocarbon processing steps, iron-based surfaces may encounter sulfur-containing fluids. At the temperatures present in the various reactors, and during long periods of contact, iron sulfide deposits (generally FeS, but sometimes FeS_2) will form. The cleaning of FeS deposits is such a dangerous operation that a separate section is devoted to the literature of the known technologies.

The low sulfur ratio sulfides (close to stoichiometric ratio of Fe and S) can be easily dissolved using mineral acids:

$$FeS + 2H^{+} = Fe^{2+} + H_2S \qquad \text{(Equation [7.5])}$$

The rates of reaction depend on the exact mineral present as well as the temperature; however, if solvent temperatures cannot be maintained at 150°F, chemical dissolution may not be an option. HCl and HCl with additives are commonly used to clean parts of surface-based gas removal facilities.[23] Mackinawite and pyrrhotite react rapidly; other sulfides such as pyrite are essentially insoluble in acid.

In addition to producing H_2S, some iron sulfide deposits are pyrophoric and will combust when in contact with air.

$$\text{A) } 4FeS + 3O_2 = 2Fe_2O_3 + 4S + \textit{heat} \qquad \text{(Equation [7.6])}$$
$$\text{B) } 4FeS + 7O_2 = 2Fe_2O_3 + 4SO_2 + \textit{heat}$$

7.7.2. Chemistries and Mechanisms for Dissolving Iron Sulfides

While the acid reaction (Equation [7.5]) for dissolving FeS is a very effective procedure, it produces large amounts of toxic H_2S. This gas produces severe safety and operational problems. Lawson[87] reviewed the major procedures for safely removing iron sulfide deposits:

- Mineral acid with an acid/gas scrubber
- Mineral acids with hydrogen sulfide suppression chemicals
- Multiple stages of oxidizing agents with acids
- Alkaline cleaners

If none of the suppression/control technologies are employed, the surface vessel being cleaned must be vented through a NaOH scrubber with enough capacity to absorb all the hydrogen sulfide during the shortest possible reaction time. As an alternative, the vessel can be vented to a flare. In both cases, it is advisable to vent through a knockout tank to remove any entrained liquid. During cleaning, the H_2S evolved makes the acid fluid very corrosive to the steel and presents a hazard when the there is a release. Fatalities to service crews have occurred in industrial plants and at well sites.

Lawson[87] suggested alternate stages of potassium permanganate and sulfuric acid for removing FeS. The proposed reactions are

$$\text{A) } 9FeS + 26KMnO_4 + 4H_2O \rightarrow +3Fe_3O_4 + 26K^{+} + 9SO_4^{2-} + 3OH^{-} \qquad \text{(Equation [7.7])}$$
$$\text{B) } Fe_3O_4 + 4H_2SO_4 \rightarrow 2Fe^{3+} + Fe^{2+} + 4SO_4^{2-} + 4H_2O$$

This process (especially the initial oxidizing stage) is commonly performed during refinery cleaning operations to control pyrophoric iron sulfides. See the article by Otzisk.[88.]

Spurrell[89] discusses the use and safety of different oxidizers in refinery cleaning that include NaOCl, H_2O_2, and $KMnO_4$. $KMnO_4$ is considered to be the most useful. His reasons are noted:

- The solution doesn't get hot when mixed
- One can add KMnO4 to water or water to $KMnO_4$

- It doesn't change the pH of the water
- It doesn't give off any gases (under normal conditions)
- Even a few ppm of $KMnO_4$ in water is visible
- It is its own indicator (turning from purple to brown) when spent
- Sophisticated lab equipment is not needed to determine concentration
- Unused material is easily stored, reused, or removed
- The manganese dioxide residual is relatively non-toxic and is not an environmental hazard

Several different technologies are available to suppress the reformation of H_2S and are used as additives to mineral acids, such as HCl or H_2SO_4. Ball,[90] Buske,[91] and Frenier[92] developed suppression agents that contain aldehydes. The most efficient agent is formaldehyde that reacts stoichiometrically with hydrogen sulfide to produce trithiane, a very insoluble material. Sometimes, a back-up scrubber system also is used to insure complete removal of sulfide gas.

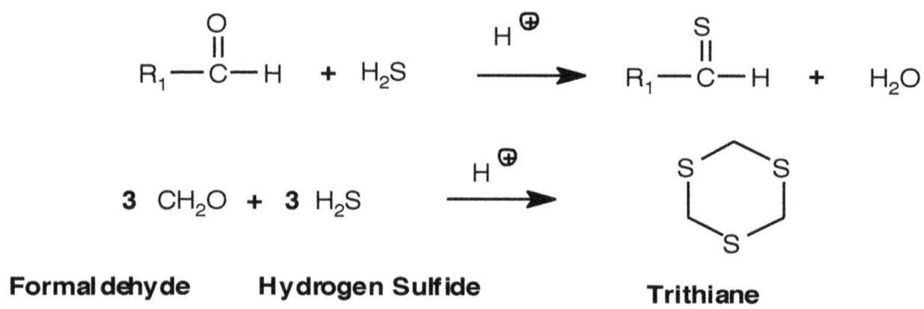

(Equation [7.8])

Tests conducted in the laboratory (backed up with treatments) demonstrated that formaldehyde used with sulfuric acid can dissolve FeS with essentially 100% suppression of H_2S evolution. There are several problems with the use of formaldehyde. This chemical is listed in several compilations of suspected carcinogens. In addition, formaldehyde can react with hydrochloric acid to form chloromethyl ethers,[93] which are known human carcinogens. In addition, sulfuric acid usually cannot be used in downhole operations because of concerns with formation of calcium sulfate deposits. In addition, the major reaction product of the reaction is trithiane (Equation [7.8]), a white solid that has very low solubilities. It usually can be flushed out of a unit, but this also could be a disposal issue.

$$HCl + 2\,CH_2O \longrightarrow \underset{\text{Chloromethylmethy ether}}{Cl-CH_2-O-CH_3} \text{ and } \underset{\text{Bischolormethyl ether}}{Cl-CH_2-O-CH_2-Cl}$$ (Equation [7.9])

Thus, formaldehyde usually is replaced by glyoxal in many treatments. The reaction product is more oily and easier to flush from a unit. Glyoxylic acid[91] also can be used with hydrochloric and sulfuric acids. Formaldehyde also can be generated in situ by adding hexamethylenetetramine to strong acid:

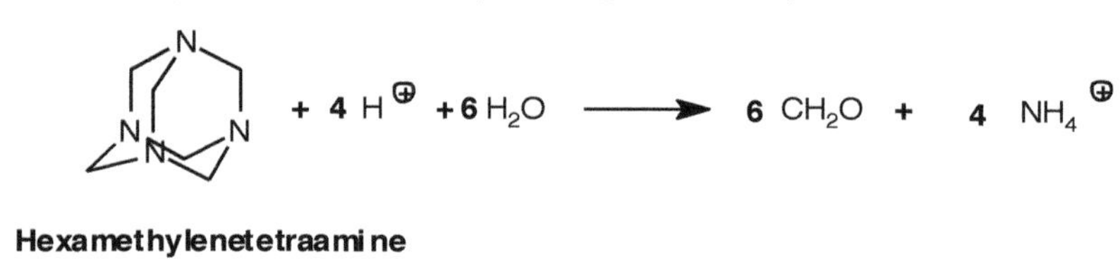

(Equation [7.10])

Additional less toxic suppression agents that include mixtures of glyoxylic acid, cinnamaldehyde, and glyoxal (Figure 7.19) also have been used in some situations in the petroleum industries. Frenier and Hill[94] developed a scavenger system that did not significantly inhibit FeS dissolution and that contained glyoxylic acid and an unsaturated aldehyde. This system uses glyoxal (preferably glyoxylic acid and cinnamaldehyde) to dissolve FeS, capture the H_2S, and lower the corrosion rate to acceptable values. These materials have been formulated in HCl.

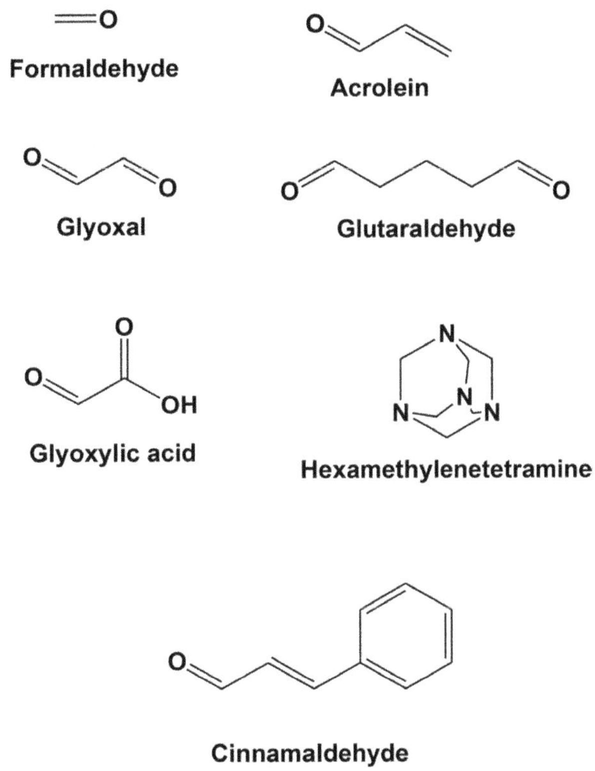

Figure 7.19 Alternative Aldehydes for H_2S Control

Dilute aqueous solutions of sulfuric acid (H_2SO_4 at about 10-15 wt.%) were used in conjunction with the aldehydes. Frenier[26,92] described the use of this solvent with aldehydes described above to safely remove FeS-containing deposits from refinery and oilfield gas-treating facilities. These solutions can be used on metals (such as 300-series stainless steels) that may be attacked by HCl-based solutions.

In an example described by Ball,[90] a heat exchanger in a refinery fouled by FeS deposits was cleaned using a solution of 10% H_2SO_4 and an aldehyde at about 125°F. The metal analyses from the solutions showed that more than 4000 lb. of FeS was removed without H_2S being detected. An example of a field treatment of piping fouled with FeS is shown in Figure 7.20. In this figure, part (a) is the fouled pipe section, and part (b) is the Fe/time dissolution curve for this treatment. From these data, the system was cleaned in about 7 hours of circulation.

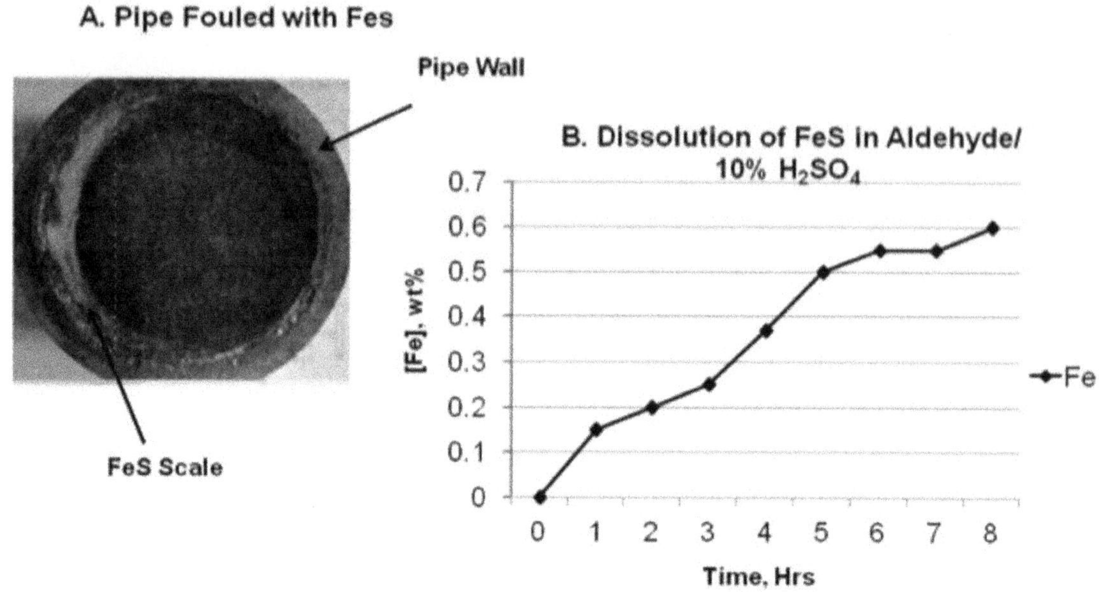

Figure 7.20 Field Cleaning of FeS Fouled Piping at 125°F[90]

Note that the setup included a gas scrubber containing a dilute NaOH solution as a backup. Similar jobs (unpublished) have been accomplished in oilfield gas plants. It is very important to note that the sulfuric acid descaling solutions require a different corrosion inhibitor family compared with an HCl inhibitor. Chemistries like those used for organic acids and chelating agents (Section 6.3) must be used in H_2SO_4 solvents. Note that flushing, passivation, and disposal processes described for HCl also must be conducted for this acidic solvent.

Miller[95] describes a FeS solvent cleaning composition that consists of formaldehyde or glutaraldehyde, methanol, 2-butoxyethanol, isopropyl alcohol, hydrochloric acid, surfactant intermediate, surfactant/corrosion inhibitors, glacial acetic acid, and water. This formulation is a combination of previous technologies but adds an organic acid (acetic acid).

The following paragraphs describe FeS removal methods that have been proposed for use in oil/gas wells but not in IC. These methods are included for a complete description of known technologies.

Nasr-El-Din et al.[96] suggest that a drawback of aldehyde-based scavengers in acidizing treatments is the formation of oily material, which may adversely affect the performance of water injectors.

This paper discusses a scavenger that is based on triazines (Figure 7.21). Laboratory work by these authors included compatibility of the old and new hydrogen sulfide scavenger with live acids. The impact of the new scavenger on corrosion of low-carbon steel was also addressed. They claim that on the basis of encouraging laboratory results, field trials were conducted on several wells. Post acidizing flowback samples were collected and analyzed to confirm laboratory results. Finally, injectivity results from the field trial were compared with offset well performance to validate the technical benefits of the new hydrogen sulfide scavenger.

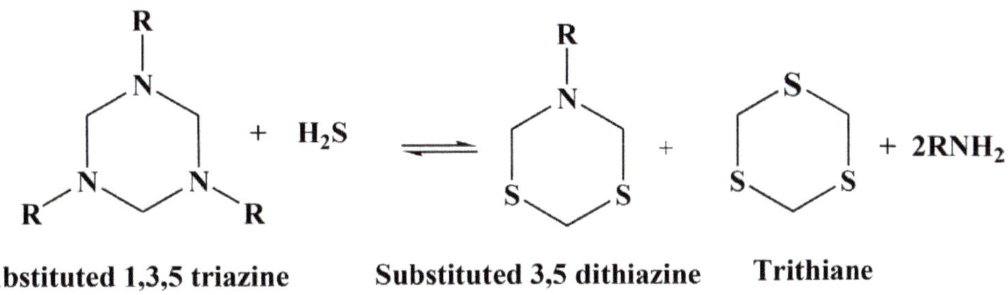

Figure 7.21 Reactions of Triazines with H_2S

Non (or low) acidic FeS removal formulations also have been proposed and tested in the lab. Mattox[97] and Talbot[98] claim that tetrakishydroxymethylphosphonium sulfate (THPS), a biocide, also can be used to dissolve FeS. These authors suggest that a chelating agent or ammonium chloride should also be in the formulation. Mattox[97] also claims that the solutions of THPS/ammonium chloride react with and complex soluble and insoluble iron(II) sulfide compounds by a series of chemical reactions (not described) at ambient temperature, resulting in the solubilization of the iron(II) compounds. This author also claims that the formulation can be pumped through pipelines as slugs between pigs (see Section 8.2) as a pig train.

However, a more recent publication on the use of THPS as a solvent suggests that a polymeric additive with THPS may also be effective. A patent application by Talbot and Jones[99] claims that adding an acidic polymer, such as a polyacrylate terminated with vinylidene-1,1-diphosphonic acid (VDPA), to THPS will dissolve FeS. Figure 7.12 shows the structure of THPS (a) and the proposed structure of an iron complex of THPS (b). Part (c) of this figure shows the reactivity of THPS formulations with different additives. The data from the applications indicate that the more acidic additive dissolved more FeS than THPS alone.

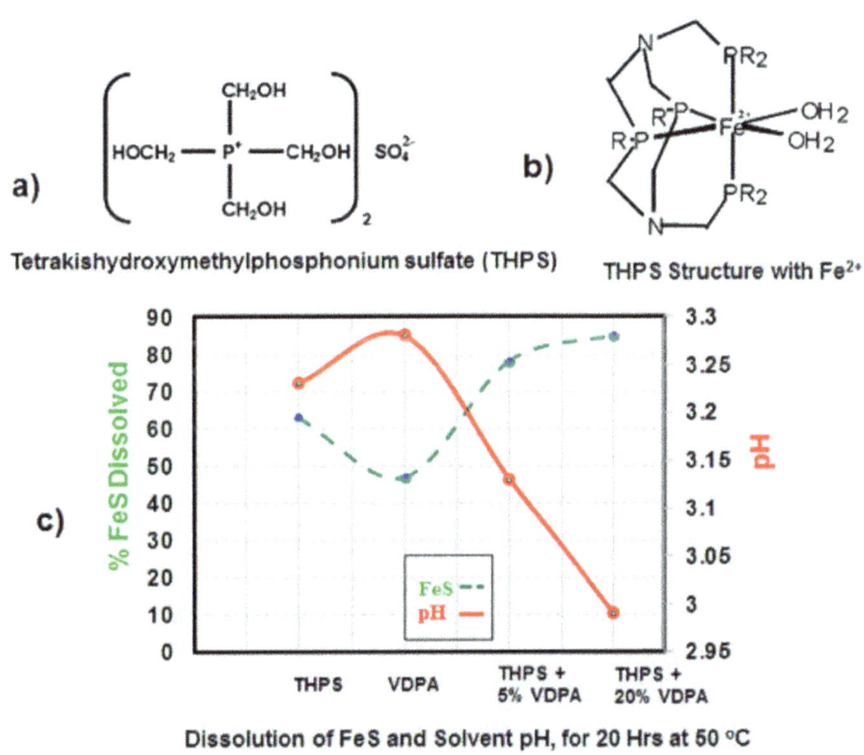

Figure 7.22 Structure and Reactivity of THPS with FeS[99]

Chelating agents like those proposed for calcite and iron oxide dissolution have been suggested to dissolve FeS.[100–102] A non-acidic technique uses alkaline chelating agents, such as ammonium EDTA[100] and oxalic acid, to dissolve as much as 3% FeS (in 100 mL) at a temperature of 300°F (148°C). The pH was about 9. The high alkalinity suppressed the evolution of hydrogen sulfide and allowed the solvent to dissolve most of the FeS.

Yap[101] and Wang[102] claim that lower pH chelate formulations can remove FeS at temperatures as low as 250°F. Wang[102] also claims that the technologies described in Mattox[97] may be very corrosive. Although these alternative higher (than mineral acids) pH formulations appear to be promising in lab tests, no field tests/commercialization (or relative costs) of these alternative formulations on FeS removal in the pipeline/facilities environments have been documented. The assertion that the THPS formulations are corrosive also has not been confirmed.

A system using maleic acid was invented by Lawson[103] to suppress hydrogen sulfide during removal of iron sulfide. Tests showed that >90% suppression of H_2S could be achieved while dissolving 65% of the FeS. Review papers on alternate oilfield FeS removal technologies are in the reports by Chen et al.[104] and Wylde.[105] Possibly some of this technology could be applied to IC processes. However, these papers do not claim that the sulfide part of FeS has been converted to an insoluble compound, so possibly re-acidification of the cleaning waste would evolve H_2S.

Although procedures for cleaning the tubulars of oil wells are physically different than those used in an industrial environment, the chemistry required is very similar. Several sulfide control techniques used in the oilfield are reviewed. Hall[106] considered several precipitation control methods for sour and sweet wells. This type of precipitation is not a significant problem while cleaning industrial equipment because the acids usually are not completely spent on the deposit. He advocated the use of a sulfide modifier (NTA and ethyleneglycolmonobutyl ether [EGMBE]) to treat sour wells containing iron. Dill[107] claimed a combination of a chelating agent (NTA) and an aldehyde (aldol) for controlling iron and sulfur precipitation. Walker[108] expanded on this concept in an SPE paper and noted that iron control agents are not required during tube clean-out procedures, but he advocated use of a sulfide scavenger. Ford[109] discusses the various types of iron sulfide found in wells and notes that the FeS (nominal) minerals are soluble in HCl, but the FeS_2 types are not acid soluble. He also advocates well clean-out procedures using only the acid and the sulfide scavenger. Williamson[110] claimed the use of a chelating agent and a ketone for the same types of treatments used by Walker.

7.7.3. Removing Pyrophoric Sulfides

If the deposits are not pyrophoric or if oxygen can be excluded, the processes described above are appropriate. The problems of specifically removing pyrophoric iron sulfide have been described by several researchers. Vella[111] claims that potassium permanganate ($KMnO_4$) is unique among the oxidants in that it can oxidize pyrophoric iron sulfide and other sulfide compounds, but it is safe to use and easy to apply. Under normal conditions, it is applied as a 1%-4% solution that is relatively harmless if it contacts skin. It requires no special materials of construction and does not form harmful or potentially explosive by-products. He also claims that at the discharge end, potassium permanganate's purple color gives a positive visual indication when the pyrophoric iron sulfide has been destroyed. The by-product of the reaction is manganese dioxide. This material is biologically inert and can be directly discharged to a waste water facility. If the manganese dioxide must be removed from the equipment or discharge stream, simple reducing agents such as sodium thiosulfate or citric acid can be used. Note that Lawson[87] suggested alternate stages of potassium permanganate and sulfuric acid for removing FeS.

McCartney[112] has described the use of stabilized H_2O_2 for neutralizing pyrophoric iron sulfides by converting the sulfide to sulfate. Unfortunately, the presence of even small amounts of soluble Fe, Cu, Mn, or Ag will cause an exothermic decomposition of this oxidant. The possible reactions are

$$
\begin{aligned}
&A)\ 2H_2O_2 + Fe, Cu, Mn, Ag(sol) \rightarrow 2H_2O + O_2(g) + \textit{heat} \\
&B)\ 2FeS + 9H_2O_2 + H^+ \rightarrow +Fe^{3+} + 2SO_4^{2-} + 5H_2O \qquad \text{(Equation [7.11])} \\
&C)\ 2FeS_2 + 15H_2O_2 \rightarrow +2Fe^{3+} + 4SO_4^{2-} + 14H_2O + 2H^+
\end{aligned}
$$

The reaction modeled by Equation (7.11A) occurs in the presence of dissolved metals unless the peroxide is stabilized using the methods from McCartney.[112] Stabilizer used in this paper was not described; however, several patents disclose the use of phosphonates,[113] high surface nanoparticles,[114] Sn(II) salts,[115] and colloidal Ag.[116] Thus, in the presence of an appropriate stabilizer, reactions B and C reduce the dangers of the pyrophoric iron sulfides. This authors claims that a Hastelloy C276 heat exchanger in a crude unit that was fouled with pyrophoric poly iron sulfide was cleaned successfully in an ultrasonic bath of stabilized H_2O_2. See Section 8.3.2 for other examples of the use of ultrasonic processes for cleaning heat exchanger units.

Otzisk[88] contends that a special formulation is an alternative to conventional cleaning agents for pyrophoric FeS and was developed to meet customer demands. It is claimed to be a unique combination of mild surfactants and gives excellent service for numerous types of cleaning and decontamination. The author says that the formulation contains an alkyl-dimethyl amine oxide.

Otzisk[88] claims that this chemical supplies oxygen to FeS in a rather slow and reduced manner (compared to many other oxidizers). This allows it to oxidize pyrophoric iron during the application without creating the dangerous exothermic problems of common oxidizers, such as hydrogen peroxide, potassium permanganate, and bleach. In addition, formulations are claimed to work as an oxidizer in both liquid and vapor phase applications. The chemistry is reported by the authors to be non-toxic, non-hazardous, and completely biodegradable, to contain no volatile hydrocarbons, and to have no flammability concerns.

7.8. Summary and Best Practices

7.8.1. Summary of Refinery and Chemical Plant Cleaning Technologies

The refinery/chemical plant environment is very complex, and all or any of the solvents described in Chapter 7 or Chapter 4 may be required.

Cleaning procedures for refinery/petrochemical equipment will continue to emphasize less toxic chemicals, high-pressure water blasting (with water recovery; see Section 8.3), and considerations for the disposal of the waste materials (Chapter 9). Continued use of sophisticated analytical methods (such as the reflectance FT-IR described in Section 2.6.1) may reduce the time necessary to develop an appropriate cleaning sequence.

7.8.2. Best Practices

Ramirez[117] evaluated, at laboratory scale, different chemical treatments commonly used in cleaning refinery components and equipment and developed a method. This method consisted of exposing deposit samples to the cleaning solutions by periods corresponding to those used in real cleaning procedures and then quantifying the weight loss of such specimens. This was related to the performance of treatments studied. In addition to the efficiency as cleaning products, the corrosivity of these chemical treatments was measured through weight loss tests.

Otzisk[88, 118] has reported on cleaning refinery units in Europe. These are very useful documents and provide details of planning and execution of refinery chemical cleaning operations.

- They note several conventional cleaning procedures that include acids and chelating agents.
- They describe solvents that remove FeS (such as $KMnO_4$) that must be used with care owing to the issues of pyrophoric sulfides. The issues of pyrophoric FeS had special emphasis.
- Green solvents containing biodegradable surfactants were described.

The web site that is the source of the Otzisk report[118] has links to several other refinery chemical cleaning articles but requires a subscription to the eptq.com library.

Al-Darwish et al.[32] describe best practices for cleaning several categories of crude oil processing units. Note that these are the recommendations of the authors of the current paper. References to mechanical processes in this book that are described in Chapter 8 are also noted in the descriptions and in Table 7.8.

- **Cleaning of Crude Oil and NGL Re-boilers**
  The ultrasonic cleaning technique is recommended, which utilizes a combination of ultrasonic transducers and aqueous cleaning fluid within a specially engineered vessel. This is to remove hydrocarbon contaminants from the surface of a steam reboiler in a swift, safe, and environmentally friendly way compared with the conventional cleaning method. Consequently, it will prolong the equipment integrity and reliability. See Section 8.3.2.
- **Cleaning of Crude Oil Stabilizer Columns**
  Vapor phase decontamination techniques (Section 7.6) for the crude oil stabilizers can be used instead of the conventional hydrojetting and grit blasting method. This technique will get the lower explosive limit (LEL) and H_2S levels to zero, neutralize the pyrophoric materials, and remove hydrocarbons to permit personnel entry to perform regular maintenance work.
- **Cleaning the Tubes of Crude Oil and NGL Fin-Fan Coolers**
  The use of aqua-drill (See Section 8.3.1) types of mechanical cleaning for the tubes of crude oil fin-fan coolers and NGL coolers can be used instead of the conventional hydrojetting method whenever fully plugged tubes with hard scales are encountered. The aqua-drill process is one of the most advanced high-pressure water-jet cleaning techniques that can be used for cleaning heat exchangers of fully plugged tubes with hard scales. It also utilizes one twister house system to rotate a high-pressure water hose. This component remotely controls all parameters of the tube internal cleaning heads, resulting in precise overlapping cutting tracks and a full surface cleaning coverage.
- **Cleaning of Crude Oil Spheroids**
  The use of spray nozzles using a decontamination solution to perform the cleaning operation can be used instead of the conventional cleaning method. Partial solution filling of the decontamination solution with a combination of spray nozzle techniques is a safe, cost-effective, and realistic approach to decontaminate and remove hydrocarbon sludge materials from crude oil spheroids. Decontamination is a technique that is used to de-oil and degrease process equipment. Moreover, decontamination will get the LEL and H_2S levels to zero to permit personnel entry to perform regular maintenance work.
- **Cleaning of Oil Storage Tanks**
  The conventional cleaning method (solvents or jetting) for the crude oil storage tanks is a viable option unless there is a necessity to minimize storage tanks' downtime during the cleaning activities. If so, the vapor phase decontamination technique can be used as described in Section 7.6.
- **Cleaning of NGL Stripper/Deethanizer Columns**
  The alternative cleaning technique is the same as vapor phase decontamination. The cost variance between the vapor phase decontamination technique and the conventional hydrojetting and grit blasting method is 82% higher with the advantage of saving 84% of the water and minimizing the equipment downtime by 72%.

- **Cleaning of Oily Water Sump (OWS) Pits**
  On the basis of the actual field experience (the authors of this paper), the alternative cleaning technique for OWS should be determined on a case-by-case basis. This is because the optimum cleaning technique is directly dependent on the amounts of hydrocarbon sludge and the possibility of totally isolating the OWS. In other words, as an alternative to the conventional cleaning method, the chemical cleaning circulation technique can be used, provided that the OWS cannot be totally isolated and that the amounts of hydrocarbon sludge cannot be easily removed by hydrojetting. On the basis of an actual field trial of this technique, field inspection following chemical cleaning (without man-entry) revealed that the OWS was free of sludge with zero LEL and zero ppm H_2S gas concentration.

A summary of all the recommendations from this paper is in Table 7.8.

Table 7.8 Summary of Recommendations from Al-Darwish et al.[32]

Equipment Type and Numeral	Service	Materials-of-Construct-ion	Type of Scales & Deposits	Optimum Cleaning Technique	Savings Variance	Notes
Crude Oil Stabilizer Columns NGL Stripper/ Deethanizer Columns	Oil or NGL	Carbon Steel	Stabilizers: hydrocarbon contaminants/sludge and H_2S gas. Stripper/ Deethanizer: black powder, H_2S gas	Vapor phase de-contamination	82% higher cost 72% water saving 84% time saving	Savings' variance is based on contractor's proposal to be verified by a field trial
Crude Oil Fin-fan Coolers	Oil	Carbon Steel	Hard hydrocarbon scales	Aqua-drill	85% higher cost 90% more water 31% time saving	Savings' variance is based on an actual field implementation
Crude Oil Re-boilers	Oil	Carbon Steel	Hydrocarbon contaminants/sludge	Ultrasonic	33% cost saving 83% water saving 67% time saving	Savings' variance is based on an actual field implementation
Crude Oil Storage Tanks (ATF-1, ATF-2)	Oil	Carbon Steel	H_2S gas, hydrocarbon contaminants/ sludge	Hydrojetting & grit blasting	85% higher cost 12% more water 62% time saving	Savings' variance is based on the use of the vapor phase decontamination
Crude Oil Spheroids	Oil	Carbon Steel	Oily sludge hydrocarbon plus solids, pyrophoric materials, and H_2S gas	3-D de-contamination	25% higher cost 26% more water 57% time saving	Savings' variance is based on an actual field implementation
NGL Reboilers, Chillers, Economizers, and Condensers	NGL	Carbon Steel	Black powder, FeS_2 (pyrite)	Ultrasonic	33% cost saving 83% water saving 67% time saving	Savings' variance is based on similar equipment to be revised based on a field trial
NGL Cooler Tubes, and Combiner Tubes	NGL	Carbon Steel	Black powder, FeS_2 (pyrite)	Aqua-drill	85% higher cost 90% more water 31% time saving	Savings' variance is based on similar equipment to be revised based on a field trial
Oily Water Sump Pits	Oil	Carbon Steel	Solids of sludge slurry rich with heavy hydrocarbons, H_2S gas, and silica mixture	Chemical Cleaning Circulation	98% higher cost 67% more water 86% time saving	Savings' variance is based on an actual field implementation
Boilers	Steam	Carbon Steel	Water scales and corrosion products	Chemical Cleaning Circulation	96% higher cost 91% more water 64% more time	Savings' variance is due to the chemical cleaning requirements

Shank et al.[6] studied the benefits and limitations of mechanical steam out and high-pressure water blasting (HPWB) versus vapor phase chemical cleaning of various parts of vacuum distillation units (VDU-Section 7.3).

Their analyses concluded that mechanical cleaning (such as hydrojetting; see Section 8.3.1) increased efficiency in the system; however, many health, safety, and environmental factors must be considered. Chemical cleaning of the system by recirculation using a two-stage chemical blend approach was unable to produce applicable results during this case study. The authors noted that several mechanical as well as onsite equipment issues affected the efficiencies of the added cleaning chemicals to reach the treatment goals. Note that onsite issues that involve several different entities often affect the job outcomes. See Section 11.2 for further discussion of onsite planning and execution.

Of the techniques that were performed in this case study, vapor-phase chemical cleaning (Section 7.6) using a two-stage chemical approach resulted in the greatest increase in efficiency within the system.

References

[1]W.W. Frenier, M. Ziauddin, R. Venkatesan, *Organic Deposits in Oil and Gas Production* (Richardson, TX: Society of Petroleum Engineers, 2010).

[2]W.W. Frenier, M. Ziauddin, *Formation, Removal and Inhibition of Inorganic Scale in the Oilfield Environment* (Richardson, TX: Society of Petroleum Engineers, 2008).

[3]W.W. Frenier, *Chemical and Mechanical Methods of Pipeline Integrity* (Richardson, TX: Society of Petroleum Engineers, 2017).

[4]M.A. Kelland, *Production Chemicals for the Oil and Gas Industry,* 2nd ed. (Boca Raton, FL: CRC Press, 2013).

[5]J.G. Speight, *The Chemistry and Technology of Petroleum* (New York, NY: Marcel Dekker, Inc, 1991).

[6]R.A. Shank, S. Gharaibeh, T.R. McCarthy, M. Panunzio, "Decontamination of a Vacuum Distillation Unit: Mechanical Versus Chemical Cleaning; a Case Study," CORROSION/2017, paper no. NACE C2017-9701 (Houston, TX: NACE, 2017).

[7]West Texas Intermediate, Wikipedia, May 5, 2009, http://en.wikipedia.org/wiki/West_Texas_Intermediate (July 28, 2009).

[8]V. Simanzhenkov, R. Idem, *Crude Oil Chemistry* (New York, NY: Marcel Dekker, Inc., 2003).

[9]M. Kayhan, "Proposed Classification and Definitions of Heavy Crude Oils and Tar Sands," 2nd UNITAR Conference on Heavy Crude and Tar Sands, Caracas, Venezuela, Feb. 7–17, 1982.

[10]C. Beal, "Viscosity of Air, Water, Natural Gases, Crude Oil and Its Associated Gases at Oil Field Temperature and Pressures," *Trans AIME* 165 (1946): 114–127.

[11]C. Beal, "Viscosity of Air, Water, Natural Gases, Crude Oil and Its Associated Gases at Oil Field Temperature and Pressures," *Trans AIME* 165 (1946): 94-95.

[12]F.J. Nellensteyn, "Constitution of Asphalt," *J Inst. Petro. Techn.* 10 (1924): 311-325.

[13]F.J. Nellensteyn, R. Loman, "Asfaltbitumen En Teer," in *Theorie En Practijk Der Bitumineuze Wegdekken*, ed. D.B. Centens (Amsterdam: Uitgevers-MILJ NV, 1932).

[14]S.I. Andersen, J.G. Speight, "Petroleum Resins: Separation, Character, and Role in Petroleum," *Petroleum Science and Technology* 19, 1 (2001): 1-34.

[15]A. Hammami, J. Ratulowski, "Precipitation and Deposition of Asphaltenes in Production Systems: A Flow Assurance Overview," in *Asphaltenes, Heavy Oils, and Petroleomics,* eds. O.C. Mullins, E.Y. Sheu, A. Hammami, A.G. Marshall (London: Springer, 2006).

[16]A. Hammami, J. Ratulowski, "Precipitation and Deposition of Asphaltenes in Production Systems: A Flow Assurance Overview," in *Asphaltenes, Heavy Oils, and Petroleomics,* eds. O.C. Mullins, E.Y. Sheu, A. Hammami, A.G. Marshall (London: Springer, 2007).

[17]K. Akbarzadeh, A. Hammami, A. Kharrat, D. Zhang, S. Allenson, S. Kabir, "Asphaltenes: Problematic but Rich in Potential," *Schlumberger Oilfield Review* 19, 2 (2007): 22-43.

[18]G. Porte, H. Zhou, V. Lazzeri, "Reversible Description of Asphaltene Colloidal Association and Precipitation," *Langmuir* 19, 1 (2003): 40-47.

[19]A.M. Zellou, "Optimizing Flow and Increasing Efficiencies in a Low Oil Price Environment," Pipeline+Energy Expo, Tulsa, OK, Aug. 24-26, 2015.
[20]W.F. Bland, R.L. Davison, *Petroleum Processing Hand Book* (New York, NY: McGraw-Hill, 1967).
[21]S.R. Brown, I.A. McConnell, "Effective Deposit Removal from Refinery Process Equipment," CORROSION/84, paper no. NACE 84004 (Houston, TX: NACE,1984).
[22]W.W. Frenier, S.J. Barber, "Choose the Best Heat Exchanger Cleaning Method," *Chemical Engineering Progress* 94, 7 (1998): 37-45.
[23]W.W. Frenier, *Technology for Chemical Cleaning of Industrial Equipment* (Houston, TX: NACE International, 2001).
[24]M. Kuno, "Crude Unit and Treater," *Hydrocarbon Processing* 60, 9 (1981): 124.
[25]M. McCoy, "Chevron Embraces Ionic Liquids," *C&E News* 94, 39 (2016): 16.
[26]W.W. Frenier, M.D. Coffey, J.D. Huffines, D.C. Smith, "Composition and Method for Removing Sulfide-Containing Scale from Metal Surfaces," U.S. Patent 4220550, 1980.
[27]Ionic Liquid. Wikipedia, Oct. 3, 2016, (Oct. 3 2016).
[28]Sulfur Recovery, U.S. Environmental Protection Agency, Washington, DC, http://www.epa.gov/ttnchie1/ap42/ch08/final/c08s13.pdf (January 1995).
[29]S.D. Kapusta, F.V.D. Berg, R. Daane, M.C. Place, "The Impact of Oil Field Chemicals on Refinery Corrosion Problems," CORROSION/2003, paper no. NACE-03649 (Houston, TX: NACE, 2003).
[30]H. Muller-Steinhagen, *Heat Exchanger Fouling-Mitigation and Cleaning Technologies* (Rugby, U.K.: Institution of Chemical Engineers, 2000).
[31]B.J. Little, P. Wagner, "Microbiologically Influenced Corrosion," in *Peabody's Control of Pipeline Corrosion*, 2nd ed., ed. R.L. Bianchetti (Houston, TX: NACE International, 2001).
[32]H.M. Al-Darwish, S.A. Al-Dossary, N.M. Al-Abdulmohsin, "Chemical and Mechanical Cleaning Case Study," CORROSION/2017, paper no. NACE C2017-9311 (Houston, TX: NACE, 2017).
[33]J.C. Cowan, D.J. Weintritt, *Mineral & Organic Scale Deposits* (Lafayette, LA: Cowan & Weintritt Publishing Company, 2004).
[34]W.W. Frenier, M. Ziauddin, *Chemistry for Enhancing the Production of Oil and Gas* (Richardson, TX: Society of Petroleum Engineers, 2014).
[35]M. McCoy, "Going Head to Head in the Laundry Room," *Chemical and Engineering News* 95, 4 (2017): 15-18.
[36]W.C. Griffin, "Calculation of Hlb Values of Non-Ionic Surfactants," *Journal of the Society of Cosmetic Chemists* 5 (1954): 235-249.
[37]M.A. Kelland, *Production Chemicals for the Oil and Gas Industry* (Boca Raton, FL: CRC Press, 2009).
[38]T. Tadros, "Polymeric Surfactants in Disperse Systems," *Advances in Colloid and Interface Science* 147–148 (2009): 281-299.
[39]Sasol 2012. Alfoterra® Surfactants, Matching Surfactant to Reservoir with HLB Concept, Houston, TX.
[40]TDW, 77-442-0213-ENG, "Pipeline Cleaning Services" (Brochure) (Tulsa, OK: T.D. Williamson, Inc., 2012).
[41]A.V.D. Werff, "The Importance of Pipeline Cleaning: Risks, Gains, Benefits, Pease of Mind," Paper 224, Pipeline Technology Conference, Hannover, Germany, April 25, 2006.
[42]W.W. Frenier, "Low Hazard Corrosion Inhibitors and Cleaning Solutions Using Quaternary Ammonium Salts," U.S. Patent 6521028, 2003.
[43]C. Sitz, W. Frenier, C. Vallejo. "Acid Corrosion Inhibitors with Improved Environmental Profiles," Paper SPE 155966, SPE International Conference & Exhibition on Oilfield Corrosion, Aberdeen, U.K., May 28-29, 2012.
[44]G.B. Matta, "D-Limonene Based Aqueous Cleaning Compositions," U.S. Patent 4511488, 1985.
[45]N.K. Mehta, R.W. Krajieck, "Decontamination of Hydrocarbon Process Equipment," U.S. Patent 5389156, 1995.
[46]R.W. Krajieck, N.K. Mehta, J.R. Duffy, "Method for Quick Turnaround of Hydrocarbon Processing Units," U.S. Patent 5425814, 1995.
[47]P.A. Mestetsky, "Method of Separating Oily Materials from Wash Water," U.S. Patent 5459066, 1995.
[48]B.R. Jansen, S. Edwards, "Heat Exchanger Cleaning Process," U.S. Patent 6936112, 2005.
[49]M.B. Lawson, K.J. Snyder, "Method for Dissolving Asphaltic Material," U.S. Patent 4108681, 1978.
[50]G.S. Penney, "Method of Increasing Hydrocarbon Production by Remedial Well Treatment," U.S. Patent 4565639, 1986.

[51]G.G. McClaflin, K. Yang, "Composition and Method for Treatment of Wellbores and Well Formation Containing Paraffin," U.S. Patent 4668408, 1987.
[52]V.R. Kruka, "Process for Removal of Wax Deposits," U.S. Patent 4646837, 1987.
[53]D.W. Jennings, S. Asomaning, M.E. Newberry, "Heat Generation Process for Treating Oilfield Deposits," U.S. Application 20110114323, 2011.
[54]I.R. Collins, I. Vervoort, "Water-in-Oil Microemulsions Useful for Oil Field or Gas Field Applications and Methods for Using the Same," U.S. Patent 6581687, 2001.
[55]M.G. Bernadiner. "Advanced Asphaltene and Paraffin Control Technology," Paper SPE 25192, SPE International Symposium on Oilfield Chemistry, New Orleans, LA, March 2-5, 1993.
[56]V.G. Levich, *Physicochemical Hydrodynamics* (Englewood Cliffs, NJ: Prentice-Hall, 1962).
[57]Y. Fan, F.M. Llave, "Chemical Removal of Formation Damage from Paraffin Deposition- Part I- Solubility and Dissolution Rate," Paper SPE 31128, SPE Formation Damage Control Symposium, Lafayette, LA, Feb. 14-15, 1996.
[58]G. Zhang, C.-Z. Yu, S.-H. Su, K.L. Kalra, D. Zhou, "Deparaffinization Compositions and Methods for Their Use," U.S. Patent 6632598, 2003.
[59]R.J. Dyer, "Method for Simultaneous Removal of Asphaltene, and/or Paraffin and Scale from Producing Oil Wells," U.S. Patent 7296627, 2007.
[60]R.J. Dyer, "Method for Simultaneous Removal of Asphaltene, and/or Paraffin and Scale from Producing Oil Wells," U.S. Application 20080047712, 2008.
[61]D. Tukenov, Nanochemistry Drives New Method for Removal and Control of Wax. Society of Petroleum Engineers, Richardson, TX, http://www.spe.org/jpt/article/7788-technology-update-1-7/ (Dec. 31, 2014).
[62]P. Andrecola, "Biodegradable Non-Reactive Oil-Well Stimulation Fluid and Method of Use," U.S. Application 20140121137, 2014.
[63]D.C. Thomas, H.L. Becker, R.A. Del Real Soria, "Controlling Asphaltene Deposition in Oil Wells," *SPE Production and Facilities* 10, 2 (1995): 119-123.
[64]A. Del Bianco, F. Stroppa, "Effective Hydrocarbon Blend for Removing Asphaltenes," U.S. Patent 5382728, 1995.
[65]G.A. Mansoori, "Arterial Blockage/ ouling Prediction and Prevention," Heavy Organics Program, Department of Chemical Engineering University of Illinois at Chicago, 2008.
[66]G.A. Mansoori, "Arterial Blockage/Fouling Prediction and Prevention," Heavy Organics Program, Department of Chemical Engineering University of Illinois at Chicago, 2001.
[67]F. Garcia-Hernandez, "Estudio Sobre El Control De La Depositacion Organica En Pozos Del Area Cretacica Chiapas-Tabasco," *Ingeniería Petrolera* July (1989): 39-45.
[68]A. Del Bianco, F. Stroppa, "Composition Effective in Removing Asphaltenes," U.S. Patent 5690176, 1997.
[69]D.A. Thorssen, D.N. Loree, "Oil and Gas Well Operation Fluid Used for the Solvation of Waxes and Asphaltenes, and Method of Use Thereof," U.S. Patent 5795850, 1998.
[70]D.N. Loree, "Oil and Gas Well Operation Fluid Used for the Solvation of Waxes and Asphaltenes, and Method of Use Thereof," U.S. Patent 6093684, 2000.
[71]A.K.M. Jamaluddin, T.W. Nazarko, "Process for Removing and Preventing near-Wellbore Damage Due to Asphaltene Precipitation," U.S. Patent 5425422, 1996.
[72]R. Cimino, A.D. Correra, T.P. Lockhart, "Solubility and Phase Behavior of Asphaltenes by Hydrocarbon Media," in *Asphaltenes, Fundimentals and Applications,* eds. E.Y. Sheu, O.C. Mullins (New York, NY: Plenum Press, 1995).
[73]L.O.S. Buzelin, C.B.d.C. Lima, "Innovative Methodology for Cleaning Pipes - Key to Environmental Protection," Paper SPE 111612, SPE International Conference on Health, Safety, and Environment in Oil and Gas Exploration and Production, Nice, France, April 15-17, 2008.
[74]S.-R. Park, Y.-K. Ahn, S.-G. Oh, K.-H. Lee, S.-J. Kim, "Chemical Cleaning Process for Removing Fouling," U.S. Patent 6485578, 2001.
[75]J.J. Wylde, J.J.L. Slayer, "Development, Testing and Field Application of a Heavy Oil Pipeline Cleaning Chemical: A Cradle to Grave Case History," Paper SPE 119688, SPE Western Regional Meeting, San Jose, CA, Mar. 24-26, 2009.

[76]C.T. Montgomery, J. Cuesta, D. Teggin, F. Fragachán, "Organic Deposits: Best Practices and Lessons Learned from the Experience 1992-1996 in Villahermosa, Mexico" (Sugar Land, TX: Schlumberger Technology Corporation, 1996).
[77]M.D. Coffey, J.L. Thompson, M.J. Carney, "Solvent-Acid Dispersions Solve Difficult Stimulation and Clean-up Problems," SPE Rocky Mountain Regional Meeting, Billings, MT, May 15-16, 1974.
[78]W.C. Lyons, G.J. Plisga, *Standard Handbook of Petroleum and Natural Gas Engineering*, 2nd ed. (Burlington, MA: Gulf Professional Publishing, 2005).
[79]W.W. Frenier, M. Brady, "Treating Composition," U.S. Patent 7427584, 2008.
[80]M. McCoy, "Cleaning up a Solvent Problem," *C&EN* 93, 3 (2015): 16-19.
[81]D. Mason, Part 5: Decontamination of the Crude Unit. (Houston, TX, FOE Chemicals blog, 2015).
[82]S.L. Kokal, "Crude Oil Emulsions," in *Petroleum Engineering Handbook, Volume I: General Engineering*, ed. J.R. Fanchi (Richardson, TX: Society of Petroleum Engineers, 2006).
[83]W.B. Brown and C.M. Loucks, "Chemical Cleaning of Metal Surfaces Employing Steam" U.S. Patent 3084 076, 1960.
[84]S.E. Sears, K.L. Roberts, "Method for Cleaning a Vessel in a Refinery," U.S. Patent 6893509, 2005.
[85]HydroChem, 2016, Cfe (Cleared for Entry). HydroChem, LLC, Deer Park, TX, http://www.hydrochem.com/services/cfe-cleared-for-entry (Sept. 29, 2016).
[86]"Hydrocarbon Decontamination" (Brochure) (Houston, TX: Halliburton, 2006).
[87]M.B. Lawson, L.D. Martin, G.D. Arnold, "Chemical Cleaning of Fes Scales," CORROSION/80, paper no. NACE 219 (Houston, TX: NACE, 1980).
[88]B. Otzisk, M. Urschey, "Chemical Cleaning and Decontamination of Refinery and Petrochemical Distillation Equipment," Proceeding of International Conference on Heat Exchanger Fouling and Cleaning, Dublin, Ireland, June 7-12, 2015.
[89]C.H. Spurrell, R.A. Kenyon, "Oxidizer Use in Refinery Chemical Cleaning: Selection Considerations and Case Histories," CORROSION/2006, paper no. NACE 06529 (Houston, TX: NACE, 2006).
[90]C.L. Ball, W.W. Frenier, "Improved Solvent for Iron Sulfide Deposits," CORROSION/84, paper no. NACE 84002 (Houston, TX: NACE, 1984).
[91]G.R. Buske, "Method and Composition for Removing Sulfide-Containing Scale from Metal Surfaces," U.S. Patent 4289639, 1981.
[92]W.W. Frenier, "Method and Composition for Removing Sulfide-Containing Scale from Metal Surfaces," U.S. Patent 4310435, 1982.
[93]L.S. Frankel, K.S. McCallum, L. Collier, "Formation of Bis(Chloromethyl) Ether from Formaldehyde and Hydrogen Chloride," *Environmental Science and Technology* 8, 4 (1974): 356.
[94]W.W. Frenier, D.G. Hill, "Well Treatment Fluids Comprising Mixed Aldehydes," U.S. Patent 6399547, 2002.
[95]A.R. Miller, "Iron Sulfide Clean-up Composition and Method," U.S. Application 20040142845, 2004.
[96]H.A. Nasr-El-Din, A.Y. Al-Humaidan, B.A. Fadhel, W.W. Frenier, D. Hill. "Investigation of Sulfide Scavengers in Well Acidizing Fluids," Paper SPE 58712, SPE International Symposium on Formation Damage Control, Lafayette, LA, Feb. 23-24, 2000.
[97]M.A. Mattox, "Method to Reduce Iron Sulfide Deposits in Pipelines," U.S. Patent 6866048, 2005.
[98]R.E. Talbot, P.D. Gilbert, M.A. Veale, K. Hernandez, "Tetrakishydroxymethylphosphonium Sulfate (THPS) for Dissolving Iron Sulfides Downhole and Topsides: A Study of the Chemistry Influencing Dissolution," CORROSION/2002, paper no. NACE-02030 (Houston, TX: NACE, 2002).
[99]R.E. Talbot, C.R. Jones, "Synergistic Compositions," U.S. Application 2015/0141377, 2015.
[100]L.D. Martin, "Methods and Compositions for Removing Deposits Containing Iron Sulfide from Surfaces Comprising Basic Aqueous Solutions of Particular Chelating Agents," U.S. Patent 4276185, 1981.
[101]J. Yap, M.J. Fuller, L. Schafer, S.K. Kelkar, "Removing Iron Sulfide Scale: A Novel Approach," Paper SPE-138520, International Petroleum Exhibition and Conference, Abu Dhabi, UAE, Nov. 1-4, 2010.
[102]Q. Wang, H. Ajwad, T. Shafai, J.D. Lynn. "Iron Sulfide Scale Dissolvers: How Effective Are They?" Paper SPE-168063, SPE Saudi Arabia Section Technical Symposium and Exhibition, Khobar, Saudi Arabia, May 19-22, 2013.
[103]M.B. Lawson, "Method for Removing Iron Sulfide Scale from Metal Surfaces," U.S. Patent 4351673, 1982.
[104]T. Chen, Q. Wang, F.F. Chang, Y.T. Al-Janabi, "New Developments in Iron Sulfide Scale Dissolvers," CORROSION/2016, paper no. NACE-2016-7264 (Houston, TX: NACE, 2016).

[105]J.J. Wylde, C. Okocha, R. Smith, A. Mahmoudkhani, C.J. Kelly, "Dissolution of Sulfide Scale: A Step Change with a Novel, High Performance, Non-Mineral Acid Chemical," SPE-179880-MS, SPE International Oilfield Scale Conference and Exhibition, Aberdeen, Scotland, May 11-12, 2016.

[106]B.E. Hall, W.R. Dill, "Iron Control Additives for Limestone and Sandstone Acidizing of Sweet and Sour Wells," Paper SPE 17157, SPE Formation Damage Control Symposium, Bakersfield, CA, Feb. 8-9, 1988.

[107]W.R. Dill, M.L. Walker, "Composition and Method for Controlling Precipitation When Acidizing Sour Wells," 1989.

[108]M.L. Walker, W.R. Dill, D.G. McFatridge, "Iron Control in West Texas Sour Gas Wells Provides Sustained Production Increase," SPE 20122, 1990 Permian Basin Oil and Gas Recovery Conference, Midland (1991).

[109]W.G.F. Ford, K.H. Hollenbeak, "Composition and Method for Reducing Sludging During the Acidizing of Formations Containing Sludging Crude Oils," U.S. Patent 4663059, 1987.

[110]C.D. Williamson, "Precipitation Control," U.S. Patent 5126059, 1992.

[111]P.A. Vella, "Improved Cleaning Method Safely Removes Pyrophoric Iron Sulfide," *Oil & Gas Journal* February (1997).

[112]T.R. McCartney, R. Shank, "Application of Stabilized Hydrogen Peroxide for the Removal of Polysulfice Deposits," International Conference on Heat Exchanger Fouling and Cleaning, Budapest, Hungary, June 9-14, 2013.

[113]A.G. Leigh, "Stabilization of Hydrogen Peroxide," U.S. Patent 4305762, 1981.

[114]H.A. Ketelson, N.D. McQueen, "Use of Inorganic Nanoparticles to Stabilize Hydrogen Peroxide Solutions," U.S. Patent 7344725, 2008.

[115]E.F. Dougherty, Q.G. Hopkins, J.E. Sanborn, A.T. Weibel, "Tin(II) Stabilizer for Hydrogen Peroxide," U.S. Patent 5078672, 1992.

[116]H. Hungerbach, W. Struzina, A. Hoburg, "Oral Hygiene Agent Containing Hydrogen Peroxide Stabilized by Colloidal Silver," U.S. Patent 5437858, 1995.

[117]M. Ramirez, A. Viloria, "A Methodology for the Evaluation of Chemical Treatments Used in Cleaning of Refinery Components and Equipment," CORROSION/2003, paper no. NACE 03650 (Houston, TX: NACE, 2003).

[118]B. Otzisk, "Chemical Cleaning and Degassing Refinery Equipment," *PTQ*, 2016, Q1 (2008): 77-81, http://www.eptq.com/articles/kurita_28012008132918.pdf

8 Mechanical Equipment and Special Fluids for Descaling Units with/without Chemicals

In 1983, Al Look[1] (at that time the President of the Dowell Division of the Dow Chemical Co.) noted that each day, hundreds (possibly thousands) of small chemical plants are set up in the world's oil/gas fields, and hundreds of thousands of pounds of chemicals are pumped into the earth to enhance and maintain the flow of oil and gas. Frequently at the end of the treatment, these "mini-plants" are broken down and transported to a new site.

Chemical IC originated from the oilfield services, and the same types of small and usually temporary chemical plants are erected in industrial plants to use equipment, chemistry, and various fluids to clean surfaces. Various types of equipment are used in all the operations described in this book. This includes conventional power generation, nuclear power generation, and refinery/chemical plants, paper mills, and numerous other industrial operations.

The author of this book is not aware of any chemical cleaning fluid that can be applied without using mechanical equipment in some manner. At a minimum, the chemicals must be pumped into a unit even if the operator's pumps provide circulation. Many different types of mechanical units often are needed, so mobile units are delivered to the treatment site and then removed for use at other sites.

Chapter 8 describes currently used mechanical units that facilitate chemical cleaning. Sections 8.1 and 8.2 introduce the equipment types. Section 8.3 describes *mostly* mechanical cleaning methods (jetting and ultrasound) and highlights the uses of jetting to clean heat exchangers, but other refinery and chemical plant units are also cleaned with this versatile technology. Section 8.4 describes the use of foam for applying chemicals to clean specific types of units. The use of vapors (steam) for cleaning refinery vessels is described in Section 7.6.

8.1. General Mechanical Equipment for Descaling Industrial Units

None of the chemical processes described in Chapters 1 and 7 can be conducted without various (and frequently massive) amounts of equipment. Some of the units requiring cleaning are described in Sections 1.3 and 7.3.

1. Al Look, private communication, 1983

Figure 8.1 is a diagram of a generic chemical cleaning setup that illustrates some of the most important and necessary components of the equipment.

- The vessel or unit must be rigged for cleaning by connecting all parts by secure hoses or hard piping. The connections must be able to accept the pressure and flow rate and the chemicals that will be pumped. That is, the chemicals cannot significantly corrode the materials of construction.
- There must be systems for introducing the chemicals, mixing them, and then circulating or agitating them. The motive system may be pumps or possibly an inert gas. Fill, soak, and dump systems and heat cycling also have been used. Because the chemicals will eventually spend themselves on the scale, fresh solvent must be constantly supplied in some manner to completely remove the deposits.
- The fluids often must be heated above ambient temperatures to obtain an adequate cleaning rate. Several systems of heating include
  - Natural heat in a unit as it comes offline or low rate "firing" for units such as power boilers
  - Injection of steam into the fluid to heat it
  - Placing a heat exchanger in the circuit and circulating hot water or steam on one side
- Toxic, pressurizing, or flammable gases must be vented, scrubbed, or otherwise removed from the fluids. Common examples are H_2S, CO_2, H_2, and hydrocarbon gases.
- There must be fittings/systems to empty the solvents to waste collection and treatment systems and to allow flushing and possibly passivation treatments. All the used fluids must be collectable, including volatile organic compounds (VOC).
- Often, the system has connections so that the fluid can be sampled to determine the rate of scale removal and to provide an end-point for the treatment.

Also, refer to Figure 4.2, which shows the cleaning hook-ups for cleaning a boiler and economizer.

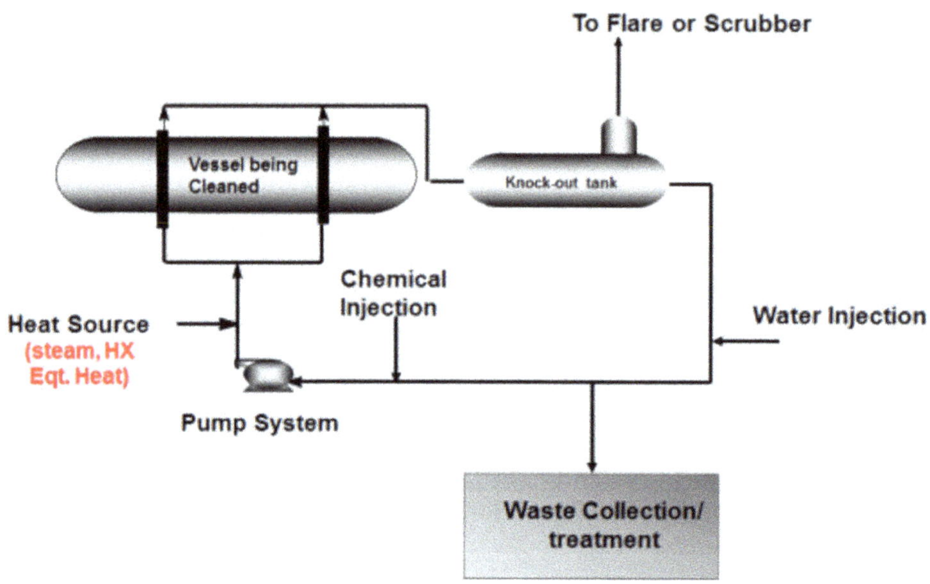

Figure 8.1 Simplified Chemical Cleaning Setup

8.1.1. General Pumping Equipment

Several examples of the very wide variety of units used are noted in the following figures. Figure 8.2 shows the evolution of pumping and chemical transport equipment over about 70 years (Frenier[1]). In the 1940s, a single truck with a tank and pump (Figure 8.2 [a]) could be used to circulate chemicals in industrial and oilfield situations. At that time, service companies may have performed both services with the same equipment.

Part b shows a modern super pumper that could be used to circulate large units such as utility power generating equipment. Part c shows the chemical transport units to support large jobs. Especially note the extensive safety rails on the decks of the modern units. Since the tanks and the fittings on the transport and pumping units must contact corrosive chemicals, the materials of construction must be carefully selected. The tanks frequently have a polymer lining to isolated the fluids from the metal shells because these units may hold very concentrated cleaning chemicals.

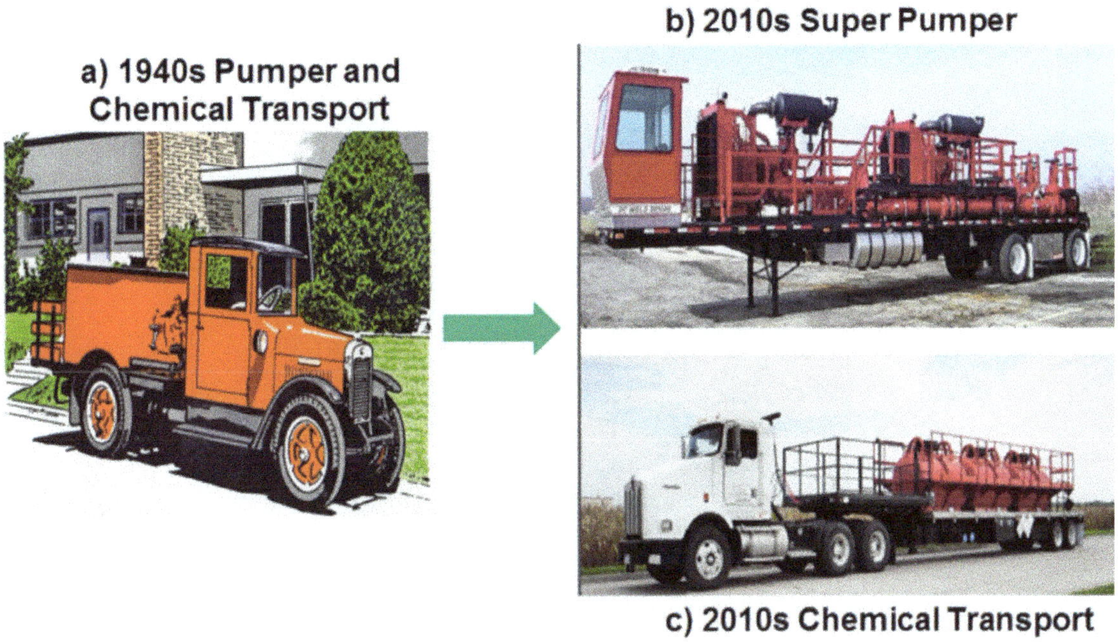

Figure 8.2 Evolution of Pumping Equipment

Chemical cleaning activities occur at such a wide variety of sites that cleaning situations require a wide range of equipment types for specific operations. The necessary portable equipment depends on the facilities, space, and utilities present at the customer's site. In many cases, steam, water, electricity, and possibly pumping are available from the plant operator. In other situations, the cleaning vendor must provide more (or all) of the equipment.

The setup at an industrial plant for chemical or mechanical cleaning usually takes place during a shut-down of the plant. At the same time, other types of maintenance activities may take place. Therefore, the storage and setup area may be congested, and safety must be emphasized. Figure 8.3 (a and b) from McGraw[2] shows the pumping and waste storage equipment during the cleaning of an HRSG (Figure 1.4) unit at a combined cycle plant.

Figure 8.4 and Figure 8.5 show the setup of some of the equipment needed to service a nuclear steam generator in Asia (Little[3]). Figure 8.4 shows mixing tanks and fill and drain pumps.

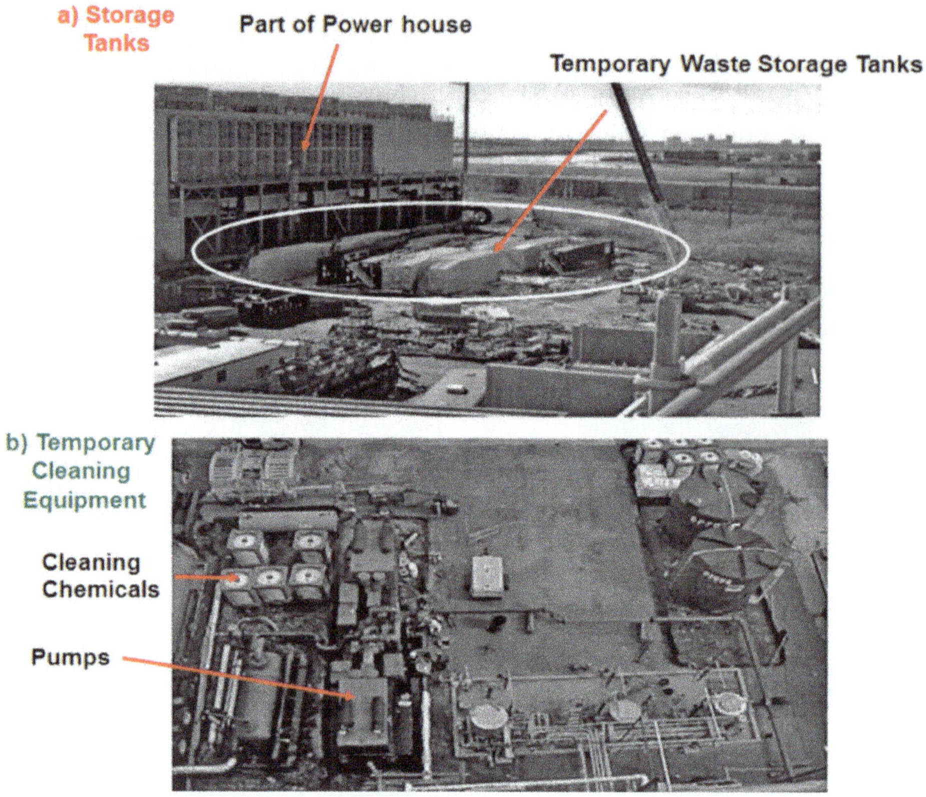

Figure 8.3 Equipment Layout and Storage for Chemical Cleaning of HRSG Units[2]

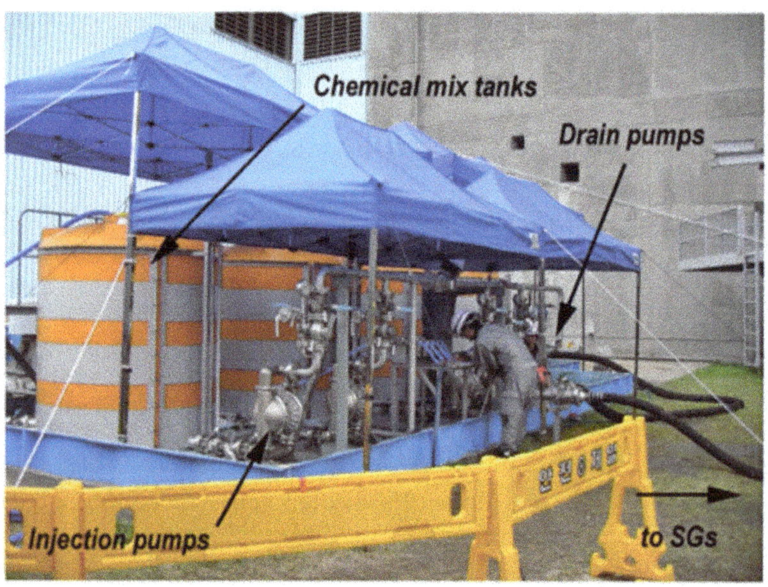

Figure 8.4 Part of Setup for Steam Generator Cleaning[3]

8.1.2. Waste Collection Equipment

As noted in Figure 8.1, the waste fluid must be collected and often stored before disposal treatments (see Chapter 9 for methods and processes). Figure 8.5 depicts waste storage tanks at a steam generator cleaning site. Note that these are just some of the activities and equipment needed for these treatments.[3] An almost limitless range of storage and/or transfer equipment may be needed, including trucks, pipelines, internal plant piping, and vacuum units.

Figure 8.5 Waste Collection from Steam Generator Cleaning[3]

Specialized vacuum truck units are designed to remove wastes from different parts of a plant (especially a refinery) and can empty waste storage tanks. An important current specification is that the air that is sucked into the vacuum truck is filtered to remove VOC before release to the atmosphere. In some states, such as in Texas, there must be two barriers between the VOCs and the atmosphere, and this can be accomplished in different ways.

A truck with VOC-activated carbon filters as well as a diluent tank is shown in Figure 8.6.[4] The diluent tank contains a liquid that absorbs most of the volatile molecules, and most of the remainder is adsorbed by activated carbon filters. This setup provides two barriers between the chemicals and the atmosphere. Although the diluent usually is a proprietary fluid, Tatina[5] describes the use of an oil/water emulsion as a method for scrubbing VOCs.

A different process for providing two barriers is described in patents by Harvey et al.[6,7] Per these documents, the apparatus includes a vacuum truck with at least one vacuum tank that has a hollow interior region. A flexible bladder is placed within the hollow interior region of the vacuum tank. The bladder is impermeable to and compatible with the collected material and capable of containing the materials pumped into it. Figure 8.7 shows an example from the patents of a bladder in the vacuum truck providing the primary isolation. The pump's exhaust is further scrubbed using carbon filters or other measures.

In all cases, the trucks are emptied into approved waste and recycle facilities. This includes the diluent tanks and the filters that are saturated with volatile hydrocarbons.

Figure 8.6 Vacuum-Scrubber Truck with Diluent and Carbon Filter System[4]

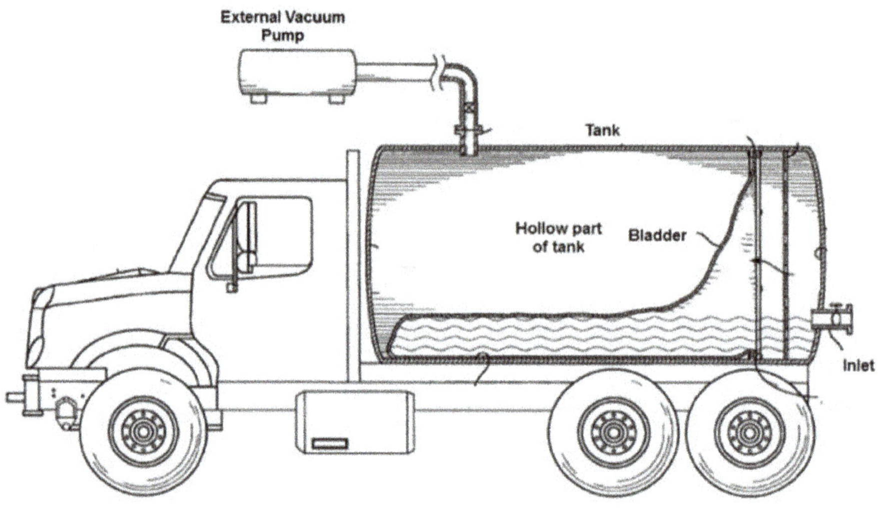

Figure 8.7 Vacuum Truck with Collapsible Bladder[6]

8.2. Pipeline Cleaning

Pipeline cleaning using pigging-type devices uses various tools to remove solids from the interiors of pipelines. Since this is not the major focus of the current book, a short introduction is provided here; much more information is available from Tiratsoo,[8] PPSA,[9] Frenier,[10] and TDW.[11] This section is included because some IC companies also provide these services or have provided them in the past. In addition, most plants have miles of piping with various diameters that may require cleaning. Some of these tools may be applicable depending on the line diameter and presence of valves and other intrusions (see Chapter 5 of Frenier[10]).

Many types of mechanical tools are used in pipeline maintenance, inspection, and treatment operations. A major maintenance device is the pipeline inspection gauge (pig; note that there may be other etymologies of the term[12]). In some places in the world, the term "scraper" also is used to describe the same types of pipeline cleaning and separation devices.

A tethered pig may also be called a "mole" in some areas of the industry. These tools are in constant use to assess the condition of pipelines and to remove various types of debris or scales and organic and mixed deposits. Pig-type devices have been used at least since 1942[2] to clear pipelines of fouling solids. Pigs are usually defined as tools that are inserted into a line and travel throughout a length of it, driven by product flow[9] or another fluid. The names of these devices are somewhat arbitrary, but the term "mole" can describe units that can be pulled by a line or are sometimes self-propelled. See additional descriptions and drawings of line moles used in downstream (refinery) operations in Cowan and Weintritt.[13]

Many pipeline tools described in this section are designed so that sealing elements provide a positive interference with the pipe wall. This seal enables gas or liquid pressure to move the unit and to provide a cleaning action (if this is part of the design). See Figure 8.8 for a view of various utility pigs (Mosher et al.[14]) that have various cleaning parts, including resin discs, various wire brushes, and items used for cleaning deposits from pits.

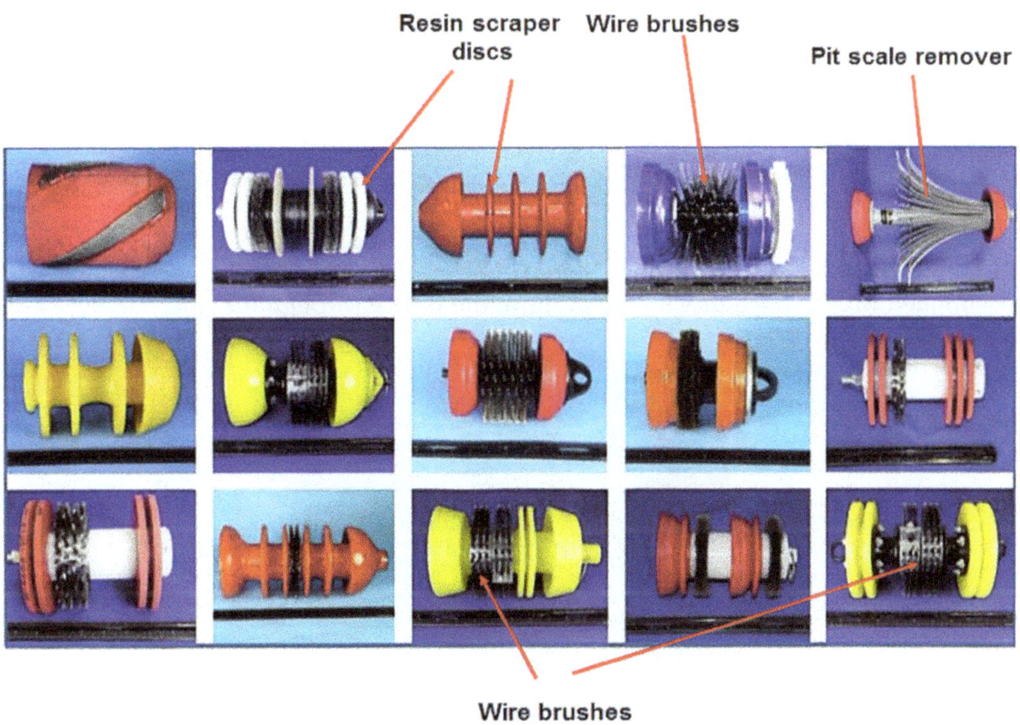

Figure 8.8 Various Types of Utility and Cleaning Pigs[14]

Davidson[15] describes the basic mechanics for the use of pigs. He states that once inserted into a line, pigs are driven through the line by applying pressure in the direction of required movement. A pressure differential is created across the pig, resulting in movement in the direction of the pressure drop. In an operational line, the pressure is usually applied using the fluid in the line.

The typical differential pressure (ΔP (lb./in^2-psi) required to drive a pig ranges from 50 psi to 300 psi. The range depends on the geometry of the line, pig types, and the drag characteristics.

2. Personal communication, Brooks Williamson, TD Williams, Inc. to Wayne Frenier, Tulsa Ok 2013

An expression provided by Wint[16] simplifies the calculation:

$$\Delta P = \frac{K}{d} \qquad \text{(Equation [8.1])}$$

Here, *K* is the type of pig factor (lb./in), and d is the nominal diameter (in) of the pipe. In one example, a 10-in disk brush pig (*K* factor = 15) is used in a pipeline. From Equation (8.1), at least 43.5 ΔP (psi-3 bars) would be required to drive this pig. Table 8.1 from Wint[16] shows approximate *K* values that reflect the relative resistance of the device during flow in a pipe.

Table 8.1

Resistance (K) Values of Devices	
Type of Device	**K (lb./in)**
Sphere or foam Pig	1
2 cup pig	4
4 cup pig	7
Disk pig	9
Cup brush pig	12
Disk brush pig	15

Once the force behind the pig becomes greater than the opposing frictional force, the pig will move in the direction of the applied force (pressure). See Figure 8.9. This is called the *breakup* pressure. The positive interface between the pig and the pipe wall imparts a cleaning action on the pipe wall. This can be further enhanced by the addition of brushes, scrapers, or even more aggressive tools to the pig. These are shown in Figure 8.8.

Pigs also can be pulled through a line with a cable. However, now it is a pulling force (not a line pressure) that must be supplied with a winch or wireline to move the device through the line.

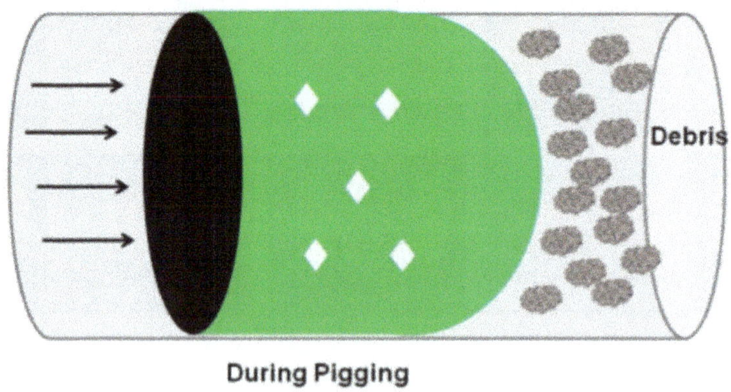

Figure 8.9 Forces on Pig

Utility pigs include cleaning, separating, and dewatering devices. Figure 8.8 shows pictures of several types of utility pigs, including cleaners and scraper devices. Davidson[15] also shows additional pictures in his paper, which describes the various designs of utility/cleaning pigs. These include foam (polymer foam not chemical foam), solid cast, and bolt-type devices. TDW[11] has photos/drawings of a large variety of pigs. As in many industries, different nomenclatures are used. Cordell[17] describes classes of utility pigs as mandrel (a solid device that is also called the body or a bolt), foam (various types), cast (various polymers), and spheres.

Cleaning chemicals can be trapped between two utility pigs and used to help clean the surfaces and to ensure that all parts of the line are contacted with the cleaner. See Figure 8.10.

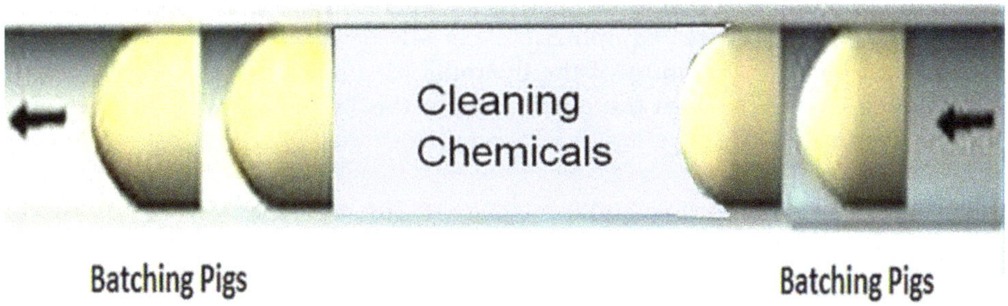

Figure 8.10 Cleaning Chemicals between Pigs

Note that moderate lengths of piping can be cleaned by circulating solvents using equipment described in Figure 8.1. In the confined areas of a refinery or power plant, this may be the preferred method of chemical application.

8.3. Heat Exchanger Cleaning Techniques and Hydraulic Jetting

A very large variety of heat exchange units are in refineries and chemical plants of all types. Section 1.3.4 describes the general function of these units, and Figure 1.15 and Figure 7.1 show drawings of some of the different configurations. This section describes cleaning methods for these vital units with emphasis on hydraulic jetting methods. As noted in Section 8.1, solvents also may be circulated through either (or both) the shell and tube sides of heat exchanger units, if the flow is not significantly blocked.

Muller-Steinhagen[18] described the fouling of heat exchangers and methods for cleaning them. The cleaning processes need to be related to the deposits (organic or inorganic solids) fouling these vital units. See Sections 3.2 and 7.1 for formation mechanisms of inorganic and organic deposits. In addition, this author[18] notes that both chemical and mechanical methods of removal have been used.

The application of various aqueous acids, chelating agents, and organic solvents have been described in Chapters 4 and 7 of this book, and Muller-Steinhagen[18] lists all of the same types of solvents. The benefits of using chemical solvents for cleaning heat exchanger units are

- Speed of cleaning
- Possibly reduced mechanical damage
- This solvent may reach less accessible areas of the units
- Possibly less labor intensive
- Performed without disassembling the unit

However, he claims that mechanical methods also may be needed if the deposits are very dense or insoluble or if flow paths are blocked and circulation is impossible or impractical. Muller-Steinhagen has an entire chapter on hydraulic jetting processes, and some of these processes will be described in the next section (8.3.1).

Thus, in some mechanical processes, chemistry plays a minor role, and high-pressure water (sometimes with chemical added) is used to clean the equipment. A recent development in mechanical cleaning (with some chemical assistance) is the use of large ultrasonic units to clean exchanger bundles. In summary, totally mechanical as well as chemical with mechanical methods are used to clean the various types of process units in addition to heat exchangers.

8.3.1. Hydraulic Water Jetting Processes

A major scope of this section is cleaning of heat exchangers and other chemical process equipment. High-pressure water cleaning of process equipment, especially heat exchangers, has a long history.[19, 20] Table 8.2 from Frenier and Barber[20] lists most of the hydraulic cleaning processes in use at the time of their report. These techniques are still in use at the publication of this book; however, several new processes also are described in this section.

During hydraulic water jet processes for cleaning exchangers, the bundles are often removed from the shell and usually transported to a specific (and protected) area, where the cleaning procedures are conducted. Various hand-operated, semiautomatic, and fully automatic jetting processes have been used to clean the bundles.

Table 8.2 Description of Hydraulic Cleaning Techniques[20]

Method	Description
Manual lance	Hand-operated stiff or flexible water lance
Shotgun	High-pressure water gun designed to clean the surface sides of a heat exchanger bundle and remove loose debris from the exterior of the tubes
Automated tube lance	Remotely operated high-pressure water lance for cleaning the inside of tubes; devices to operate multiple lances are available
Rotating tube lance	Remotely operated water lance that also rotates to ensure more complete removal of deposits from the inside of tubes
Hydraulic ramming	Remotely operated water lance that uses hydraulic energy to push the stiff lance through plugged tubes while jetting at high pressure
Hydraulic assisted tube cleaners	Devices with cutting surfaces that are propelled through a tube using water pressure to remove hard scale
Automated bundle blaster	Device that cleans the shell side of exchanger bundles by remotely rotating the bundle and indexing the jet head; some can also clean the tube side
Tube mole and rotojet	Spherical device with multiple jet heads that can transit a large tube, heat exchanger shell, or pipeline and rotates to clean the entire surface with high-pressure water
Hydraulic/abrasive	Abrasives can be fed into the water stream when additional mechanical forces are required to remove the deposits and polish the surfaces. The abrasives range in hardness from sodium bicarbonate and seed shells through garnet. Many of the techniques described above can be modified to use abrasives with the water. Chemicals such as detergents and solvents may also be added to the jet water to improve cleaning efficiencies.

Many of these types of processes are described in this section, and Figure 8.11(a) shows a diagram of a water lance that is inserted into a tube to blast out the deposits. Very high-pressure water (up to 10,000 psi) is used. Figure 8.11(b) shows the application of high-pressure water using a hand lance.

In some cases, water, with particulates added, has been used for this operation.[21] Environmentally friendly particulates such as sodium bicarbonate[22] have been added to the water stream to improve removal effectiveness. High-pressure water jetting also can be used to clean several different equipment types, such as tanks, piping, and other surfaces.

In addition, Jalalirad et al.[23] have described the use of projectiles, including sponge balls of different size and hardness, that are propelled through a heat exchanger tube to remove deposits. They researched the contact area of the projectiles and the stiffness and found that the best size is 10% bigger than the tube and a stiffness that is 1% under 0.6N force.

Figure 8.11 (a) Water Lance and Shell and Tube Heat Exchanger; (b) Person Performing Hand Jetting

The jetting process requires a high-pressure pumping unit (Figure 8.12) to provide water flow rates that will remove a specific type of deposit. The design of the jet nozzle on the end of the lance is critical to the performance of the service. The hydro jet[24] is a nozzle intended to eject gas or fluid in a coherent stream into a surrounding medium. The purpose of a jet nozzle is to increase the kinetic energy of the flowing medium at the expense of its pressure and internal energy.

Nozzles can be convergent (narrowing down from a wide diameter to a smaller diameter in the direction of the flow) or divergent (expanding from a smaller diameter to a larger one). A de Laval nozzle has a convergent section followed by a divergent section. Convergent nozzles accelerate subsonic fluids. If the nozzle pressure ratio is high enough, then the flow will reach sonic velocity at the narrowest point (i.e., the nozzle throat). In this situation, the nozzle is said to be choked. Table 8.3 from Muller-Steinhagen[18] suggests jetting operating conditions on the basis of the contamination situations. That book also displays a wide variety of jet tips for different conditions.

Figure 8.12 Hydraulic Jetting Pump Unit (20,000 psi) and Recirculation Unit

Table 8.3 Hydrojetting Suggestions for Various Heat Exchanger Foulants[18]

Industry	Contamination	Jetting Pressure bar	Flow Rate l/min
Petrochemical	Solvents, synthetic rubber, soft plastics, latex	> 800	180
Petrochemical	Plastics, petroleum derivitives	Up to 3000	30
Refineries	Crude oil, viscous petoleum, tar, asphalt	600-800	180-300
Refineries	Gasoline, diesel oil, fuel oil, petroleum coke	600-800	180-200

Patents for jet nozzles that can be used for surface cleaning are noted. Henshawa[25] describes a rotating head that mounts a nozzle and creates cavitation in a pressurized fluid to quickly and efficiently clean a surface. The rotation of the nozzle ensures a relatively wide cleaning path. Such a unit may be useful for clearing the outside of exchangers or the inside of tanks.

Steinhilber[26] claims a design of a flat spray nozzle. He also states that the spray nozzle serves the function of allowing water to be aimed with high pressure of 20 to 250 bar to produce a flat jet with a strip-form or virtually linear spray pattern.

The patent also claims that good cleaning is achieved with such high-pressure jets because the highly accelerated water droplets from the nozzle bring about a mechanical removal effect on the impingement surface to be cleaned. The aim of such cleaning apparatuses is to use as little water as possible to produce a high cleaning effect, i.e., a good removal effect.

A measure for this is the generated *jet force*, which is defined as the force that is exerted on a given surface area by the impinging jet. The exact design and materials used in jet nozzles are trade secrets for many manufacturers.

A patent by Vargas[27] describes a device that supplies pressurized fluid to inner portions of tanks for fluid storage (e.g., petroleum) to enable cleaning or the recirculation of the stored fluids. The device includes a revolving element to change the direction of the flow of pressurized fluids inside the tank. Figure 8.13, which is based on this patent, shows possible locations of the unit on the sides of a storage tank.

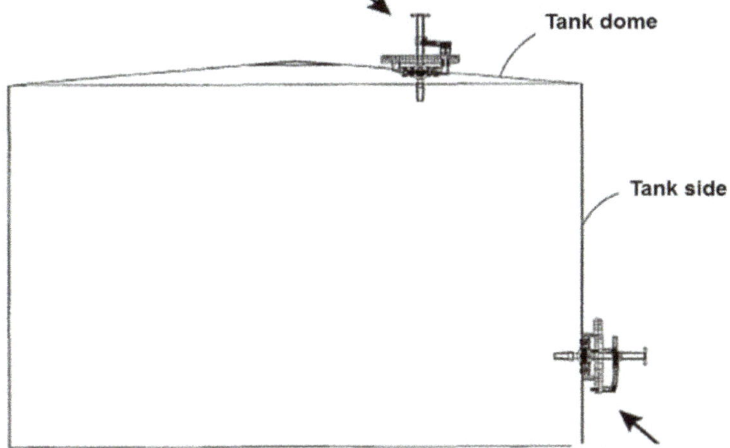

Figure 8.13 Location of Cleaning Jets on a Storage Tank[27]

An example of a high-pressure pumping system for hydraulic jetting with a water-recycling unit is shown in Figure 8.12. Note that this pumping unit provides the high pressures needed for knocking off hard scale but has low flow rates. In contrast, the high-flow rate pumps in Figure 8.2 are needed to pump large volumes of fluid to fill and circulate boilers and other equipment. However, they may not need very high pressures to overcome head pressures and friction pressures in the units being cleaned.

For many years, a hand-lance has been a major part of the operations. However, this can be very dangerous, and where a hand lance is required, the operator needs substantial personal protective equipment (see Section 10.2.1). Note the heat exchanger and lance in Figure 8.11(a) and a person (b) in required personal protective equipment (for the 1990s time frame, but the face shield is not lowered) ready to start the jet using a foot trigger. Figure 8.14 shows a photo of an operator in full protective gear jetting a large heat exchanger inside of a plant. In this case, the bundle was not removed from the shell owing to its size and weight.

Figure 8.14 Hand Jetting on Large Heat Exchanger

This type of jetting still may be required, depending on the locations and sizes of the units being cleaned, but automated (and safer) alternatives are also now available. Note that in some cases a detergent may be added to the water to improve removal of dispersible deposits. In addition, a hand-held device called a shot-gun can be used to clean the shell side of heat exchangers.

Frenier[19] and HydroChem[21] describe automated multitude lance devices (see Figure 8.15) that are used to improve safety and speed of cleaning by keeping people away from the HP water. A layout of a highly automated system has been patented,[28] and a drawing from the patent is shown in Figure 8.16. The authors claim that the system provides a heat exchanger tube cleaning assembly that can automatically (without ongoing human intervention) survey the tube sheet of a heat exchanger in three-dimensions, convert and record the survey results as a digital record in three-dimensions, and then, according to sequential parameters input via custom software, automatically coordinate via computer one or more cleaning devices to clean each desired tube of the heat exchanger. As seen in a drawing from the patent,[28] various digital cameras survey the tube bundle and then direct the computer to control the insertions of cleaning water lances into the tubes.

Figure 8.15 Automated Tube Lancer[29]

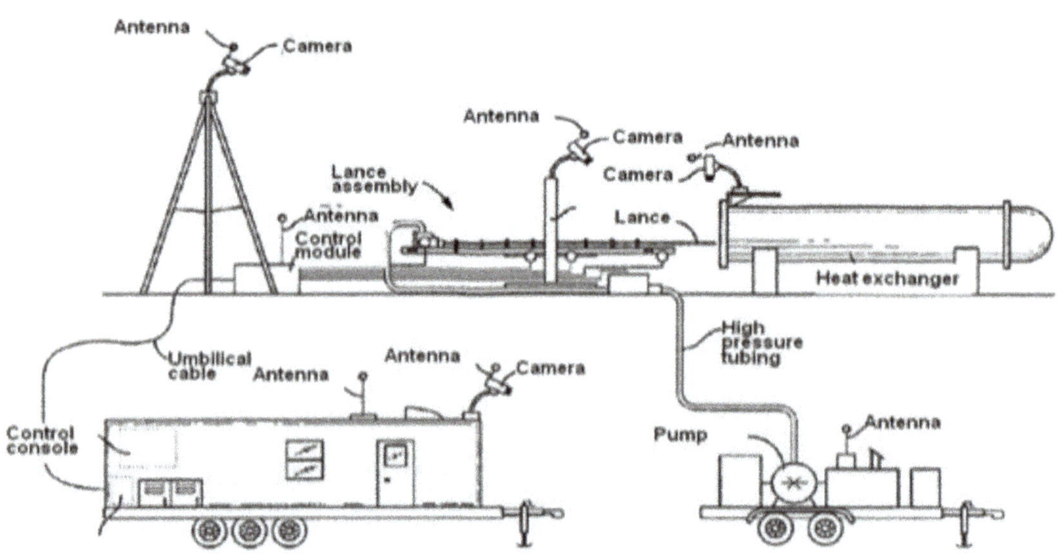

Figure 8.16 Automated Lancing System[28]

These systems are designed for cleaning the tube side of the exchangers. Several images of different automated shell side blasters are shown in Figure 8.17. These illustrations show fixed site units in which the heat exchanger is removed from its shell and placed in a rack (usually with rollers that will turn the bundle). Here, HP water jets on an overhead carriage will clean the outsides of the tubes.

Figure 8.18 shows a single person-operated shell-side blaster[30] that can be maneuvered to clean smaller units that cannot accommodate the larger overhead jetting systems. For complete coverage, the bundle may be on a set of rollers that can move the unit under the powerful jets. However, the image shown describes different equipment that clean both the tubes and the shell side of a small bundle.

Figure 8.17 Automated Shell Side Blasters[21]

Hydraulic jetting can be applied to a very wide range of process equipment, including piping and flow lines. To clean the insides of these types of units, a jet head on a long flexible high-pressure cable is threaded into the line. Figure 8.19(a) shows a cable feed system that advances the jet head (b) through the line at a programmed speed to clean the deposits.

Hydrojetting is a very flexible method that can be customized to work in a wide range of process units. In addition to tube bundles that are cleaned after complete removal from the shell, HP jetting is used to clean vertical heat exchangers, remove coke, and clean many other types of process units.[30]

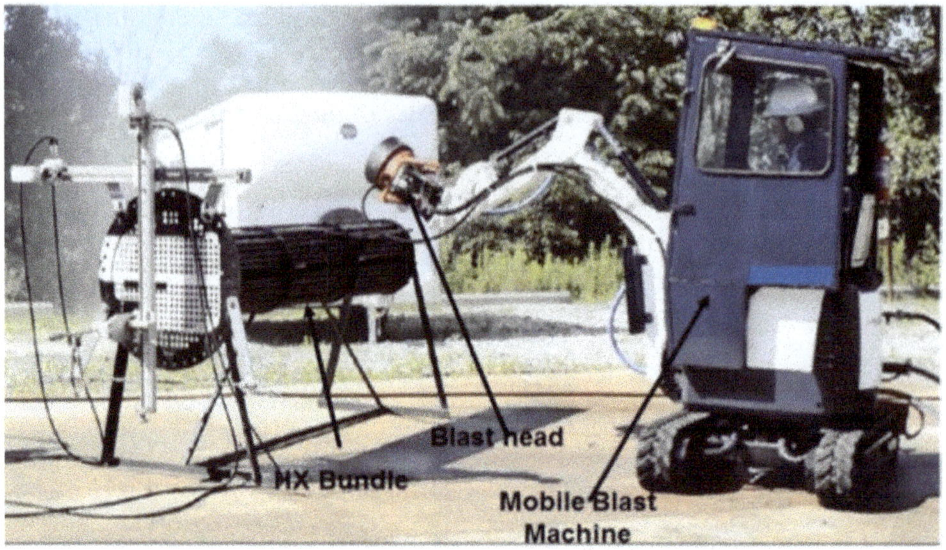

Figure 8.18 Moveable Shell Side Heat Exchanger Water Blaster

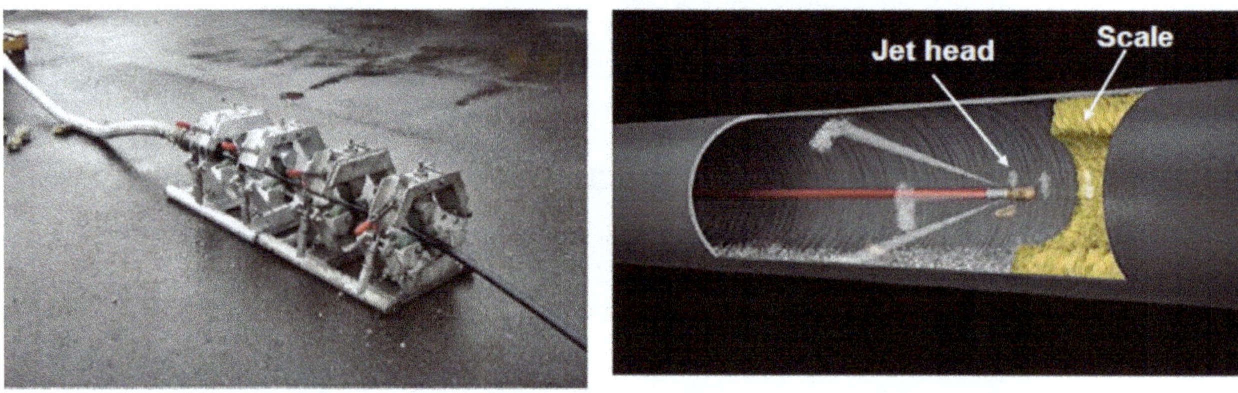

a) Jetting cable feeder

b) Jetting cable and head inside line

Figure 8.19 Jet Cleaning of a Flow Line: (a) Jet Cable Feeder and (b) Jet Head in Line[30]

An additional technology for cleaning the interior of heat exchanger tubes includes propelling small cleaning devices through a clogged tube using a jet of moderate pressure water. Saxon[31, 32] describes spring-loaded tube cleaners that are sent through the tube. The blades are mounted on a spindle; at one end of the spindle is a serrated plastic disk that allows a jet of water to propel the cleaners through a tube with greater hydraulic pressure than air. A drawing by the author of this book shows (Figure 8.20) the general shape of the device, and the papers[31, 32] give more details of the designs and uses. Saxon[32] also asserts that this method uses lower water pressures (~ 600 psi) and much less water than high-pressure hydraulic jetting processes shown in Figure 8.17, Figure 8.18, and Figure 8.19. The reason stated is that the spring-loaded cleaning tools are propelled by a water jet that only operates while the cleaner is in the tube.

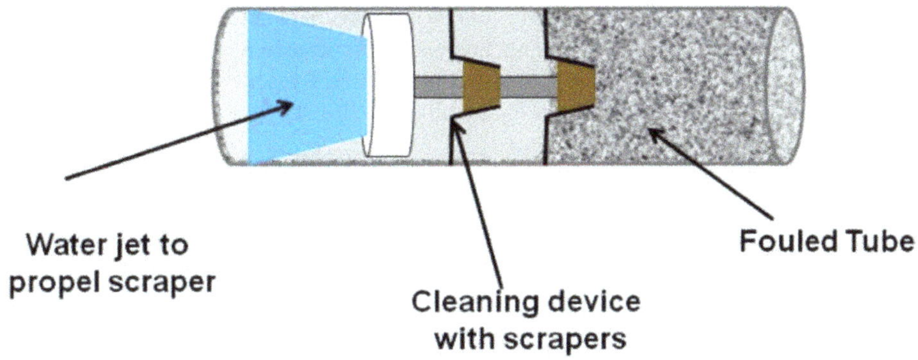

Figure 8.20 Drawing of Tube Cleaning Tool Being Propelled by Water Jet

8.3.2. Cleaning Heat Exchangers with Ultrasound

Ultrasonic processes are used to clean heat exchange bundles. Patents[33,34] and a report from Tubetech[35] describe devices and processes that use ultrasonic units inserted into the tubes or between the tube bundles to improve cleaning efficiencies. A case history[35] notes that the shell side of a heat exchanger in a Texas refinery was cleaned with a combination of water jetting assisted by an ultrasonic device. The authors claim that the combination of mechanical, ultrasonic, and polymeric water ensured that the deposits were removed effectively.

Kieser[36] reported cleaning complete bundles by immersing them in a bath of chemicals (detergents; details not revealed) that may remove tube and shell side deposit. A cleaning bath is depicted in Figure 8.21 and Figure 8.22. Although effective, cleaning of large bundles may require 40,000 L of fluid that requires disposal.

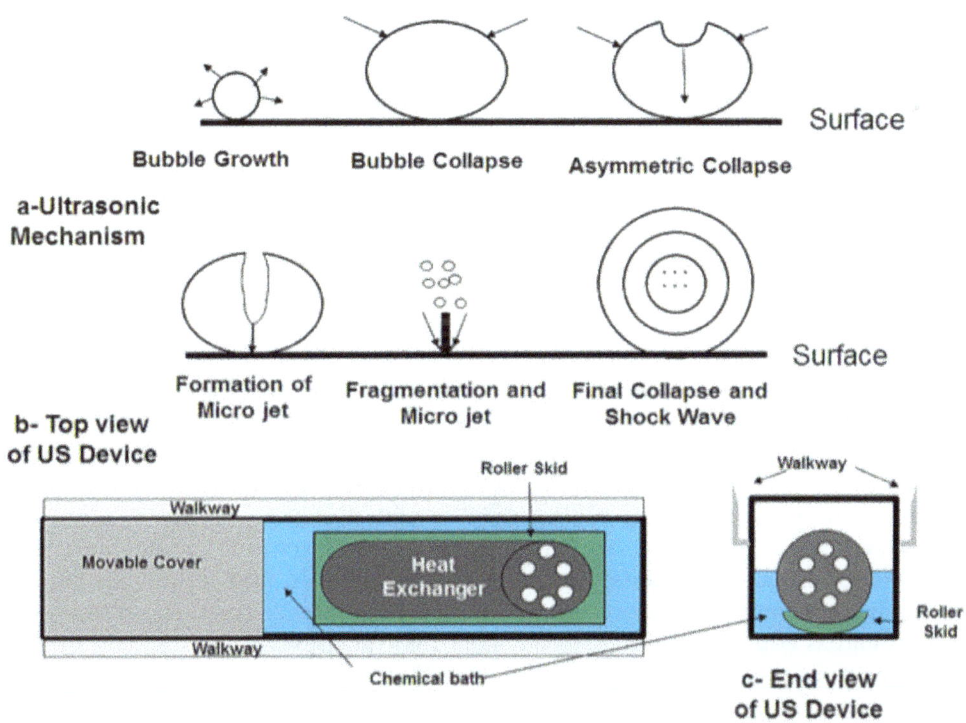

Figure 8.21 Ultrasonic Shock Wave Mechanism (a) and Views (b, c) of Ultrasonic Bath

A general theory of how ultrasonic processes remove solids from surfaces is in the report by Fuchs.[37] The theory and practice of using ultrasound to clean large objects like heat exchange bundles is illustrated in Figure 8.21, which was drawn on the basis of the information in Kieser[36, 38] and Naude.[39] Part (a) illustrates a mechanism of the process. Here, the collapse of bubbles at the surface produces a jet of fluid as well as a shock wave that dislodges solids from the surface.

Drawings in Figure 8.21 (b and c) show a large vat that has been equipped with strategically placed ultrasonic transducers (not shown) that generate sonic energy. This energy produces bubbles that cavitate and transmit the energy to the surfaces. Kieser[36] claims that filling the bath to about the midpoint of the exchanger and rotating it (c) in the bath removes more deposits. Note that the bath contains chemicals (not described) that increase the solids removal process. A photo of a heat exchanger being cleaned in a ultrasonic bath is shown in Figure 8.22.[40] In a similar cleaning service from Variclean,[41] ultrasonic energy provides an alternative and more efficient way to clean heat exchangers. The generated cavitation separates the contamination from the surfaces without damaging the object.

Figure 8.22 Heat Exchanger in Ultrasonic Bath Showing Fluids Draining from the Tubes[40]

A patent application by Phillips[42] describes a system that requires ultrasonic transducers that operate in the range of 20-30 kHz and deliver 2000 watts of energy. They are spaced along the bath based on 2-10 wavelengths of the energy wave form.

The application also defines some of the cleaning bath chemistries, including solvents used to remove asphaltenes, bitumen, and other heavy crude oil derivatives. The patent claims that an aqueous-based degreasing solution, with near neutral pH, of a mixture of trisodium phosphate and a nonionic surfactant (Paratene[43]) provides excellent performance and relatively simple disposal. The application also notes that, in some cases, small amounts of an organic solvent may be added to the aqueous solution to enhance the removal of certain contaminants. In other cases, it also may be necessary to use strongly acidic or basic cleaning fluids to address specific contaminants, such as polymers, epoxies, and scales. The choice of materials in construction of the container is therefore important; although normal (or carbon) steels perform well as structural elements and as container walls in strictly near neutral applications, stainless steel is preferred as a wall material to avoid corrosion in the case of non-neutral cleaning fluids.

See Section 7.5 for possible surfactants and solvents. Note that bubble formation is a necessary part of the process, so the stability of the bubbles and foam will be important experimental variables in developing an ultrasonic bath fluid.

8.4. Use of Foams for Chemical Cleaning Operations

Although chemical cleaning solvents usually are applied by pumping solvents through process equipment or power boilers, several other methods have been used, including foams and steam (or other vapors) for cleaning/decontaminating refinery vessels (see Section 7.6).

Reactive chemical foams have been used for many years to clean equipment that cannot (or should not) stand the weight of a liquid fill or if it is not economical to fill the equipment with liquid. Industrial process equipment that is successfully cleaned with foam includes utility surface condensers, towers, tanks, and utility power turbines. Section 8.4.1 briefly presents a theory about foaming methods.

8.4.1. Foaming Theory

Foam cleaning is a chemical process developed in the 1950s. One of the first foam cleaning jobs occurred because of a customer's need for a creative solution to a problem.[44] An overhead process gas line (550 ft. of 12-in. stock) required cleaning for removal of acetylides before dismantling. Acetylides explode when subjected to shock. Although chemical cleaning with an acidic solution was a known method to remove acetylides, a "fill and soak" process" could not be used because of the piping design. The piping support structure was designed to support only the weight of a gas. A chemical cleaning engineer solved the problem by reducing the density of the acidic solution by converting the liquid to a foam. The job was successful, and foam cleaning processes have been used in other situations. Foam cleaning was soon directed toward the cleaning of utility type surface condensers, which cannot tolerate the weight of water fill.

Foams are coarse dispersions of gas in a relatively small amount of liquid. A short description follows; see Rossen[45] and the discussion in Frenier[46] for more details of foam theory. In oilfield nomenclature, the quality of foam is expressed as

$$\text{Foam quality} = \frac{V_g}{V_g + V_l} \bullet 100 \qquad \text{(Equation [8.2])}$$

Here, V_g is the volume of the gas, and V_l is the volume of the liquid. The sum is the volume of the foam. Foam quality is considered to range from 52% to 95%. Above 95%, the foam usually changes to a mist, with gas as the continuous phase. Below about 52%, foam is not stable because there are no bubble/bubble interactions to provide resistance to flow or to gravity separation. Above 52% gas, the gas concentration is high enough that the bubble surfaces touch. Fluids with less than 90% quality are frequently called energized fluids, and foams are fluids in the 90%–95% range. Note that this same equation (Equation [8.2]) also can be used to calculate foam quality with a liquid/gas flow rate, as long as the volumes/time are known. It is clear from this equation and Karam[47] that foam can occupy a much larger volume of space than an equivalent mass of a liquid.

Foams cannot form from pure liquid. The thermodynamic equation is

$$\Delta F = s\Delta A - p\Delta V \qquad \text{(Equation [8.3])}$$

Here, ΔF is the change in Helmholtz free energy at constant T, s is the surface tension of the liquid, A is its surface area, p is the pressure, and V is the volume of the gas. Equation (8.3) applied to a foam shows that a decrease in free energy results from the loss of area and expansion of the gas, both of which come about by coalescence of the bubbles; hence, foam composed of a pure liquid is thermodynamically unstable. The third component necessary for stable foam is usually a solute in the liquid but also can be a finely divided solid, a liquid/crystal, or an insoluble monolayer. The stabilizer (foaming agent) must be surface active (migrate to the interface) and be positively adsorbed. The work necessary to desorb the stabilizer is added to Equation (8.3) to give

$$\Delta F = s\Delta A - p\Delta V + W(desorption) \qquad \text{(Equation [8.4])}$$

This term may make the whole equation positive (more thermodynamic stability); however, even when the foam is not completely stable (as is the case with most foams), the work of desorption reduces the decay rate of the foam. Examples of surfactants that may be used as a foamer are in Section 7.5. A major factor for picking a foamer is the compatibility with the aqueous phase.

The elasticity of the surface layer as well as the Marangoni effect[48] (explained below) help to stabilize the foam. The draining of a liquid, including the motion of a surface that carries an adsorbed layer, creates fresh surfaces of higher surface tension (surfactants usually lower the surface tension), a non-equilibrium temporary condition. The local difference of surface tension is removed by the motion of the surface from regions of lower to higher surface tension. This is called the Marangoni flow and opposes the movement of drainage and acts as a stabilizing factor to restore the thickness of the lamella layer. Thus, agents that create large differences in surface tension or increase the elasticity of the lamella will lead to greater foam stability. Although a surfactant lowers the surface tension of the liquid, lamella may lack cohesion. A second agent (foam stabilizer) may be necessary to increase cohesion. A very compressive account of foam and emulsion technologies are in Kokal.[49]

The rate of drainage can be described by an exponential equation:

$$V = V_o \exp(-kt) \qquad \text{(Equation [8.5])}$$

Here, V is the volume of liquid that drains in time (t), and V_O is the total amount of liquid in the foam. This equation was used to describe drainage rates from many different types of slow-draining foams. The constant (k) is equivalent to a first-order rate coefficient.

Materials that inhibit or "break" foam should have a low surface tension, be destructive to the cohesion of the film, be insoluble in the aqueous phase, and have a low volatility. The polydimethylsiloxanes[50] have all these characteristics and are the most universal antifoaming agents. Other agents (such as oils and fats) have more complex behavior and may stabilize some foam at certain concentrations.

The rheology of foam is complex, and viscosity must be measured under flowing conditions. The major things to know are that foams are viscous as well as shear-thinning, even when the surfactants and liquid display Newtonian rheology. This means that foam will suspend solids when the flow stops, and the apparent viscosity decreases as flow increases (when it is being pumped). This is a well-known effect and is used in hydraulic fracturing.[46] The effect in a flowing pipe is that the viscosity is increased (accompanied with a liquid phase), and the density is decreased. Also, there will be a low Reynolds number (Re), high friction factor, and higher pressure drops (ΔP) compared with multiphase gas flow. Formation of foam and uses in cleaning operations are described in the next section.

8.4.2. Foaming Agents and Foaming Process for Cleaning Surfaces

Foam used in pipelines or process vessel cleaning/treatments require

1) A liquid source
2) A gas (usually a hydrocarbon gas, N_2 or CO_2)
3) A surfactant to stabilize the foam (see Figure 7.9 and Figure 7.11 for basic surfactant structures)

Figure 8.23 shows a simple setup in which water/surfactant (or hydrocarbon/surfactant) mixes with a second fluid such as an acid or a concentrated corrosion inhibitor and then with the gas phase. The mixing section (in this case, static mixers) provide the shear to form the foam. Different surfactants are required for different aqueous fluid foams as well as hydrocarbons. With enough shear, a foam could be formed directly in the pipeline or process unit by injecting a gas into a liquid flow. Examples of the use of foam to clear liquid from lines as well as other treatments are described next.

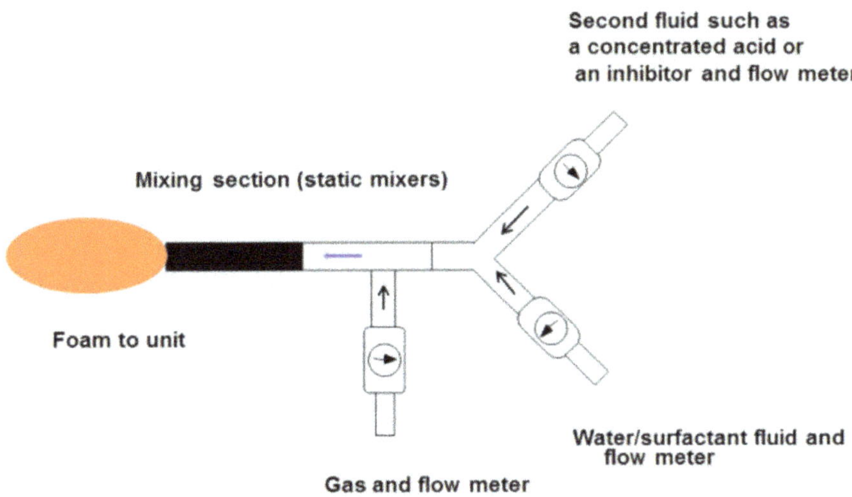

Figure 8.23 Foam Generation Unit

The use of foamed solvents for cleaning surfaces was one of the first technologies patented by the chemical cleaning industry. Early foam generators[44, 51] produced foam by adding air or a gas through spargers submerged in the cleaning solution. To obtain a large volume of foam, many spargers were necessary. The spargers could become contaminated with oil from the local air supply and cause rapid breaking of the foam as it was generated. The spargers also failed from overpressure. If the gas/liquid ratio was too great for the capacity of the generator, generating efficiency was lowered, and the result was a poor quality of foam and excess gas. The size of the foam generators made setup sometimes difficult. The first generators weighed more than 2 tons. See Figure 8.23 for a more modern design that uses static mixtures.

8.4.3. Issues with Foam Cleaning Technology

Several issues with using foam to clean industrial equipment are described.

Corrosion. Air or oxygen cannot be used in foam technologies unless the usage is for removing copper from non-corrodible surfaces. If acids or chelating agents are used, the appropriate corrosion inhibitors (see Sections 6.2 and 6.3) for the fluids must be in the liquid phase, and the inhibitor must be tested in the presence of the foaming agents.

Recycle. Early recycle processes completely ignored the gaseous component of foam. The liquid portion was recovered from natural foam breakout or foam breakers and regenerated. The half-life of the foam, or the amount of time to recover one-half of the amount of liquid used to generate the foam, became an important consideration. If a foam did not have sufficient half-life, breakout would occur before the foam filled the tube or vessel being cleaned. If the foam had too long of a half-life, natural breakout would not occur, and foam breakers had to be used to facilitate breakage. One method included the collection of effluent foam in a large tank to allow more time for breakage. Modern foam formulations are designed to meet almost any foam half-life requirements.

Frenier[19] recognized that a once-through process for foam cleaning was very wasteful, and it would be desirable to recycle and recirculate the foam through the vessel being cleaned. In a once-through process, the cost of cleaning solutions is directly proportional to the time and flow rate–not on stoichiometric quantities of solvent required to dissolve the scale. A once-through foam cleaning process requires a person on the effluent end to continually spray on a foam breaker. The foam is converted to a liquid before disposal. A new process[52] was developed for the complete recycle of foam cleaning products. Excess gas is not used because recycling uses only the available volume of gas from the closed system. Recycle foam flow is steady and continuous and does not depend on excess gas and slug flow for foam generation. The gas phase, liquid phase, and foam are recycled as a closed system or a non-pressurized system. Pressure-drop calculations are more realistic. Two-phase slug flow is reduced.

Greater contact of foam in upper level condensers could be achieved because of improved foam flow characteristics. Foamed solutions could then be used until they are completely spent without fear of excessive corrosion problems. Also, it is not necessary to chemically break the foam so it can be recycled as a liquid. Breakout liquid is regenerated into foam using the breakout gas. The process has the advantage of using an inert gas because the gas is also recovered and not wasted in a once-through process. Corrosion inhibitors (Frenier[53]) can be used to decrease the cost because of recycling. There is less exposure to toxic and irritating gases. Organic cleaning materials can be more economical because of recycle and on-site disposal by incineration.

8.4.4. Specific Solvents and Practical Examples of Use of Chemical Foams

This subsection reviews the range of solvents and equipment types that have been successfully cleaned using chemical foams.

Barrufet[54] claims that reasons for using a foamed solvent to clean different equipment, including surface condensers and utility boilers, are that air pockets can trap a liquid solvent and possibly some of these components cannot accept the weight of a liquid fill. Other reasons for using a foamed solvent are

- Water cost
- Chemical cost
- Disposal cost
- Passivation requirements
- Preparation cost owing to temporary piping

Chemicals. Crowe[55] patented a foam cleaning system that uses a fluorocarbon carrier and a stabilizing agent containing a perfluoro alkane sulfonamide. The system incorporates the concept of foam recycle, and the formulation could contain nitric acid to remove hydrocarbon deposits. HCl also is a commonly used solvent.

Kelly[56] reviewed the technology advancements in foam cleaning. These improvements include the more efficient foam generator described above and a surfactant system (based on cationic surfactants) that is less sensitive to metals removed from the deposits than earlier systems. Ball[57] reviewed foam cleaning for boiler subsystems at the 1979 International Water Conference. The highlight of the paper was the development of cost/benefit calculations that demonstrated that foam cleaning of utility surface condensers and turbines had an 8:1 benefit/cost ratio.

In 1986, Hoy[58] presented a paper to the International Water Conference that described an improved process for cleaning copper/alloy surface condensers: a solvent system of foamed HCl, followed by ammonium EDTA and a copper complexer. The complexer reduces the copper bleeding that usually follows cleaning of copper metal using hydrochloric acid. Foamed sodium EDTA was used by Flores[59] to remove calcium sulfate deposits from utility surface condensers.

Mestetsky and Davis[60] proposed a foamed solution of enzymes and nonionic (N-oxides) surfactants to clean batch cokers in refineries. The solution was heated to 210°F (99°C), and the water vapor and foam generated were passed upward into the fractionator, where cleaning took place for 12 hours before the unit was drained.

Cleaning Power Turbines. These very large structures (Figure 1.5 and Figure 8.24) can be cleaned using mechanical and chemical methods. A patent by Amrian[61] describes the use of chemical solvents to remove dirt and organics from the turbine blades. The report claims the solvent contained a hydroxylated wetting agent, a chelating agent, an emulsifier, an optional crown ether, and an optional non-aromatic solvent. The cleaning composition also is claimed to be 100% biodegradable, water-based, and able to remove adhered particles having a particle size of less than 35 micrometers. However, the solvent must be flushed through the turbine using a pump as the turbine is rotated using an external engine.

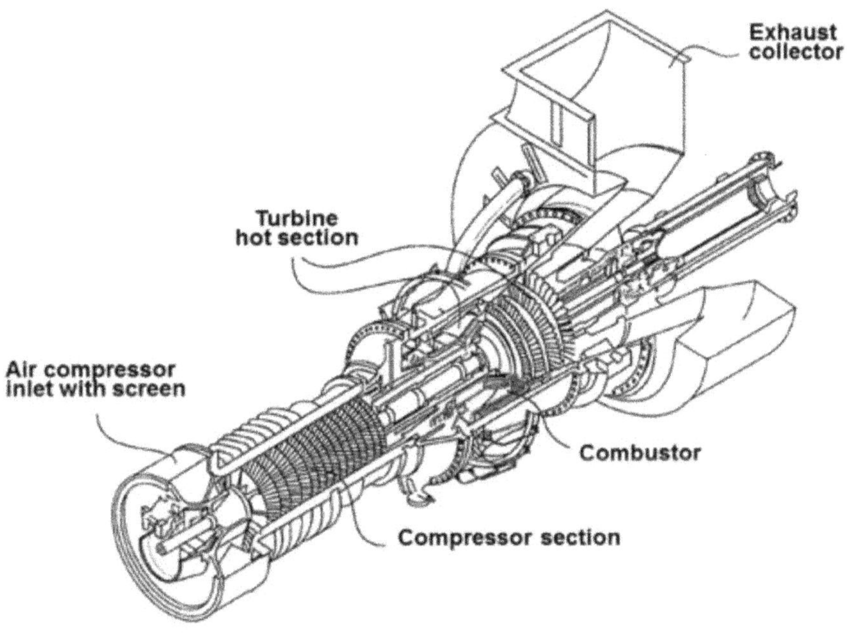

Figure 8.24 Diagram of a Gas Turbine Showing Blades in Hot Section[61]

A more efficient method of cleaning the turbine blades may include a foamed solvent. Because of its ability to fill large and unusual spaces, foam can be used to clean turbine generator blades, which become fouled with deposits containing minerals and copper (see Figure 1.5 and Figure 8.25). Very small amounts (25 lb.) can cause the turbine to lose power. The effect of the small amounts of deposits may be because

- Blades are of an airfoil design for efficiency
- Deposits change the shape of the blade
- Deposits increase the surface area
- Deposits reduce clearance between blades
- Deposits affect the flow across the surface
- Deposits reduce blade efficiency

Examples of Foam Cleaning of Turbines. Foster[62,63] and Wicker[64] claim that deposits that may foul the blades and nozzles of a turbine may include Cu, CuO, sodium phosphate, or calcium-based deposits. These authors claim that most turbine fouling is copper based. See Figure 8.25. These investigators[62, 63] claim that foam has been used more than 450 times (around the world) during the past 20 years to descale turbines and improve power production.

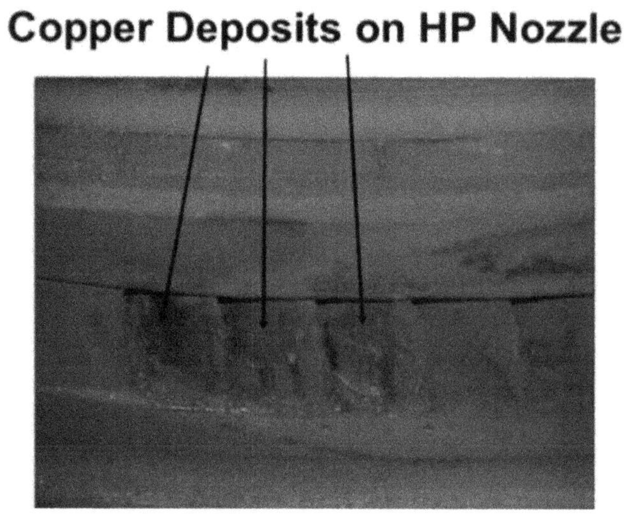

Figure 8.25 Copper Deposits on High-Pressure Turbine Nozzle

A setup of the chemical foam treatment of turbines is shown in Figure 8.26.

A patented valve,[65] solutions of ammonia ammonium carbonate and a foaming agent, and, if necessary, an oxidant (oxygen or hydrogen peroxide) have successfully removed copper deposits from turbine blades. Figure 8.27 shows how the foam fills the steam chest and then flows to the next turbine.

When cleaning turbines, the solvent is chosen for the deposit on the surface:

- Copper metal removal requires ammonium carbonate, ammonia, and an oxidizing agent.
- Copper oxide deposits require the same alkaline combination, but the oxidizing agent may not be needed.
- Calcium salts require formic acid or an EDTA solvent.
- Water-soluble salts, such as sodium phosphate, may only need deionized water for removal.

In all cases, a foaming agent appropriate and tested for that solvent is required. Foster[63] notes that the foam quality (Equation [7.1]) is important to produce good results, and a thick foam (high quality) enhances the rate of scale removal. At the end of the treatment, a water foam is flushed through the turbine, and defoamer is added at the end of the train to the foam to separate the phases. See Figure 8.26.

Figure 8.28 shows a dataset from Foster[63] that demonstrates that foam cleaning is as effective as grit blasting to remove deposits and return the power output of a fouled unit. The latter treatment does not require dismantling the turbine.

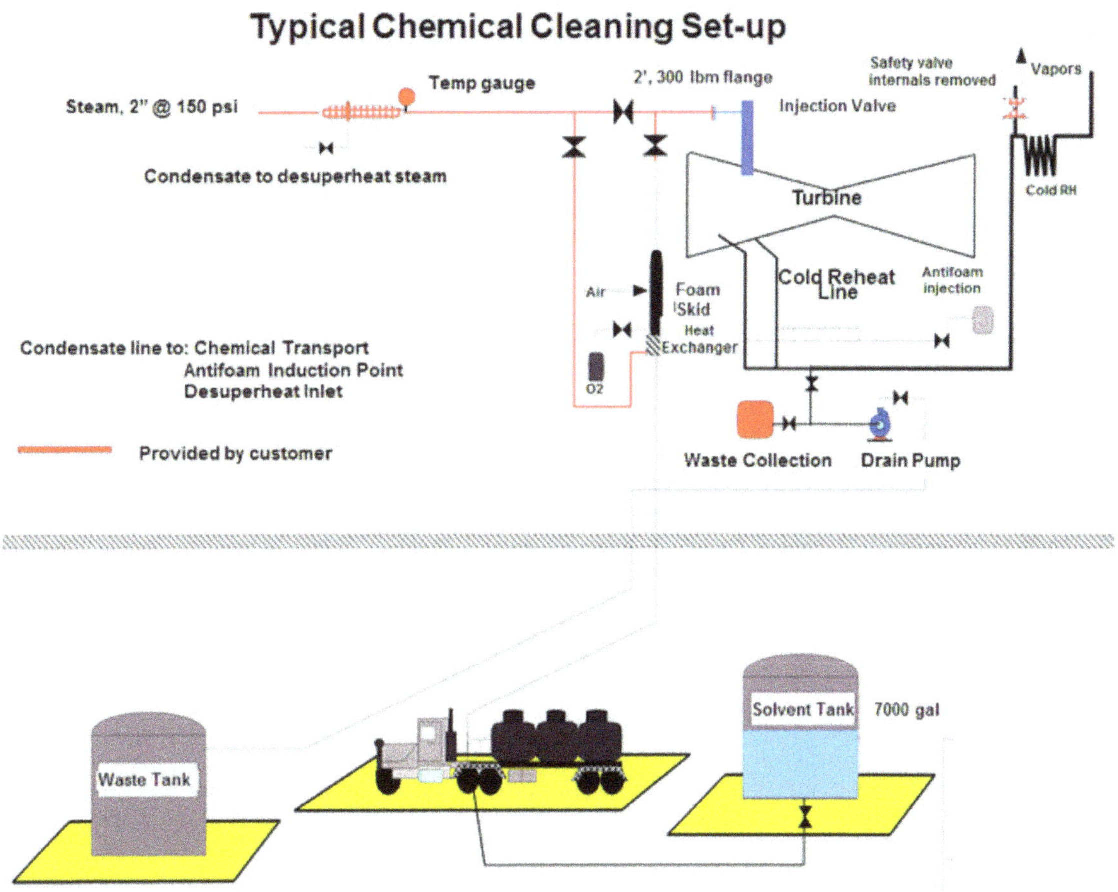

Figure 8.26 Equipment Setup for Foaming of Turbines[62, 63]

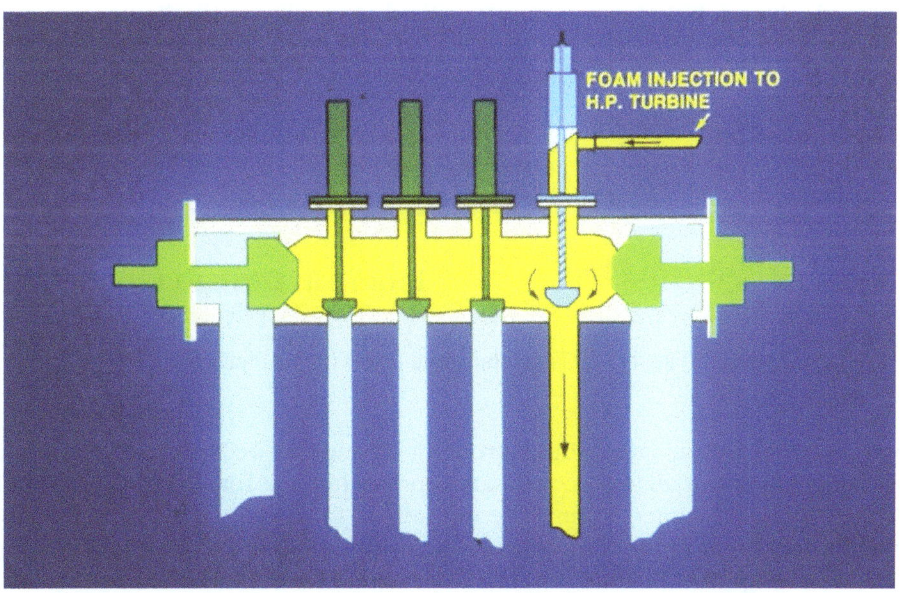

Figure 8.27 Foamed Solvent Filling the Sections of Turbine Generators

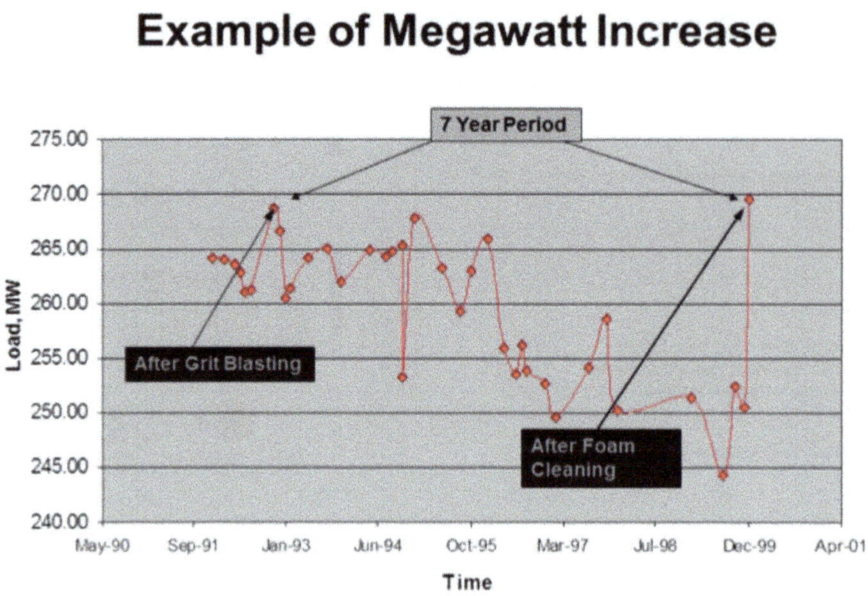

Figure 8.28 Power Increase after Grit Blasting and Foam Cleaning of a Turbine[62,63]

Cleaning Pipes and Pipelines. Karam[47] described the use of foam to remove liquids from gas lines. This paper lists several surfactants suitable for producing a pipeline foam to remove liquids for subsea gas lines. These include ammonium salts of alcohol ether sulfates and olefin sulfonates. These are both anionic surfactants. See Figure 7.9 for general surfactant structures.

Figure 8.29 depicts the action of foam removing a liquid from the line. The mechanism can include physical removal because the foam is more viscous than the liquid, and some of the liquid is incorporated into the foam. Note that the foam can be generated using an inert gas or a hydrocarbon gas. The surfactants should be compatible with the liquid, but excess amounts of fluid may collapse the foam.

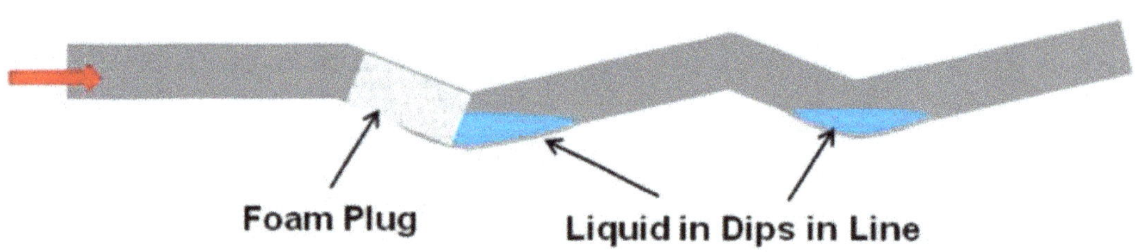

Figure 8.29 Foam Removing Liquid from a Pipeline

Yang[66] describes a method for unloading fluids from an oil and gas well, pipeline, or flowline that includes a synergistic combination of an alkyl glucoside and an amphoteric surfactant as a foaming composition. This foaming composition is claimed to improve removal of formation fluids (water or NGL), particularly production fluids with natural gas condensate and water, and therefore increase the rate of production from the oil and gas well. Examples of foaming surfactants include decyl glucoside, octyl glucoside, or dodecyl glucoside used with alkyl betaines, alkylamidopropyl betaines, alkylampho acetates, alkylampho propionates, alkyl hydroxysultaines, or alkylamidopropyl hydroxysultaines. Yang[67] also claims a wide range of anionic and amphoteric surfactants can be pipeline foaming agents.

Kouba[68] describes pipeline foams that are claimed to relieve various flow assurance problems. The authors claim that issues such as deposits (e.g., asphaltene, wax, and hydrates) and, particularly in the case of gas-dominated hydrocarbon fluids, slippage between the gas phase and the liquid phase can be partially eliminated by introducing a foamed fluid and passing it through the line. This patent application is quite detailed mechanically but does not describe suitable surfactants.

Embodiments claimed include

- Methods of (a) generating either an oil-based (hydrocarbon-based) foam and/or a water-based foam; (b) transporting the foam through a transport pipe used for hydrocarbon fluids; and (c) de-foaming the foam after it is transported. Typically, such hydrocarbon fluids are gas-dominated, but in some embodiments, they can be liquid-dominated.
- The foam is generated from foam precursor(s). In some presently contemplated embodiments, the foam comprises the hydrocarbon fluid that the pipe transports. That is, the hydrocarbon fluid serves as a foam precursor. In some such embodiments, all or part of the hydrocarbon fluid is at least partially transported as a foam within the associated pipeline.
- Typically, there is a means for introducing the foam into the transport pipe, such as with an injector. In all such embodiments, the foaming can be done either continuously or intermittently.
- Regions suitable for inducing turbulent flow include vertical flow in a well-bore or riser; flow through a mixer, valve, choke, and/or pump; and flow through appropriately sized restrictions. Also, foam is generated by adding a foaming agent such as surfactants, surface active agents, foamers, soaps, or solid particles. Note that some crude oils may contain natural surfactants that could cause foam to form (see Section 4.3.3).

8.5. Summary and Best Practices for Mechanical Technologies and Foamed Fluids for Clearing

8.5.1. Summary

This chapter highlights the use of many different types of portable equipment that must be assembled in an industrial plant to achieve cleaning of fouled units.

- Modern pumping gear, transport tanks, and waste collection units must be used on each job. However, each setup is unique. All chemical cleaning treatments must have various mechanical units to apply the chemicals.
- High-pressure water and ultrasonic processes are used to clean heat exchangers and ancillary chemical plant units fouled with organic and mixed deposits. Sometimes, safe, abrasive agents such as sodium bicarbonate or detergents may be added to the water to assist with the cleaning.
- Specialized fluids that include foams are also used to convey cleaning chemicals to areas of various types of plants that cannot accept the weight of a full water treatment. These special fluids also can be used in areas in which normal cleaning agents cannot be used. In addition, these fluids may reduce the volume of waste and can provide heat to facilitate the cleaning processes.

8.5.2. Best Practices

Frenier and Barber[20] describe general methods for cleaning heat exchange equipment, including both mechanical and chemical procedures, and give guidelines for selecting between chemical and mechanical cleaning and among the various types of chemical cleaning processes. The key factors for selecting a method are

- Goal of the cleaning
- Physical accessibility
- Short-term vs long-term costs and risks
- Metallurgical, heating, and disposal factors

This paper reviews the various factors of solvent and method choice in the planning stage for all cleaning procedures described in this chapter and in Chapters 4 and 7. The authors also note that any chemical or mechanical cleaning project should begin with thorough planning. All the physical and chemical aspects of the unit and its surroundings should be considered in the development of a cleaning plan. Well-trained and supervised crews should execute the plan. At the end of the project, the results should be evaluated and physical inspections should be conducted if possible. Did the job achieve the desired results? Record and save the evaluations with the permanent file of the unit.

The comprehensive book by Muller-Steinhagen[18] is a source for additional information on cleaning heat exchangers.

The book has sections on

- Fouling mechanisms
- Off-line cleaning (chemical and mechanical)
- On-line cleaning and fouling mitigation methods

Bademose[69] describes cleaning of pipelines to ensure product flow continuity and pipeline integrity. In the mechanical cleaning of an oil pipeline via pigging, a collection of sand, wax, scales, and other debris are successfully swept away, making room for unhindered transportation of pipeline content. However, the microbial organisms feasting on the pipe walls are missed in this cleaning process. The water content, which is also a byproduct of crude production, is not stopped in this process to eliminate the chances of internal corrosion buildup. The chemical cleaning method would adequately create pipe wall layers to resist the water to pipe reactions and would eradicate SRBs and other reactive agents.

To be competitive, this author[69] suggests that operators must appreciate the distinct advantages of both the chemical and the mechanical methods of cleaning the pipeline and recognize that these methods complement each other. This technical paper seeks to analyze the effect of each method in the light of their economic value and recognize the direct/indirect debilitating effect on product and asset value in cases in which either method substitutes the other. The author of the paper claims that using pigs (mechanical and gelled) with solvents is more economical that just pigging the line.

References

[1]W.W. Frenier, "Review of HSE Innovations in the Chemical Cleaning Industries," CORROSION/2015, paper no. NACE C2015-5553 (Houston, TX: NACE, 2015).

[2]J. McGraw, D. Fawcett, "Chemical Cleaning HRSG's, Top 10 Lessons Learned," CORROSION/2003, paper no. NACE-03580 (Houston, TX: NACE, 2003).

[3]M.J. Little, R.D. Varrin Jr., A.T. Pellman, M.A. Kreider, "Advanced Scale Conditioning Agent (ASCA) Applications: 2010 Experience Update," EPRI SGMP 2010 SG Secondary Side Management Conference, San Antonio, TX, Mar. 2–4, 2010.

[4]HydroChem 2015. Zevac Deer Park, TX, HydroChem Industrial Services.

[5]R. Tatina, L. Mouraa, N. Dietricha, S. Baig, G. Hébrarda, "Physical Absorption of Volatile Organic Compounds by Spraying Emulsion in a Spray Tower: Experiments and Modeling," *Chemical Engineering Research and Design* 104 (2015): 409-415.

[6]J.F. Harvey, P.K. Alvarez, R.S. Holland, R.S. Holland Jr., J. Bilke, T. Garcia, J. Daniel, "Mounted Bladder for Storage Tank," U.S. Patent 8763855, 2014.

[7]J.F. Harvey, P.K. Alvarez, R.S. Holland, R.S. Holland Jr., J. Bilke, T. Garcia, J. Daniel, "Bladder and Engagement Device for Storage Tank," U.S. Patent 9216885, 2015.

[8]J. Tiratsoo, *Pipeline Pigging & Integrity Technology*, 3rd ed. (Houston, TX: Clarion Technical Publishers, 2008).

[9]PPSA, "About Pigs" (Gloucestershire, U.K.: Pigging Products & Services Association, 2008).

[10]W.W. Frenier, *Chemical and Mechanical Methods of Pipeline Integrity* (Richardson, TX: Society of Petroleum Engineers, 2017).

[11]TDW, "Guide to Pigging, Brochure" (Tulsa, OK: T. D. Williamson, Inc., 2011).

[12]Pigging. Wikipedia, July 18, 2014, http://en.wikipedia.org/wiki/Pigging (July 22, 2014).

[13]J.C. Cowan, D.J. Weintritt, *Mineral & Organic Scale Deposits* (Lafayette, LA: Cowan & Weintritt Publishing Company, 2004).

[14]W. Mosher, T. Lam, H. Tsaprailis, "Methodology for the Evaluation of Cleaning Pigs on Sludge Deposits from Corrosion Pits," CORROSION/2016, paper no. NACE 7023 (Houston, TX: NACE, 2016).

[15]R. Davidson, "An Introduction to Pipeline Pigging," PPSA Aberdeen Seminar 2002, Aberdeen, U.K., Nov. 9, 2002.

[16]D. Wint, "Difficult to Pig Pipelines," Appalachian Underground Corrosion Short Course, Morgantown, WV, May 16-18, 2011.

[17]J. Cordell, "An Overview of Utility Pigs, Their Types and Uses," in *Pipeline Pigging and Integrity Technology* ed. J. Tiratsoo (Houston, TX: Clarion Technical Publishers, 2003).

[18]H. Muller-Steinhagen, *Heat Exchanger Fouling-Mitigation and Cleaning Technologies* (Rugby, U.K.: Institution of Chemical Engineers, 2000).

[19]W.W. Frenier, *Technology for Chemical Cleaning of Industrial Equipment* (Houston, TX: NACE International, 2001).

[20]W.W. Frenier, S.J. Barber, "Choose the Best Heat Exchanger Cleaning Method," *Chemical Engineering Progress* 94, 7 (1998): 37-45.

[21]"HydroChem Story," HydroChem, Deer Park, TX, http://www.hydrochem.com/ (Jan. 18, 2014).

[22]A.C. Rubey III, A.M. Taylor, J. William E. Spears, "Cleaning Method and Apparatus Using Sodium Bicarbonate Particles," U.S. Patent 5588901, 1996.

[23]M.R. Jalalirad, M.R. Malayeri, "A Criterion for the Selection of Projectiles for Cleaning Tubular Heat Exchangers," International Conference on Heat Exchanger Fouling and Cleaning, Budapest, Hungry, June 9-14, 2013.

[24]"Nozzles (Jet)," Wikipedia, Nov. 14, 2015, https://en.wikipedia.org/wiki/Nozzle (Jan. 31, 2016).

[25]T.L. Henshawa, "Rotating Cavitating Jet Nozzle," U.S. Patent 5217163, 1993.

[26]E. Steinhillber, T. Haegler, E. Steiner, "High-Pressure Jet Nozzle," U.S. Patent 5833148, 1998.

[27]J.L. Vargas Jr., "Device for Fluid Recirculation and Tank Cleaning, and Storage Tank," U.S. Patent 9061329 B2, 2015.

[28]J.E. Gardner, P.E. Amador, W.S. Howard, James A. Stoner, "Automated Heat Exchanger Tube Cleaning Assembly and System," U.S. Patent 8524011 B2, 2013.

[29]HydroChem, "Improving Heat Exchanger Efficiency," HydroChem, LLC Deer Park, TX, 2014, Exchanger-Cleaning_web.pdf.
[30]HydroChem, "Automation" (Deer Park: TX: HydroChem Industrial Services, 2015).
[31]G.E. Saxon Jr., R.E. Putman, "The Practical Application and Innovation of Cleaning Technology for Heat Exchangers," Engineering Conferences International: Heat Exchanger Fouling and Cleaning: Fundamentals and Applications, Santa Fe, NM, May 18-22, 2003.
[32]G.E. Saxon Jr., "Innovative and Efficiency: Low-Pressure Water Mechanical Cleaning of the Heat Exchanger Cases in Heat Exchanger Cleaning," International Conference on Heat Exchanger Fouling and Cleaning, Dublin, Ireland, June 7-12, 2015.
[33]S.L. Sachs, F.K. Hill, "Ultrasonic Cleaning Method and Apparatus for Heat Exchangers," U.S. Patent 4244749, 1981.
[34]E. Jukkala, "Ultrasonic Cleaner for a Heat Exchanger," WO Application 2013093183, 2013.
[35]Tubetech, "Ethylene Dichloride Heater Shell Side Cleaning" (Rayleigh, U.K.: Tubetech International Ltd, 2014).
[36]B. Kieser, A. Goode, R. Pholion, "From Practice to Theory-Practice: Advances in the Cleaning of Heat Exchangers Using Ultrasound," International Conference on Heat Exchanger Fouling and Cleaning, Budapest, Hungary, June 9-14, 2013.
[37]F.J. Fuchs, "Ultrasonic Cleaning: Fundamental Theory and Application" (Washington, DC: National Aeronautics and Space Administration, 1995).
[38]B. Kieser, R. Phillion, S. Smith, T. McCartney, "The Application of Industrial Ultrasonic Cleaning to Heat Exchangers," International Conference on Heat Exchange Fouling and Cleaning, Crete Island, Greece, June 5-10, 2011.
[39]C.F. Naude, A.T. Ellis, "On the Mechanism of Cavitation Damage by Non-Hemispherical Cavities in Contact with a Solid Boundary," *ASME J. Basic Engineering* 83 (1961): 648-656.
[40]M. Bellamy, "Ultrasonic Cleaning of Heat Exchangers" (Toronto, ON, Canada: Tech Sonic Services, 2014).
[41]"Ultrasonic Cleaning of Heat Exchangers," 2014, VARICLEAN, Enkhuizen, The Netherlands, http://www.variclean.eu/site/ultrasonic-cleaning-heat-exchangers (Jan. 22, 2016).
[42]W.L. Phillips, S. Smith, B. Kieser, "Apparatus for Cleaning Industrial Components," U.S. Application 20140230844, 2014.
[43]Paratene™ D728 Water-Based Degreaser, 2005, Woodrising Resources, Ltd., Calgary, Canada, http://www.paratene.com/d728.html (Dec. 7, 2016).
[44]J.W. Brenner, J.P. Engle, "Foam Cleaning of Surfaces," U.S. Patent 3037887, 1962.
[45]W.R. Rossen, "Theory of Mobilization Pressure Gradient of Flowing Foams in Porous Media," *Journal of Colloid and Interface Science* 136, 1 (1990): 1-16.
[46]W.W. Frenier, M. Ziauddin, *Chemistry for Enhancing the Production of Oil and Gas* (Richardson, TX: Society of Petroleum Engineers, 2014).
[47]T. Karam, "Foam...The New Liquid Accumulation Reducer in Gas Pipelines" (Trondheim, Norway: Norwegian University of Science and Technology, 2012).
[48]S. Petrovic, T. Robinson, R.L. Judd, "Marangoni Heat Transfer in Subcooled Nucleate Pool Boiling," *International Journal of Heat and Mass Transfer* 47, 23 (2004): 5115–5128.
[49]S.L. Kokal, "Crude Oil Emulsions," in *Petroleum Engineering Handbook, Volume I: General Engineering*, ed. J.R. Fanchi (Richardson, TX: Society of Petroleum Engineers, 2006).
[50]A.C. Huber, R.K. Panandiker, "Liquid Laundry Detergent Compositions with Silicone Antifoam Agent," U.S. Patent 5288431, 1994.
[51]D.B. Carroll, J.W. Rodgers, "Foam Generator," U.S. Patent 3212762, 1960.
[52]B.G. Simmons, "Method for Recycling Foamed Solvents," U.S. Patent 4849027, 1989.
[53]W.W. Frenier, *Chemical Methods for Controlling Corrosion During Oil and Gas Activities* (Houston, TX: NACE International, 2017).
[54]M.A. Barrufet, R.A. Flommerfelt, H.L. Chen, "Chemical-Foam Cleaning Processes for the Control of Power Plant Particle Erosion," *International Journal Energy Research,* 18 (1994): 665-687.
[55]C.W. Crowe, R.G. Duranleau, "Cleaning by Foam Contact and Foam Regeneration Method," U.S. Patent 3436262, 1969.
[56]M.Y. Kelly, "New Developments in Foam Cleaning," CORROSION/80, paper no. NACE 80218 (Houston, TX: NACE, 1980).

[57]C.I. Ball, B.G. Simmons, P.P. Lalena, paper IWC-79-30, 40th International Water Conference, Pittsburgh, PA, 1979.
[58]E.F. Hoy, "Enhanced Foam Cleaning of Copper Alloy Surface Condensers," paper IWC-86-1, 47th International Water Conference, Pittsburgh, PA, Oct. 27-29, 1986.
[59]O.E. Flores, "Condenser Chemical Cleaning Using Foamed Ethylenediamine Tetraacetic Acid," CORROSION/80, paper no. NACE 85185 (Houston, TX: NACE, 1980).
[60]P.A. Mestetsky, R.E. Davis, "Method of Cleaning Using a Foamed Liquid," U.S. Patent 5462607, 1995.
[61]M.C. Amiran, "Chemical Cleaning Solution for Gas Turbine Blades," U.S. Patent 6310022, 2001.
[62]C.D. Foster, "Chemical Foam Cleaning of Steam Turbines."
[63]C.D. Foster, J. Trimble, "Foam Cleaning Boosts Turbine Output," *Power Engineering* March (1999).
[64]K. Wicker, "Break out the Foam," *Power Magazine* June (2004): 59-60.
[65]C.D. Foster, "Methods and Apparatus for Periodic Chemical Cleanings of Turbines," U.S. Patent 5090205, 1992.
[66]J. Yang, "Synergistic Surfactant Compositions for Unloading Fluids from Oil and Gas Wells," U.S. Application 2007/0181307, 2007.
[67]J. Yang, "High Performance Foams for Unloading Gas Wells," U.S. Patent, 7422064, 2008.
[68]G.E. Kouba, A. Montesi, L.D. Rhyne, "Foam for Mitigation of Flow Assurance Issues in Oil & Gas Systems," U.S. Application 2008/0099946, 2008.
[69]A. Bademosi, "The Economics of Combining Chemical and Mechanical Cleaning in a High Debris Pipeline," CORROSION/2017, paper no. NACE C2017-9696 (Houston, TX: NACE, 2017).

9 Disposal of Chemical Cleaning Wastes

Chemical cleaning wastes usually contain quantities of metals and other materials that exceed liquid discharge limits of the plants in which the waste is generated. This is a complex subject and is different from state to state, country to country, and plant to plant. Several specific items that may be controlled are described in Section 10.3.

General characteristics that must be limited to various levels are

- Acidity or high alkalinity
- Prohibited organic compounds or those that are not biodegradable
- Heavy metals (Fe, Cu, Cr, Ni, Pb, Co, and Mo). See Table 9.1 for some specific elemental limits for very toxic metals.
- Radioactivity
- Total dissolved solids (TDS)

Table 9.1 RCRA[1] Waste Limits

Controlled Metal	Hazardous Waste Code	EPA Allowable Limit in a waste
Arsenic (As)	D004	5.0 ppm (mg/L)
Barium (Ba)	D005	100.0 ppm (mg/L)
Cadmium (Cd)	D006	1.0 ppm (mg/L)
Chromium (Cr)	D007	5.0 ppm (mg/L)
Lead (Pb)	D008	5.0 ppm (mg/L)
Mercury (Hg)	D009	0.2 ppm (mg/L)
Selenium (Se)	D010	1.0 ppm (mg/L)
Silver (Ag)	D011	5.0 ppm (mg/L

Although the environmental revolution of the past few decades has increased the need for better disposal and waste minimization techniques, responsible cleaning vendors have been concerned with safe disposal for many years. Cleaning procedures can generate volumes of wastes that range from a few hundred gallons to several hundreds of thousands of gallons of waste fluids, produced during cleaning of large boilers or nuclear steam generators. Many methods have been devised to dispose of these wastes. Chapter 9 reviews both mechanical and chemical methods of disposal.

Because wastes containing chelating agents and complexing agents are more difficult to treat than wastes from cleaning treatments using mineral acids, much of the literature is concerned with these more complex wastes. The procedures that are reviewed are divided into methods that are mostly mechanical and those that are mostly chemical; however, all of the methods usually contain some mechanical and chemical procedures. A general reference to boiler cleaning waste disposal is an EPRI[2] manual.

9.1. Mechanical Methods

Different mechanical methods that include thermal treatments and condensing the volume of the wastes have been applied to treat/reduce waste volume.

9.1.1. Thermal Treatments of Cleaning Wastes

The development of cleaning fluids containing organic acids, especially high-pH chelating agents (Section 4.1.2), necessitated that the industry (especially the utility power companies) develop an onsite method for safely (and economically) disposing of very large volumes of cleaning wastes. This relatively simple process consists of injecting wastes from the later (high pH) stages of cleaning (ammonium citrate or ammonium EDTA) directly into the fired box of an online (or one at reduced power) boiler unit.

Incineration (i.e., thermal treatment) of chelant wastes has proven to be a very economical method of disposal for both EDTA and citric acid wastes. The development of safe thermal evaporation techniques greatly enhanced the use of chelating agents at the expense of mineral acid formulations.

Test burns of spent EDTA solutions have been described by Sisson.[3] Following trial injections of spent EDTA chelants into the fire box of operating boilers, 65,000 gal (245,700 L) of spent EDTA chelant containing 19,000 ppm Fe and 320 ppm Cu were injected into a boiler operating at 255-260 MW. The burn lasted over 5 hours at an injection rate of 180 gal/min (660 L/min). Over the time of the burn, 1,701 lbms (770 kg) of iron and 9 lbms (4.1 kg) of copper left the stack (most of which was from the coal).

The measured values of iron and copper were less than 1% of the threshold limit values (TLV) for those two metals. Wackenhut[4] reported on the incineration of a citric acid waste in a coal-fired boiler. Over 18,000 gal (69,660 L) of an ammonium citrate waste containing 1,560 lbms (706 kg) of iron and 30 lbms (13.6 kg) of copper were successfully incinerated without adversely affecting the stack discharges.

Hallaren[5] summarized more than 10 years of successful incineration of wastes from chelant cleaning of utility boilers. Since the development of thermal treatment as a technique for treating wastes, this method usually is the least expensive procedure for disposal in localities where the technique is still legal. As of the date of this book's publication, this method can be used in some states and localities.

Daniels[6] discusses current (2014) issues of thermal treatments of boiler cleaning chemical wastes (BCCW). Several points made by this author are noted.

- For some time, utilities have used an exemption provided by the Bevill Amendment[7] and a subsequent letter from the Environmental Protection Agency (EPA) to group BCCW with other wastes that were uniquely associated with coal-fired utilities (such as fly ash and bottom ash) and treat them as exempt from hazardous waste regulations. This allowed the comingling of BCCW with fly ash or bottom ash and disposal in the ash pile without first having to determine if they were characteristically hazardous.

- Daniels notes that for many years, coal-fired utilities have evaporated characteristically nonhazardous BCCW (<5 ppm Cr) by spraying it directly into the fireball of an operating coal-fired boiler, typically at a rate of 30 to 50 gpm, depending on the amount of coal going to the boiler. This was the practice with EDTA wastes. The water is evaporated, the organic EDTA is consumed in the fireball, and metals are combined with the fly and bottom ash leaving the boiler. The small rate at which the BCCW was being added versus the coal feed made no measurable differences to any of the fly ash or bottom ash characteristics or stack gases. Some even found a slight benefit to NOx emissions during the time the BCCW was being evaporated. This practice had been successfully used for many years by coal-fired power plants.
- However, Daniels claims that changes to the definition of a nonhazardous secondary material by the EPA, together with changes to definitions of what constitutes a commercial industrial solid waste incinerator, essentially prohibit a conventional utility boiler from evaporating BCCW in the future.
- Because evaporation and comingling with fly ash will be prohibited, other options for properly disposing of BCCW need to be considered. The amount of BCCW generated with the combined solvent drains and rinses can be very large, and offsite treatment or disposal costs could double the cost of the cleaning. Onsite treatment, particularly of EDTA wastes, can be time-consuming and expensive. Other innovative recovery and beneficial reuse of these wastes has been considered in the past but never commercialized. However, in the present regulatory environment, these options may be in more demand as utilities strive to clean their boilers while controlling costs and liability.

Daniels also notes that between 4 and 8 boiler volumes of waste must be stored until disposal. See Figures 8.3-8.5 and the discussion about equipment in Section 8.1. A rendering of a portable tank (frequently called a "frac tank") that could be used for waste storage is shown in Figure 9.1.

Figure 9.1 Typical Portable Tank

9.1.2. Reduction in Volume

Fluids resulting from cleaning the secondary side of nuclear steam generators are processed by reducing the volume using a submerged combustion evaporator,[8] followed by solidifying the concentrated wastes. This process was developed in 1988 and was used to treat 180,000 gals of EDTA waste from the Oconee Nuclear Station. An 85% volume reduction was achieved before solidification.

Peters[9] described this process in more detail. Chemical cleaning solutions containing high concentrations of organic chelating wastes are difficult to reduce in volume using existing technology. Current methods for evaporating low-level radioactive waste solutions often use high-maintenance evaporators that can be costly and inefficient. The heat transfer surfaces of these evaporators are easily fouled, and their maintenance requires a significant labor investment. To address the volume reduction of spent, low-level radioactive, chelating-based chemical cleaning solutions, a liquid volume reduction system (LVRS) was developed.

The LVRS is based on submerged combustion evaporator technology that was modified for treating low-level radioactive liquid wastes. Peters[9] claimed that an improved and modularized system was used at Arkansas Nuclear One to process approximately 170,000 gal of chemical cleaning waste in early 1991. The LVRS provided nuclear utilities with an economically acceptable alternative for reducing the volume of nonstandard, difficult-to-process, low-level Class "A" radioactive solutions. After volume reduction, the concentrated waste solution is processed for burial. The procedure uses a waste solidification process that can solidify 40-48 gal of waste in a standard 55-gal drum, significantly reducing processing and disposal costs. The distillate from the evaporator undergoes a secondary processing step, if necessary, and is released after it meets all site discharge limits.

Snow[10] has reviewed (53 references) several different methods for reducing the volume of nuclear steam generator wastes. Disposal methods that are mostly mechanical include direct solidification, thermal degradation, incineration, and evaporation. Direct solidification methods include adding a cement binder and solidifying with a vinyl ester binder. The latter method was used to solidify the NS-1 waste from the decontamination of Dresden-1 after partial evaporation to reduce the volume. Thermal degradation and incineration methods can require a large amount of equipment to ensure destruction of the very stable EDTA molecule, which must be subjected to temperatures >200°C (392°F) to effect rapid degradation.

9.2. Chemical Treatments to Remove Metals

After cleaning a unit (especially a boiler with HCl), the strongly acidic cleaning wastes are generally neutralized as they leave the boiler, and then that waste is combined with neutralization and passivation steps and rinses to produce a combined waste that is not characteristically hazardous based on pH. The other chemical cleaning solvents are not characteristically hazardous by pH to begin with. Daniels[6] claims that a BCCW[11] can be classified as a hazardous waste if it contains a concentration of one of the RCRA 8 toxic metals. See Table 9.1 for details about the metals and federal limits in wastes. With the exception of Cr, which may exist as Cr(III) and the more toxic Cr(VI), the RCRA waste metals are unusual in most cleaning fluid wastes.

The primary metal that is of concern is chromium. Chromium comes from stainless steel feedwater heater and condenser tubes. This accumulates in deposits in the boiler. The regulation is specifically aimed at hexavalent chromium. Normally, utilities measure total chromium first and only address the hexavalent chromium issue if the total chromium is greater than the RCRA limit of 5 ppm.

The final waste to be chemical treated probably will be a mixture of the cleaning stages, rinses, and possibly passivation stages. Wastes also may be segregated depending on the concentrations of heavy metals.

There has been a great deal of work[12] developing methods for chemically treating wastes (usually from boiler tube cleaning fluids) either to make the waste acceptable for further disposal or to recycle a useful chemical, such as EDTA acid. Many metals can be precipitated from chelant wastes or HCl wastes using chemical treatments.

Rogoshewski[12] tested the use of lime and dilution of cleaning wastes with air preheater washing wastes. If the waste from the treatment of the boiler (ammoniated citrate or hydroxyacetic acid/formic acid) were diluted by about 7/1, lime treatment effectively reduced heavy metal concentrations to < 1 ppm. The sludge could be sent to a solids disposal site (or possibly reused) and the water released to a licensed site.

Most of the chemical waste treatment methods use some form of precipitation technology to remove the heavy metals, which are the primary problem species. The addition of lime ($CaCO_3$), slaked lime (CaO), or other alkalis (NaOH) will raise the pH to > 10 and precipitate Fe, Cu, Ni, and Cr as the hydroxides. The solubility of the hydroxides in the matrix depends on the final pH and the other ions in the fluids. For Cu(II), the sulfides are less soluble than the hydroxides, and various sulfides have been suggested as precipitation additives.

Ferrier joined the R/D section of the Dow Chemical Company (a manufacturer of EDTA chemicals) in 1970, and one of the ongoing projects at that time was aimed at precipitating EDTA acid from spent boiler cleaning wastes, cleaning it, and reusing the valuable chemicals. The method included acidifying the alkaline fluid with dilute H_2SO_4 to precipitate H_4EDTA, and it used a filter press to separate the EDTA acid. Unfortunately, it was very difficult to remove all the heavy metals (Fe, Cu, and Ni) to the levels needed for reuse. The following paragraphs describe methods for reducing the metals to acceptable disposal values.

Table 4.1 shows the characteristics of the wide variety of chemicals that are used in the various cleaning services along with the $\log K_{stability}$. To understand the precipitation chemistries, the solubility product Equations (3.11-3.19) must be compared with the stability products (Table 4.1). For example, when a precipitating chemical (NaOH) is added to a waste containing Fe(III), the reaction is

$$Fe^{3+} + 3OH^- \Leftrightarrow Fe(OH)_3(s) \qquad \text{(Equation [9.1])}$$

The [Fe] remaining in solution depends on the ratio $K_{stability}/K_{sp}$. The larger that this ratio becomes, the lower amount of residual metals will be found in solution. Thus, metals that form very insoluble salts with the precipitation chemicals will be removed from solution to values below discharge limits (frequently 1 ppm). For example, for Fe(III)EDTA, $K_{stability} = 10^{25}$ and K_{sp} for $Fe(OH)_3 = 10^{-30}$, and the result is 10^5. This is a large driving force, and the iron is easily separated from ferric EDTA.

After precipitation, the slurry can settle, and the thickened solids are dewatered using mechanical devices such as filter presses or centrifuges. Bell[13] has reviewed treatment methods for many types of wastes. Spent mineral acids are neutralized with alkyl-containing solutions. Each source of alkali (sodium hydroxide, lime, or sodium carbonate) has advantages and disadvantages. Copper-ammonia solutions can be treated by reducing the pH to about 7 to produce ammonium ions and to precipitate copper hydroxide.

Uhr[14] conducted a literature survey of disposal methods for waste streams containing citrate and gluconate from chemical cleaning and metal treating processes. His paper included the functions of citrate and gluconate in formulations, other additives possibly present, and interference caused by citrates and gluconates in alkaline precipitation of heavy metals. Citrate is recommended as easiest to treat for disposal. The process uses oxidation with ozone, peroxide, persulfate, or hypochlorite, followed by addition of lime and NaOH to raise the pH and precipitate citrate as the calcium salt. Additives discussed include ammonia, fluorides, reducing agents, surfactants, and inhibitors. Although gluconate may be precipitated with lime at pH 12 or destroyed by boiling with about 300 ppm ferric ion in NaOH solutions, neither technique reduces gluconate concentration below guidelines. The chemistries of chelation, buffering, precipitation, and oxidation of citrates and gluconates are described in this report.

Holcombe[15] tested several cleaning solutions, including inhibited hydrochloric acid containing thiourea, ammonium bromate, EDTA, citrate, and hydroxyacetic-formic acid (HyFor). See Sections 4.1 and 4.2 for details about these solvents.

The raw field wastes from different boiler cleaning jobs[15] were treated using conventional lime and caustic precipitation and by co-ponding with coal fly ash. Analytical results were compared with federal effluent discharge guidelines (NPDES) for copper and iron. The caustic/lime treatments were effective for the bromate and HyFor wastes and removed both Fe and Cu. The HCl waste required caustic lime and sulfide to remove both Fe and Cu.

Table 9.2 shows the results of treatments with standard chemicals or adding the waste to a fly ash waste pond (done in the lab). The wastes were diluted 1/100 with water before the fly ash exposure tests. The neutral ash effectively treated the nonchelating boiler cleaning wastes: hydrochloric acid, HyFor, and ammonium bromate. It considerably reduced the iron content of the acid boiler cleaning waste but had little effect on the EDTA boiler cleaning waste. The two alkaline ashes successfully reduced all of the wastes to the detection limit, which is 1/10 of the regulatory concentration.

Table 9.2 Tests Using Chemicals of Coal Ash Waste to Treat Boiler Cleaning Waste[15]

Lab Results, Chemical Treatment of Field Waste				Lab Results, Co-ponding with Coal Ash			
Treatment	**pH**	**Fe (mg/L)**	**Cu (mg/L)**	**Treatment**	**pH**	**Fe (mg/L)**	**Cu (mg/L)**
	HCl Waste				**Metal in Ash Slurries**		
None	1.8	2300	160	Acidic Ash	3.6	1.1	0.9
Lime	6.0	1180	72	Neutral Ash	7.0	<0.1	0.5
Lime	8.0	11.7	57	Alkaline Ash #1	11.1	<0.1	<0.1
Lime	10.0	1.5	29	Alkaline Ash #2	11.2	<0.1	<0.1
Lime	12.0	1.1	12				
NaOH	8.0	5.2	37				
NaOH	12.0	0.5	43				
NH_4BrO_3 +NaOH	12.0	9.3	43				
	HyFor Waste				**HCl Waste**		
None	3.7	830	< 0.5	After 1/100 Dilution	-	41.4	0.2
Lime	7.1	< 0.5	< 0.5	Acidic Ash	3.7	31.8	1.5
Lime	8.0	< 0.5	< 0.5	Neutral Ash	7.0	<0.1	<0.1
Lime	9.8	< 0.5	<0.5	Alkaline Ash #1	10.4	<0.1	<0.1
				Alkaline Ash #2	10.1	<0.1	<0.1
	NH_4BrO_3 Waste				**HyFor Waste**		
None	10.6	< 0.5	120	After Dilution	-	8	<0.1
Lime	11.0	< 0.5	1.4	Acidic Ash	4.2	7	1.1
Lime	12.3	< 0.5	3.4	Neutral Ash	7.7	<0.1	<0.1
				Alkaline Ash #1	10.4	<0.1	<0.1
				Alkaline Ash #2	10.1	<0.1	<0.1
	NH_4BrO_3 Waste, Diluted 1/10				**NH_4BrO_3 Waste**		
None	10.3	< 0.5	8	After Dilution		<0.1	3.5
Lime	10.8	< 0.5	< 0.5	Acidic Ash	4.2	0.2	3.0
Lime	12.2	< 0.5	< 0.5	Neutral Ash	7.0	<0.1	<0.1
				Alkaline Ash #1	10.0	<0.1	<0.1
				Alkaline Ash #2	10.8	<0.1	<0.1
	Citric Acid Waste				**Citric Acid Waste**		
None	10.0	4200	4	After Dilution		69.5	< 0.1
Lime	11.1	140	0.7	Acidic Ash	4.6	6.1	0.4
Lime	12.2	2.3	0.8	Neutral Ash	7.9	1.4	0.9
Na_2S	12.0	2.3	0.14	Alkaline Ash #1	10.0	<0.1	<0.1
				Alkaline Ash #2	10.8	<0.1	<0.1

	Low Cu EDTA Waste				**Ammoniated EDTA Waste**		
None	9.7	3200	1.2	After Dilution		54.5	2.8
Lime	12.0	62.7	<0.5	Acidic Ash	4.3	53.9	1.3
Na_2S	12.0	544	1.2	Neutral Ash	8.1	49.8	1.1
				Alkaline Ash #1	9.7	<0.1	<0.1
				Alkaline Ash #2	10.6	<0.1	<0.1
	High Cu EDTA Waste						
None	9.7	3010	200				
NaOH	12.0	1.7	160				

The conclusions of the tests[15] are summarized:

- The solvent chemistry that promotes the dissolution of scale and deposits during boiler chemical cleaning also makes treatment of the spent solutions to reduce metal ion concentrations difficult.
- This study identified several treatment or pretreatment steps for these wastes. For example, treatment may require dilution before pH adjustment. Conventional physical-chemical treatment consisting of pH adjustment with lime or caustic followed by solids precipitation and separation is capable of treating hydroxyacetic-formic acid and hydrochloric acid wastes.
- Ammonium bromate solutions may require dilution before treatment to precipitate copper.
- Citrate and EDTA-based wastes are more difficult to treat and require additional processing steps.
- Co-ponding of boiler cleaning solutions in alkaline ash ponds can successfully reduce the dissolved metal concentrations. The highly alkaline ashes were successful in treating all wastes tested. One method of predicting the success of co-ponding includes the screening experiments used in this study. However, positive laboratory results do not guarantee successful field applications owing to differences in many physical parameters. The neutral ash did not perform acceptably with citrate or EDTA wastes. The acidic ash did not achieve 1 mg/L copper and iron levels with any waste type.

Treatment of more concentrated chelant wastes (containing EDTA) requires the use of a more aggressive chemistry. Kennedy[16] demonstrated that iron and copper could be removed from EDTA by a combination of lime, NaOH, and sodium-N,N-diethyldithiocarbamate. This process removed >99.9% of the iron and copper. However, only 82% of the nickel was precipitated. Samuelson[17] described the use of this technology to treat several waste streams from operational cleaning of utility power boilers. See Figure 9.2 for an example of the precipitation of metals from EDTA wastes.

The problems of treating wastes containing high concentrations of nickel were addressed by Hoy.[18] This metal can be removed by heating the waste to temperatures of about 180°F (82°C) in the presence of sodium-N,N-diethyldithiocarbamate and calcium. After being heated to 180°F (82°C), then cooled to room temperature, Ni concentrations in EDTA fell from 35 ppm to 2 ppm in 120 hours. This process is difficult to perform, but it can remove most of the metals. However, if the EDTA is precipitated by adding an acid and then precipitated as H_4EDTA, it may not be of commercial quality for reuse. The equipment needed to process wastes containing EDTA is significant. See Figure 9.3.

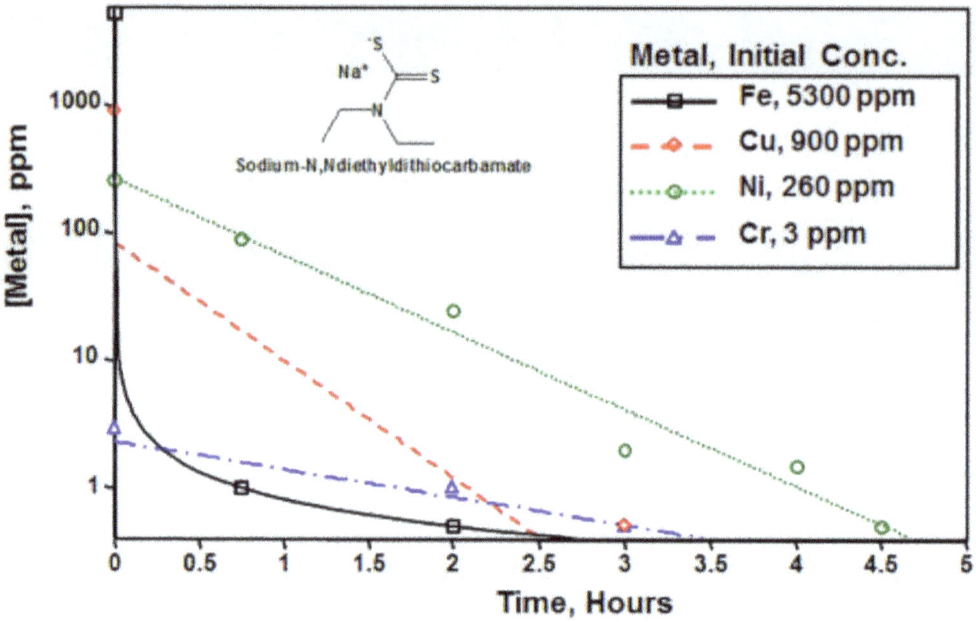

Figure 9.2 Precipitation of Metals form EDTA Waste with Lime and Sodium-N,N-diethyldithiocarbamate at 150°F and pH 10.8

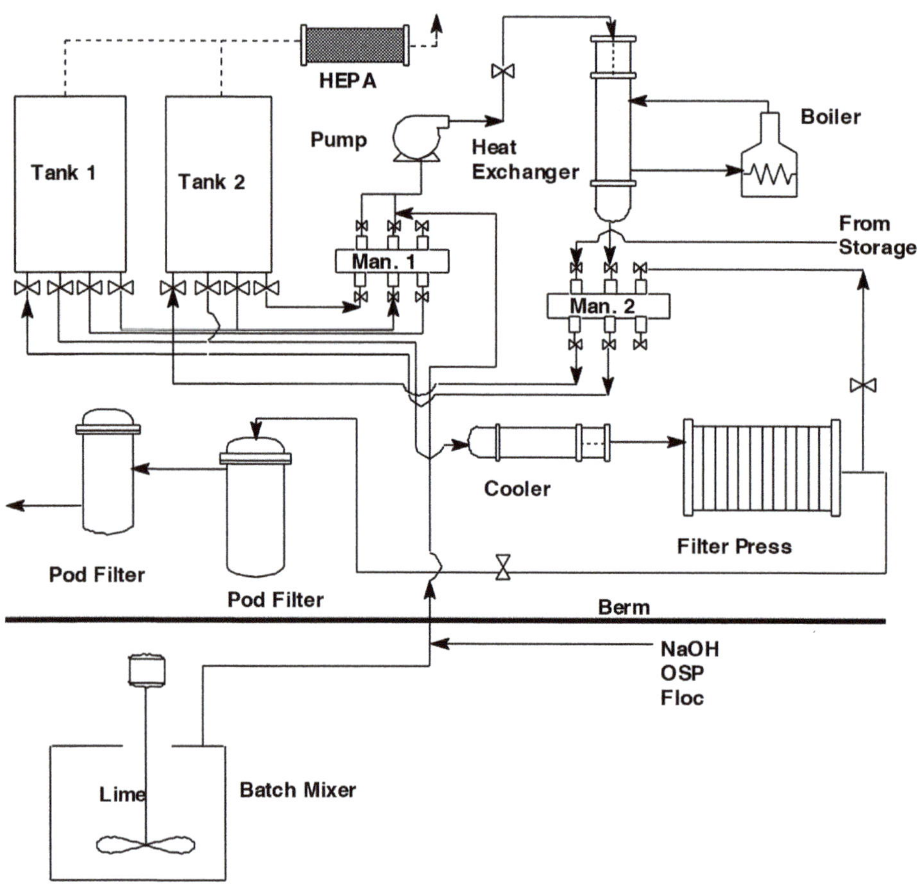

Figure 9.3 Schematic of Waste Treatment at Byron One Nuclear Plant[17,19]

More recently, electrochemical methods have been attempted in lab settings. Huang et al.[20] claim that an electroreduction process can separate metal ions from their EDTA complexes. The free EDTA is claimed to be recovered from a field power plant and reused as a cleaning agent. The results from controlled, direct current, electrolysis experiments showed that the electroreduction process can successfully remove metal ions from the synthetic solution and field boiler cleaning waste samples and can deposit elemental metals on the cathodic plate. At a current density of 15.5 mA/cm^2 and a temperature of 40°C, the percentage metal removal and current efficiency were 94.16% and 8.27%, respectively, with a cadmium cathodic plate. Moreover, results also indicated that the percentage of metal removal and overall current efficiency increase with increasing current density and temperature under optimal conditions (15.5 mA/cm^2 and 40°C). The electroreduction process was able to separate metal ions from their EDTA complexes. The free EDTA recovered from the field power plant can thus be recovered from chemical cleaning wastes. The author of this book does not know if this method has been used commercially.

Additional treatment methods for removing metals from EDTA-containing wastes have been described in the patent literature. Frost[21] described a process that included addition of a sulfide, separation of metal sulfides, then oxidation by sodium nitrite at 575°F. Frost claimed that the metals, including chromium, were completely removed. A process for removing heavy metals from solutions containing chelating agents was proposed by Guess.[22] This process includes the addition of ferrous ions followed by addition of the sodium dithionite to precipitate the metals. A reverse osmosis technique was developed by Terao[23] for treating boiler cleaning wastes containing citric acid.

A very different approach for disposal of chelated iron boiler wastes made use of these materials to oxidize hydrogen sulfide at the Geysers Power Plant in California.[24] The spent EDTA-iron waste was used to oxidize the sulfide gases produced with the geothermal steam. The chemical reaction is

$$2Fe(III)EDTA + H_2S + 2OH^- \rightarrow S^0 + 2H_2O + 2Fe(II)EDTA \qquad \text{(Equation [9.2])}$$

Varrin and Little[25] patented a process that stores and reuses chemical cleaning solutions associated with steam generator treatments. The patent states that an aqueous cleaning solution that has been previously used to remove deposits from a nuclear steam generator (or other vessel) is reused after being transferred from the steam generator into an external vessel. The spent cleaning solution may then be reconditioned and reused to further clean the same steam generator or a different steam generator. The chemistry adjustments described can include the addition of chemicals to enable the solvent to dissolve more iron oxide or to do a different process such as copper removal. Note that the boiler cleaning solvents described in Sections 4.1 and 4.2.3 are often replenished by adding additional solvent chemicals or are significantly modified by changing the pH or by adding oxidizers while the fluids are still in the boiler. These changes have similar purposes as those described by the Varrin and Little[25] patent.

9.3. Chemical Degradation of Organic Content

Several methods that chemically degrade some or all of the organic content of a cleaning solution have been suggested and may be in commercial use. These chemical reactions may occur after the removal of heavy metals or may also cause the complexes to release the metals.

9.3.1. General Chemical Methods

Most of the previous discussion was concerned with removing the metals from cleaning wastes. The organic portion of the waste also may require treatment.

Hydrochloric acid solutions may contain thiourea if this solvent is used to remove copper and iron from a boiler. Since thiourea is a known animal carcinogen, treatment methods must include a step to reduce the concentration of this chemical. McConnell[26, 27] has developed an oxidation method for removing thiourea from acid solutions. Alkaline solutions can be oxidized to carbon dioxide and nitrogen in a two-step procedure.

$$\text{A) } H_2N-(CS)-NH_2 + 4H_2O_2 \rightarrow H_2N-(CO)-NH_2 + SO_4^{2-} + 5H_2O \qquad \text{(Equation [9.3])}$$
$$\text{B) } H_2N-(CO)-NH_2 + 3NaOCl + 2NaOH \rightarrow N_2 + 3NaCl + Na_2CO_3 + 2H_2O$$

Disposal of EDTA may be a problem because it is not biodegradable. EDTA and other polyaminopolycarboxylic acid chelating agents undergo hydrolysis and other degradation reactions when heated to moderately high temperatures. Venesky[28] showed that EDTA was not stable above 200°C (392°F) and postulated that the reactions were initiated by dissolved oxygen. Martell[29] studied NTA and EDTA in aqueous solution and showed that EDTA is cleaved at the C-N bond at 260°C (554°F), but NTA was much more stable and did not decarboxylate rapidly until 290°C.

Motekaitis[30] determined the kinetics of hydrolysis and ammonolysis of EDTA at temperatures of interest for chemical cleaning activities. The rate increases with decreasing pH in the presence of ammonia. The rates are high enough at temperatures above 160°C (320°F) to affect the commercial use of ammonium EDTA. The molecule is cleaved at the C-N and C-C bonds. EDTA is rapidly oxidized by ferric ions. Motekaitis[31] also determined that ferric EDTA would oxidize other EDTA molecules rapidly at temperatures as low as 125°C (257°F). The major reaction cuts off carboxylate groups in a sequential fashion. NTA is oxidized more slowly than EDTA.

The oxidation of EDTA, NTA, and DTPA by various peroxo compounds has been investigated. Snow[10] reviewed various oxidation techniques for degrading EDTA. The most effective methods are oxidation by ozone or hydrogen peroxide and UV light. Karell[32] studied ways for degrading EDTA in the presence of dissolved ferric iron and concluded that peroxide and UV light at 254-nm was the most effective method. However, he concluded that this method was too expensive to use to degrade nuclear cleaning wastes because the degradation products also are chelating agents and could stabilize radioactive metals.

Iron dissolved in the EDTA can significantly influence the degradation process by means of Fenton chemistry.[33] In the Fenton reaction, a ferrous chelate is oxidized to a ferric chelate by hydrogen peroxide and liberates a hydroxyl radical that causes further oxidation of chelate molecules in solution. See Equation (9.4).

$$\text{Fe(II)Chelate} + H_2O_2 \rightarrow \text{Fe(III)chelate} + OH^- + \bullet OH \qquad \text{(Equation [9.4])}$$

Chen[34] has determined that peroxide degrades chelating agents via the Fenton reaction in the following sequence: NTA<EDTA<HEDTA. The chelating agent DTPA is the most chemically stable material in the presence of peroxide and iron. The major reaction products are iminodiacetic acid (IDA), hydroxyethyliminodiacetic acid (HEIDA) and glycine. These materials are still capable of complexing many metals, albeit not as strongly as EDTA.

Alvino[35] patented a method for treating nuclear decontamiantion fluids such as the CANDECON solvents (see Section 5.2). The patent specifics that an oxidation method effectively destroys organic chemicals, such as EDTA, HEDTA, NTA, oxalic acid, citric acid, picolinic acid, or formic acid or similar salts.

The patented process is conducted on solutions that contain less than about 1% by weight of a carboxylic complexing agent and that are contaminated principally by iron, cobalt, nickel, and other cations. The solution is pumped out of the reactor coolant system after it has circulated throughout the system for 30 minutes to 1 hour or longer as needed to decontaminate the unit. An oxidizing agent is then added to the contaminated solution to oxidize the complexing agent and thus destroy the complex. Oxidants such as ozone, hydrogen peroxide, or a chlorite solution may be added to the contaminated solution, which reacts with the complexing acid to form carbon dioxide and water. This oxidation step may be enhanced with UV light. The temperature of the solution is preferably maintained from about 140°F to about 212°F during the treatment.

The treatment results, shown in Figure 9.4, demonstrate that adding hydrogen peroxide continuously is more efficient than a periodic addition of the oxidant for reducing the carbon content.

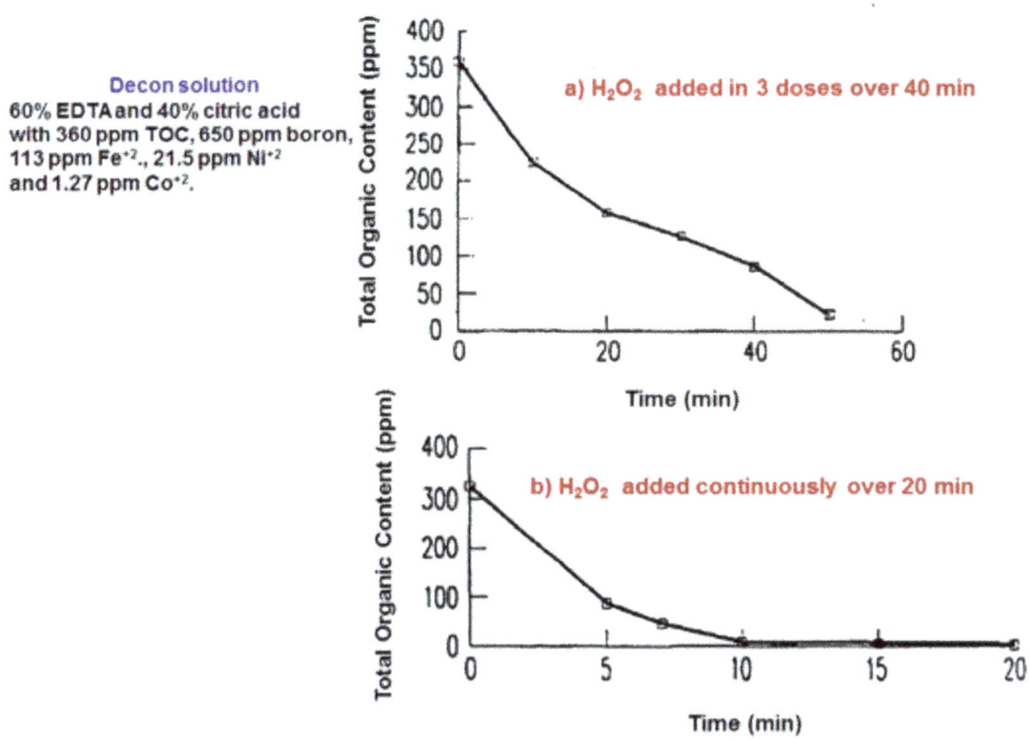

Figure 9.4 Treatment of a Decon Effluent with H_2O_2

An additional oxidation method that uses supercritical water to rapidly destroy most organic molecules is described in Section 9.3.2.

9.3.2. Supercritical Water Oxidation (SCWO)

Under normal conditions, water is present in any of three states: steam, liquid water, or ice. If water is heated and compressed to sufficiently high temperature and pressure, water enters a fourth state known as the supercritical state. In the case of water, this occurs above 374°C and 221 bar (Figure 9.5).

SCWO efficiently destroys organic compounds because it has unique properties between those in its liquid and gas state.

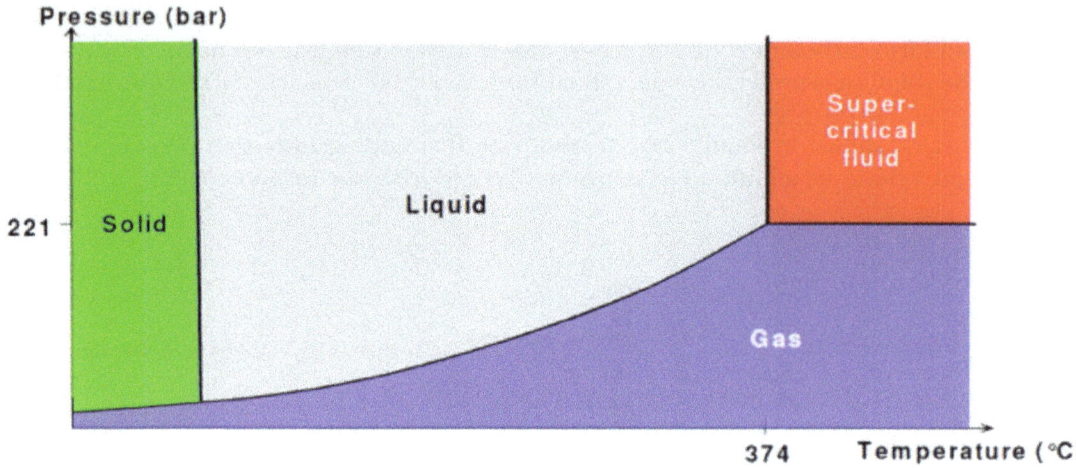

Figure 9.5 Simplified Phase Diagram of Water

Although it is currently unclear whether the process has been used commercially to treat chemical cleaning wastes, the following short review provides some recent information.

Modell[36] patented a supercritical water method of oxidizing organic materials to obtain useful energy, and/or detoxifying or destroying noxious and toxic organic materials, and/or removing unwanted salts from water. This method also enables efficient oxidizing of conventional and unconventional organic fuels with maximized energy recovery in a simplified apparatus for use in a wide variety of heating and power cycles.

O'Reagan[37] notes that the density of supercritical water is comparable to liquid water densities and is high enough for reasonable throughputs in a process. However, Modell[36] notes that the viscosity and diffusivity in the supercritical region are more like that of a gas. Owing to the low dielectric constant of supercritical water, the solubility of organic compounds and gases is high. This low dielectric constant, together with the high diffusivity, means that there is an insignificant mass transfer resistance, thus enabling very fast reaction rates. High molecular weight organic compounds are destroyed or transformed almost immediately, whereas smaller molecules such as acetic acid are generally slower and are typically the rate-controlling compounds in the process. Nevertheless, reaction time is extremely fast (typically less than 60 seconds). Nitrogen-containing compounds will revert to elemental nitrogen without production of NOx.

Yesodharan[38] notes that under supercritical conditions, organic substances are completely soluble in water. The critical point occurs at a temperature of 374°C and a density of 0.3 g/cm^3. The supercritical region lies above 374°C. The technique has been tested for the decomposition of heterocyclics such as quinoline and isoquinoline, which are frequently encountered in the environment because of their presence in fossil and synthetic fuels (oil shales, coal, oil sands, and shale oils) and some pesticide mixtures, such as creosote.

Yesodharan[38] also claims that studies on the SCWO of quinoline and isoquinoline suggest that the process is effective for the complete degradation of these pollutants in water. Here, water acts as a chemical reagent as well as a solvent. The decomposition was more effective when compared to pyrolysis. These studies make it likely the chelating agent/organic acid cleaning chemicals also could be treated using SCWO.

Jimenez-Espadafor[39] reports a study that treated industrial oily wastes using SCWO in a pilot plant (25 kg/h), and a simulation was performed to evaluate the viability of energy production on an industrial scale. The SCWO pilot plant effluent is good for producing hot water or steam by recovering the heat of waste organics. Both alternatives were evaluated for a SCWO industrial plant design; with 1000 kg/h, it was possible to recover a maximum of 118 kW, that is, 71% of the energy content of the wastewater.

9.3.3. Biodegradable Chelating Agents

Biodegradable chemicals is a very complex subject. Each chemical gives different biodegradation data depending on the specific test. So, a claim that a chemical is biodegradable must be considered on the basis of the type and intent of the test method. See Frenier[40] and OECD.[41]

However, alternate approaches to address ecological issues include developing biodegradable chelants for use in cleaning formulations or developing microorganisms capable of feeding on EDTA. Note the caution in the prior paragraph. Lauff[42] found that *Agrobacterium* can degrade EDTA or PDTA complexes in wastewater. Optimal Fe^{3+} EDTA degradation occurred below pH 7. Lauff[42] also described a pure culture of *Agrobacterium sp*, which was isolated from a waste treatment facility. The isolate grew on ferric EDTA as the sole carbon source. The degradation produced carbon dioxide, ammonia, and an unidentified metabolite. The rate of degradation was 24 mM/day. In a similar vein, Bunch[43] studied settled sewage sludge, which was used to inoculate BOD dilution water. After 7 days, the test solutions were tested for the chelating agents. NTA and HEIDA lost their chelation ability, whereas EDTA, DTPA, HEDTA, and CYDTA were unchanged. It was concluded that chelants with single nitrogen atoms are biodegradable, but those with two nitrogen atoms are not biodegradable.

The following paragraphs review information on chemicals that have *not* been well known in the chemical cleaning industry but may have significant uses in the future due to improved biodegradable characteristics, compared with EDTA-type chemicals. These materials are IDS, GLDA, MGDA, HEIDA, and EDDS (See Figure 4.1 and Figure 9.6). The biodegradable chelating agents citric acid and NTA have been used for several years along with non-biodegradable phosphonates and EDTA, HEDTA, and DTPA. Uses of some of these chelating agents for scale removal have been described in Frenier.[44]

Bayer[45] claims that sodium iminodisuccinate (IDS) belongs to the aminocarboxylate class of chelating agents and that all monocarboxylates used today are acetic acid derivatives produced from amines, formaldehyde, sodium hydroxide, and hydrogen cyanide. In contrast, sodium IDS can be produced from maleic anhydride water, sodium hydroxide, and ammonia. The only solvent used in the production process is claimed to be water, and the only side product formed, ammonia dissolved in water, is recycled back into IDS production or used in other Bayer processes. They claim that because sodium IDS is a readily biodegradable, nontoxic, and nonpolluting alternative to other chelating agents, it can be used in various applications that use chelating agents. For example, it can be used as a builder and bleach stabilizer in laundry and dishwashing detergents to extend and improve the cleaning properties of the eight billion pounds of these products that are used annually.

AkzoNobel[46] states that GLDA is based on the food-approved natural amino acid salt, monosodium L-glutamate (MSG). This report also claims that GLDA, compared with various alternative chelating agents, is the most environmentally benign chelating agent. Uses for GLDA in calcite dissolution treatments were described in Section 4.4.1.

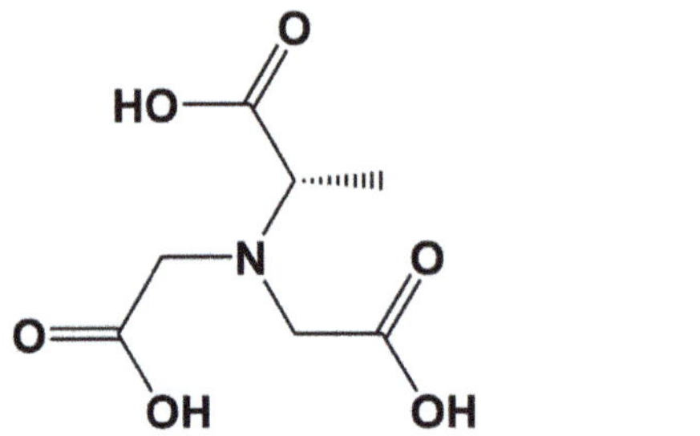

methylglycine-n,n-diacetic acid-MGDA

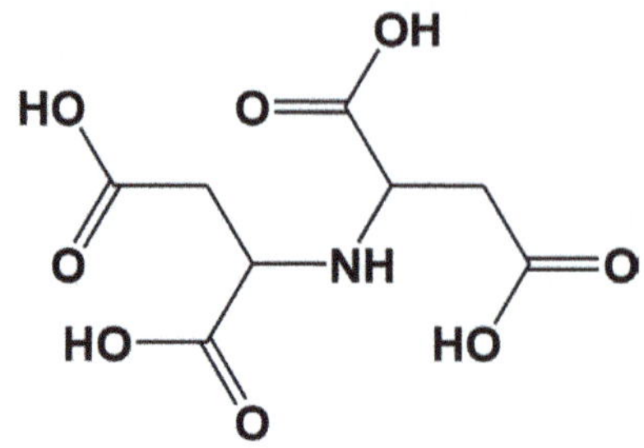

iminodisuccinic acid- IDS

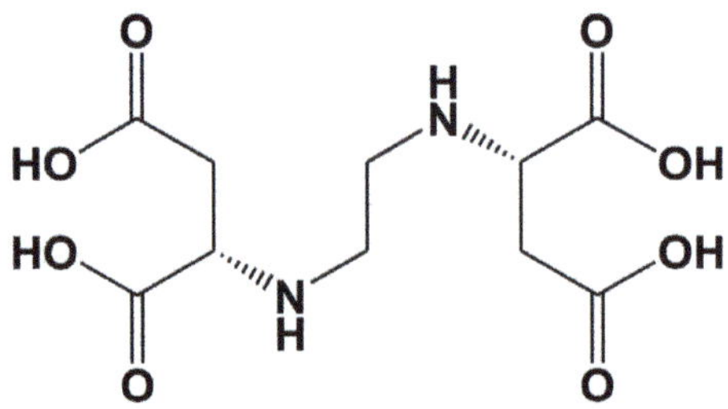

s,s-ethylenediamine n,n'-disuccinic acid-EDDS

Figure 9.6 More Biodegradable Chelating Agents

Al-Mayouf[47] claims that pressure is mounting for the removal and replacement of non-biodegradable chelating agents with alternative, readily biodegradable chelates. Alternatives, such as MGDA and GLDA, are claimed to possess good ecological and toxicological profiles. The authors[47] claim that MGDA and GLDA meet all of the requirements listed above and thus can be applied as new effective and environmental friendly chelating agents. They also note that MGDA is a strong chelating agent. See the stability constants in Table 1.2.

HEIDA also has been shown to be readily biodegradable.[48,49] Dow[48] also has found that the potassium salts of HEIDA were more effective in removing calcium scale compared with the sodium salts. Tandy[50] contends that EDTA and phosphonates are non-biodegradable chelants that are used extensively in domestic and industrial applications. They claim that an environmental concern is that once the chemical is in river systems, the chelants can extract heavy toxic metals from mud and sediment and cause remobilization of the metals in the environment that may have long-term adverse effects. EDDS is claimed to be readily biodegradable before discharge into river systems and to be completely mineralized in the environment; therefore, EDDS cannot remobilize heavy metals and is classified as nonhazardous.

Charkhutian et al.[51] investigated the biodegradable chelating agents EDDS, IDS, MGDA, and NTA as potential alternatives and compared them with EDTA for effectiveness for extracting metals from soils.

Sato et al.[52] used electrochemical methods to study the galvanic coupling between magnetite and iron in EDDS solutions both with and without added iron(II) ions. Galvanic coupling accelerates the corrosion of iron owing to the small shift in its potential in the anodic direction. At the same time, the potential of magnetite is cathodically polarized away from the potential range in which the only faradic process—the reductive dissolution of magnetite—takes place and results in a considerable decrease in its dissolution. Magnetite dissolves faster at the galvanic potential when [EDDS] ≥ [Fe^{2+}], whereas iron is affected to a much lesser extent. The ratio between the rates of dissolution of magnetite at the galvanic potential to those at its steady state potential tends to decrease at higher temperatures and at higher EDDS and Fe^{2+} concentrations. This study shows that temperature plays a decisive role in the dissolution of magnetite coupled to iron. For complete removal from the iron surface, high temperatures should be used. Corrosion inhibitors have to be considered when high losses of the base metal cannot be tolerated. They suggest that EDDS is a promising substitute for EDTA in the nuclear energy industry because of its composite chelating power.

Crump and Wilson[53] made preliminary investigations into the replacement of widely used chelating agents for industrial cleansing of radionuclide-contaminated items using readily biodegradable alternatives that have equivalent effectiveness. The chelation of radionuclide cations by [S,S′]-EDDS has been compared with traditional decontamination agents EDTA and citrate. Models indicate that in many respects, [S,S′]-EDDS compares favorably with EDTA .

Wilson and Crump[54] tested the dissolution of iron with EDDS. A sample (1.0 g of EDDS) was dissolved in deionized water, and sufficient aqueous ammonia solution was added to obtain a pH of 9.0. Deionized water was added to obtain a 5 wt% EDDS solution. The resulting solution was heated at 100°C for 3 hours with 0.2 grams of iron oxide (Fe_3O_4) and was cooled to room temperature. The amount of soluble iron was then determined using inductively coupled plasma spectroscopy. After 3 hours, approximately 31 wt% of the iron was dissolved. In a similar experiment, an EDDS solution was adjusted to a pH of 4.5 with an aqueous ammonia solution. After 3 hours at reflux, approximately 70 wt% of the iron oxide had been dissolved. These data show that ammoniated polyamino disuccinic acids are effective in dissolving iron oxide scales.

LePage et al.,[55] Boonstra et al.,[56] Charkhutian et al.,[51] Mahmoud et al.,[57] and Rabie et al.[58] have extensively evaluated the use of Na_4GLDA solutions. This chemical is a commercial raw material,[59] and the data show it to be soluble in several fluids of interest in reactive stimulation, including HCl (15% and 28%), acetic acid (15% and 28%), and formic acid (15% and 28%). These authors report that the material also shows high biodegradation values.

LePage et al.[55] and Mahmoud et al.[57] describe calcite dissolution tests using GLDA and other chelants over a large range of pH values in which the pH was adjusted by adding HCl. The high dissolution values are attributed to the acid reaction of the added HCl as well as a chelate-type reaction. See Table 4.13.

The use of more biodegradable chelating agents may affect chemical cleaning formulation choices. However, at the time of this book's publication, citric acid is still the only biodegradable agent in wide use in IC treatments.

9.4. Summary of Waste Disposal Technologies

The continued use of chemicals to descale industrial equipment requires waste minimization and waste disposal techniques that are cost-effective with other cleaning methods such as high-pressure water and on-stream cleaning. These are:

- Mechanical reduction in volume
- Thermal treatments in coal-fired boilers
- Chemical precipitation and filtering
- Oxidation of organic wastes
- Development of safer and biodegradable chemicals

The development of safe, biodegradable cleaning agents is the number one challenge of the cleaning industry.

References

[1]"Resource Conservation and Recovery Act," Title 40 of the Code of Federal Regulations (CFR), parts 239 through 282 (Washington, DC: U.S. Environmental Protection Agency, 1976).

[2]EPRI, "Boiler Chemical Cleaning Waste Management Manual" (Palo Alto, CA: Electric Power Research Institute, 2013).

[3]A.B. Sisson, G.V. Lee, "Incineration Safely Disposes of Chemical Cleaning Solvents," American Power Conference, Chicago, IL, 1972.

[4]E. Wackenhuth, L.W. Lamb, E.J.P., "Incineration Safely Disposes of Chemical Cleaning Solvents," *Power Engineering* 76 (1972): 68.

[5]T.F. Hallaren, "Incineration of Boiler Cleaning Chelate Wastes," *MP* 21, 4 (1982): 30-32.

[6]D. Daniels, "Boiler Chemical Cleaning: Doing It Correctly," *Power Magazine,* Sept. 1, 2014, http://www.powermag.com/boiler-chemical-cleaning-doing-it-correctly/?pagenum=1 (2014).

[7]Subtitle C in the Solid Waste Disposal Act Amendments of 1980 (Bevill Amendment) (Washington, DC: U.S. Environmental Protection Agency, 1980).

[8]D.B. Semmler, R.M. Propst, J.M. Storton, K.S. Mayo, "Processing of Chemical Cleaning Solutions at the Oconee Nuclear," *Nuclear Plant J.* 7, 4 (1989): 46-69.

[9]G.A. Peters, American Power Conference, Chicago, IL, 1981: pp. 1302-1307.

[10]A.L. Snow, "Chemical Cleaning Waste Disposal, Final Report" (Palo Alto, CA: Electric Power Research Institute, 1986).

[11]"Resources Conservation and Recovery Act," 40 USC parts 239 through 282 (Washington, DC: U.S. Environmental Protection Agency, 1976).

[12]P.J. Rogoshewski, D.D. Carstea, "Evaluation of Lime Precipitation for Treating Boiler Tube Cleaning Wastes" (Washington, DC: U.S. Environmental Protection Agency, 1980).

[13]W.E. Bell, J.E. Burquet, "Considerations in Disposal of Spent Cleaning Wastes," CORROSION/77, paper no. NACE 77076 (Houston, TX: NACE, 1977).

[14]D.R. Uhr, "Removal of Heavy Metals from Spent Cleaning and Finishing Solutions Which Contain Citrate and Gluconates," CORROSION/81, paper no. NACE 176 (Houston, TX: NACE, 1981).

[15]L.J. Holcombe, G.P. Behrens, W.C. Micheletti, "Methods for Removal of Copper and Iron from Boiler Chemical Cleaning Wastes," *Environmental Progress and Sustainable Energy* 6, 2 (1987): 74-81.

[16]W.C. Kennedy, "Removal of Iron from Chelant Solutions," U.S. Patent 4629570, 1986.

[17]M.L. Samuelson, S.B. McConnell, E.F. Hoy, "An On-Site Chemical Treatment for Removing Iron and Copper from Chelant Cleaning Wastes," Paper IWC-88-39, 49th International Water Conference, Pittsburgh, PA, Oct. 24-26, 1988.

[18]E.F. Hoy, S.B. McConnell, "Process for Removal of Nickel from Effluent Containing EDTA," U.S. Patent 4971775, 1990.

[19]D. Radtke, W. Frenier, R. Rusek, J. Carmen, D.J. Herrmann, "Waste Treatment of the Effluent for Chemical Cleaning of the Steam Generator at Byron One Unit," Paper IWC-95-38, 56th International Water Conference, Pittsburgh, PA, Oct. 30–Nov. 1, 1995.

[20]C.P. Huang, M.-C. Hsu, P. Miller, "Recovery of Edta from Power Plant Boiler Chemical Cleaning Wastewater," *Journal of Environmental Engineering* 126, 10 (2000): 919-924.

[21]J.G. Frost, K.J. Snyder, "Method for Treating Chelant Solutions to Remove Metals in Solution," U.S. Patent 5160631, 1992.

[22]R.G. Guess, "Ferrous Dithionite Process and Compositions for Removing Dissolved Heavy Metals from Water," U.S. Patent 5122279, 1992.

[23]T. Terao, A. Miyzaki, "Chemical Cleaning Waste Treatment by Reverse Osmosis," CORROSION/93, paper no. NACE 93364 (Houston, TX: NACE, 1993).

[24]E. Rodríguez, W.S. Harvey, E.J. Ásbjörnsson, "Review of H_2S Abatement Methods in Geothermal Plants," SGP-TR-202 2014, Thirty-Eighth Workshop on Geothermal Reservoir Engineering, Stanford University, Stanford, CA, Feb. 24-26, 2014.

[25]R.D. Varrin Jr., M.J. Little, "Method of Reusing a Cleaning Solution," U.S. Patent 9498802, 2016.

[26]S.B. McConnell, "Chemical Removal of Thiourea from Hydrochloric Acid," U.S. Patent 4822494, 1989.

[27]S.B. McConnell, "Chemical Treatment of Thiourea in Acidic Cleaning Wastes," CORROSION/90, paper no. NACE 90149 (Houston, TX: NACE, 1990).
[28]D.L. Venesky, "Thermal Stability of EDTA and Its Salts," 32nd Int. Water Conference, Pittsburg, PA, Nov. 2-4, 1971.
[29]A.E. Martell, R.J. Motekaitis, A.R. Fried, J.S. Wilson, D.T. MacMillan, "Thermal Decomposition of EDTA, NTA, and Nitrilomethylenephosphonic Acid in Aqueous Solution," *Canadian J. Chem.,* 53, 22 (1975): 3471-3476.
[30]R.J. Motekaitis, D. Hayes, A.E. Martell, W.W. Frenier, "Hydrolysis and Ammonolysis of EDTA in Aqueous Solution," *Canadian Journal of Chemistry* 57, 9 (1979): 1018-1024.
[31]R.J. Motekaitis, A.E. Martell, D. Hayes, W.W. Frenier, "The Iron(III)-Catalyzed Oxidation of EDTA in Aqueous Solution," *Can. J. Chem.* 58, 19 (1980): 1999-2005.
[32]E.J. Karell, "Photodegradation of iron(III) ethylenediaminetetraacetic acid: Application to treatment of nuclear steam generator chemical cleaning waste solutions," PhD dissertation, University of Virginia, 1993.
[33]D. Chen, A.E. Martell, D. McManus, "Studies on the Mechanism of Chelate Degradation in Iron-Based, Liquid Redox H_2S Removal Processes," *Canadian Journal of Chemistry* 73, 2 (1995): 264-274.
[34]D. Chen, A.E. Martell, D. McManus, "Oxidation of H_2S to S by Air with Fe(III)-NTA as a Catalyst Degradation," *Canadian J. Chem.* 53, 22 (1993): 1524-1531.
[35]W.M. Alvino, D.C. Grant, E.J. Laboda, W.A. Byers, R.D. Burack, "Method of Treating a Contaminated Aqueous Solution," U.S. Patent 5564105, 1996.
[36]M. Modell, "Processing Methods for the Oxidation of Organics in Supercritical Water," U.S. Patent 4338199, 1982.
[37]J. O'Regan, S. Preston, A. Dunne, "Supercritical Water Oxidation of Sewage Sludge" (Cork, Ireland: SCFI Group, 2007).
[38]S. Yesodharan, "Supercritical Water Oxidation: An Environmentally Safe Method for the Disposal of Organic Wastes," *Current Science* 82, 9 (2002): 1112-1122.
[39]F. Jimenez-Espadafor, J.R. Portela, V. Vadillo, J. Sánchez-Oneto, J.A.B. Villanueva, M.T. García, E.J.M.d.l. Ossa, "Supercritical Water Oxidation of Oily Wastes at Pilot Plant: Simulation for Energy Recovery," *Ind. Eng. Chem. Res.* 50, 2 (2011): 775–784.
[40]W.W. Frenier, *Chemical Methods for Controlling Corrosion During Oil and Gas Activities* (Houston, TX: NACE International, 2017).
[41]OECD, "OECD Guidelines for Testing Chemicals Part I Degradation of Chemicals" (Paris, France: Organization for Economic Cooperation and Development, 2005).
[42]J.J. Lauff, D.B. Steele, L.A. Coogan, J.M. Breitfeller, "Degradation of Ferric Chelate of EDTA by a Pure Culture of an Agrobacterium Sp," *Appl. Environ. Microbiol* 56, 11 (1990): 3346-3353.
[43]R.L. Bunch, M.B. Ettinger, "Biodegradability of Potential Organic Substitutes for Phosphates," *Purdue Univ. Eng. Bull., Extn. Srs,* 129, 1 (1967): 393-396.
[44]W.W. Frenier, M. Ziauddin, *Formation, Removal, and Inhibition of Inorganic Scale in the Oilfield Environment* (Richardson, TX: Society of Petroleum Engineers, 2008).
[45]Bayer, "An Environmentally Friendly and Readily Biodegradable Chelating Agent," 2001 Greener Synthetic Pathways Award, Bayer Corporation, Munich, Germany, 2001.
[46]AkzoNobel, "Dissolvine® Gl Technical Brochure" (Chicago, IL: AkzoNobel Functional Chemicals LLC, 2009).
[47]A.M. Al-Mayouf, "Dissolution of Magnetite Coupled Galvanically with Iron in Environmentally Friendly Chelant Solutions," *Corrosion Science* 48, 4 (2006): 898-912.
[48]Dow, "Versene (Tm) HEIDA Chelating Agent Technical Data," Midland, MI, 2007.
[49]Xus40855.01, "Developmental Chelating Agent," Brochure (Midland, MI: Dow Chemical Co., 2007).
[50]S. Tandy, K. Bossart, R. Mueller, J. Shauser, L. Shauser, R. Schulin, B. Nowack, "Extraction of Heavy Metals from Soils Using Biodegradable Chelating Agents," *Environ. Sci. Technol.* 38, 3 (2004): 937-944.
[51]K.B. Charkhutian, B.L. Libutti, F.L.M. De Cordt, J.S. Ruffini, "Process for Inhibiting Scale and Fouling on the Metal Surfaces Exposed to an Aqueous System," U.S. Application 20040011743, 2004.
[52]M. Saito, T. Yamamoto and S. Soya, "Cleaning Agent Composition," U.S. Patent 6013612, 2000.
[53]D.K. Crump, D.A. Wilson, "Formulations with Unexpected Cleaning Performance Incorporating a Biodegradable Chelant," U.S. Application 20090099058, 2009.

[54]D.A. Wilson, D.K. Crump, "Succinic Acid Derivative Biodegradable Chelants and Uses and Compositions Thereof," U.S. Patent 5859273, 1999.
[55]J.N. LePage, C.A. DeWolf, J.H. Bemelaar, H.A. Nasr-El-Din, "An Environmentally Friendly Stimulation Fluid for High Temperature Applications," paper SPE 121709, SPE International Symposium on Oilfield Chemistry, The Woodlands, TX, Feb. 20-22, 2009.
[56]T. Boonstra, Oedse, M. Heus, A. Carstens, J. LePage, "Glutamic Acid N,N-Diacetic Amide, Glutamic Acid N-Acetic Amide N-Acetonitrile, Alkali Metal Salts Thereof, Process to Prepare Them and Their Use," WO Application WO/2009/024519, 2009.
[57]M.A. Mahmoud, H.A. Nasr-El-Din, C.A. DeWolf, J.N. LePage, J.H. Bemelaar, "Evaluation of a New Environmentally Friendly Chelating Agent for High-Temperature Applications," Paper SPE 127923, SPE International Symposium on Formation Damage, Lafayette, LA, Feb. 10-12, 2010.
[58]A.I. Rabie, M.A. Mahmoud, H.A. Nasr-El-Din, "Reaction of GLDA with Calcite, Reaction Kinetics, and Transport Studies," Paper SPE 139816, SPE International Symposium on Oilfield Chemistry held in The Woodlands, TX, April 11–13, 2011.
[59]AkzoNobel, "Dissolvine Gl-38 Safety Data Sheet" (Deventer, NL: AkzoNobel, 2006).

10

Health, Safety, and Environmental Issues in IC

10.1. Introduction

Four interrelated issues concerning the handling and use of chemicals in IC environments are reviewed in Chapter 10:

1. The *health and safety* of all persons who may contact the chemicals that are being used, including the primary pipeline/facilities products and the reaction products of any treatment and waste streams. These concerns and possible solutions are described in Section 10.2.1.
2. The *effects* of the chemicals and reaction/waste products on organisms in any environment that may be affected by the treatment processes.
3. The plans and techniques to *manage* and *safely dispose/reuse* chemicals and wastes, including methods for recycling the chemicals and the water used to generate them (see Section 10.3).
4. *Emergency* plans/operations for controlling spills also must be considered in each pipeline maintenance phase (see Section 10.4).

10.2. Health and Safety Considerations during Chemical/Mechanical Cleaning Operations

10.2.1. General Considerations for All Cleaning Operations

Developing a corporate safety attitude to reduce and hopefully eliminate injuries, accidents, and releases of toxic chemicals has been practiced for many years. The author of this book recalls a "Life is Fragile: Handle with Care" program practiced at the laboratories and production plants of the Dow Chemical Co. in the 1970s. This program was based partially on the ideas of a safety pyramid (see Figure 10.1), with the goal of eliminating the root causes of injuries and accidents by eliminating the un-safe acts themselves. The metric was a reduction in treatments of individuals.

Smith[1] has reviewed these programs and contends that traditional industry standards driven by the U.S. Occupational Safety and Health Administration[2] are used to classify safety events based on the treatment (see Figure 10.1[a]) and/or restrictions of the injured person.

However, they claim that this treatment-based approach has limitations. They claim that with intense focus on administrative reporting (i.e., is the incident recordable or not?) and incident escalation management, the approach does not naturally resonate with workforce members to enable desired cultural changes. These authors also note that historical approaches have not included a potential *consequence assessment,* which is a critical element in preventing future injuries. Thus, incident classifications are often inconsistent because work restrictions and medical treatments are subject to individual medical provider judgment.

The authors'[1] company (ExxonMobil) is claimed to have a Hurt-based approach (see Figure 10.1[b]) to mitigate limitations of the treatment-based approach. The benefits are claimed to include

- Natural safety language of humans that will protect families and prevent injuries
- Consistent description of actual injury severity
- Integral assessment of potential injury severity
- Resonates with workforce to enable cultural changes based on caring for people
- Alignment with the corporation's safety vision of "Nobody Gets Hurt"

Both of these approaches are designed to educate the employee (and family in most programs) to eliminate the causes of accidents/incidents. The "Nobody Gets Hurt" process has an additional layer to highlight the multiple fatality incidents that affect many people long-term and can greatly affect the health of a whole corporation. The explanations and examples are given in Table 10.1.

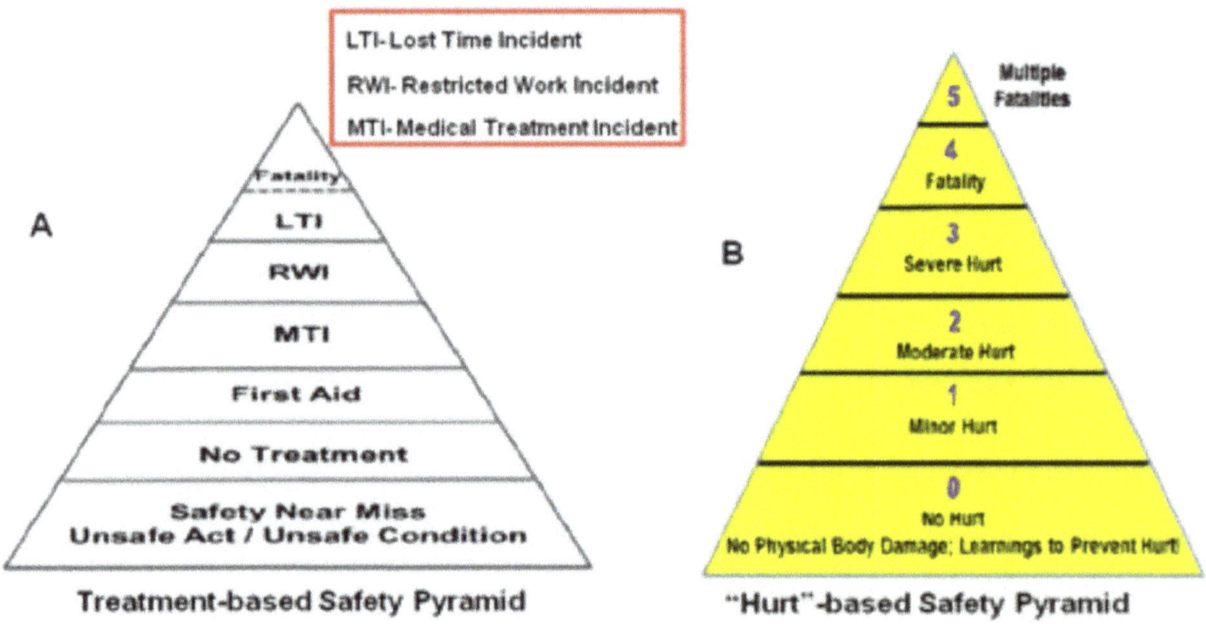

Figure 10.1 Treatment-Based and Hurt-Based Safety Pyramids[1]

Table 10.1 Hurt Pyramid Levels[1]

Severity Levels	Duration of Injury	Examples
4 & 5 Fatality (ies)	Forever	5- Multiple fatalities 4- Single fatality
3 Severe hurt	Long-term and life-altering	Amputation/severe disfigurement Total loss of organ/vision/hearing
2 Moderate hurt	Weeks to months	Bone fractures, significant lacerations Moderate hearing loss/vision loss
1 Minor hurt	Hours to days	Minor cuts/bruises/sprains/strains Moderate hearing loss/corneal abrasions
0 No hurt	No body damage	Object/chemical in eye removed by flushing Slips, trips, fall with no bruising or swelling General soreness

Health, safety, and the environment (HSE) awareness and activities continue to evolve. The control of HSE must shift from upper management, who proclaim rules and regulations, to the operational teams, who buy into the idea of eliminating accidents and spills.

The activities and technologies described in the introductory paragraphs are interconnected; however, it is useful to consider them separately since they are primarily associated with different parts of most jobs. Section 10.2.1 considers additional safety-related issues.

These concerns are related to mechanical and chemical hazards and are chiefly associated with human contact with chemicals and equipment during the storage, transport, and execution of a treatment. As such, contact with the treating chemicals is mostly a concern for the employees and contractors of the service companies and the employees and contractors of the plant owners. Public exposure to chemical hazards also must be considered, but these issues usually are associated with waste products of the treatments and are described in Sections 8.3 and 8.4 and in the PHMSA rupture/release regulations.[3]

Chemical hazards to personnel are a result of the inherent chemistries of the individual products, the final formulations, and the treating fluids. Equipment-related hazards also mostly affect the maintenance company and operator employees. Methods to *plan for* and to *reduce* hazards are also considered in Section 10.3.

Direct exposure to chemicals can affect people through contact with the skin and eyes and through inhalation and ingestion. The *toxicity* of most chemicals has been determined through laboratory animal tests,[4] but these data are mostly associated with ingestion or inhalation of the chemical. At the work site, hazards owing to acute *contact* (chemical burns) are a frequent source of concern, but people who handle large amounts of chemicals (such as bulk plant operators and shippers) also must be concerned with long-term and delayed affects owing to periodic contact.

Inhalation of toxic gases such as H_2S and CO_2 also must be considered, and testing equipment, breathing support, and wind direction indicators should be available at the job site and at a mixing site. H_2S may come from the unit being cleaned or from FeS reacting with an acid. CO_2 is unusual in IC but could be present when calcite is dissolved with HCl.

This gas is not highly toxic but excludes oxygen and may thus cause asphyxiation. Both of these gases are heavier than air and accumulate in low spots at the site. N_2 liquid (a very cold fluid) or compressed gas may also be present. Very large quantities also could cause asphyxiation, but it is slightly lighter than air (density of air at standard temperature and pressure is 1.292 kg/m^3, and the density of nitrogen is 1.251 kg/m^3) and usually would not cause problems unless it entered an enclosed space, such as a van or an enclosed well site structure.

The *reactive* hazards of some chemicals used in treatments, during transportation, onsite, and during and after a treatment also should be controlled. Some treating chemicals are flammable or combustible. The flash point (described in Section 10.3.1) is the major metric for determining flammability hazards. Some industrial chemicals contain oxidizing agents that could potentially react (explode or cause a fire) if they come in contact with organic compounds (including fuels and other organic materials on site). Storage and handling practices must be in place to prevent this contact until the oxidizers are needed. Other chemicals, such as HCl or NaOH are *corrosive* on contact to metals, human flesh, and other materials.

Material Safety Data Sheets (MSDS) are the major document for communicating chemical safety risks to all personnel at the pipeline/facilities sites, at a supplier's area, and during transit (DOT hazard placards also are required on transport vehicles). An MSDS book for all onsite chemicals or other MSDS compilations (such as digital/internet) should be available and used as part of the planning and during onsite safety meetings. The first page and an interior page of a MSDS for HCl is shown in Figure 10.2. These documents may be as many as 10 pages, so only a small set of the data are reproduced in this figure. These documents contain health, safety, and toxicity information of every chemical onsite. A very important part of the document is the list of personal safety equipment that is needed to safely handle the chemical. A list of typical safety equipment includes

- Hard hats
- Appropriate eye protection
  - Safety glasses with side shields
  - Chemical goggles
- Face shields
- Hearing protection
- Flame/chemical-resistant clothing
- Respirators/air packs
- Steel toe/chemical-resistant boots
- Chemical-resistant gloves

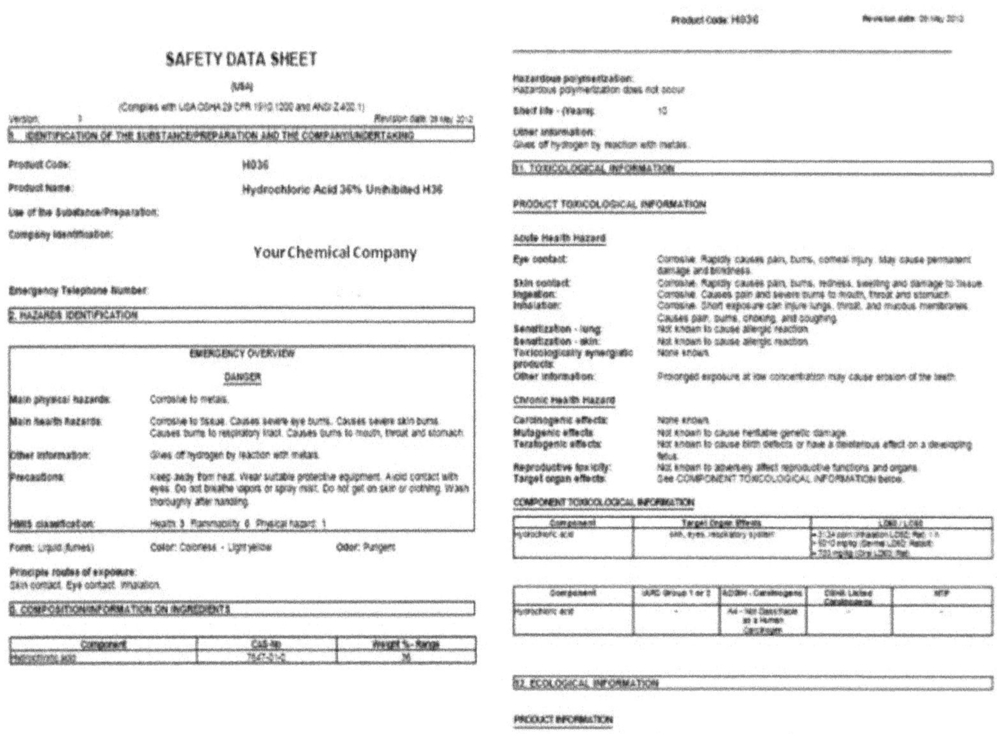

SAFETY DATA SHEET

Product Code: H036

Product Name: Hydrochloric Acid 36% Uninhibited H36

Your Chemical Company

Emergency Telephone Number:

2. HAZARDS IDENTIFICATION

EMERGENCY OVERVIEW

DANGER

Main physical hazards: Corrosive to metals.

Main health hazards: Corrosive to tissue. Causes severe eye burns. Causes severe skin burns. Causes burns to respiratory tract. Causes burns to mouth, throat and stomach.

Other information: Gives off hydrogen by reaction with metals.

Precautions: Keep away from heat. Wear suitable protective equipment. Avoid contact with eyes. Do not breathe vapors or spray mist. Do not get on skin or clothing. Wash thoroughly after handling.

Form: Liquid (fumes) Color: Colorless - Light yellow Odor: Pungent

Principle routes of exposure:
Skin contact. Eye contact. Inhalation.

Hazardous polymerization:
Hazardous polymerization does not occur

Shelf life - (Years): 10

Other information:
Gives off hydrogen by reaction with metals.

11. TOXICOLOGICAL INFORMATION

PRODUCT TOXICOLOGICAL INFORMATION

Acute Health Hazard

Eye contact: Corrosive. Rapidly causes pain, burns, corneal injury. May cause permanent damage and blindness.
Skin contact: Corrosive. Rapidly causes pain, burns, redness, swelling and damage to tissue.
Ingestion: Corrosive. Causes pain and severe burns to mouth, throat and stomach.
Inhalation: Corrosive. Short exposure can injure lungs, throat, and mucous membranes. Causes pain, burns, choking, and coughing.
Sensitization - lung: Not known to cause allergic reaction.
Sensitization - skin: Not known to cause allergic reaction.
Toxicologically synergistic products: None known.
Other information: Prolonged exposure at low concentration may cause erosion of the teeth.

Chronic Health Hazard

Carcinogenic effects: None known.
Mutagenic effects: Not known to cause heritable genetic damage.
Teratogenic effects: Not known to cause birth defects or have a deleterious effect on a developing fetus.
Reproductive toxicity: Not known to adversely affect reproductive functions and organs.
Target organ effects: See COMPONENT TOXICOLOGICAL INFORMATION below.

COMPONENT TOXICOLOGICAL INFORMATION

12. ECOLOGICAL INFORMATION

PRODUCT INFORMATION

Figure 10.2 MSDS for 36% HCl

Brown[5] has provided a list of safety considerations required during hydraulic fracturing operations that also should apply to any application of chemicals or intrusive work around industrial equipment. Here, this list has been modified to fit IC operations.

Brown[5] notes that at no time should the safety aspects of a treatment be compromised. Safety guidelines have been developed from experience derived from *previous incidents/accidents*. Many of these incidents have had great potential to seriously injure personnel or destroy valuable equipment. The inherent risk of dealing with high pressures can be greatly minimized by following simple safety procedures. Therefore, cleaning treatments can never be considered a success if an accident results in the destruction of equipment or injury to personnel.

- **Personal safety equipment**. Each person on location should wear appropriate safety equipment to minimize the risk of personal injury. Hard hats, hard-toed shoes, and safety glasses should be the minimum level of safety equipment worn on location. Other equipment such as hearing protection, goggles, fire-retardant fabrics, and filter masks should be worn if exposure to the conditions they protect against is a possibility. High-pressure water jetting treatments may require additional gear, such as face masks and water-resistant suits. Wearing safety equipment is a simple step that creates a positive safety atmosphere on location. The minimum personal protective safety equipment is noted in the MSDS example (Figure 10.2); however, individual contractors and plant owners may have more comprehensive requirements.
- **Safety meeting**. Holding a pretreatment safety meeting ensures that all personnel on location are aware of specific dangers of and required procedures for the treatment. Each person on location should clearly understand his or her role during the treatment and individual responsibility during emergency situations. A head count must be taken to account for everyone on location. An escape route and meeting place should be agreed upon, where all personnel will gather in the event of an emergency situation. Personnel who are not directly involved in the treatment should have limited location access during the actual pumping/treating operations. Everyone should be aware of the unique dangers of each process. Some locations may be in an area with hydrogen sulfide (H_2S), or the fluids being pumped may be highly flammable. The treatment site may have windsocks to indicate the wind direction, so people can evacuate if needed in a direction opposite to or perpendicular to the indicated wind direction.

  All potential safety problems or concerns should be brought to the attention of everyone. Maximum pressure limits should be set at this time, and every high-pressure pump operator must be aware of these limits. Instructions for pressure testing the lines and connections must also be covered. The high-pressure treating lines should be tested; a properly tested line includes tests of each pump in addition to the main treating line. The pressure rating of the plant facility's units should be checked to make sure it exceeds the treating pressure. The pretreatment safety meeting is the principal communication to all personnel. A well-organized safety meeting helps ensure that the treatment is an operational success without threatening human safety. The safety meeting should be a part of the pre-operational planning document and also should include operator personnel so all communications are in place.

  Many industrial plants have their own safety controls and may require contractors to attend periodic classes and show certification before entering the plant.
- **Pressure control**. Ensure that the equipment pressure control is always maintained. Many types of valves are used in cleaning operations and in process plants and have to be maintained and operated by trained personnel using approved procedures.
- **Precautions for flammable fluids**. Note the MSDS and the company manuals requirements for handling the cleaning fluids. Oil-based fluids should be tested for volatility before they are accepted as a fracturing fluid. An oil is generally considered safe to pump if it has a Reid vapor pressure < 1, API gravity $< 50°$ and open-cup flash point of 10°F [-12°C]. However, even if the fluid is considered safe to pump, several additional safety rules should be followed when pumping oil. Storage tanks for flammable fluids should be diked and spotted at least 150 ft. from a unit.

  Many other treating chemicals have low flash points and may be flammable. Any fluid that contains any alcohol (methanol or isopropyl alcohol) may be flammable. Spotting the fluids in this manner helps minimize problems during pumping. Also, all low-pressure hoses should be enclosed in a hose cover to prevent oil from spraying on hot engine components of the trucks, should a hose leak. Care must be taken to ensure that there is *no smoking on location*. It is a good idea to have all personnel check matches and lighters when they arrive on location. Finally, fire-fighting equipment should be on location and ready to be operated. In this way, a small fire may be contained before it has a chance to spread and become a major disaster. To further reduce the risk for an on-location fire, equipment should be both bonded and grounded. Bonding forms an electrical conduit between vessels containing flammable fluids and the equipment that transfers this fluid and pumps it. It controls the electrical potential from electrostatic buildup between all the units handling the flammable fluid. The requirements of the plant owner always must be part of the procedures.

- **Environmental considerations**. Treatment operations should be conducted using sound environmental practices to minimize the potential for contaminating air, water, and soil. All operations should comply with all applicable environmental laws and regulations. Hazardous material spills should be cleaned up quickly in accordance with a spill plan. All waste and unused materials should be handled and disposed of in accordance with locally approved regulations. More details of ecotox management and waste disposal are in Chapter 9 and Section 10.3.1.

10.2.2. Onsite HSE and Process Control

At the job site, a significant task is to monitor the process to allow addition of new chemicals and to determine when the treatment is complete. Previously, and now on small jobs, the progress, based on changes in the solvents, has been conducted on the tailgate of the pick-up using wet chemical methods. See Section 2.2.6 for more details. Some contractors now deploy mobile labs to the job site where more sophisticated equipment can be housed and used.

A depiction of the interior of a modern lab is shown in Figure 10.3. The monitoring procedures vary from plant to plant, and the rules of the plant and contractor must be applied. These labs frequently include an atomic absorption spectrophotometer for metals analyses (right side of Fig. 10.3). A wide range of electrochemical monitoring devices (Section 6.6.2) also can be used in a mobile lab. These units can significantly improve job process control since a sample line from the unit being cleaned usually comes into the lab. It also provides a safe, comfortable space for onsite personnel in a possibly dangerous industrial site.

Figure 10.3 Mobile Onsite Lab[6]

10.3. HSE Management

10.3.1. Chemical Selection to Enhance HSE Compliance

Developing green, low-toxicity formulations or less hazardous chemicals has become the goal of many formulators; however, there is no consensus (see Gupta[7]) on what constitutes green chemistry, even though many proposed green formulations are shown in sections throughout this book.

Darling[8] of the U.S. Environmental Protection Agency defines green products as "pollution prevention at the molecular level, designed to reduce or eliminate the use or generation of chemicals that are hazardous to human health or the environment." Here, this definition is used where possible, but many of the references cited define "green" in relationship to more limited criteria, such as a governmental regulation or list of regulations. Some of these are listed in Section 10.3.1 and in Chapter 1 of Kelland.[9]

10.3.2. Chemical Development Processes

A major service company[10] described a quantitative assessment of HSE impacts of prospective new well production chemicals that includes four areas of concern. The program[10] considers the following criteria (with scores in parentheses).

The total scores are added, and the worst (with the highest scores) products are then slated for elimination or modification. The four areas of concern are listed as follows with the specific items that were evaluated or tested. Note that this program was developed in the mid-1990s and some of the individual control items may have changed; however, the methodology can be used with appropriate updates.

1. Global "Restricted Use" Considerations–this consideration notes if the product and its components are listed on any of the following:

EU controlled substances	(4)
EU land-banned material	(4)
North Sea banned list	(3)
Organohalogens (nonpolymeric)	(4)
CERCLA/CWA Hazardous substance table 302.4 with RQ 100 lb. or less	(3)
Environmental Protection Agency extremely hazardous materials	(3)
Carcinogenic, mutagenic, or toxic to reproduction (CMR) substances	(3)
Marine pollutants	(2)
Resource Conservation and Recovery Act waste code U or P	(4)
CA Prop 65	(2)
Priority Pollutants Clean Water Act 2010	(4)
Process safety management chemicals list	(4)
Arms Control and Disarmament Agency (ACDA) schedule	(2)
U.S. Health Department Annual Carcinogen Report	(3)
Toxic metals list	(4)

2. Physical/Chemical Properties—this assessment scores physical and chemical properties that are based on the following values.

Flammability, based on flash point:

>141°F (60.5°C)	(0)
73–141°F (23–60.5°C)	(1)
10–73°F (-12–23°C)	(2)
< 10°F (-12°C)	(3)
Flammable Solid	(Yes = 2, No = 0)

Corrosivity:

pH < 2 or > 12.5	(2)
pH 2–12.5	(0)

Reactivity: Reactivity scores are based on the material's National Fire Prevention Association (NFPA) rating (0–4) or if the material's product structure has an energy-containing functional group. If yes, score = 3, if no, score = 0.

3. Environmental Screening—three parameters are evaluated under environmental screening: biodegradation, bioaccumulation, and toxicity. Ecotox scoring is based on the following values:

Biodegradation:

28-Day oxygen depletion test:

Very low biodegradation (<20%)	(3)
Low biodegradation (20–70%)	(2)
High biodegradation (>70%)	(0)

Bioaccumulation:

Log $P_{O/W}$ > 3 and MW < 600	(2)
Log $P_{O/W}$ < 3 and/or MW > 600	(0)

Toxicity:

EC_{50}	1 mg/L	(4)
LC_{50}	1–10 mg/L	(3)
	10–100 mg/L	(2)
	>100 mg/L	(0)

4. Exposure Hazards—Exposure hazards are assessed using the following data sources and data provided on vendor's MSDS. The testing procedures and nomenclature were described by Frenier[4]:
   - Registry of toxic effects of chemical substances
   - MDL Information Systems, Inc. MSDS database
   - CCINFO MSDS database
   - Other resources as applicable

   Score assignments are as follows:

Inhalation:

Highly toxic:	$LC_{50} < 0.5$ mg/L	(4)
Toxic:	$0.5 < LC_{50} < 2$ mg/L	(3)
Low-toxic:	$LC_{50} > 2$ mg/L	(0)

Ingestion:

Extremely toxic:	$LD_{50} < 5$ mg/kg	(4)
Highly toxic:	$5 < LD_{50} < 50$ mg/kg	(3)
Toxic:	$50 < LD_{50} < 500$ mg/kg	(2)
Low-toxic:	$LD_{50} > 500$ mg/kg	(0)

Skin absorption:

Highly toxic:	$LD_{50} < 40$ mg/kg	(4)
Toxic:	$40 < LD_{50} < 200$ mg/kg	(3)
Low-toxic:	$LD_{50} > 200$ mg/kg	(0)

Carcinogen: If the product or its components are classified, the following International Agency for Research on Cancer group rating (or equivalent rating from other agencies) applies:
1 = 4, 2A = 3, 2B = 2, 3 or none = 0
If the product or its components is a mutagen: Yes = 2, No = 0

The assessment process will score the potential product and will attempt to produce a product with the lowest overall score. A score of zero is the goal. This initial assessment probably will not be the final consideration of toxicity. If the material is to be used in North Sea countries or if the material is not registered in various potential use countries, additional testing may be required.

Jenkins[11] has provided a review of the current U.K. off-shore regulations and the protocols for screenings. See Figure 10.4 for a diagram of the Harmonized Mandatory Control Scheme (HMCS). This chart identifies the steps to determine if additional data are needed before a new chemical can be introduced. The acronyms are

- OSPAR, Convention for Protection of the Environment (Europe)
- PLONAR, Pose Little or No Risk to the Environment
- HOCNF, Harmonized Offshore Chemical Notification Format
- OEDC, Various Methods of Biodegradation Determination
- CHARM, Chemical Hazard Assessment and Risk Management Model

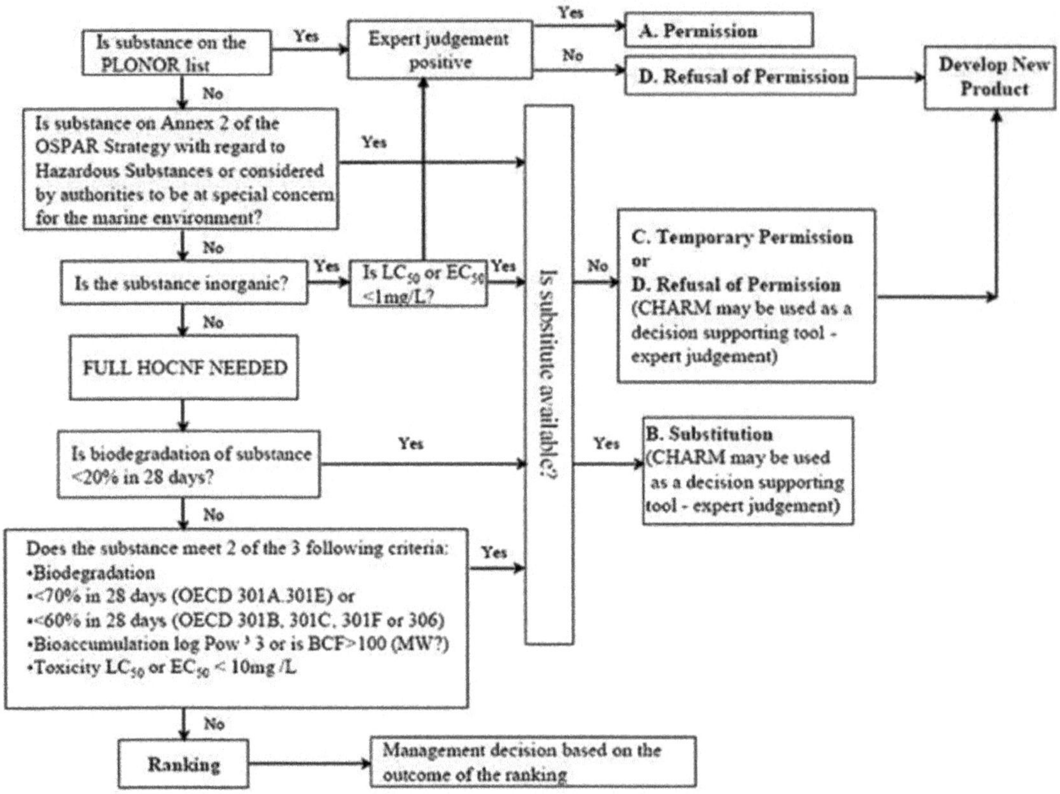

Figure 10.4 Harmonized Mandatory Control Scheme (HMCS)[11]

The European Union's Parliament has passed new regulations for the safe use of chemicals in the EU countries. The regulations[12] are called REACH:

- All substances manufactured or imported into the EU at 1 ton per year or more per manufacturer/importer must be tested and registered.
- Includes substances supplied in their pure form or in preparations.
- Includes substances supplied in products intended to be released from the product.
- Certain substances that have some adverse effect to health or environment present in products.
- The aim is to ensure a high level of protection of human health and the environment
- It went into force on June 1, 2007.
- It will be implemented stepwise over 11 years.
- It has a very broad scope, affects all products and markets, and will include 2-3 million substances, chemicals, and materials.
- Although the European Regulation only applies to the EU, they are reflecting themselves into all geographies.

Substances of Very High Concern (SVHC) under REACH are

- CMR Cat. 1, 2 (carcinogenic, mutagenic, or reprotoxic; see ECHA[13])
- PBTs (persistent, bioaccumulative, and toxic)
- vPvBs (very persistent, very bioaccumulative)
- Substances for which there is scientific evidence of probable serious effects to human health or the environment (e.g., endocrine disrupters, which affect the hormone system)

More details and references are in Kelland.[14] He notes that the big issues will be producing toxicity and ecological data on these chemicals/substances. Most of the types of tests required are noted in the previous paragraphs. Also see Frenier.[4]

There are additional ways to look at the development of new chemicals. Jordan[15] summarizes industry chemical activities and contends that production chemical development should focus on the development of new chemicals that have higher activity but lower viscosity than currently used chemicals, which would allow deployments at colder temperatures and over longer distances. HSE issues are also affected by these considerations.

Figure 10.5 shows the principal factors that affect the selection of suitable scale, wax, and hydrate inhibitors for an oilfield (especially subsea) application per these authors. These general rules apply to any IC chemical, and the approach can be modified where needed. Some of the very specific characteristics may not apply to all IC fluids.

The chemical and physical properties of importance include

- Physical properties
- Fluid compatibility
- Ability to monitor
- Material compatibility, including corrosion of metals
- Cost/performance characteristics
- Application, including injection and sampling
- Effects of system conditions, including temperature and viscosity
- Compatibility with aqueous and hydrocarbon fluids

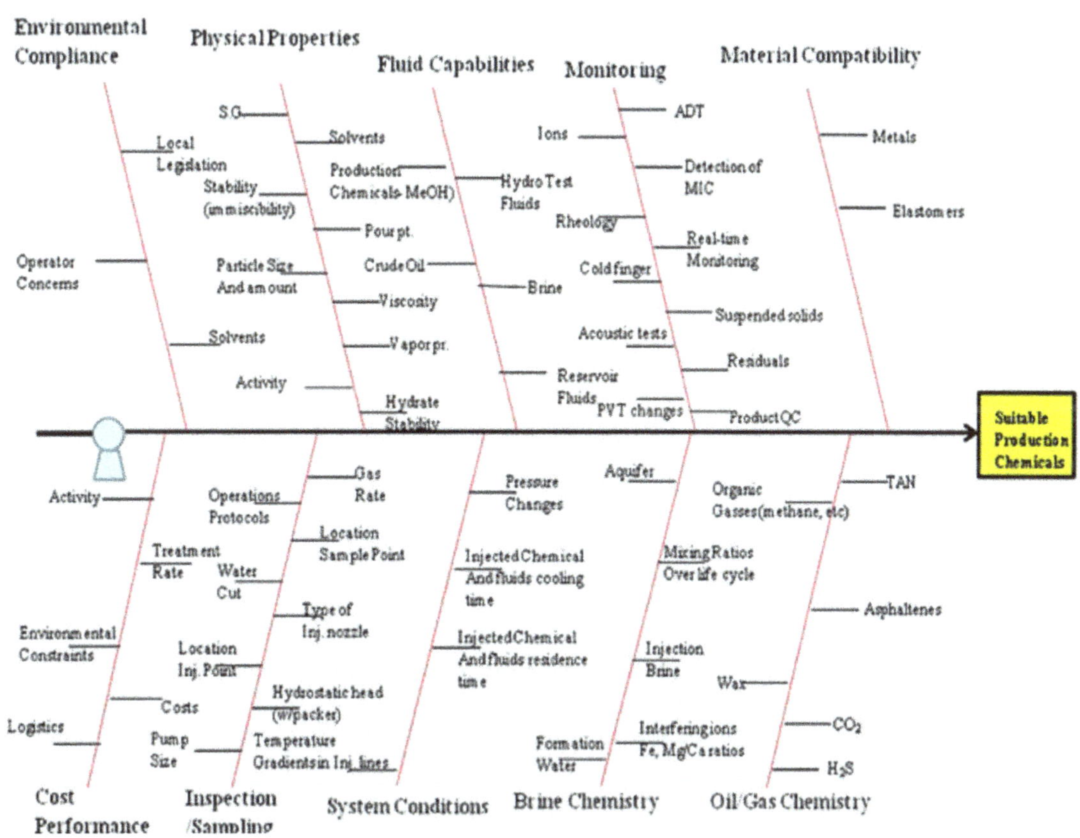

Figure 10.5 Additive Selection Properties[15]

Hill[16] has provided details on a process for developing more environmentally acceptable products. This requires working with regulatory agencies, clients, and chemical suppliers to achieve regulatory compliance for MSDS, labels, and environmental testing protocols. The paper recognizes that international environmental statutes and guidelines combined with clear corporate goals to minimize environmental impact are key requirements in regulating industrial use and discharge of environmentally harmful chemicals. The development of more environmentally acceptable products requires a thorough screening of environmental fitness and status of compliance with relevant laws and regulations.

Details are given by the authors[16] on the aggressive programs developed to address a wide range of environmental opportunities and issues. International environmental statutes and guidelines are the strongest driving force to restrict industrial use and discharge of environmentally harmful chemicals. The primary tool of environmental risk reduction is an aggressive product development program incorporating more environmentally friendly chemical solutions. HSE activities and considerations critically affect each step of the product development process. See Figure 10.6. This represents an update of the program developed by Purinton.[10]

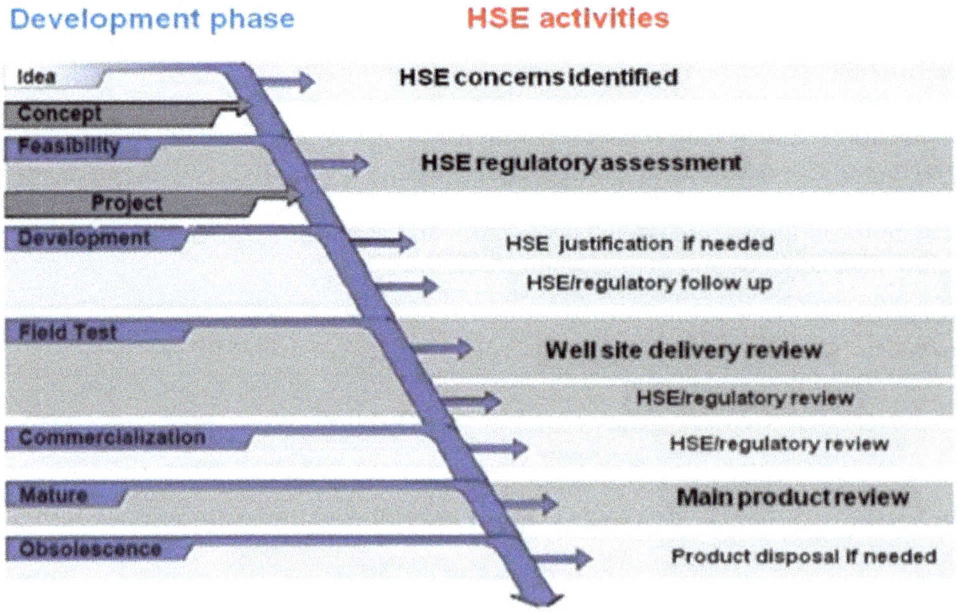

Figure 10.6 HSE Considerations during Product Development[16]

Cleaning chemicals have primary active ingredients that may be harmful if discharged to the environment and other chemicals that are part of an onsite formulation. Improving the characteristics of these products to reduce risk or damage to marine life requires changes in previously acceptable products, such as eliminating restricted materials and incorporating components with improved ecotoxicity values. Also, HSE issues are a major concern for the use of certain chemicals, especially in the North Sea operations area. See Kelland[17] and Del Villano.[18]

Kelland[9] begins his book with a chapter on selecting chemicals and the environmental issues associated with the OSPAR regulations for using production chemicals in the northeast Atlantic (and North Sea) region. This chapter has 54 references. The second edition (Kelland[14]) has 99 references to environmental/safety documents, reflecting the continued worldwide concerns about these issues.

The Society of Petroleum Engineers held a workshop on risk assessment planning.[19] An organizer of the event (Kristin M. Koblis, Nobel Energy) notes that there is a need to identify chemicals in which no additional concern is warranted and to also identify the chemicals where additional risk characterization and management may be needed. This process also should be specific to field operations and identify the field-specific exposure likelihood. The process must evaluate not only the potential human health and environmental hazards but also the potential for people or ecological API receptors to be exposed to the chemicals in the products. Chemical risk is based on both the chemical's inherent hazard to human or ecological receptors and the potential for such receptors to be exposed to the chemical throughout the life cycle of the chemical.

10.4. Spill Control

If there is a spill in an industrial plant during IC operations, the plant's management will have their own plan. The following information is provided only as an illustration.

Notification, control, removal, and final disposal of the fluid or solid may require the cooperation and participation of multiple governmental agencies and commercial entities. In the U.S. territories and waters, the first response to a spill should be the National Response System.[20] Initially, this system was focused on oil spills and select hazardous polluting substances discharged into the environment. It has since been expanded by other legislation to include hazardous substances and wastes released to all types of media.

The NRS functions through a network of interagency and inter-government relationships, which were formally established and described in the 13T of the National Oil and Hazardous Substances Pollution Contingency Plan (NCP).[21] The website includes the emergency contact number that should be used in the case of a spill. This system is part of the National Incident Management System,[22] a part of the U.S. Homeland Security Department. The NCP document includes details of the planning and control of a spill incident. The National Incident Management System (NIMS) provides a systematic, proactive approach to guide departments and agencies at all levels of government, nongovernmental organizations, and the private sector to work seamlessly to prevent, protect against, respond to, recover from, and mitigate the effects of incidents, regardless of cause, size, location, or complexity, to reduce the loss of life and property and harm to the environment.

The system[23] provides spill response and vessels for control of spills in European waters. Individual countries may have more procedures, but the author of this book has not accessed them.

API[24] provides a comprehensive document that gives guidance and highlights industry-recommended practices for the construction and integrity of wells that are hydraulically fractured. The guidance provided here will help to ensure that shallow groundwater aquifers and the environment are protected, while also enabling economically viable development of oil and natural gas resources. Although this plan was designed for oilfield situations, the basic information can be modified for in-plant operations. The plant managers also will have their own guidelines.

The document proposed that in the event of a spill, it is extremely important for all responsible operating personnel to know how to respond quickly and effectively to control, contain, and clean up the spill. A contingency plan should be prepared for inland areas and for areas near water. The plans should provide use of oil spill cooperatives, whenever advantageous. Spill plans should address the need to advise the public about significant releases. Also review the National Response System,[20] described earlier in this section.

The plan should include procedures to advise government officials and provide appropriate information and access to the press. Section 8.7.5 of Control and Containment of API[24] states that in the event of a spill:

1. The source of the spill should be stopped or reduced as much as possible in a safe manner. The spread of the spilled substance should be controlled or contained in the smallest possible area to minimize adverse effects. Some methods that can be used to control and contain discharged substances, particularly oil, include retaining walls or dikes around tanks and other spill-prone equipment.
2. Secondary catchment basins are designed to prevent the spread of oil if it escapes the primary wall or dike.
3. Permanent booms should be constructed in water basins adjoining the facility.
4. Temporary booms should be deployed in the water after the spill occurs.
5. Special chemicals should be used to gel or biodegrade the oil to prevent the spread of the spill into or on water. Operators should evaluate the potential for spills and damages and use this information to determine the type and size of primary and secondary containment necessary. The type and footage of containment booms installed or stored for deployment will vary with the type, size, and location of the facility and spill potential. This information should be developed for each main area or facility and be stated in the facility contingency plan. In addition, the contingency plan should list where emergency equipment is located.

Spill containment is the primary responsibility of the plant owner, and their requirements must be applied for all IC jobs.

References

[1]R.M. Smith, M.L. Jones, "A Hurt-Based Approach to Safety," Paper SPE 163757, SPE Americas E&P Health, Safety, Security & Environmental Conference, Galveston, TX, Mar. 18–20, 2013.

[2]OSHA 3132, Occupational Safety and Health Administration Information Booklet (Washington, DC: U.S. Department of Labor, 2000).

[3]Pipeline and Hazardous Materials Safety Administration, 2012, U.S. Department of Transportation, Washington, DC, http://phmsa.dot.gov/pipeline (Feb. 5, 2012).

[4]W.W. Frenier, "Introduction to Toxicity Testing," CORROSION/96, paper no. NACE 96154 (Houston, TX: NACE, 1996).

[5]E. Brown, R.W. Thrasher, L.A. Behrmann, "Fracturing Operations," in *Reservoir Stimulation*, 3rd ed., eds. M.J. Economides, K.G. Nolte (Chichester, U.K.: John Wiley & Sons, 2001).

[6]"HydroChem Story," 2014, HydroChem, Deer Park, TX, http://www.hydrochem.com/ (Jan. 18, 2014).

[7]D.V.S. Gupta, "What Is Green?" CORROSION/98, paper no. NACE 98206 (Houston, TX: NACE, 1998).

[8]D. Darling, R. Rakshpal, "Green Chemistry Applied to Corrosion and Scale Inhibitors," CORROSION/98, paper no. 207 (Houston, TX: NACE, 1998).

[9]M.A. Kelland, *Production Chemicals for the Oil and Gas Industry* (Boca Raton, FL: CRC Press, 2009).

[10]R.J. Purinton, T.S. Manning, "The New Role of HSE in Chemical Product Development," Paper SPE 35822, SPE Health, Safety, and Environment in Oil and Gas Exploration and Production Conference, New Orleans, LA, June 9-12, 1996.

[11]A. Jenkins, "Performance of High-Temperature, Biodegradable Corrosion Inhibitors," CORROSION/2011, paper no. NACE 11272 (Houston, TX: NACE, 2011).

[12]"Registration, Evaluation, Authorization and Restriction of Chemicals (Reach)," Regulation (EC) No 1907/2006 (Brussels, Belgium: European Parliament and Council, 2006).

[13]"Classification, Labeling, and Packaging of Substances and Mixtures," (EC) No 1272/2008 (Helsinki, Finland: European Chemical Agency (ECHA), 2008).

[14]M.A. Kelland, *Production Chemicals for the Oil and Gas Industry*, 2nd ed. (Boca Raton, FL: CRC Press, 2013).

[15]M.M. Jordan, N.D. Feasey, "Meeting the Flow Assurance Challenges of Deepwater Developments: From Capex Development to Field Start Up," Paper SPE 112472, SPE North Africa Technical Conference & Exhibition, Marrakech, Morocco, Mar. 12-14, 2008.

[16]D.G. Hill, K. Dismuke, W. Shepherd, I. Witt, H. Romijn, W. Frenier, M. Parris, "Development Practices and Achievements for Reducing the Risk of Oilfield Chemicals," Paper SPE 80593, SPE/EPA/DOE Exploration and Production Environmental Conference held in San Antonio, TX, Mar. 10-12, 2003.

[17]M.A. Kelland, T.M. Svartaas, L. Dybvik, "A New Generation of Gas Hydrate Inhibitors," Paper SPE 30695, SPE Technical Conference, Dallas, TX, Oct. 22-25, 1995.

[18]L. Del Villano, R. Kommedal, M.A. Kelland, "Class of Kinetic Hydrate Inhibitors with Good Biodegradability," *Energy Fuels* 22, 5 (2008): 3143-3149.

[19]SPE, "Risk Assessment Workshop," Focus Sharpens on Chemical Risk Assessment, San Antonio, TX, Feb. 24-25, 2015.

[20]National Response System, 2009, U.S. Environmental Protection Agency, Washington, DC, http://www.epa.gov/ceppo/web/content/nrs/

[21]National Oil and Hazardous Substances Pollution Contingency Plan (NCP), 2014, Environmental Protection Agency, Washington, DC, http://www2.epa.gov/emergency-response/national-oil-and-hazardous-substances-pollution-contingency-plan-ncp-overview (Nov. 20, 2014).

[22]"National Incident Management System (NIMS)," 2008.

[23]EMSA, "Network of Standby Oil Spill Response," European Maritime Safety Agency, 2010.

[24]API Guidance Document HF1, Hydraulic Fracturing Operations-Well Construction and Integrity Guidelines, 1st ed. (Washington, DC: American Petroleum Institute, 2009).

11

General Conclusions

11.1. Summary

During the past 30 years, many new solvents for inorganic deposits have been developed. Low hazard corrosion inhibitors have replaced some of the more toxic materials used to protect the structural surfaces during cleaning. Development of nuclear cleaning methods has extended the life of some very valuable nuclear generating plants. New waste disposal methods have enabled the industry to continue to operate but have required new chemicals to be developed. Improvements for cleaning chemical/refinery equipment include ways to remove FeS, and more environmentally friendly solvents have been developed.

These trends will continue, and the need for less hazardous chemicals in all areas and disposal concerns will drive much of development. There may be much improved procedures for cleaning CP/R equipment, including improved analytical methods for deposit analyses. There will be more application of computational methods to cleaning technology, as computers and software for simulating chemical reactions become more powerful. Industrial cleaning changes have included

- Chemistry Evolution
  - Solvents: HCl + thiourea of oxidizers, toxic organic solvents
  - Organic acids, chelating agents, and new oxidizing agents
  - FeS control technologies
  - New, less toxic organic solvents, additives (glycol ethers, surfactants)
  - Inhibitors for HCl and organic acids; evolution of structures and formulations
- Equipment/Application Evolution
  - Automated jetting technologies
  - Foams/gels
  - Vapor phase fluids (steam)
- Disposal Evolution
  - Dumping in the river is not allowed
  - Thermal treatments
  - Neutralizations, precipitations of metals
  - Biodegradable agents

- HSE Evolution
  - Chemical development based on HSE plan
  - Mobile labs
  - PE and onsite safety

11.2. General Job Design Considerations

The preface of this book states the basic approach to any cleaning job. A successful design of a treatment requires the operator to

- Understand the problem (Sections 3, 3.8, and 7.4)
- Evaluate solution options (Sections 2.6.4, 4, 5.2, 5.3, and 7.5)
- Design and execute cleaning procedures (Sections 2.6, 8.1, and 8.4)
- Perform a safe treatment (Sections 9 and 10.2)
- Evaluate the results (Section 2.6.6 and 10.3.2)

This short section applies to all the cleaning services described in Chapters 3–9. If a chemical cleaning or mechanical job fails to achieve the desired results, the cause usually is poor job design and improper chemistry or equipment. The chemistry cannot be effective unless the solvent reaches the deposits at the proper temperature to allow dissolution to take place and is constantly replenished. Important job design considerations include

1) Understanding the deposits, the solvents, and mechanical equipment
2) Determining the flow rate. There must be enough flow to fill and circulate all portions of the unit being cleaned. The hoses, inlets, and outlets must be sized appropriately with the available pumps.
3) Determining the appropriate pressure and water flow rate for hydrojetting for the equipment being cleaned. Higher forces may damage the surfaces, and lower forces may not remove the solids.
4) Venting. If the cleaning process can produce gas (such as H_2S while dissolving FeS), the system must be able to vent this gas.
5) Flushing and draining. Most solvents will leave undesirable films on the cleaned surfaces unless they can be flushed with fresh water or a passivating solution.
6) Heating and cooling. During cleaning, the heat fluid system can be heated with an external heat exchanger, process fluids, by direct firing, use of injected steam, or plant-supplied hot water. Any of these methods can be successful. The key is to maintain the solvent at its optimal cleaning temperature for the duration of the job without overheating, which could cause excessive corrosion or solvent degradation.
7) Analyzing the progress using chemical analyses and monitoring pressure, temperature, and flow rates.
8) Choosing disposal options. Each plant location (and some individual units within a plant) can have different chemical disposal requirements. See Chapter 9 for some general options. The disposal options must be considered and chosen for the cleaning plan *before* the solvent (or mechanical method) is chosen. The cost of the disposal frequently exceeds the cost of the cleaning operations. For example, Radtke[1] described the treatment procedures required to process chemical cleaning wastes generated during the cleaning of a nuclear steam generator. The cleaning phase lasted about 2 weeks, but the waste processing took 3 months. This is an extreme example but illustrates the importance of considering waste disposal during the planning phase of any cleaning operation. See Chapter 9.
9) Analyzing the cleaning effectiveness and reporting to the client. Surfaces may be examined physically; the job logs are evaluated and tube samples may be examined. This closes the loop of planning, executing, evaluating, and improving.

11.3. Example of a Specific Job Plan

This is a job plan for cleaning units at a chemical plant (see Gutzite[2]). See the various sections of this book for more details.

PLAN

1. **Description of Unit**
   a. Type (Function and product)
   b. Size
   c. Metallurgy
   d. Connections and flow paths, restrictions, and location in plant
   e. Relation and isolation from other units or operations
   f. Temperature restrictions
   g. Product side, water side, or both
   h. Goal of cleaning
   i. Cleaning history
   j. Utilities available
   k. Plant disposal options
2. **Sample of Deposit or Tube**
   a. Cut tube before present cleaning
   b. Obtain sludge sample before present cleaning
   c. Tube or sludge sample from previous cleaning or earlier opening of unit
   d. Submit sample to contractor lab as early as possible (2 weeks minimum is optimal)
   e. Description of distribution of fouling deposits (estimated thickness and uniformity)
   f. Analyze deposit obtained and test solvent
   g. Choose one or more solvents for detailed testing
   h. Conduct advanced tests including loop tests if necessary
   i. Conduct custom corrosion tests if necessary
3. **Choose Solvent System**
   a. Technical considerations: solvency, corrosion, metallurgy, temperature, flow, draining, disposal, equipment, and chemical availability
   b. Cost considerations: current budget or long-term gain and cost of disposal options
   c. Scheduling considerations: critical path, time required, effect on other operations
   d. Safety: toxicity, flammability, personnel exposure
4. **Plan Solvent Systems**
   a. Estimate chemical requirements, including corrosion inhibitors and passivation stages
   b. Estimate excess requirements
   c. Source and order chemicals early enough for job placement
5. **Plan Physical Aspects**
   a. Pumping: client or contractor. Estimate capacity needed and types of pumps and metallurgy
   b. Heating: customer hot water, customer heat exchanger, contractor heat exchanger, direct fire, steam sparge, other
   c. Tankage: Client chemical and disposal storage, contractor trucks and tanks, rental tanks. Metallurgy, capacity, and connections
   d. Layout and connections: temporary piping, contractor piping and hoses, valves, fittings, metallurgy. Pressure requirements and ratings. Size for flow requirements. Multiple loops. Other utility requirements including power and steam
   e. Venting and valuing: heating connections. Sample locations: Draining and flushing
   f. Monitoring points: Temperature, flow, corrosion probes
   g. Special equipment: general analytical procedures, mobile lab, filter press, chemical mixing, and control

h. Make drawing of layout and hook-up showing sizes, orientation, access, and egress. Location of safety equipment and potential physical and chemical hazards

6. **Plan Staffing**
   a. Client personnel per shift
   b. Contractor personnel: category and number per shift. Source: local or offsite. Special personnel (lab, safety)
7. **Produce Detailed Plan**
   a. Written plan including all people, chemicals, and equipment requirements
      1. Time line required including long lead-time items
      2. Set responsibilities of contractor and client to supply equipment, services, personnel, and physical changes needed
   b. Client and contractor approve plan

EXECUTE

1. **Client Preparation**
   a. Take unit offline
   b. Isolate, install blinds
   c. Install temporary piping
   d. Empty and/or purge unit of product and toxic gases (unless that is the goal of the job)
   e. Make space available for contractor equipment and chemical storage tanks
   f. Install utility and power lines, steam and water lines, and any required gas lines
2. **Contractor Onsite Preparations**
   a. Obtain necessary permits and clearances from the client
   b. Deliver all contractor equipment including tanks to the site according to drawings
   c. Hook-up according to drawing
   d. Pressure check all lines with water
   e. Install all utility and power connections
   f. Set up analytical equipment, including mobile lab and special equipment (corrosion monitoring)
   g. Site chemical trailers
   h. Safety tape area
   i. Perform any onsite chemical blending
   j. Fill unit with water if required by plan
   k. Conduct tail-gate safety meetings and include customer personnel
3. **Chemical Injection**
   a. Start chemical injection, with heating if required by plan
   b. After unit is full of chemicals, heat/cool and circulate as required by plan
   c. Start sampling procedures
   d. Analyze samples and record values in log and in computer if available
   e. As job proceeds, replenish chemicals if necessary
   f. Change valves to circulate alternate loops if required. Verify temperatures, flow rates, and pressures with secondary devices if possible
   g. When agreed-upon parameters are stable for the required time period, start drain to temporary tanks or to designated facility or site
   h. Refill and rinse if required
   i. Inject next stage of chemicals if required and repeat steps d through h
   j. Complete final drain and rinse if required
   k. Send drain samples to contractor lab for detailed analyses
4. **Waste Treatment**
   Waste treatment procedures vary too much to provide general guidelines, but they must be planned, set up, and executed with the same care as the cleaning part of the job. In addition, coordination with plant, local, state, and federal agencies may be required before, during, and after the job.

5. **Completion of the Job**
    a. Using the plan as the guide, disconnect contractor piping, drain, and flush all lines
    b. Secure contractor equipment and remove from site
    c. Remove temporary piping, valves, and blinds

EVALUATE

1. **Inspection of Unit**
    a. Ensure that unit is cool, has been purged of toxic gases, and has a safe oxygen level
    b. Obtain vessel entry permits and comply with contractor vessel entry requirements, including all PPE
    c. With client representative present (if possible), inspect critical areas that are accessible
    d. Look for scale, sludge, copper, after-rust, and evidence of corrosion damage. Look in pits if accessible. Use any available fiber-optic probes, mirrors, etc.
    e. Photograph surfaces if allowed by client
    f. Obtain samples of any remaining scale or deposits and send to HydroChem lab for analyses
    g. Inspect as many critical areas as feasible, especially high-heat zones or any identified trouble areas
    h. Cut tubes if possible and send to contractor lab for analyses
2. **After Job Evaluation**
    a. Evaluate the job results versus the plan
    b. Determine and document lessons learned for future jobs

11.4. Future Quality Improvements

- Accelerate the search for safer, more environmentally acceptable chemicals as well as improved application methods. This process will offer economic, environmental, and public perception benefits. This will include automating many of the more hazardous operations (such as high-pressure water cleaning).
- Quality also can be enhanced via a thorough understanding of the entire industrial supply chain. This is especially true of the petroleum product supply chain. This includes production (upstream), transportation (midstream of multiple fluids, including oil, gas, and aqueous phases and finished products), and refining/petrochemical (downstream) manufacturing. This book has attempted to emphasize that chemical and physical principles are the same in any environment in which the conditions are also the same. In addition, it must be emphasized that each phase of the supply chain affects the next, and there are also feedback loops, in which one phase affects earlier phases. For example, the quality requirements for pipeline transportation of fluids affects the well products that are needed as well as the facilities and refineries required for the treatments' final products.
- All chemical cleaning fluids are potential products, *not* wastes for disposal, which provides a different perspective for resource conservation. Thus, the products are seen to include the hydrocarbon liquids, gases, water, and the dissolved salts. This view also includes CO_2 capture products for any sources and even problem gases such as H_2S. Some industry sectors already use these valuable chemicals; gases and brines are used in enhanced oil recovery, and HF water is being recycled. With the exception of possible use of the salt cake from evaporation of brines for road salt,[3] however, the author of this book is not aware of routine brine uses as a chemical feed stock. Veil[4] has shown several examples of methods (including distillation and evaporation for producing the salt cakes). Detailed analyses of some brines may reveal useful amounts of rare or costly chemical species.

    In other industries, fresh brines from selected formations have a long history as chemical feed stocks. For example, the Dow Chemical Company[5] was founded on extraction of bromine (and then NaOH and chlorine) from brines. There is also the example of the great overlap with the chemical/petrochemical industries that routinely use other firms' "waste streams" as their raw materials. There has been a long history of attempts to reuse cleaning wastes (Chapter 9), but more work will produce more economical and safer processes.

References

[1]D. Radtke, W. Frenier, R. Rusek, J. Carmen, D.J. Herrmann, "Waste Treatment of the Effluent for Chemical Cleaning of the Steam Generator at Byron One Unit." Paper IWC-95-38, 56th International Water Conference, Pittsburgh, PA, Oct. 30–Nov. 1, 1995.

[2]J. Gutzite, "Cleaning of Process Equipment and Piping," MTI Publication No. 51 (St. Louis, MO: MTI, 1999).

[3]S. Rassenfoss, "From Flowback to Fracturing: Water Recycling Grows in the Marcellus Shale," *JPT* 63, 7 (2011): 48-51.

[4]J.A. Veil, "Produced Water Management Options –One Size Does Not Fit All," 2009, Society of Petroleum Engineers, Richardson, TX, http://www.spe.org/dl/docs/2009/Veil.pdf

[5]D. Whitehead, *The Dow Story: The History of the Dow Chemical Company* (New York, NY: McGraw-Hill, 1968).

Index

B

C

F

G

H

O

P

Q

R

S

www.ingramcontent.com/pod-product-compliance
Ingram Content Group UK Ltd.
Pitfield, Milton Keynes, MK11 3LW, UK
UKHW051959041026
15590UKWH00009B/2386

9 781575 903675